PETROLEUM FUELS MANUFACTURING HANDBOOK

PETROLEUM FUELS MANUFACTURING HANDBOOK

Including Specialty Products and Sustainable Manufacturing Techniques

Surinder Parkash, Ph.D.

McGraw Hill

New York Chicago San Francisco Lisbon London Madrid
Mexico City Milan New Delhi San Juan Seoul
Singapore Sydney Toronto

The McGraw-Hill Companies

Library of Congress Cataloging-in-Publication Data

Parkash, Surinder.
Petroleum fuels manufacturing handbook : including specialty products and sustainable manufacturing techniques / Surinder Parkash.
p. cm.
Includes index.
ISBN 978-0-07-163240-9 (alk. paper)
1. Petroleum as fuel. 2. Petroleum products. I. Title.
TP355.P28 2010
665.5—dc22 2009017071

1 2 3 4 5 6 7 8 9 0 DOC/DOC 0 1 5 4 3 2 1 0 9

ISBN 978-0-07-163240-9
MHID 0-07-163240-9

Sponsoring Editor
Taisuke Soda

Editing Supervisor
Stephen M. Smith

Production Supervisor
Pamela A. Pelton

Project Manager
Smita Rajan, International Typesetting and Composition

Copy Editor
Anne Lesser

Proofreader
Eina Malik, International Typesetting and Composition

Art Director, Cover
Jeff Weeks

Composition
International Typesetting and Composition

Printed and bound by RR Donnelley.

This book is printed on acid-free paper.

To my wife, Rita

ABOUT THE AUTHOR

SURINDER PARKASH, PH.D., has over three decades of experience in petroleum refining and the related fields of process design, refinery operational planning, international marketing, and project planning. He has worked with many well-known companies and organizations such as Indian Institute of Petroleum, Iraq National Oil Company, Bahrain National Oil Company, and Kuwait National Petroleum Company. He is the author of *Petroleum Refining Handbook*, published by Gulf Professional Publishing. At present, Dr. Parkash is president of NAFT-ASIA (www.naft-asia.com), an independent consulting firm.

CONTENTS

Chapter 5. Diesel Fuels 65

Chapter 6. Residual Fuel Oils 81

Part 2 Petroleum Specialty Products

Chapter 7. Bitumen 101

Chapter 8. Petroleum Coke 141

Chapter 9. Carbon Black 165

Chapter 10. Lube Base Stocks 189

Chapter 11. Lubricating Oil Blending 211

PREFACE

Petroleum products are everywhere around us. They appear in visible forms, such as gasoline, diesel, kerosene, and aircraft fuels, and in less visible forms over the entire spectrum of industry, such as automobile lubricants, greases, carbon black for truck tires, bitumen for road building, the waterproofing in house roofs, feedstock for petrochemicals, synthetic fibers, and plastics. Petroleum feedstock is used in the manufacture of white mineral oils in eye ointment, hair oils, cosmetics, petroleum solvents, and pest control sprays. Transportation fuels, however, remain the most important use of petroleum.

The consumption of petroleum products throughout the world is ever-increasing to meet the rising energy needs of countries. But this rapid rise has led to undesirable air and water pollution levels. Environmental pollution affects everyone on the planet. During the last two decades, the manufacture and blending of petroleum products has changed rapidly, with a view to reduce atmospheric pollution and conserve petroleum feedstock. The lead phaseout from gasoline, sulfur reduction in all transportation fuels, and new lube-making technologies that produce longer-lasting engine oils or lower fuel consumption are a few illustrations of these changes.

This book surveys the manufacture, blending, properties, specifications, and uses of petroleum fuels and specialty products (products made out of petroleum feedstock for nonfuel use except petrochemicals). There are a very large number of specialty products—petroleum solvents, bitumen for paving and industrial uses, lubricating oils, greases, white mineral oils, carbon black, petroleum coke, spray oils, and so on—to meet the requirements of industry. Possibly far more technical personnel are engaged in petroleum specialty manufacture and the handling of petroleum products than are found in refineries. Although petroleum fuels are generally made in refineries out of crude oil distillation, petroleum specialty products are made in relatively smaller downstream units starting with refinery streams as feedstock. A refinery may produce five or six basic products, such as liquified petroleum gas (LPG), naphtha, kerosene, diesel, and fuel oils, but specialty manufacturers may produce a large number of their products from these basic refinery products. There is very little published information on specialty manufacturing processes. The selection of a petroleum product for a specific job has become more challenging. Specifications and the test methods used on petroleum products are important for the proper selection of a petroleum product for a given end use.

Part 1, the first six chapters, is devoted to petroleum fuels. Part 2, the remaining chapters, deals with petroleum specialty products. The book presents manufacturing processes, product blending, and specifications of various petroleum products. To make the book useful to the professional in the petroleum industry, an in-depth treatment of each subject not normally found in textbooks is provided. It is hoped that this book will be of direct interest to students and all those engaged in the manufacture, blending, storage, and trading of petroleum products.

Surinder Parkash, Ph.D.

PETROLEUM FUELS MANUFACTURING HANDBOOK

P • A • R • T • 1

PETROLEUM FUELS

CHAPTER 1
LIQUEFIED PETROLEUM GAS

Liquefied petroleum gas, commonly called LPG, is also known by the names of its principal generic components, propane and butane. The normal constituents of LPG are propane (C_3H_8), propylene (C_3H_6), butane (C_4H_{10}), and butylenes (C_4H_8). These are commercial products and may contain other impurities such as smaller quantities of C_5+ hydrocarbons. LPG as a liquid is colorless, and in vapor form it cannot be seen. Pure LPG has no smell, but for safety reasons an odoring agent, usually a mercaptan, is added during manufacture to aid detection at very low concentrations. LPG exists in a gaseous form under normal atmospheric pressure and temperature. It is easily liquefied by moderate pressure at ambient temperatures, which means it can be easily and conveniently stored as a liquid, a big advantage over natural gas, which can be liquefied only at a very low temperature and high pressure. LPG as a liquid is 250 times denser than LPG as vapor, so a large quantity can be stored in a relatively small volume.

Table 1-1 shows the physical properties of LPG constituents. The boiling point at atmospheric pressure of n-butane is 31.08°F and for propane is –43.7°F. Thus propane can be stored in liquid form in tanks exposed to the atmosphere without the danger of freezing in cold winter ambient temperatures. The calorific value of LPG on a volume basis is significantly higher (propane, 95 MJ/m^3; butane, 121 MJ/m^3) compared with that of natural gas (38 MJ/m^3). For this reason, natural gas appliances and LPG appliances cannot be interchanged.

LPG has the following main uses:

1. LPG is the most versatile fuel used in domestic applications. It is used like natural gas and can do everything that natural gas can do. LPG is used for cooking, central heating, space heating, and hot water supply, as well as in a large number of appliances, such as ovens, stovetops, and refrigerators in homes, hotels, and restaurants.
2. LPG is increasingly being used as automobile fuel because of its cost advantage over gasoline and diesel. LPG is a clean-burning fuel. The absence of sulfur and very low levels of nitrogen oxides (NOx) and particulate emissions during its combustion make LPG a most environmentally friendly source of energy. The disadvantage is that LPG has a lower calorific value per unit volume, and thus the vehicle has to refuel more frequently. In industry, LPG is used to power industrial ovens, kilns, furnaces, and for various process heating applications. LPG is used in brick kilns and aluminum die casting, in ceramics, and in glass manufacture. LPG is used to heat bitumen for road building. It has other diverse uses, such as the following:
   - In agriculture, for crop drying, waste incarnation, greenhouse heating, and running power equipment.
   - As a feedstock for chemical manufacture, in water desalination plants, and in aerosol manufacture as a propellant.
   - As a standby fuel for natural gas. LPG is used as automobile fuel in forklift trucks.

In developed countries, most of the LPG demand (more than 80 percent) is for the industrial sector; less than 20 percent of the demand is for the domestic market. In the developing countries of Asia, Africa, and South and Central America, the largest demand for LPG is in the domestic sector. The rural communities that earlier were using biomass (e.g., wood and charcoal) as domestic fuel are now switching over to LPG as the supply available is more.

TABLE 1-1 Properties of LPG Gases

Constituent	Formula	Boiling point 1 ATM °F	Critical temperature °F	Critical pressure lb/in^2	Specific gravity 60/60°F	Vapor pressure at 100°F lb/in^2
Propane	C_3H_8	–43.75	206.06	616.00	0.5070	188.64
Propylene	C_3H_6	–53.86	196.90	669.00	0.5210	227.607
n-Butane	C_4H_{10}	31.08	305.62	550.60	0.5840	51.706
Isobutane	C_4H_{10}	10.78	274.46	527.90	0.5629	72.581
1-Butene	C_4H_8	20.73	295.59	583.00	0.6005	63.2775
Cis-2-Butene	C_4H_8	38.70	324.37	610.00	0.6286	45.7467
Trans-2-Butene	C_4H_8	33.58	311.86	595.00	0.6112	49.8821
Isobutene	C_4H_8	19.58	292.55	580	0.6013	64.583

AUTOMOTIVE LPG

Automotive LPG, or autogas, refers to the LPG used in automotive applications. LPG consists mainly of propane, propylene, butane, and butylenes in various proportions. The composition of autogas varies from country to country depending on the prevailing ambient temperatures. In moderate ambient temperatures, autogas typically consists of 60 to 70 percent propane and 30 to 40 percent butane. The addition of butane slows down combustion speed in an engine and reduces NOx emissions. Components of LPG are gases at normal ambient temperature and pressure but can be easily liquefied for storage by an increase in pressure from 8 to 10 bar or a reduction in temperature. LPG used in automobiles is stored in liquid form in an onboard steel cylinder. LPG has a long and varied history in transportation applications. It is estimated that more than 4 million automobiles use LPG worldwide at present. It has been used in rural farming areas as fuel for farm machinery. LPG is used for some special applications such as forklifts in warehouses. The use of LPG can result in lower vehicle maintenance costs, lower emissions, and fuel cost savings compared with conventional gasoline or diesel fuels. LPG is considered a particularly suitable fuel for heavy vehicles, buses, and delivery vehicles because of its significantly lower particulate emissions compared with diesel-powered buses. The use of LPG as automotive fuel varies from country to country depending on the relative cost of alternative fuels such as gasoline and diesel.

LPG STORAGE

For domestic applications, LPG is stored in 15-kg cylinders. Domestic bulk LPG tanks vary in size from 200 to 2000 kg. They are installed outdoors on customer premises and LPG is delivered from road tankers. The amount of gas delivered is recorded via an onboard meter and charged to the customer. Storage tanks are usually installed aboveground. Propane is stored in a tank as a liquid under a pressure of 7 to 10 bars (100 to 150 PSIA). The gas pressure is reduced in two stages to bring it to a safe working pressure of 37 millibar (0.53 lb/in^2), for which the gas appliances are usually designed to operate.

LPG MANUFACTURE

LPG from Field Gases

About 60 percent of the world supply of LPG comes from associated gas processing, and 40 percent of the LPG is produced in oil refineries from crude distillation, fluid catalytic cracking units (FCCUs), delayed cokers, hydrocrackers, and other conversion processes. The worldwide estimated production of LPG in 2005 was estimated at 250 million tons per year.

Acid Gas Removal

The raw natural or associated gases from a group of wells are received in a knockout drum where gas and liquid phases are separated. The gas is disentrained with the aid of a mist eliminator pad incorporated in the knockout drum and then compressed by a gas compressor for pipeline transport to an acid gas removal plant. Condensate separated in a knockout drum is injected back into the gas stream after water separation. Water separated in the knockout drum is disposed of as wastewater.

The oil field gases contain carbon dioxide and hydrogen sulfide, together known as acid gases. Because these gases are corrosive, poisonous, or both, they are removed first before further processing or LPG separation. Acid gases are separated from the gas stream by amine treating or by the Benfield process in which gases are treated with a solution of potassium carbonate containing some additives. The Benfield process uses an inorganic solution containing 25 to 35 wt % (percentage of weight) K_2CO_3. The absorption is chemical not physical. Figure 1-1 shows the reactions.

The Benfield solution has vanadium pentoxide (V_2O_5), which results in higher gas loading, lower circulation rate, and less corrosion. The absorber operates at 200 to 400°F.

Figure 1-2 shows a process flow diagram of acid gas (CO_2 and H_2S) removal based on the Benfield process. The gases and liquid coming from the field enter feed surge drum V-101, which removes any entrained water. The gas and liquid feed are recombined, and the two-phase mixture is heated in E-101 by heat exchange with sweet gas coming from the top of acid gas absorber V-103. It is further heated with 50 lb/in^2 steam in E-103. It is next fed to absorber V-103 near the bottom. A lean potassium carbonate solution is fed to the absorber at its top and middle sections. The rich solution reaching the bottom of absorber is pumped to regenerator column V-104 via flash drum V-107. The sweet gas from absorber V-103 overhead is cooled in heat exchanger E-101 and next by cooling water in E-102 on its way to separator drum V-102 where the condensate is separated. Sweet gas exits the separator drum V-102 for further processing in an LPG extraction unit. Water separated in the drum is returned to flash drum V-107. Sweet hydrocarbon product is pumped out to mix with sweet gas from drum V-102. Potassium carbonate solution rich in acid gas is regenerated in V-104. The solution is fed to the top of a packed column. The rich solution is regenerated by reboiling with steam in reboiler E-106. The lean solution is collected at the bottom of the column and returned to the absorber. Any makeup potassium carbonate solution required by the absorber is drawn from carbonate storage drum V-106. The regenerator overheads are condensed by air cooler E-105 and collected in regenerator accumulator V-105. Acid gases remain uncondensed and exit V-105 to the sulfur plant.

Extraction Plant

The combined feed to extraction plant typically comprises associated gases and condensate from oil-producing areas plus refinery gases after treating for acid gas removal. The extraction process involves feed compression, feed/effluent heat exchange, dehydration, absorption, and stripping. Three product streams are produced; a liquid stream rich in propane, butane, and gasoline that is sent to the fractionation plant and two overhead gas streams that supply gas to the fuel system. Absorption oil is provided by a recycled gasoline product. A closed cycle propane refrigeration system supplies low-temperature chilling.

Referring to the process flow diagram in Fig. 1-3, oil field gases and refinery gases from acid gas removal plant are received in knockout drum V-201 at 336 lb/in^2 where gas and liquid phases are separated. The gas is compressed by gas compressor K-201 to 571 lb/in^2 and after-cooled in after-cooler E-201 while liquid separated is pumped by pump P-201 to accumulator drum V-202.

$$K_2CO_3 + CO_2 + H_2O = 2\,KHCO_3$$

$$K_2CO_3 + H_2S = KHS + KHCO_3$$

FIGURE 1-1 Absorption and regeneration of acid gas.

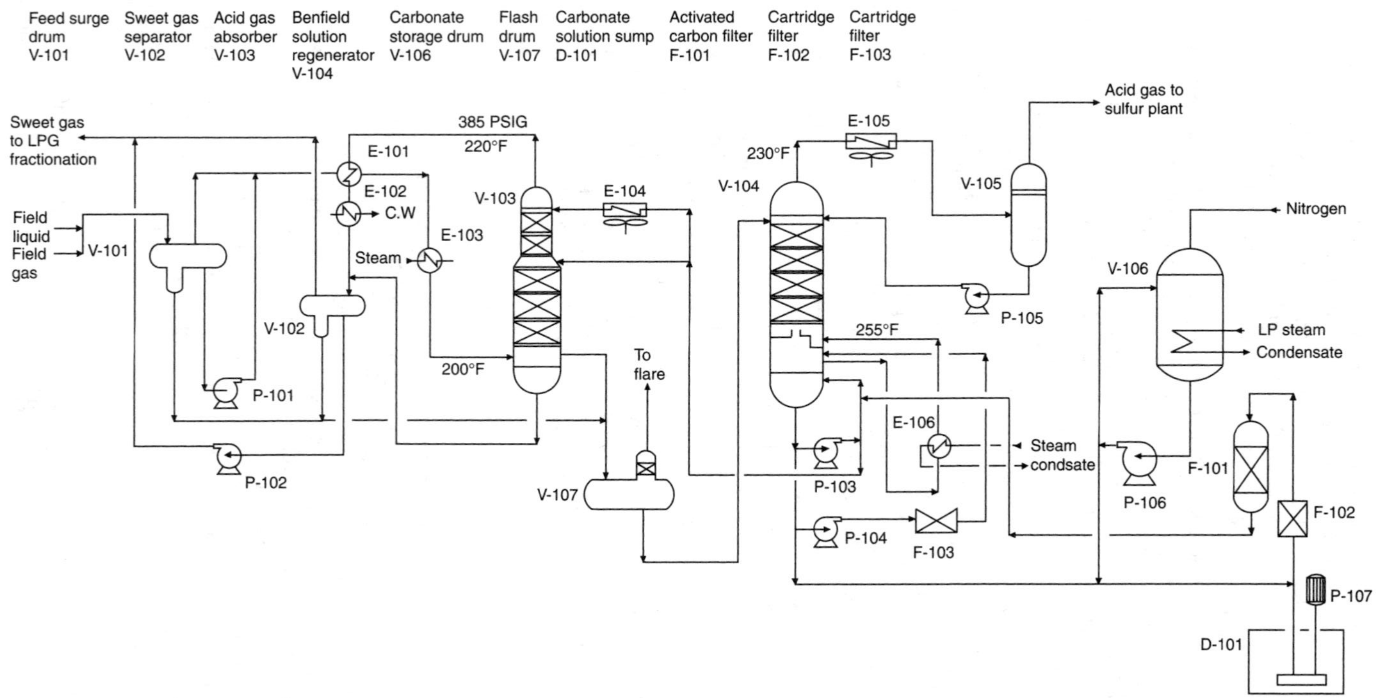

FIGURE 1-2 Acid gas removal plant.

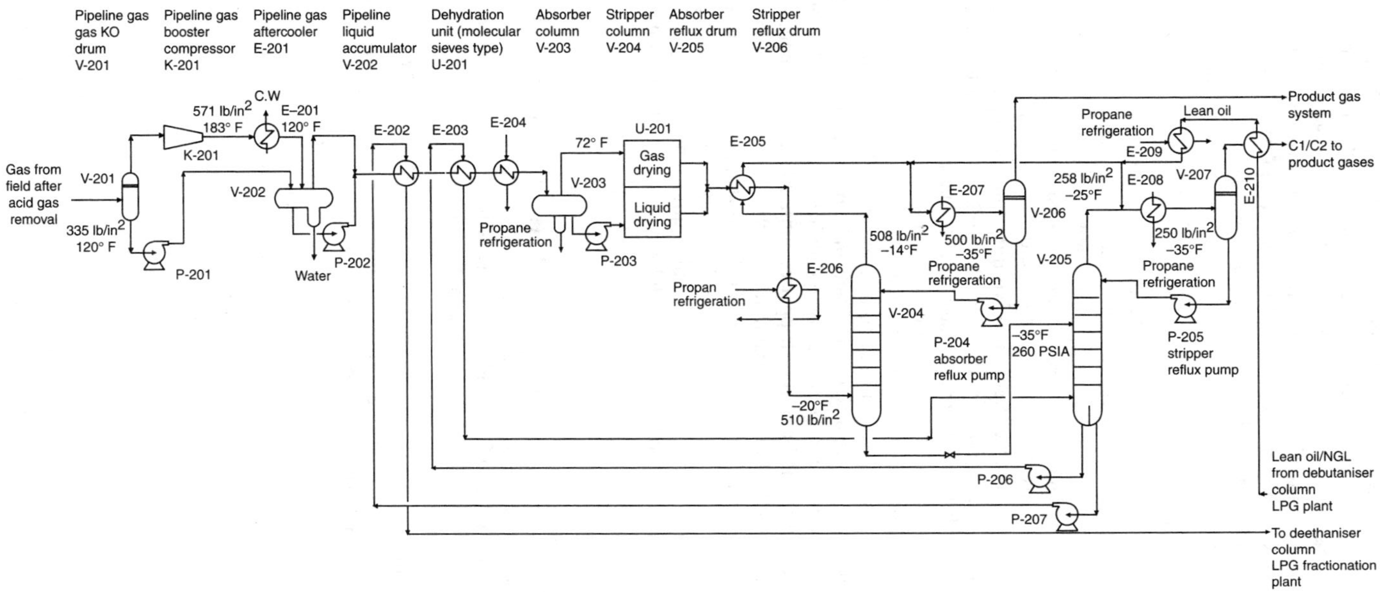

FIGURE 1-3 LPG extraction plant.

The mixed-phase feed from V-202 exchanges heat with stripper (V-204) bottoms in feed/stripper bottom exchanger E-202 and then reboils the stripper reboiler E-203. The feed gas is further cooled by chilling with high-level refrigerant propane in E-204. The condensed hydrocarbons are separated from gas in V-203. Gases that leave V-203 go to gas dehydration unit U-201 while liquid hydrocarbons are pumped out by P-203 to a liquid dehydration unit.

Dehydration units are provided to remove moisture from gas and liquid and thus prevent freezing in the cold end of the plant. The gas enters the gas dehydration unit at 544 lb/in^2 and 72°F. When it leaves the unit, the water content is reduced to 1 ppm maximum. Similarly, water content of liquid phase is reduced to 4.5 ppm maximum.

The dried gas and liquid streams from dehydration unit U-201 are combined for further chilling in exchangers E-205 and E-206 and cooled from 72 to –20°F at the absorber column V-204 inlet. The absorber column V-204 recovers propane, butane, and heavier hydrocarbons, from the feed with a minimum loss of these components. The absorbent for this operation is natural gasoline recycled from fractionation plant debutanizer column bottoms. The two-phase feed at –20°F and 510 lb/in^2 enters the bottom of absorber V-204 where liquid and vapor are separated. The ascending vapor contacts the descending liquid absorbent on valve trays, and absorption of heavier components take place. The overhead vapor is mixed with chilled lean oil and cooled to –35°F by heat exchange with low-level propane in absorber oil presaturator E-207. The effluent from E-207 is phase separated in absorber reflux drum V-206. The liquid from reflux drum is pumped by reflux pump P-204 to absorber column as reflux. Absorber overhead vapor leaves the plant to product gas/fuel systems. The rich liquid from absorber bottom is transferred to a stripper V-205 via a throttle valve. The function of stripping column V-205 is to reduce the methane and ethane content of the absorber bottoms. Stripping is done at reduced pressure, approximately 260 lb/in^2. The absorber bottoms are let down to stripper bottom pressure and flashed into the stripper column. Most of the methane and some ethane are flashed off and ascend to the top of the column contacting the descending reflux on valve trays where some of the heavier components are reabsorbed. The stripper overheads are mixed with chilled lean oil and cooled to –35°F by low-level propane in stripper oil presaturator E-208. The cold mixture is separated in stripper reflux drum V-207, and the liquid is pumped by reflux pump P-205 to the stripper column. The overhead vapor from V-206 leaves the plant to a gas distribution/fuel system.

Fractionation Plant

The stripper bottom product from the LPG extraction plant is comprised of propane, butane, and natural gasoline with some associated ethane and lighter components. This stripper bottom constitutes feed to the LPG fractionation plant where it is separated into a gas product, propane, butane, and natural gasoline in three fractionation columns.

Deethanizer. Referring to the process flow diagram in Fig. 1-4, the stripper bottoms from the extraction plant enter deethanizer column V-101 near the top. The overhead vapor is partially condensed in deethanizer condenser E-101 by heat exchange with medium-level propane at 20°F. Condensed overhead product in overhead reflux drum V-104 is pumped back to the deethanizer by reflux pump P-101. The noncondensed vapor, mainly ethane, leaves the plant to fuel the gas system. Heat is supplied to the column by forced circulation reboiler E-104. The deethanizer column operates at approximately 390 lb/in^2. Approximately 98 percent of the propane in the deethanizer feed is recovered in the bottom product. The residual ethane concentration is reduced to approximately 0.8 mol % (mole percentage) in the bottom product. The bottom product from deethanizer pressure drains into depropanizer column V-102.

Depropanizer. Deethanizer bottoms are expanded from 390 to 290 lb/in^2 and enter depropanizer V-102 as mixed-phase feed. The depropanizer fractionates the feed into a propane-rich overhead product and a bottom product comprised of butane and natural gasoline. Tower V-102 overhead vapor is totally condensed in the depropanizer condenser E-102 by cooling water, and condensate is collected in depropanizer column reflux drum V-105. A part of the condensed overhead product is sent back to the column as reflux via pump P-103 while the remaining part is withdrawn as a liquid propane product.

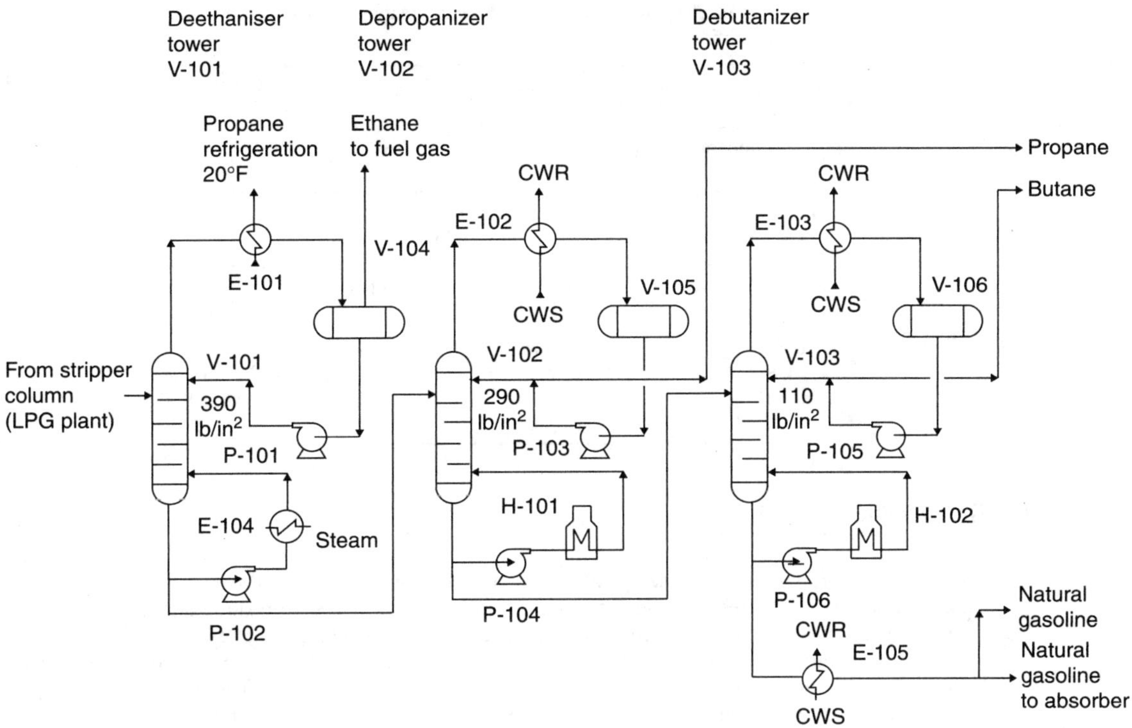

FIGURE 1-4 LPG fractionation system.

Column V-102 reboil heat is supplied by direct-fired heater H-101. Reboiler circulation is aided by reboiler circulation pump P-104. The bottom product is sent to debutanizer column V-103.

Debutanizer. The depropanizer bottoms are expanded from approximately 290 to 110 lb/in^2 and enter the debutanizer column as a mixed-phase feed. The column feed is fractionated into a butane-rich overhead product and natural gasoline bottoms. The columns overhead are totally condensed in the debutanizer condenser E-103 by heat exchange with cooling water, and condensate is collected in reflux drum V-106. The debutanizer reflux and product pump P-105 serve the dual purpose of supplying reflux to the column and allowing withdrawal of column overhead product butane from the reflux drum. The column reboil heat is supplied by a direct-fired debutanizer reboiler H-102, and the boiler circulation is aided by debutanizer reboiler circulating pump P-106. The bottom product leaving the column is cooled in product cooler E-105. A part of the gasoline product is recycled to the LPG extraction unit and serves as lean oil for the absorber column.

Product Treatment Plant

Propane and butane products from the fractionation plant contain impurities in the form of sulfur compounds and residual water that must be removed to meet product specifications. The impurities are removed by adsorption on molecular sieves. Each product is treated in a twin fixed-bed molecular sieve unit. Regeneration is done by sour gas from the stripper overhead followed by vaporized LPG product. Operating conditions are listed in Table 1-2 and impurities to be removed are listed in Table 1-3.

TABLE 1-2 Molecular Sieve Product Treating Process Operating Conditions

Operating variable	Units	Propane	Butane
Pressure	lb/in^2	325	155
Temperature	°F	110	110
Phase		Liquid	Liquid

TABLE 1-3 Typical Contaminant Level in Untreated LPG

Contaminants	Units	Propane	Butane
H_20	wt ppm	10	Trace
H_2S	wt ppm	100	Trace
COS	wt ppm	34	Trace
C_3SH	wt ppm	100	40
C_2H_5SH	wt ppm	Trace	220

LPG SPECIFICATIONS

Commercial propane and butane specification conforming to U.S. Gas Processor Association standards are listed in Tables 1-4 and 1-5. Indexes for "R" and "O" give residue and oil stain results, respectively, in whole numbers. In these specifications, under residual matter, "R" refers to residue volume in milliliters multiplied by 200. "O" refers to 10 divided by oil stain observation in millimeters. Specifications for autogas conforming to EN 589 are listed in Table 1-6. The most important specifications for auto LPG are motor octane number and vapor pressure. Commercial butane-propane (BP) mixtures used for domestic uses contain varying amounts of C_3 and C_4 hydrocarbons as per the ambient conditions (Table 1-7).

TABLE 1-4 Commercial Propane Specifications

Property	Units	Limit	Value	Test method
Composition				ASTM D 2163
C_2 and lighter	Mol %	Max.	2.0	
C_3 hydrocarbons	Mol %	Min.	96.0	
C_4 and heavier	Mol %	Max.	2.5	
Cu corrosion strip, 1 h	@ 37.8°C	Max.	No. 1	ASTM D 1838
Hydrogen sulfide			Negative	ASTM D 2420
Moisture content			Pass	ASTM D 2713
Residual matter				ASTM D 2158
"R" number		Max.	10	
"O" number		Max.	33	
Relative density	60/60°F		Report	ASTM D 1657/D 2598
Sulfur	ppm	Max.	60	ASTM D 2784/D 3246
Vapor pressure @ 37.8°F	lb/in^2	Max.	200	ASTM D 1267
Ammonia	ppm	Max.	Report	Drager tubes
Carbonyl sulfide	ppm	Max.	Report	UOP 212
Diene	Mol %	Max.	0	ASTM D 2163
Hydrogen sulfide (H_2S)	ppm		Report	UOP 212
Unsaturates	Mol %	Max.	1.0	ASTM D 2163
Volatile residue				
Temperature @ 95 % evaporation	°C	Max.	–38.3	

TABLE 1-5 Commercial Butane Specifications

Property	Units	Limit	Value	Test method
Composition				ASTM D 2163
C_4 Hydrocarbons	Mol %	Min.	95.0	
C_5 and heavier	Mol %	Max.	2.0	
Free water content	Visual		None	
Cu corrosion strip, 1 h	@ 37.8°C	Max.	No. 1	ASTM D 1838
Hydrogen sulfide			Negative	ASTM D 2420
Relative density	60/60°F		Report	ASTM D 1657/D 2598
Sulfur	ppm	Max.	60	ASTM D 2784/D 3246
Vapor pressure @ 37.8°F	lb/in^2	Max.	70	ASTM D 1267
Ammonia	ppm	Max.	Report	Drager tubes
Diene	Mol %	Max.	0	ASTM D 2163
Hydrogen sulfide (H_2S)	ppm	Max.	Report	UOP 212
Unsaturates	Mol %	Max.	1.0	ASTM D 2163
Volatile residue				
Temperature @ 95% evaporation	°C	Max.	2.2	

TABLE 1-6 Autogas (LPG for Automobiles) Specifications

Characteristics	Units	Limit	Value	Test method
Vapor pressure, 40°C	kPa	Min.	800	ISO 8973
		Max.	1530	
Volatile residue (C5 and heavier)	Mol %	Max.	2.0	ISO 7941
Diene	Mol %	Max.	0.5	ISO 7941
Total volatile sulfur	mg/kg	Max.	100	ASTM D 2784
Motor octane (Mon)		Min.	90.5	ISO 7941/EN 589
Cu strip corrosion test, 38°C			No. 1	EN ISO 6251
Residue on evaporation	mg/kg	Max.	100	JLPGA-S-03
Moisture content, @ 0°C			NIL	EN 589
Hydrogen sulfide			NIL	EN ISO 8819

TABLE 1-7 Commercial LPG (B-P Mixture)

Property	Units	Limit	Value	Test method
Composition				ASTM D 2163
C_3 Hydrocarbons	Mol %	Min.	22.0	
		Max.	24.0	
C_4 Hydrocarbons	Mol %	Min.	76.0	
		Max.	78.0	
Cu corrosion strip, 1 h	@ 37.8°C	Max.	No. 1	ASTM D 1838
Hydrogen sulfide			Negative	ASTM D 2420
Odorant				
Tetrahydrothiophene			*	
Relative density	60/60°F		Report	ASTM D 1657/D 2598
Residual matter				ASTM D 2158
"R" number		Max.	10	
"O" number		Max.	33	
Sulfur	ppm	Max.	60	ASTM D 2784/D 3246
Vapor pressure @ 37.8°C	lb/in^2	Max.	93	ASTM D 1267

* Sufficient to odorize product.

CHAPTER 2
NAPHTHA

Naphtha is the lightest liquid distillate product of crude distillation consisting of C5 to C10 hydrocarbons boiling in the 100 to 310°F range. It is produced from the atmospheric distillation of crude oil and from many secondary processing units in the refinery. Unlike other petroleum fuels such as kerosene, diesel, or fuel oil, naphtha is not a direct petroleum fuel but is used as a feedstock for the manufacture of plastics and polymers, synthetic fiber, petrochemicals, fertilizer, insecticides and pesticides, industrial solvents for making specialty solvents such as food grade hexane, dyes, and chemicals. In refineries, naphtha is one of the basic feedstocks for the manufacture of gasoline. At locations where natural gas is not available, naphtha is used as a feedstock for producing hydrogen required for hydroprocessing units in refineries. Naphtha is sometimes used as fuel in gas turbines or boilers for power generation units. The worldwide naphtha demand in 2006 was estimated at 900 million tons.

NAPHTHA PRODUCTION

Naphtha is produced from the following units:

- Crude distillation units in the refinery.
- Secondary processing units in the refinery.
- Gas-processing units separating LPG from field gases. Naphtha thus separated is known as natural gas liquid.

Crude Distillation Unit

The yield of naphtha cut from crude distillation depends on the crude oil processed. Lighter crude oils yield larger volumes of naphtha on processing. Table 2-1 lists the yield of naphtha from some Middle Eastern crude oils. Naphtha produced in the refinery is typically a straight C5-310°F cut from the crude distillation unit. Naphtha cut withdrawn from crude column is not a sharp cut because it contains lighter as well as heavier components such as LPG and kerosene.

Naphtha production in the refinery is a two-step process:

1. Production of a broad cut from a crude distillation unit (CDU).
2. Refractionation of the broad naphtha cut to remove light and heavier components.

In the CDU (Fig. 2-1), crude oil is preheated by heat exchange with product streams and enters preflash tower V-100. The preflash tower is a small distillation column with four to five plates that removes most of the LPG gases and some light naphtha as overhead product. The preflash tower top vapors are cooled in exchangers E-101 and E-102 and collected in reflux drum V-103. A part of this preflashed naphtha is sent back to column V-100 as reflux, and the rest is routed to naphtha refractionation section via V-102. The topped crude from the preflash tower is fed to main atmospheric

TABLE 2-1 Yield of Naphtha from Various Crude Oils

Crude	Arab light	Kuwait export	Bahrain	Arab heavy	Bombay high	Safania	Dubai
Crudei API	34.2	30.5	30.4	28.3	39.5	27.1	31.78
Yields, Vol %							
LSR	7.60		4.80	7.30	8.70	4.30	7.10
HSR	10.40		7.60	8.10	15.60	6.90	9.80
Total naphtha	18.00	15.10	12.40	15.40	24.30	11.20	16.90
Kerosene	16.00		19.40	14.60	19.70	20.60	14.40

distillation column V-101. Naphtha is withdrawn from the crude distillation column's reflux drum V-102 and routed to the naphtha refractionation unit. Naphtha liquid withdrawn from the CDU column reflux drum V-102 contains heavy ends that must be removed. Similarly, the LPG gas product from V-102 reflux drum contains some naphtha vapor that must be recovered. Naphtha vapors from V-102 are compressed in compressor C-101 and cooled in a series of water-cooled heat exchangers.

Naphtha Refractionation Unit. The condensed naphtha is collected in naphtha feed drum V-500 (Fig. 2-2). The uncondensed vapors from V-500 enter absorber V-501 near the bottom and are absorbed in a stream of kerosene that enters V-501 near the top. The rich kerosene stream

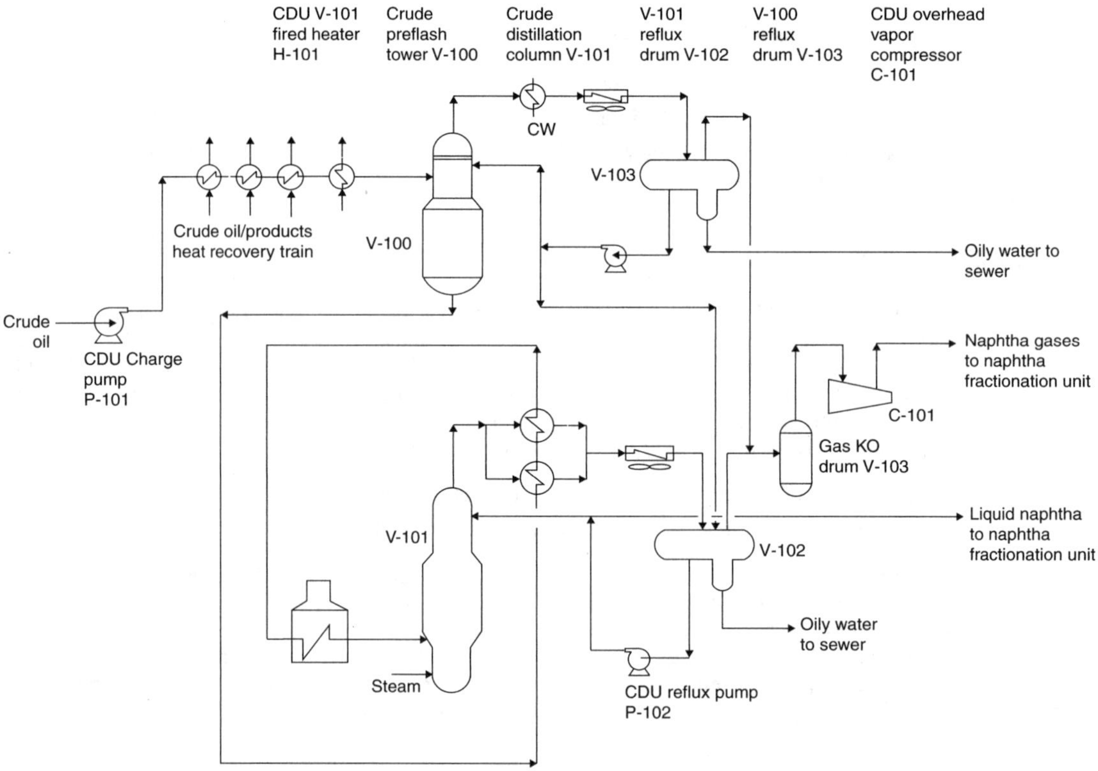

FIGURE 2-1 Simplified process flow diagram for naphtha production from crude distillation unit in refinery.

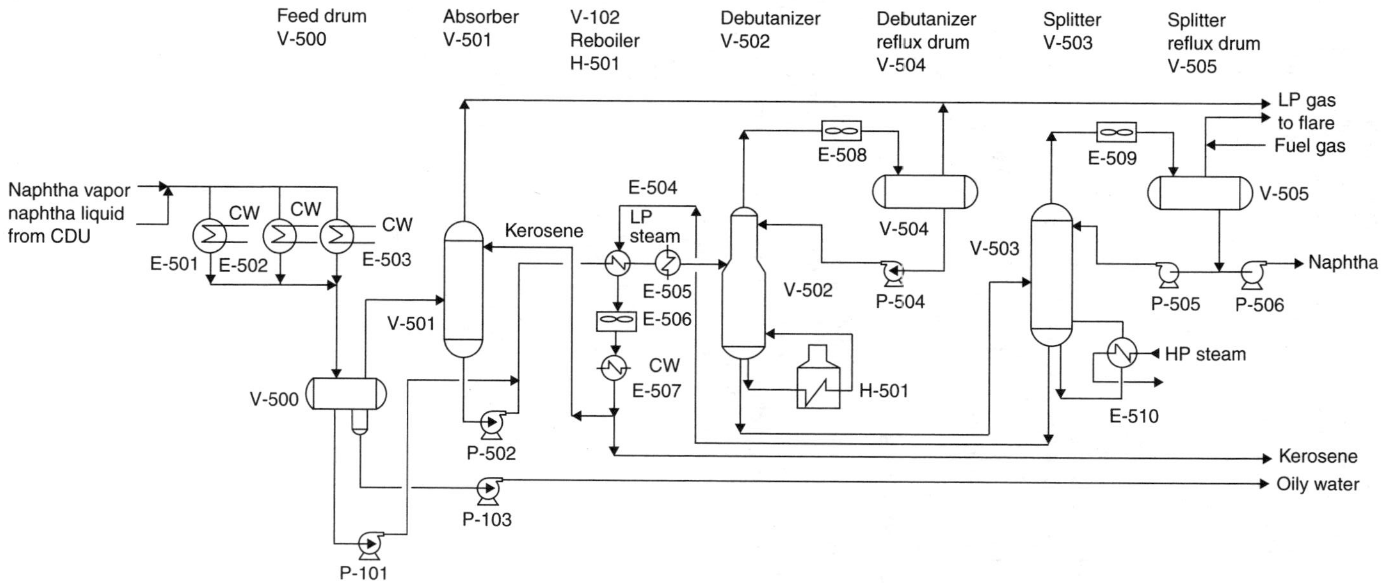

FIGURE 2-2 Naphtha refractionation.

leaving V-501, along with condensed naphtha from V-500 after heating with steam in E-505, enters debutanizer column V-502, which removes all C4 and lighter product from naphtha as overhead product. The bottom product from debutanizer V-502 is sent to a splitter column V-503 where naphtha is removed as a top product and kerosene as a bottom product. A part of kerosene is recycled to absorber V-501 as sponge oil.

Production from Secondary Processing Units

Naphtha is also produced from secondary conversion units such as distillate hydrocrackers, delayed coker units, and resid hydrocrackers. Small quantities of naphtha are also produced by distillate desulfurizer units. However, the distillate hydrocracker is the most important conversion unit, which produces approximately 31 vol % (percentage of volume) naphtha on feed. Compared with straight run naphtha, hydrocracker naphtha has a lower paraffin and higher naphthene content. Hydrocracker heavy naphtha, because of its high naphthene content, is a preferred feedstock for catalytic reformer units. Feed with high naphthene content gives a higher reformate and hydrogen yield.

Production from Associated Gas

Almost 10 percent of total naphtha production comes from associated gas processing. A large quantity of associated gas is also produced as a by-product during crude oil production. Gas separated from oil may contain carbon dioxide, hydrogen sulfide, methane, ethane, propane, normal butane and isobutane, and C5+ hydrocarbons. The typical associated gas composition from a Middle Eastern oil field is listed in Table 2-2. The gas is first processed to remove acid gases (CO_2 and H_2S). Next C3+ components such as propane, butane, and natural gasoline are separated from C1 and C2 gases by cooling with a propane refrigeration system to a low temperature. C3+ hydrocarbons condense as liquid and are separated in a flash drum. The separated hydrocarbons are further separated into propane, butane, and natural gasoline by fractionation in a series of columns. The separated C1 and C2 gases are stripped of any heavier hydrocarbons

TABLE 2-2 Typical Associated Gas Composition

Component	Weight %
H_2	0.00
N_2	0.22
CO_2	2.61
H_2S	0.04
C_1	37.40
C_2	20.97
C_3	19.42
I C_4	3.31
N C_4	8.16
I-C_5	2.00
N C_5	2.63
C_6	2.44
C_7	0.51
C_8	0.20
C_9	0.05
C_{10}+	0.01
Water	0.03
Total	100.00

TABLE 2-3 C_4+ Natural Gasoline Composition and Properties*

	Vol %
Isobutane	0
Normal butane	0
Isopentane	25.6
Normal pentane	37.7
Cyclopentane	0
2,2 Dimethyl butane	1.2
2,3 Dimethyl butane	2.2
2 Methyl pentane	9.7
3 Methyl pentane	6.3
Normal hexane	2.2
Methyl cyclopentane	5.2
Cyclohexane	5.8
Benzene	1.8
C_7+	2.3
Density, g/mL	0.6568
PONA, vol %	
Paraffins	73.9
Naphthene	15.1
Aromatics	11
Sulfur, ppmw	0.5

*Separated from field gases.

they may contain by absorbing in natural gasoline liquid in an absorber. Naphtha produced from associated gas is called light naphtha. Table 2-3 lists its composition and properties. Light naphtha consisting mainly of C5 and C6 hydrocarbon components is a preferred isomerization unit feed. Isomerization unit isomerizes C5 and C6 normal paraffins to branched chain hydrocarbons and increases the research octane number (RON) from 70 to 83. Isomerate is an important gasoline blend component for controlling the Reid Vapor Pressure (RVP) and distillation specification of blended gasoline.

SECONDARY PROCESSING UNITS

Table 2-4 lists the typical naphtha yield from various secondary processing units. Naphtha properties from secondary processing units such as the distillate hydrocracker and delayed coker are presented in Tables 2-5 and 2-6. Naphthas produced from coker or resid hydrocrackers usually have high nitrogen, sulfur, and olefin content, and they require hydrotreating before blending into the naphtha pool.

TABLE 2-4 Naphtha Yield from Various Refinery Units

Units		Naphtha yield
Distillate hydrocracker	Vol %	31.5
Delayed coker	Vol %	1.9
Resid hydrocracker (H oil)	Vol %	7.3
Resid desulfurizer	Vol %	3.0
Diesel desulfurizer	Vol %	0.9
Kerosene desulfurizer	Vol %	1.3

TABLE 2-5 Light and Heavy Naphtha Properties Ex Hydrocracker

Property	Units	C_5-180°F	180–320°F
Gravity	°API	79	52.5
Density	g/mL	0.672	0.769
Aniline point	°F		107
Distillation	°F		
IBP		110	215
10%		115	225.0
30%		125	245.0
50%		135	260.0
70%		150	270
90%		170	295
EP		195	325
PONA	Vol %		
Paraffins		74	27
Naphthenes		22	58
Aromatics		4	15
Sulfur	ppm	<10	<10
Octane number	Clear, F-1	77	61

TABLE 2-6 Light Coker Naphtha Properties

Property	Units	Value
Gravity	°API	82.1
Density, 60/60°F	kg/L	0.662
Distillation		
IBP		
10 vol %	°F	118
50 vol %	°F	126
90 vol %	°F	147
RON clear		81.8
PONA		
Paraffins	Vol %	36
Olefins	Vol %	56
Naphthenes	Vol %	0
Aromatica	Vol %	8.0
Bromine number		70.0
Nitrogen content	ppm	100.0
Sulfur	Wt %	0.10
Reid vapor pressure	kPa @100°F	91

NAPHTHA DESULFURIZATION

Naphtha produced from crude oil distillation, coking units, or from field gases may contain sulfur, mercaptan, and H_2S as impurities that must be removed or reduced to a low level before naphtha can be used as feedstock in any downstream catalytic process. The naphtha hydrodesulfurization (HDS) unit serves to make naphtha feed suitable for catalytic conversion processes by removing sulfur, nitrogen, trace metals, or other catalytic poisons from feed. This is done by reacting feed

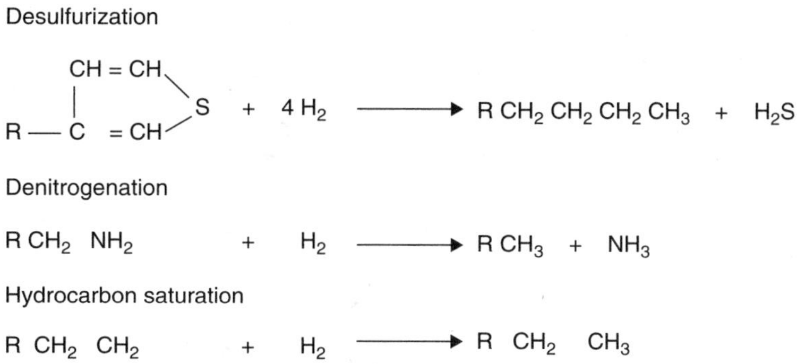

FIGURE 2-3 Hydrodesulfurization reactions.

with hydrogen at high temperature and pressures. Sulfur is converted to H_2S and nitrogen to NH_3, which are removed by distillation. Typical desulfurization reactions occurring in the HDS reactor are shown in Fig. 2-3.

Naphtha feed to cat reformers must meet 1 ppm or lower sulfur level specifications to protect noble metal catalyst in the reforming unit. In the naphtha steam reforming process for the production of hydrogen, naphtha sulfur must be reduced to less than 0.5 ppm in order to prevent poisoning of the nickel catalyst in the reactor. Sulfur is removed from naphtha in a naphtha desulfurization unit. However, if only H_2S is present in the feed, vapor feed is passed over a guard reactor containing ZnO, which absorbs H_2S.

NAPHTHA HDS UNIT

Referring to the process flow diagram in Fig. 2-4, naphtha feed from storage tanks is pumped by charge pump P-101 through reactor effluent-feed exchangers E-101 to E-103 and fired heater H-101 into the top of reactor V-101. The reactor is loaded with a desulfurization catalyst consisting of cobalt-molybdenum (Co-Mo) metals on an alumina base. Hydrogen from the catalytic reformer or hydrogen plant is compressed by centrifugal compressor C-101 and sent to the feed stream upstream of the reactor effluent-feed exchanger. The reactor effluent is cooled in effluent-feed exchangers E-101, E-102, E-103, in air cooler E-104, and in trim water cooler E-105 before flowing into high-pressure (HP) separator V-102. Hydrogen-rich vapors from a HP separator are routed back to compressor C-101. A small part of this stream is purged off to prevent buildup of H_2S in the hydrogen stream. HP separator liquid is flashed in low-pressure separator V-103 where H_2S and lighter hydrocarbons are separated from liquid. The liquid from V-103 is pumped to stabilizer column V-104 through feed/bottom exchanger E-106. Stabilizer column V-104 overhead vapors are condensed in air cooler E-107 and in water trim cooler E-108 and flow into accumulator V-105. The vapor from the accumulator (C4 and lighter) along with gas from the flash drum is sent to an amine unit for H_2S removal and gas recovery. The liquid from the V-105 is returned to the stabilizer column as total reflux. Stabilizer bottoms flow through E-106 to dehexanizer column V-106. The objective of dehexanizer column V-106 is to split desulfurized naphtha into light and heavy naphtha. Heavy naphtha is used as a feed for the catalytic reforming unit. The dehexanizer overhead vapors are condensed in E-109 and flow into reflux drum V-107. A part of the condensed liquid is sent back to the column as reflux, and the remainder is pumped through water trim cooler E-113 to storage as light naphtha (C5/C6) product. Dehexanizer bottoms (heavy naphtha) are cooled by pumping through air cooler E-111 and water trim cooler E-112 and routed to storage tanks. Table 2-7 lists the key operating conditions for a naphtha HDS unit.

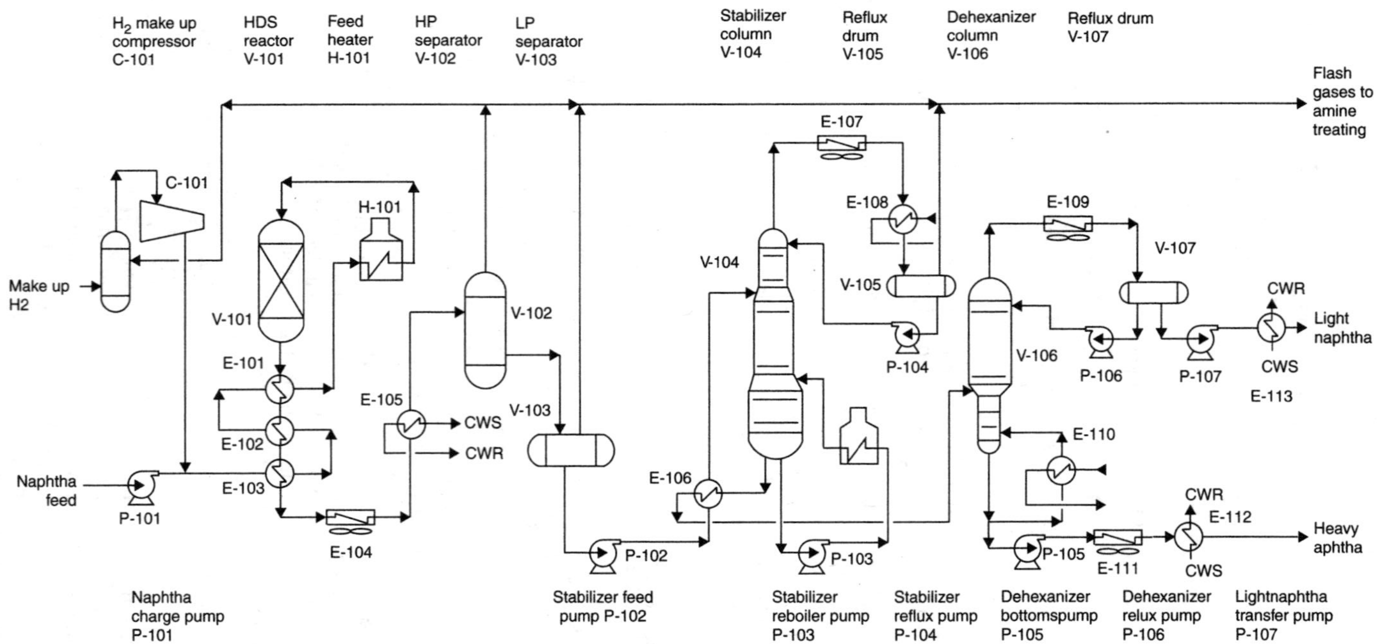

FIGURE 2-4 Naphtha desulfurization and fractionation.

TABLE 2-7 Naphtha HDS Unit Operating Conditions

Property	Units	Value
Reactor inlet temperature @ SOR/EOR	°F	608/698
HP separator pressure	kg/cm^2	21.3
Hydrogen partial pressure at reactor outlet	kg/cm^2	11.2
VVH (m^3/h naphtha/m^3 catalyst)	15°C	4.00

NAPHTHA SPECIFICATIONS

Naphtha may be classified by its boiling range or by its end use:

- Light straight run (LSR) naphtha
- Wide straight run (WSR) naphtha
- Petrochemical naphtha

LSR NAPHTHA

LSR naphtha is a light naphtha cut produced from crude oil distillation with a boiling range of C5 to180°F. It consists mainly of C5 and C6 hydrocarbons. It is highly paraffinic. The paraffin content of light naphtha is greater than 80 vol %. Table 2-8 lists the specifications of light naphtha, which is typically blended from hydrocracker light naphtha, meroxed coker light naphtha, and natural gasoline separated from associated gas. Specifications limit blending of light cracked naphtha such as light coker naphtha in LSR blends to an olefin content of 1.0 vol % maximum. LSR naphtha, because of its volatility, is a preferred feedstock for refinery isomerization unit to make a light gasoline blending component. LSR naphtha has a low RON of approximately 60. In the isomerization unit, feed is vaporized, mixed with hydrogen, and passed over a platinum-impregnated chlorinated

TABLE 2-8 Light Naphtha Specifications

Property	Units	Limit	Value	Test method
Color, saybolt			+20	ASTM D 156
Density	kg/L	Min.	0.645	ASTM D 1298
		Max.	0.700	
Distillation				ASTM D 86
IBP			Report	
10 vol %	°F	Max.	131	
50 vol %	°F	Max.	149	
90 vol %	°F	Max.	239	
End point	°F	Max.	320	
Lead content	ppb	Max.	50	IP 224
PONA				Chromatography
Paraffins	Vol %	Min.	80	
Olefins	Vol %	Max.	1.0	
Naphthene	Vol %	Max.	18.0	
Aromatics	Vol %	Max.	5.0	
Sulfur	Wt %	Max.	0.03	ASTM D 1266
Vapor pressure, Reid	kPa @100°F	Max.	91	ASTM D 323

alumina catalyst in an isomerization reactor. Isomerization of C5 and C6 normal paraffins to isoparaffins increases RON by 18 to 22 numbers. Isomerate is a very useful blend component to control gasoline distillation and to reduce high aromatic content reformate blending in gasoline.

WSR Naphtha

WSR is a (C5-310°F) cut from crude distillation units. Typical product specifications for WSR naphtha are listed in Table 2-9. A significant part of WSR is used as petrochemical naphtha or as a feedstock for the catalytic reforming unit for the production of a motor gasoline blend component.

TABLE 2-9 Wide Straight Run (WSR) Naphtha Specifications

Property	Units	Limit	Value	Test method
Color, Saybolt			Report	ASTM D 156
Density	kg/L	Min.	0.690	ASTM D 1298
		Max.	0.735	
Distillation				ASTM D 86
IBP			Report	
10 vol %	°F	Min.	109	
		Max.	210	
50 vol %	°F	Min.	174	
		Max.	270	
90 vol %	°F	Min.	230	
		Max.	351	
End point	°F	Max.	399	
Lead content	ppb	Max.	200	IP 224
Olefins	Vol %	Max.	1.0	ASTM D 1319
or				
Bromine number		Max.	1.0	ASTM D 1159
PONA			Report	
Sulfur	Wt %	Max.	0.07	ASTM D 1266
Vapor pressure, Reid	kPa @100°F	Max.	75	ASTM D 323

Petrochemical Naphtha

In refineries, petrochemical naphtha is blended from the straight run naphtha ex-crude distillation column and hydrocracker naphtha. Naphtha used as petrochemical feedstock must be highly paraffinic (minimum 70 vol %) with low aromatic content (less than 11 vol %). Table 2-10 lists the typical specifications of petrochemical naphtha.

Naphtha is widely used in the fertilizer and petrochemical industries (naphtha crackers for production of ethylene, propylene, etc.) as feedstock. The most important criteria for the selection of naphtha as feedstock in petrochemical and fertilizer plants are its paraffin and aromatic content. This is because high-paraffin naphtha is an excellent cracking feedstock in naphtha cracker and yields higher volumes of ethylene and other olefinic products. In the fertilizer plant, naphtha is mixed with steam passed over a nickel catalyst at high temperature (steam reforming) to synthesis gas ($CO + H_2$). The synthesis gas is further converted to ammonia and urea. Paraffinic naphtha requires lower steam reformer severity, whereas high aromatic feed is basically resistant to the steam reforming reaction requiring a higher severity operation. Aromatics also produce more coke lay-down on catalyst with a resultant shorter catalyst life.

TABLE 2-10 Petrochemical Naphtha Specifications

Property	Units	Limit	Value	Test method
Color, Saybolt		Min.	+20	ASTM D 156
Density, 60/60°F	kg/L	Min.	0.680	ASTM D 1298
		Max.	0.725	
Distillation				ASTM D 86
IBP			Report	
10 vol %	°F	Min.	Report	
		Max.		
50 vol %	°F	Min.	122	
		Max.	248	
90 vol %	°F	Min.	167	
		Max.	320	
End point	°F	Max.	356	
Lead content	ppb	Max.	200	IP 224
Olefins	Vol %	Max.	1.0	ASTM D 1319
PONA				
Paraffins	Vol %	Min.	70	Chromatography
Olefins	Vol %	Max.	1.0	
Naphthene	Vol %		Report	
Aromatics	Vol %		Report	
Sulfur	Wt %	Max.	0.07	ASTM D 1266
Vapor pressure, Reid	kPa @100°F	Max.	91	ASTM D 323

NAPHTHA USES

Catalytic Reforming

Catalytic reforming is an important refinery process for the manufacture of gasoline from naphtha. Straight run naphtha from crude oil distillation consists mainly of paraffins and naphthenes. It has a low octane number and cannot be used for gasoline blending.

The catalytic reformer unit converts low-octane heavy naphtha to high-octane catalytic reformate. Reformate is used as a blend component for gasoline blending. The reformate octane number can be increased by increasing the process severity, that is, increasing the reactor temperature or decreasing the space velocity. Increasing severity decreases reformate yield. Most cat reformers for gasoline manufacture operate in the 95 to 100 RON severity range. The manufacture of aromatics requires catalytic reforming units with continuous catalyst regeneration, operating in a higher severity RON range (100 to 102).

Figure 2-5 presents the chemical reactions that take place during catalytic reforming. Important chemical reactions are dehydrogenation, dehydrocyclization, isomerization and hydrocracking. Dehydrogenation reactions convert naphthenes to aromatics with higher octane number. Dehydrocyclization reactions convert normal paraffins to naphthenes, and isomerization of normal paraffins convert n-paraffins to branched hydrocarbons with high RON. Hydrocracking of paraffins decreases molecular weight and increases the volatility of reformate. The yield of reformate is a function of feed paraffin, olefin, naphthene, and aromatic (PONA) and cat reformer operating conditions. Naphthas with high naphthene content are the preferred feed for the catalytic reformer unit. The cat reforming process also produces hydrogen gas, which can be used for various desulfurization units in the refinery.

Naphtha feed for the reformer unit comes from the naphtha desulfurization unit where naphtha, sulfur, and nitrogen are both reduced to below 0.5 ppm.

Referring to the process flow diagram in Fig. 2-6, in the reforming unit, naphtha feed from the naphtha HDS unit or storage tank is mixed with hydrogen from recycle compressor K-101

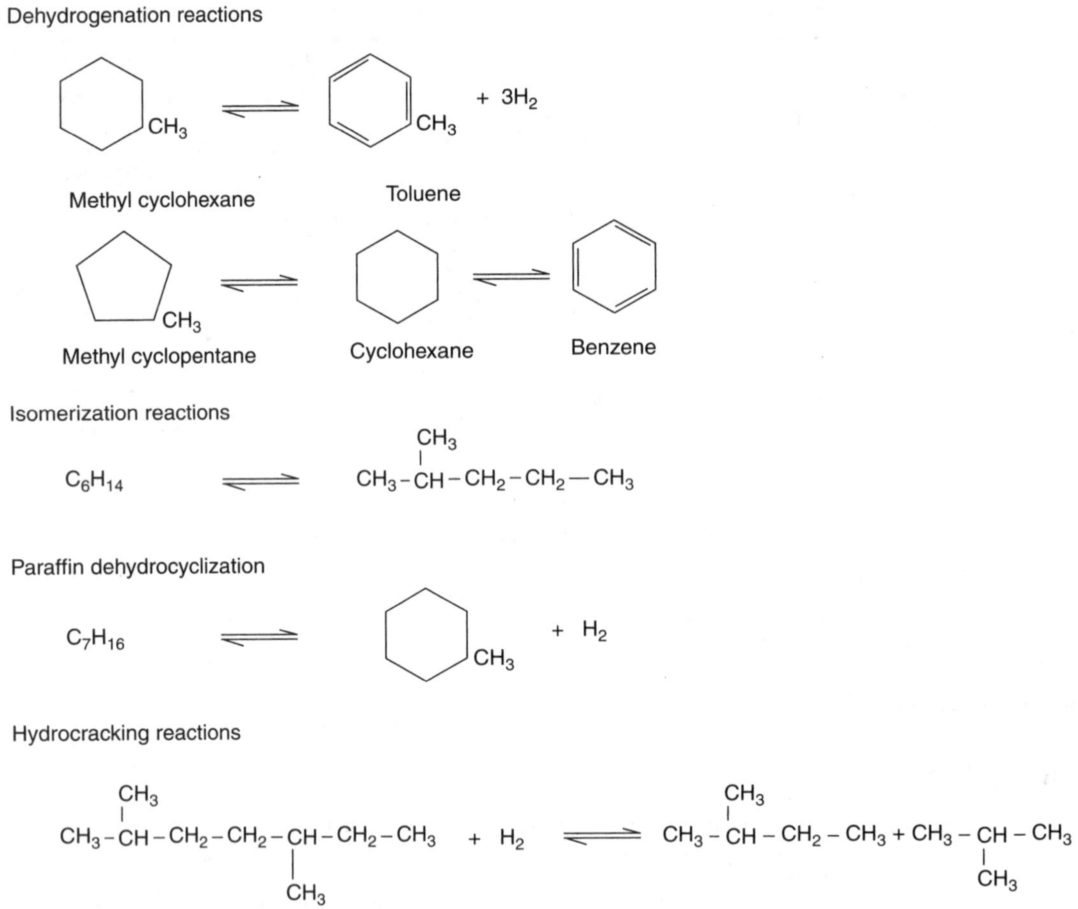

FIGURE 2-5 Catalytic refforming reactions.

and is preheated in feed-effluent exchanger E-101. The feed-hydrogen mix is next heated to reaction temperature by first fired heater H-101 and passed over reforming catalyst in first reactor R-101. Due to the endothermic nature of reactions, the temperature of the effluent coming out of the first reactor drops. It is reheated in the second fired heater H-102 after which the effluent from the first reactor enters the second reactor R-102. Effluent from the second reactor is reheated in third fired heater H-103 after which it enters the third reforming reactor R-103. Effluent from third reforming reactor is cooled first by heat exchange with incoming feed in E-101, next in air cooler E-102, and finally in trim water cooler to 104°F before entering HP separator drum V-103. Pressure in V-103 is controlled at 170 lb/in^2. High-pressure gas, mainly hydrogen, is sent to recycle compressor K-101 and recirculated to incoming fresh feed. Reforming reactions produce large volumes of hydrogen. Excess hydrogen from a flash drum is sent to other refinery units via knockout (KO) drum V-105 and compressor K-102.

Liquid effluent from HP Separator V-103 is pumped via P-102 to stabilizer column V-106, which separates fuel gas and LPG as top product; reformate is the bottom product. Reformate is cooled by heat exchange with incoming feed in E-105, E-106, and next cooled in water cooler E-108 before sending it to storage.

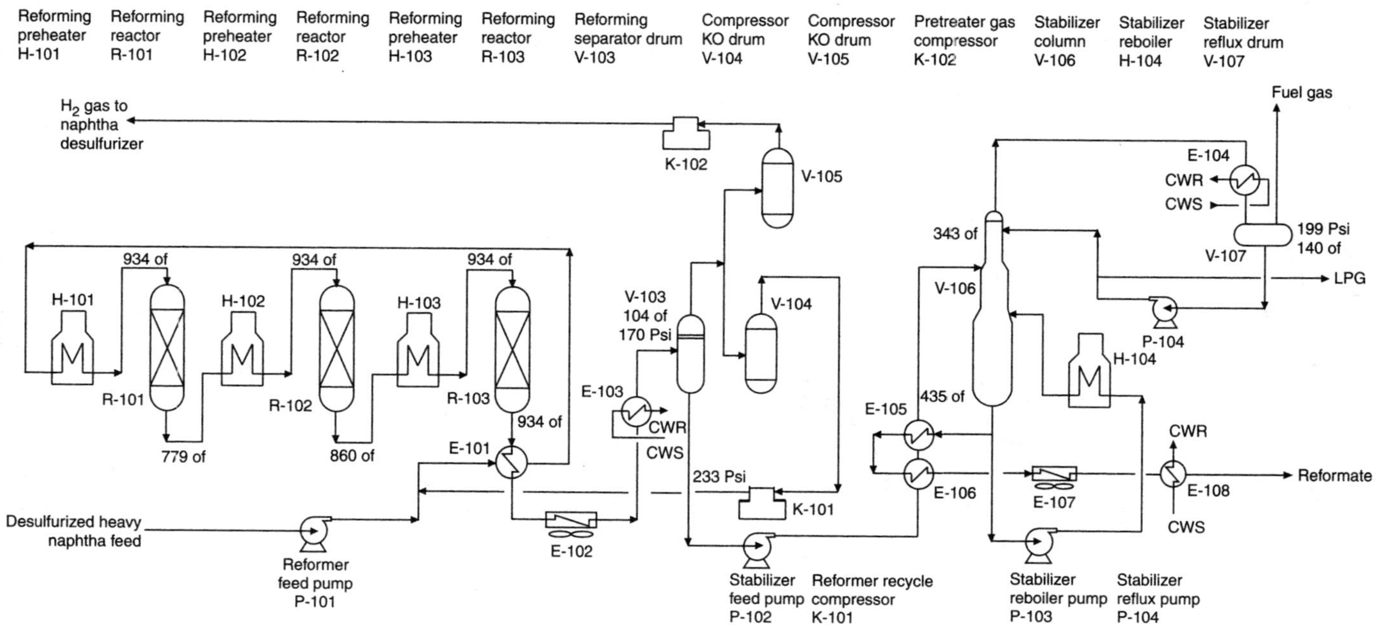

FIGURE 2-6 Process flow diagram: Catalytic reforming unit.

Operating Conditions

Reforming catalyst is a high-purity alumina catalyst impregnated with platinum and other noble metals. Reforming reactions are favored by high temperature and low pressure. The yield of reformate is a function of feed PONA and process severity. Typical operating conditions, feed, and product properties are shown in Tables 2-11 and 2-12. Motor gasoline blends typically contain 30 vol % or more reformate. Considering the large volumes of motor gasoline blended, the volumes of heavy naphtha feedstock consumed are a significant fraction of total naphtha production.

TABLE 2-11 Catalytic Reformer Operating Conditions

Operating conditions	Units	Value
Reactor inlet temperature (SOR/EOR)	°F	934/1013
HP separator temperature	°F	104
HP separator pressure	lb/in^2	170
Space velocity	WHSV	2.75
Recycle ratio (moles of H2/moles of feed)		4.5
Catalyst distribution	Wt %	
First reactor		15
Second reactor		25
Third reactor		60
Stabilizer column		
Column top/Bottom temperatures	°F	180/435
Column top pressure	lb/in^2	205

TABLE 2-12 Catalytic Reformer Feed and Product Properties

Property	Units	Feed	Reformate C5+
API gravity	°	61.3	47.8
Specific gravity		0.734	0.789
Octane number		38	96
TBP distillation	°F		
IBP		194	140
10%		203	
30%		221	194
50%		239	230
70%		257	257
90%		275	
FBP		284	311
PONA	Vol %		
Paraffins		69	
Naphthenes		20	
Aromatics		11	
Sulfur	Wt %	0.015	
Mercaptan	Wt %	0.008	
Nitrogen	ppm	1.00	

Aromatics Production

Benzene, toluene, and xylene are important feedstocks for the petrochemical industry. Benzene ranks in the top-20 chemicals worldwide in terms of production volumes. Benzene is used in the manufacture of plastics and resins, nylon and synthetic fiber, in rubber, lubricants, dyes, detergents, drugs, and pesticides, and in many other industries. Paraxylene, a basic raw material for the polyester fiber industry, is

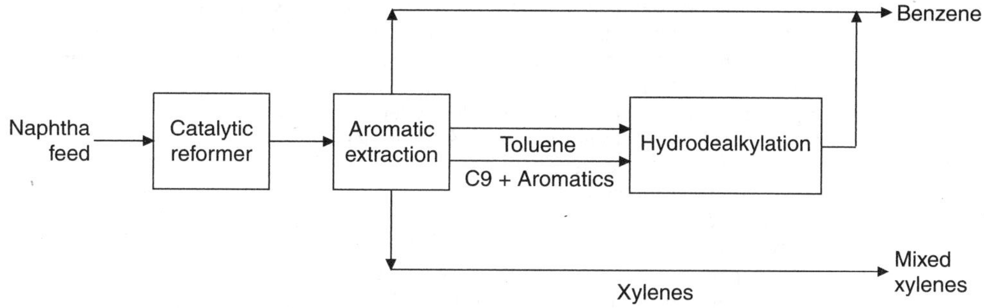

FIGURE 2-7 Aromatics production from naphtha.

used for the manufacture of terephthalic acid (TPA) and dimethyl terephthalate (DMT). Both TPA and DMT are used in the manufacture of polyethylene terephthalate (PET) resin. The major uses of PET are fiber, film, and engineering resins.

The catalytic reforming of naphtha is also used for the production of aromatics such as benzene, toluene, and xylenes (Fig. 2-7). The only difference from the cat reformer for motor gasoline (mogas) is that reforming operations are carried out at a higher severity (98 to 102) in a continuous catalyst regeneration configuration. Feed is 183 to 313°F cut. It is used for the production of benzene, toluene, and mixed xylenes. Aromatics from reformate are separated from nonaromatics by solvent extraction. Aromatic extract is separated into benzene, toluene, and xylene by fractionation or other separation processes. Because there is little demand for toluene, it is converted to benzene by a hydrodealkylation process.

Fertilizer/Petrochemical Industry

Naphtha is used as a feedstock for the manufacture of nitrogen fertilizers such as urea, ammonium nitrate, ammonium sulfate, and so on. In regions where natural or associated gas is not available, naphtha is the preferred feedstock. Naphtha is also used for the manufacture of methyl alcohol with a large industrial demand.

Naphtha is gasified and next reacted at high temperatures (1575°F) with steam over a nickel catalyst that converts naphtha to carbon monoxide and hydrogen (Fig. 2-8). Carbon monoxide reacts with more steam to produce more hydrogen and carbon dioxide. Hydrogen is reacted with nitrogen from the air to ammonia. Ammonia in turn is reacted with CO_2 to manufacture urea. Methanol is manufactured by reacting carbon monoxide with hydrogen over a catalyst.

Steam reforming reaction

$$C_5H_{12} + 5\,H_2O = 5\,CO + 11\,H_2$$

Low-temperature shift reaction

$$CO + H_2O = CO_2 + H_2$$

FIGURE 2-8 Steam reforming of naphtha.

Naphtha Steam Cracker

Naphtha is one of the basic raw materials for the production of ethylene and propylene. Ethylene is the largest volume petrochemical produced worldwide (estimated at 95 million tons in 2002). Ethylene has no direct use but is used exclusively as a chemical building block. Ethylene, propylenes, and C4 olefins are used for the production of plastics, fibers, films, textiles, pharmaceuticals, detergents, and so on.

The bulk of the worldwide commercial production of olefin is based on the thermal cracking of naphtha or heavier hydrocarbons with steam. The process is called pyrolysis, or steam cracking.

Naphtha is pumped through the convection section of the pyrolysis furnace where it is heated by heat exchange with flue gases. It is then mixed with steam and further heated to incipient cracking temperature (approximately 950°F) depending on feedstock. This stream next enters the radiant heating section of the furnace where it is heated under controlled conditions (residence time and temperature profile) to 1380 to 1607°F for 0.1 to 0.5 seconds. Final heating temperature depends on feed properties. Pyrolysis converts heavier hydrocarbons into lighter fractions, primarily ethylene and propylene, by removing hydrogen. Ethylene yield is roughly 33 wt % of the naphtha feed. The hot gas effluent from the furnace is then passed through a quench section where it is rapidly cooled to retard further cracking and to condense heavy fractions. Heavy fractions are subsequently processed into fuel oil, light cycle oil, and pyrolysis gasoline. The pyrolysis gasoline produced from naphtha cracker has a high benzene, toluene, and xylene content (Table 2-13). It also has a high olefin and diolefin content. It is first hydrotreated to saturate diolefins and monolefins. Aromatics are next extracted by a selective solvent such as sulfolane and separated by fractionation. Steam generated in the quench section is recycled back to the furnace for reuse.

TABLE 2-13 Pyrolysis Gasoline Properties and Composition Ex Naphtha Steam Cracker

ASTM distillation		°F
IBP		95
10	Vol %	133
30		160
50		187
70		221
90		298
EP		392
Density	g/mL, 60°F	0.770
Bromine number		71
Diene number		57
Total sulfur	ppm	800
Mercaptan sulfur	ppm	100
Molecular weight		85.73
Organic nitrogen	ppm	10
Existing gums	ppm	33

Composition	Wt %
C_4 diolefins	0.50
C_4 mono olefins	0.40
C_4	0.10
C_5 cyclopentane	1.00
N-C_5	5.30
I -C_5	3.60
C_5 mono olefins	4.50
C_5 diolefins	8.30
N C_6	1.30
Methyl cyclopentane	0.10
Cyclohexane	0.10
I-Hexane	1.50
C_6 mono olefins	1.90
C_6 diolefins	3.70
Benzene	22.30
C_7 naphthene	0.60
N-Heptane	0.55
ISO heptane	0.26
C_7 mono olefins	0.90
C_7 DI Olefins	1.70
Toluene	14.40
N-Octane	0.25
Iso octane	0.25
C_8 monolefins	0.30
C_8 diolefins	0.60
Xylene+ethyl benzene	6.80
Styrenes	3.10
C_9+	15.69

Power Plant Fuel

Naphtha is increasingly being used as a fuel in gas-based power plants in place of natural gas, associated gas, or liquefied natural gas (LNG). Naphtha as a fuel cannot compete with natural gas due to its higher price, but because of the shortage of natural gas feedstock, power producers in many regions of the world use naphtha as fuel to meet electric power output requirements. Naphtha with high aromatic content (less than 20 vol %) can be used. Net heat of combustion is 20,000 Btu/lb or 11,300 kcal/kg, which is much more than natural gas on a per unit volume basis. Using naphtha in place of gas requires adjustment of operating conditions to avoid high boiler temperatures of 2000°F or more. Also, use of naphtha results in higher corrosion rates due to the sulfur in naphtha, which in turn causes shorter run length or more plant maintenance.

CHAPTER 3
GASOLINE

Gasoline may be defined as the fuel derived from crude oil, in the boiling range of 100 to 400°F, for use in spark-ignited internal combustion engines. Gasoline is one of the most important petroleum products. Most passenger cars use gasoline as fuel, and the demand for gasoline is directly linked to the growth of the automobile industry. It is estimated that production of gasoline worldwide in 2007 was approximately 1150 million tons, which corresponds to an estimated 27 percent gasoline yield from crude oil. The automobile industry is witnessing explosive growth. In 2006, estimated production of passenger cars was 49 million units, and demand for gasoline is growing in proportion to the number of automobiles fueled by motor gasoline (mogas) on the roads. About half a century ago, gasoline was also a fuel for aircraft, but with the advent of aviation turbine engines, most military and commercial aircraft switched to kerosene fuel, and demand for gasoline for this use almost disappeared. However, a small number of piston engine aircrafts for commercial and military use still use gasoline fuel.

GASOLINE ENGINE

The gasoline engine is an internal combustion engine. Fuel is supplied by a pump from the automobile fuel tank. Gasoline is mixed with air, and the gasoline-air mixture enters the cylinder during the intake stroke. It is compressed by a piston on the compression stroke and ignited by a spark plug. As the fuel burns, the expanding hot gases force the piston down on the power stroke. Unburned gases are finally ejected from the cylinder on the exhaust stroke through an exhaust valve.

The stoichiometric ratio of air to fuel for complete combustion is 15:1 by weight. A richer mixture is used under full throttle where maximum power is desired. Gasoline is a mixture of hydrocarbons. In a perfect combustion process, oxygen in the air would convert all hydrocarbons to carbon dioxide and water, releasing energy for moving the car. In actual practice, the combustion process produces many types of pollutants that are emitted from the tailpipe of the automobile: unburned hydrocarbons, nitrogen oxides, carbon monoxide, and carbon dioxide.

GASOLINE PROPERTIES

Octane Number

The octane number is a relative measure of knocking, or the tendency to self-ignition of a fuel in a spark-ignited internal combustion engine. In a normal cycle, the fuel-air mixture is ignited at the end of the compression stroke by a spark plug. If fuel has too low an octane number, it may spontaneously ignite during the compression stroke itself due to the temperature rise during adiabatic compression or from hot spots in the engine. This kind of ignition, called preignition, can quickly damage the engine. Fuels with a high normal paraffin content have a low octane and can cause preignition. The efficiency of an engine is directly proportional to its compression ratio; the higher the compression ratio, the higher the power output. For higher compression ratio engines, gasoline with a higher octane or resistance to knocking is required because the tendency to preignition is lower. The antiknock quality of motor gasoline is expressed in terms of octane numbers. In various test methods, the test fuel is compared to blends of two pure hydrocarbons: normal heptane and isooctane (2,2,4, trimethyl pentane). Normal heptane is low in its resistance to knock and assigned an octane number of 0. Isooctane is quite high in its resistance to knock and assigned an octane number of 100. Blends of the two serve as a reference fuel system. Octane number is measured with a standard American Standard Testing Method (ASTM) Cooperative Fuel Research (CFR) single-cylinder engine with a variable compression ratio. To measure the fuel's octane, the engine is operated with test fuel at 900 r/min (revolutions per minute) and the compression ratio is increased until onset of knocking. Next the test fuel is replaced with a mixture of isooctane and normal heptane in varying proportions until a mixture is found that matches the knocking characteristics of the test fuel. The percentage of isooctane in the blend is the octane number of the fuel. Both the research octane number (RON) and the motor octane number (MON) are measured with the same test engine but with different test conditions. RON is determined as per ASTM D 2699, the test is run at 600 r/min, input air at 20 to 52°C, depending on barometric pressure, and timing is fixed at 13 degrees. In the MON method (ASTM D 2700), the test is run at 900 r/min, input air temperature is maintained at 38°C, and ignition timing varies between 14 and 26°C depending on compression.

The RON of most commercial unleaded gasoline fuels varies between 83 and 95. Motor octane (MON) is generally 5 to 10 numbers less than RON.

Reid Vapor Pressure

A gasoline engine needs a fuel that is sufficiently volatile to allow easy formation of the fuel vapor-air mixture required for combustion. If fuel vaporizes too much, it will not flow to the engine due to vapor lock. If fuel is less volatile, it will not evaporate and will remain in liquid form. Specifications are set to control these properties based on their vapor pressure and distillation. The Reid Vapor Pressure (RVP) is vapor pressure of gasoline at 100°F. RVP and boiling range determine the ease of starting, the engine warmup, and the vapor lock temperature. The RVP of various gasoline grades varies between 35 and 84 kPa. The RVP of gasoline is adjusted according to the ambient conditions of the region where the fuel is being used. RVP is adjusted to a lower value during the summer and a higher value during the winter months. RVP is adjusted by increasing or lowering the volume of volatile components such as normal butane, light straight run (LSR) naphtha, or isomerate in gasoline blends. Vapor emissions from automobiles are a major component of VOCs (volatile organic compounds) in the atmosphere. RVP specifications of gasoline are being reduced to the minimum level possible to reduce hydrocarbon emissions from the storage and handling of gasoline. Table 3-1 lists the octane number and RVP of some commercial gasoline grades.

TABLE 3-1 Octane Number and RVP of Gasoline Grades

Property	Units	Limit	1	2	3	4	5	6
Research octane number (RON)			83	90	91	91	94	95.3
Motor octane number (MON)					81	81	84	85.5
Vapor pressure at 100°F	kPa	Min.		49	49	35	44	50
		Max.	66.5	77	84	60	54	83.3
Benzene	Vol %					3	4.5	4.8
Olefins	Vol %	Max.		20	20	21		
Aromatics	Vol %	Max.			50			
Oxygenates	Vol %	Max.						10.0
Sulfur	Wt %	Max.	0.2	0.01	0.03	0.015	0.01	0.01

Distillation

Gasoline's tendency to vaporize is also characterized by determining a series of temperatures at which various percentages of the fuel have evaporated as described in ASTM D 86. The temperatures at which 10, 50, and 90 percent evaporation occur define the volatility of gasoline.

The 10 percent evaporated temperature is directly affected by the seasonal blending of gasoline. This temperature must be low enough to provide easy cold starting but also high enough to minimize vapor lock and hot weather derivability problems. Problems can occur in cool weather by the use of summer season gasoline. The 50 percent evaporated temperature must be low enough to provide good warmup and cool weather drivability without being too low to cause hot weather drivability and vapor lock problems. The 90 percent and final boiling point (FBP) must be low enough to minimize crankcase and combustion chamber deposits, spark plug fouling, and dilution of engine oil. The higher boiling fractions of gasoline have a significant effect on the emission levels of undesirable levels of hydrocarbons and aldehydes from automobile exhaust.

Vapor Lock Temperature

Vapor lock is the disruption of gasoline supply to the internal combustion engine's carburetor or fuel injection system caused by excessive vaporization of gasoline. Vapor lock occurs when the automobile fuel pump, which is designed to pump liquid, has a suction pipe full of vapor instead of liquid and is thus unable to deliver gasoline to the engine, resulting in stalling. The fuel can vaporize due to heating by the engine or local climatic conditions or because of the lower boiling point at high altitudes. In regions with large variations in summer and winter temperatures, the use of winter grades with high RVP and low 10 percent distillation temperature can cause vapor lock problems if used in the summer months. An empirical index, the vapor lock index (VLI) is defined in some gasoline specifications for defining limits of volatility for seasonal grades in terms of vapor pressure and vol % (percentage by volume) distilled at 70°C. VLI is calculated by the relation shown in Fig. 3-1.

$$VLI = 10^{*} VP + 7^{*} E70$$

Where:

VLI = Vapor lock index

VP = Vapor pressure in kPa, 38°C

E70 = Vol % distilled at 70°C

FIGURE 3-1 Vapor lock index of gasoline.

Gasoline Blending

All gasoline blend components except n-butane must be liquids at ambient temperature falling in the naphtha boiling range, that is, between 100 and 400°F so that the fuel must vaporize rapidly in a stream of air to yield engine charge. Also gasoline components must have a high octane number to prevent autoignition of fuel during compression. The octane number of a hydrocarbon is a function of its chemical composition. Hydrocarbons with a high percentage of straight chain paraffins have a very low octane number. Thus n-heptane (C_7H_{16}) has a boiling point of 208°F and octane (RON) of 0. Branched chain normal paraffins, olefins, and aromatics have high octane numbers. Straight run naphtha from crude distillation has the correct boiling range for gasoline blending but because of the high percentage of n-paraffins, it has low octane number: 65 to 70. Target RON of gasoline is typically in the range of 85 to 95 depending on gasoline specifications in the region of usage.

Gasoline is blended from the following refinery streams:

- N-butane
- LSR
- Isomerate
- Light cat naphtha
- Heavy cat naphtha

TABLE 3-2 Properties of Mogas Blending Components

Property	Units	N-Butane	Isomerate	Light cat naphtha	Heavy cat naphthas	Cat reformate	Alkylate	Coker light naphtha	MTBE	LSR
SG (specific gravity)		0.5844	0.641	0.7083	0.8441	0.7811	0.700	0.69	0.74	0.667
RON		95	84	92.9	95.0	96.4	96.0	74	116.0	68.0
MON		92	81.3	81.3	80.4	84.8	94.0	68.8	100	63
RVP	lb/in^2	51.92	13.5	8.4	0.4	9.5	6.2	13.1	9	10.1
Aromatics	Vol %	0	0	15	59	62.5	0	8	0.6	2.4
Olefins	Vol %	0	0	35	10	1	0	56	0	0

- Cat reformate
- Alkylate
- Coker light naphtha
- Methyl tertiary butyl ether (MTBE)

Table 3-2 lists the typical properties of these streams. A typical refinery blending scheme for gasoline is shown in Fig. 3-2.

The key gasoline blend components are light and heavy cat naphtha ex fluid catalytic cracking (FCCU) unit, catalytic reformate, n-butane, and LSR. Coker naphtha is only available to refineries with a delayed coker unit. Isomerate stream originates from the LSR isomerization unit. Alkylate is a very valuable gasoline blending component with high octane and low RVP. MTBE is not strictly a refinery stream. It is an oxygenate with a very high octane number that may be imported by a refinery for gasoline blending.

GASOLINE BLEND COMPONENTS

FCCU Cat Naphtha

Feed to the FCCU unit is vacuum gas oil from the vacuum distillation of reduced crude. The vacuum gas oil (VGO) may be desulfurized to reduce sulfur in FCCU products. FCCU is the catalytic cracking of VGO to produce gases, light cat naphtha (LCN), medium cat naphtha, light cycle oil, heavy cycle oils, and decant oil. Approximately 43 vol % of VGO feed is converted to LCN, and another 16 vol % is converted to heavy cat naphtha. LCN is an excellent blendstock for gasoline blending with a RON of 93 and a MON of 81. Medium cat naphtha (MCN) also has good RON and MON values but consists of higher boiling components with an ASTM distillation range of 265 to 365°F compared with a distillation range of 105 to 250°F for LCN.

Catalytic Reformate

Catalytic reforming is one of the key processes for increasing the octane number of straight run naphtha. Feed to the unit is straight run heavy naphtha cut from 185 to 340°F. Feed is first hydrotreated to remove sulfur, nitrogen, and other catalyst poisons. Reformate has a very high aromatic content with high RON and MON. The octane number of reformate can be increased by increasing the severity of the reforming reactions. Most reforming units for gasoline manufacture are designed for 95 to 100 reformate octane. Vol % yield of C5+ reformate is typically 80 vol %. Yield decreases with the increase in octane number. Because the reformer feed is not limiting, reformer capacity can be chosen independently according to gasoline requirements.

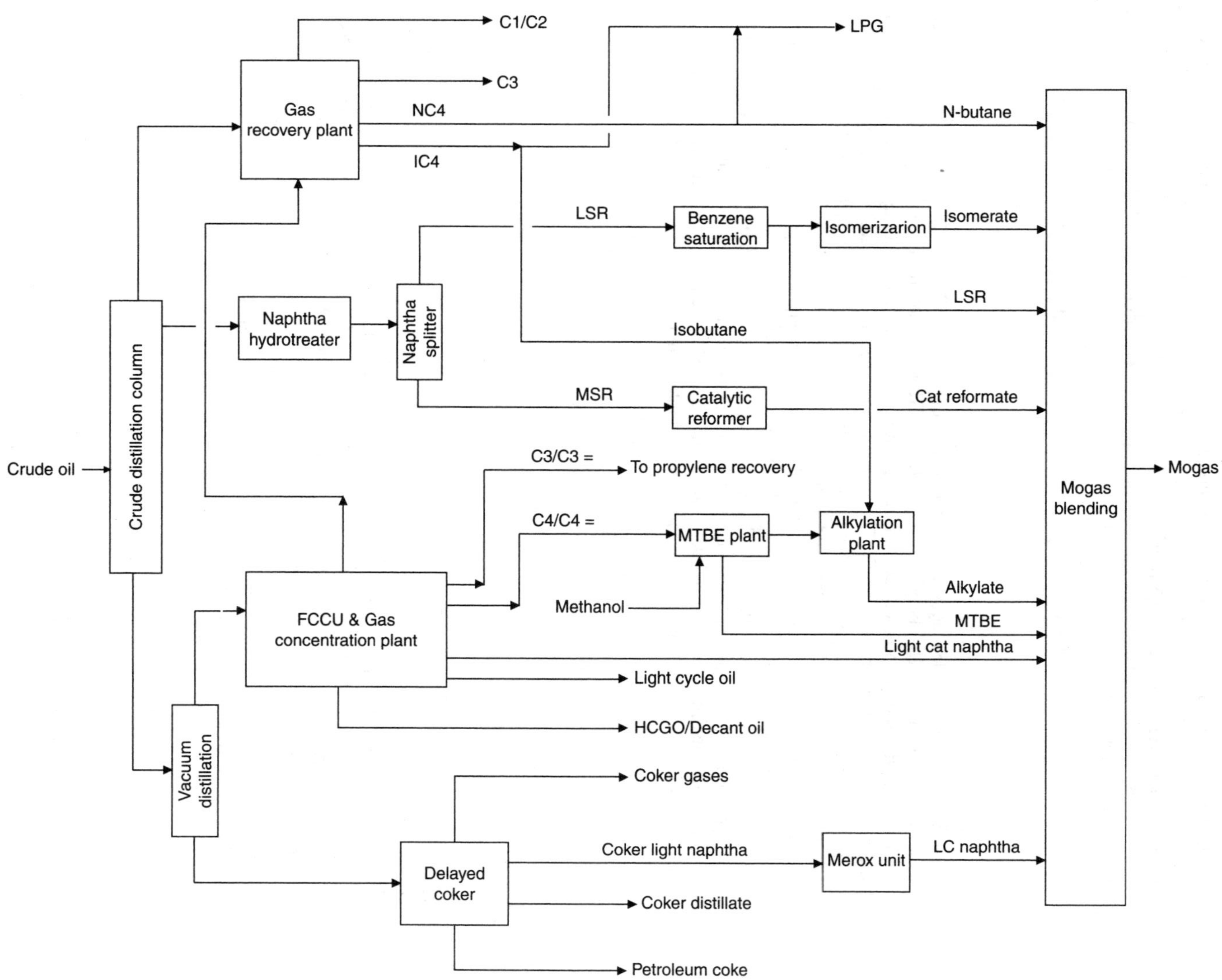

FIGURE 3-2 Refinery mogas blending.

Isomerate

Light straight naphtha contains mainly C5 and C6 paraffins. LSR is included in the gasoline pool to adjust its RVP and distillation. However, LSR stream has a low RON, generally less than 70. It is possible, however, to increase the RON of this stream by 15 to 20 points by the modern isomerization process. In the conventional once-through process, desulfurized feed is heated, mixed with hydrogen, and passed over an alumina or zeolite catalyst impregnated with platinum. Feed is isomerized to an equilibrium mixture of normal and highly branched isomers of C5 and C6 hydrocarbons. The mixture has higher octane by 15 to 20 points. The yield is essentially 100 percent.

MTBE

MTBE is produced by the reaction of isobutylene with methanol (Fig. 3-3). The C4 cut from FCCU gases with 10 to 20 vol % isobutylene is used for MTBE manufacture. The etherification reaction of isobutylene with methanol is conducted in the liquid phase at a moderate temperature of 156°F in the presence of ion exchange resins. Oxygenates are superior to aromatics as an octane source. MTBE is most suited as a gasoline blend component, and it being used in gasoline blending to partly replace aromatics. MTBE can be used up to 11 vol % in the blend, and it gives an oxygen content of 2 wt % (percentage by weight) to gasoline blend. However because oxygen does not contribute to energy release, the net calorific value of fuel is reduced, requiring more gasoline per kilometer compared to gasolines blended without MTBE. The U.S. Clean Air Act of 1990 introduced a requirement of oxygen content of gasoline for a reduction in emissions of carbon monoxide and reactive organic gases and thus improved air quality. To meet the oxygen content requirement, gasoline blenders used MTBE, ethanol, and other blend components containing oxygen.

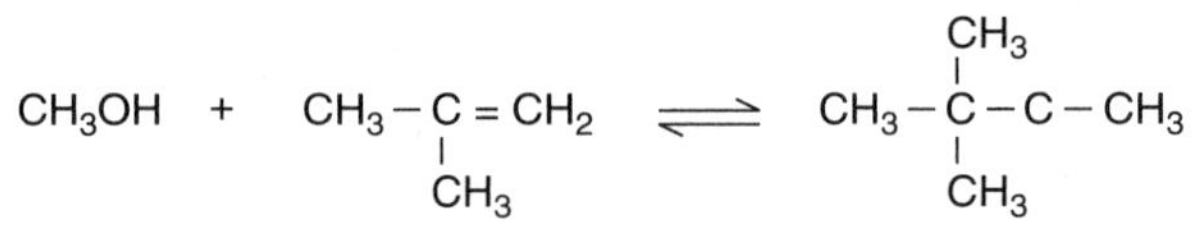

Methanol Isobutylene MTBE

FIGURE 3-3 MTBE manufacture.

California was the first state to enforce this regulation, and refiners extensively used MTBE in gasoline blends. Under the 1990 Clean Air Act Amendment, refiners were required to meet 2 percent oxygen requirements for reformulated gasolines. Refiners are allowed to choose from a variety of oxygenates, such as MTBE, ethanol, but MTBE emerged as the choice of refiners.

A negative effect of MTBE use in gasoline is ground and surface water pollution, which can occur through leaking underground storage tanks, pipeline, spills, and emissions from marine engines. MTBE is much less biodegradable than petroleum products. The use of MTBE as an oxygenate is currently under investigation due to health and environmental concerns.

Alkylate

Alkylate is an excellent motor gasoline and aviation gasoline blend stock because of its high octane and low volatility. It is manufactured by the reaction of isobutylene with isobutane in the presence of sulfuric acid (H_2SO_4) or hydrofluoric acid (HF) catalyst (Fig. 3-4).

The alkylation unit feed originates from the FCCU unit. FCCU gases are fractionated in a C3/C4 splitter to remove propane and lighter constituents. The bottom from the C3/C4 splitter is charged to the MTBE unit, which mainly depletes isobutylene from this feed. The raffinate from the MTBE unit

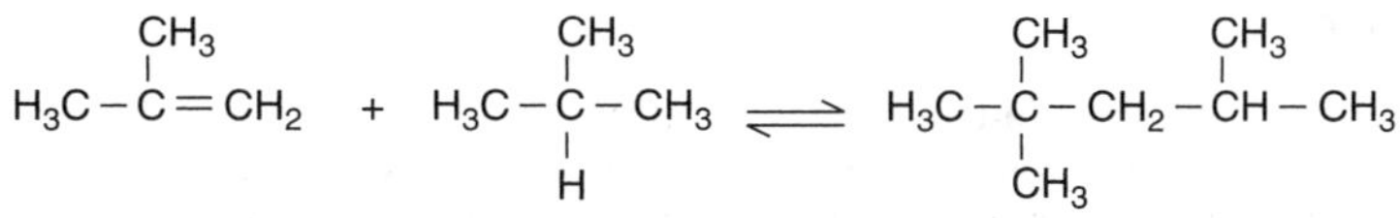

FIGURE 3-4 Alkylate manufacture.

containing ISO butylene and isobutane becomes feed for the alkylation unit. In case the MTBE unit is shut down, feed to the unit will not contain adequate isobutane to alkylate the olefin. So provision is made to supply isobutane separated from field butane. Table 3-3 lists the typical composition of alkylation unit feed.

The alkylation reaction is almost instantaneous and highly exothermic and is favored by low temperature conditions. The process is carried out at 44 to 50°F. The alkylate product is a mixture of branched hydrocarbons in the gasoline boiling range. Alkylate has an RON of 93 to 98 and a MON of 90 to 95. The reactors in which alkylation reactions occur are horizontal pressure vessels containing a mixing impeller, an inner circulation tube, and a tube bundle to remove heat generated by the alkylation reaction. The feed is injected into the suction side of the impeller inside the circulation tube. The impeller rapidly disperses hydrocarbon feed into the acid catalyst to form an emulsion. The emulsion is circulated by the impeller at high rates within the contractor. A portion of the emulsion is withdrawn from the contractor on the discharge side of the impeller. The emulsion flows to an acid settler where acid and hydrocarbon phase separate out. The acid, being heavier, settles to the bottom and is returned to the suction side of the impeller. Thus the impeller acts as an emulsion pump between the reactor and settler. Sulfuric acid acts as a catalyst for the alkylation reaction and theoretically remains unchanged. In reality, however, a certain amount of acid is consumed as a result of side reactions and feed contaminants. To maintain desired spent acid strength (90 wt % H_2SO_4), a small amount of spent acid is withdrawn and an equivalent amount of 98.5 wt % fresh acid is charged to the reactor. The impure alkylate stream contain some esters, which are removed by reacting with fresh sulfuric acid followed by washing with a dilute alkaline water stream. The treated contractor effluent is fed to an isostripper tower to remove isobutene as overhead product, normal butane as side draw, and alkylate as bottom product. For aviation gasoline blending, alkylate may be rerun to remove heavy ends from aviation alkylate. The typical composition of alkylate product is listed in Table 3-4.

TABLE 3-3 Alkylation Unit Feed Composition Ex MTBE Plant

Composition	Vol %
Propane	0.07
Isobutane	47.31
n-Butane	11.37
Isopentane	1.43
Normal pentane	0.02
Propylene	0.01
Butylene	38.91
Amylene	0.67
Butadiene	0.21
Water	0
Methanol	0
MTBE	0

TABLE 3-4 Alkylate Product Composition Isobutane, Isobutylene Feed

Composition		Mass %
i- Butane		0.19
n-Butane		7.86
i-Pentane		5.31
n-Pentane		0.03
C6 + alkylate		86.61
Properties		
Density		0.695
API gravity		72.1
RVP	lb/in^2	8.9
RON		96.4
MON		93.9

Normal Butane

Normal butane (n-butane) has an RON of 95 and RVP of 52 lb/in^2. It is added to gasoline to meet RVP specifications. An important consideration is that vapor pressure of blended gasoline must be less than 14.7 lb/in^2, the atmospheric pressure, on the hottest summer day. In winter months it can be blended in gasoline in a higher proportion because winter RVP specs are higher than those of summer. Butane is the cheapest gasoline blend component and plentiful. Refiners aim to maximize n-butane inclusion in gasoline[1] to increase refinery profitability. Gasoline blends typically contain 2 to 4 vol % butane. Vapor pressure of gasoline is a compromise between high RVP to improve economics and lower RVP to prevent vapor lock and reduce evaporation losses

Tetraethyl Lead (TEL)

Tetraethyl is an organometallic compound with the formula $(C_2H_5)_4$ Pb. TEL (Fig. 3-5) has been used to suppress knock in gasoline engines since the 1920s. In the following 50 years, TEL was almost exclusively used in gasoline to increase its octane number. In the internal combustion engine, during fuel combustion, TEL decomposes completely into lead, lead oxides, and short-lived ethyl radicals that scavenge radical intermediates in the combustion process. This in turn prevents ignition of unburned fuel during the exhaust stroke. Combustion of TEL produces not only carbon dioxide and water but also metallic lead, which can oxidize further into PbO. Pb and PbO can quickly accumulate in the engine, making it inoperable after a short time. To prevent Pb or PbO accumulation in the cylinder, some other chemical additives, such as 1,2 dibromoethane and 1,2 dichloroethane, are added to TEL to make TEL fluid (Table 3-5). These additives, on combustion, form volatile lead bromide and lead chloride, which exit the engine as exhaust gases.

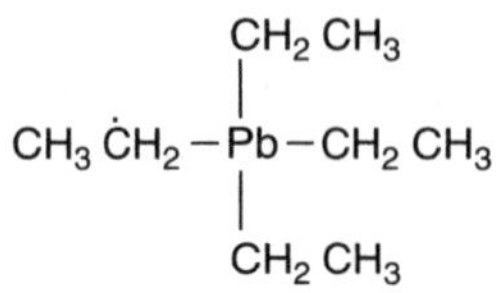

Formula	$C_8H_{20}Pb$
Molecular weight	323.44
Density at 66°F	1.653
Melting point °F	–212.8
Refractive index	1.519

FIGURE 3-5 Tetraethyl lead antiknock.

The widespread use of lead alkyl additives in motor gasoline followed the 1921 discovery by General Motor engineers of their ability to increase octane rating and hence the antiknock properties of a gasoline to which it is added. The maximum concentration of elemental lead in 1 gallon of motor gasoline was fixed at 4.23 g. The limit for aviation gasoline was fixed at 4.6 g/gallon. It was found that the octane number of gasoline could be increased by 12 to 15 numbers within this lead limit. Despite advances in refining technology, lead alkyls remained the most cost-effective option to increase the octane number of gasoline. One of the greatest advantage of TEL over other antiknock agents is the low concentration needed, approximately 1 part TEL to 1260 parts untreated gasoline. Alternative options to produce high-octane gasoline blend components, such as catalytic reforming, alkylation, isomerization, and FCCU units require large capital investment and were not considered economic compared with lead additives for the same objective. In addition to increasing gasoline octane, an unexpected benefit of lead alkyl additives is that lead acts as a lubricant

TABLE 3-5 Typical Composition of TEL Fluids for Motor and Aviation Gasolines

Components	Motor gasoline % mass	Aviation gasoline % mass
Tetraethyl lead	61.49	61.41
Ethylene dibromide	17.86	35.68
Ethylene dichloride	18.81	
Solvent, antioxidant, dye, inerts	1.84	2.91

between the contact surfaces of exhaust valves and valve seats on the cylinder head. This proved to be an advantage because the valve seat could be machined directly into the engine body with a lead additive providing all the protection against wear. Due to the high temperature of exhaust gases, exhaust valves operate in a hot environment, valve seats are susceptible to recession, and valve seats recede into the cylinder head. When unleaded fuel is used in an engine with a susceptible valve seat, valve seat recession may occur.

For more than 50 years, refiners added lead to gasoline in the form of TEL or tetramethyl lead (TML) to increase the octane number. Lead is a relatively inexpensive source of incremental octane (both RON and MON). For example, 0.53 g Pb/L in the form of TML increases the RON of a gasoline blend from 91 to 98. At the same lead level, RON of another gasoline blend stock increased from 79 to 90. In practice, gasoline lead susceptibility is a function of gasoline composition and blend properties. In general, the higher the octane of the base gasoline, the lower is the lead susceptibility.

The low concentrations present in gasoline and exhaust were not perceived as dangerous. In the engine, exhaust compounds of lead are dispersed into the air in the vicinity of roads.

Lead is a poisonous metal that has neurotoxic effects on humans even at very low levels. The most significant reason for the lead phaseout was that it enables the introduction of catalytic converters, which are the single most effective method of reducing harmful exhaust emissions from vehicles. Lead in automotive engine exhaust can be deposited on the catalyst in the converter and thus permanently deactivate the catalyst.

The lead phaseout in the United States began in 1976 and was completed in 1986. By 2000, leaded gasoline was withdrawn from the European Union and many other countries. However, TEL remains an ingredient of high-octane motor racing fuels and aviation gasolines.

POLLUTION FROM GASOLINE COMBUSTION

Combustion pollutants from millions of gasoline-fueled automobiles on the roads end up in the atmosphere. This in turn adversely affects the environment, human health, and contributes to global warming. All hydrocarbon-fueled vehicles increase air pollution because the combustion of the air-fuel mixture is never 100 percent complete. Hydrocarbon emissions result when fuel molecules do not burn in the engine or burn only partially. When hydrocarbons and nitrous oxides from auto emissions are exposed to sunlight, a photochemical reaction takes place that results in the formation of ozone, a major component of ground-level smog. Ozone irritates the eyes, damages the lungs, and aggravates respiratory problems. Some of the hydrocarbons in the auto exhaust are known carcinogens. The main emissions from a car engine are carbon dioxide, water vapor, and nitrogen. Air contains approximately 78 percent nitrogen, and it passes unchanged through the engines. However, a smaller amount of other pollutants are generated during the combustion process, discussed next

Carbon Monoxide

Carbon monoxide (CO) is a product of incomplete combustion and forms when combustion takes place in an air-deficient condition. CO is a poisonous gas and can cause death if humans are exposed to automobile exhaust in closed space.

Nitrogen Oxides

Under high pressure and temperature condition of the engine, nitrogen in the air reacts with oxygen in the air to form nitrogen oxides (NOx). Like hydrocarbons, NOx are precursors to ozone formation. Oxides of nitrogen contribute to smog and acid rain.

Unburned Hydrocarbons

These are hydrocarbons that pass through the engines unburned or hydrocarbon is released to the air due to evaporation from fuel storage tanks. All these pollutants end up in the atmosphere.

CATALYTIC CONVERTER

In 1950, an average new car in the United States emitted 8.1 g HC, 2.3 g NOx, and 54.4 g CO per kilometer of travel. These were reduced to 0.156 g HC, 0.25 g NOx, and 2.125 g CO per kilometer. All this was achieved because of fundamental improvement in engine design plus the addition of catalytic converters to automobiles. Catalyst converters reduce the output of these three types of emissions by 90 percent or more, converting them into relatively harmless gases such as CO_2, N_2, and water (Fig. 3-6). To reduce emissions, modern cars have been designed to carefully control the air-to-fuel ratio very close to stoichiometric, which is the ideal ratio of air to fuel. For gasoline, the stoichiometric ratio is approximately 14.7:1, signifying that for every kilogram of fuel burned, 14.7 kg of air is required. During actual drive the fuel-to-air ratio varies significantly from the ideal ratio.

Reduction of nitrogen oxides to nitrogen and water

$$2NO + 2CO = N_2 + 2CO_2$$

Oxidation of carbon monoxide to carbon dioxide

$$2CO + O_2 = 2CO_2$$

Oxidation of unburnted hydrocarbons to CO_2 and water

$$2C_XH_Y + \frac{(2X+Y)}{2} O_2 = 2XCO_2 + YH_2O$$

FIGURE 3-6 Catalytic converter reactions.

Most modern cars are equipped with three-way catalytic converters that help regulate three types of emissions: NOx, CO, and HC. The converter uses two different types of catalysts: a reduction catalyst that reduce the oxides of nitrogen and carbon monoxide to nitrogen and carbon dioxide and one oxidation catalyst that oxidizes CO to carbon dioxide and unburned HCs to carbon dioxide and water. Both types of catalysts consist of an alumina support impregnated with noble metals platinum, rhodium, and/or palladium. The reduction catalyst is in the first stage of the converter. The oxidation catalyst is the second stage of the catalytic converter. A control system monitors the oxygen content of exhaust gases upstream of the converter and adjusts the air-to-fuel ratio, to ensure that exhaust gases have sufficient oxygen for the converter to function efficiently.

The European Union emission standards for gasoline-fueled cars were first introduced in 1992. The permissible levels of CO, HC, and NOx emissions in grams per kilometer have been revised downward with every passing year (Table 3-6) to control atmospheric pollution while the number of gasoline-powered cars on the roads continue to rise.

TABLE 3-6 Emission Standards for Passenger Cars M1 Class (>2500 kg), Gasoline Powered*

Standard	Date	Carbon monoxide (CO)	Hydrocarbons (HC)	HC + NO_x	NO_x	Particulate matter (PM)
Euro 1	July 1992	2.72	—	0.97	—	—
Euro 2	Jan. 1996	2.20	—	0.5	—	—
Euro 3	Jan. 2000	2.30	0.20	—	0.15	—
Euro 4	Jan. 2005	1.00	0.10	—	0.08	—
Euro 5	Sept. 2009	1.00	0.1†	—	0.06	0.005
Euro 6	Sept. 2014	1.00	0.1†	—	0.06	0.005

*All figures in g/km.
†And nonmethane hydrocarbons (NMHC) = 0.068.

GASOLINE SPECIFICATIONS

Table 3-7 lists the typical gasoline specifications for some commercial gasoline grades. In 1990, the U.S. Environmental Protection Agency issued regulations that would require gasoline specifications to be "reformulated" by inclusion of oxygenates in the blend so as to result in significant reduction in vehicle emissions of ozone-forming and toxic air pollutants. This cleaner burning gasoline is called Reformulated gasoline (RFG). RFG is required to be used in nine major metropolitan areas of the United States with the worst ozone air pollution problems. Gasoline specifications in other countries are also gradually moving toward the RFG gasoline model.

Blend components used to make RFG are no different from those used in blending conventional gasolines. RFG and conventional gasoline differ only in the volume percentage of these components and reducing the use of components that contribute to air pollution. However, like other gasolines, RFG is formulated to burn in the engine like any other conventional gasoline. RFG has no adverse effect on vehicle performance or engine life.

RFG composition has upper limits for certain organic compounds that contribute to air pollution. These are discussed next.

TABLE 3-7 Unleaded Premium Motor Gasoline 91 and 95 RON

Property	Units	Limit	91 RON	95 RON	Test method
Acid number total	mg KOH/g	Max.	0.03	0.03	ASTM D 974
Appearance, visual			Bright and clear	Bright and clear	
Aromatics	Vol %	Max.	50		
Benzene	Vol %	Max.	4.0	4.0	ASTM D 3606
Corrosion, Cu strip, 3 h, 50°C		Max.	NO 1	NO. 1	ASTM D 130
Density, 15°C	kg/m^3	Min.	720	725	ASTM D 4052
		Max.	783	775	
Distillation					ASTM D 86
Evaporated @ 70°C	Vol %	Min.	20	20	
		Max.		48	
Evaporated @ 100°C	Vol %	Min.	80	46	
		Max.		70	
Evaporated @ 180°C	Vol %	Min.	90	85	
End point	°C	Max.	210	220	
Residue	Vol %	Max.	2.0	2.0	
Existent gum	mg/100 mL	Max.	3	3	ASTM D 381
Flash point	°C	Max.	21	21	ASTM D 56
Induction period	Min	Min.	360	360	ASTM D 525
Lead	gPb/L	Max.	0.005	0.005	ASTM D 3237
Octane number					
MON		Min.	81.0	85.5	ASTM D 2700
RON		Min.	91.0	95.3	ASTM D 2699
Odor			Marketable	Marketable	
Olefins	Vol %	Max.	20.0		ASTM D 1319
Sulfur	Wt %	Max.	0.03	0.10	ASTM D 4294
Sulfur , mercaptan	mg/kg	Max.	15	15	ASTM D 3227
MTBE	Vol %	Max.	10	10	
Vapor pressure, Reid @ 37.8°C	kPa	Min.	49.0	50	ASTM D 323
		Max.	84.0	83.3	
Vapor lock index		Max.	620–700	1100	

Benzene

Benzene is a low molecular weight volatile aromatic compound (boiling point 176.2°F) present in gasoline. Benzene is also a known carcinogen. Human exposure to benzene can occur during vehicle refueling at a service station or in refineries, petrochemical plants, or other facilities handling petroleum solvents. As the lightest aromatic, it is always a part of evaporative hydrocarbon emissions. Most of the ambient benzene concentration results from automotive emissions. Catalytic reformate is the major source of benzene in the gasoline pool, typically contributing 65 percent of the benzene in the pool. The other major source of benzene is FCCU LCN, which contributes 25 percent. The remaining 10 percent comes from various other streams, such as coker light naphtha and light straight run gasoline.[2] In a refinery, naphtha is fractionated into two streams: an LSR naphtha stream going to the isomerization unit or for direct blending into the naphtha pool and a bottom cut containing all C6+ going to the cat reformer.[3] For removal of benzene from catalytic reformer feed, the operation of the naphtha splitter is adjusted to remove all benzene and benzene precursors such as cyclohexanes and methyl cyclopentane as overhead product. Removal of benzene and its precursors is a difficult fractionation because the splitter has to accomplish this separation while maximizing C7+ recovery as bottom product, which is cat reformer feed. California reformulated gasoline specifications (CaRFG3) limit benzene concentration in gasoline to 1.1 vol % maximum.

Aromatics

Aromatics are reactive hydrocarbons that contribute to the formation of ozone in the atmosphere. Catalytic reformate and FCCU light and medium cat naphthas are the main gasoline blending components, and both of these have a high concentration of aromatic compounds. Cat reformate may typically contain 63 vol % aromatics, and FCUU gasoline may contain 29 percent aromatics.[4] The control of aromatics in gasoline is affected by lowering the end point of reformer feed. The temperature at which 90 percent gasoline is distilled is controlled at 290°F. Thus cat reformate coming from reactors may be distilled to remove heavy ends and adjust the FBP of reformate at 290°F. Similarly, FBP of cat naphtha may be adjusted to reduce aromatics in the blendstocks. Limiting the aromatic content of gasoline and distillation temperature are the most important parameters for controlling vehicle emissions. Aromatics oxidize to large condensable molecules that lead to increased combustion chamber deposits and increased particulate matter in exhaust. Unburned aromatics also contribute to ozone formation. Adding oxygenates to gasoline is one of the most effective means for both decreasing aromatics and distillation temperatures. Reformulated gasolines typically limit aromatics in gasoline blends to 35 vol % maximum.

Olefins

Olefins have long been known to be a major contributor of smog in the atmosphere.[5] Light olefins have a high vapor pressure and are extremely photochemically reactive with respect to ground-level ozone formation. These hydrocarbons react with nitrogen oxides in the presence of sunlight to form ozone. For most refiners, FCCU gasoline is the only blending component with significant olefins. Gasoline olefins are controlled by limiting the percentage volume of cat naphthas in gasoline blend. RFG limit olefins in the gasoline pool to 10 vol %.

Sulfur

The requirements for low-sulfur gasoline are important. Some of the gasoline blend components, such as cat reformate, MTBE, and alkylate, are sulfur free. Cat reformer feed is hydrotreated to reduce sulfur in the 1 to 2 ppm range in the cat reforming unit itself. The major source of sulfur in gasoline is FCCU cat naphthas. Typically feed to FCCU is vacuum gas oil, which may contain

2.5 to 3 wt % sulfur. Sulfur in gasoline[6] may range from 0.08 to 0.3 percent. Feed hydrotreating to 0.15 percent sulfur will produce a light cat naphtha stream with less than 50 ppm sulfur. Feed hydrotreating has other benefits, such as increasing gasoline yield, reduced coke make, reduced sulfur level in light and heavy cycle oils, and reduced SOx level in regeneration gases. RFG limits gasoline sulfur to 30 ppm sulfur maximum.

RVP

Gasoline vapor pressure must be high enough so that at a low temperature enough fuel gets into the cylinder but not so high that oxygen cannot enter the cylinder. At high elevations, atmospheric pressure is low and vapor pressure must not be so high that the fuel vaporizes in the fuel system. Gasoline RVP varies between 5 and 15 lb/in^2 depending on the ambient temperature. RVP is so set that at maximum expected ambient temperature in the area of use, vapor pressure of gasoline should be less than the atmospheric pressure or 14.7 lb/in^2. RVP needs to be higher during cold weather and lower during hot weather. The RVP of gasoline is adjusted by blending butane or other high volatility components such as LSR or isomerate in gasoline. In reformulated gasoline specifications, to minimize hydrocarbon emissions, RVP of gasoline during the summer is reduced to the 6.4 to 7.2 lb/in^2 range.

Oxygenates

Oxygenates blending add oxygen to fuel via oxygen-bearing compounds such as MTBE, ETBE, and ethanol, and so reduces the amount of CO and unburned HCs in exhaust gases. RFG regulations allow gasoline formulation with oxygenates so as to contain at 0 to 3.7 wt % oxygen. Adding oxygenates to gasoline in the form of ethanol is one of the most effective ways to reduce aromatics and distillation temperature. MTBE is no longer favoured in gasoline because of its low biodegradability and concerns over ground and surface water contamination. Ethanol is a common replacement. An ethanol-gasoline mix containing 10 vol % ethanol and 90 percent gasoline is called gasohol, or E10. An ethanol-gasoline mix containing 85 percent ethanol is called E85. The most extensive use of ethanol in gasoline blending takes place in Brazil where ethanol is derived from sugarcane. In the United States, in 2004, 10 million tons of ethanol was produced from corn for fuel use.

Gasoline Distillation

In reformulated gasoline specifications, T50 and T90, or 50 percent distillation point and 90 percent distillation point, have been reduced to 200°F and 290°F, respectively. T90 reduction is done essentially to remove the tail end of gasoline, more specifically C_{10} to C_{12} hydrocarbons. Reduction of T90 is generally controlling, and no specification changes are needed to meet T50 specification. Inclusion of oxygenates such as ethanol, which has a low boiling point, also lowers T50 temperature. The source of most heavy aromatics is naphtha reformate. The heavy aromatics from the cat reformer can be controlled by limiting the end point of feed.

AVIATION GASOLINE

In the 1930s and 1940s, before the advent of aviation turbine or jet engines, all military and commercial planes used piston engines similar to those in automobiles. These planes were fueled by a special type of gasoline (aviation gas, or avgas). Both automobile and piston engine planes use spark-ignited internal combustion engines, and specifications for motor gasoline and avgas are similar to some extent. The specifications of gasoline for these planes were developed jointly by engine manufacturers and refineries, resulting in a fuel that could deliver the premium performance required for military and commercial applications at that time. Demand for aviation gasoline fuels virtually ceased with the advent of turbine engines toward the end of World War II. Aviation gasoline was no longer needed for large military or commercial aircraft. At present, avgas is used in small piston engine powered aircrafts for activities such as crop spraying, flight training, and flying clubs. Worldwide, the total avgas volumes consumed are low, estimated at 0.5 percent of total mogas demand. Avgas-fueled aircraft are much smaller, but these planes outnumber jet-fueled aircraft. Piston engine planes operate using the same basic principles as spark ignition engines of the cars, but they have a much higher performance requirement. An important difference between the airplane engine and automobile engine is the use of supercharger in the airplane engine. Supercharger is simply a compressor that increases the pressure and density of either air alone or the air-fuel mixture. This enables a larger charge to be admitted to the cylinder on the intake stroke. The aircraft engine has very high power output per unit engine weight compared to that of automobile, and also it has lower specific fuel consumption. This is achieved by higher compression ratio of the engine. Engine efficiency is directly proportional to its compression ratio. Raising the compression ratio requires higher octane fuel to protect against knocking. Thus aviation gasoline requires much higher octane compared with automobile gasoline. The specifications of aviation gasoline for use in piston-type engines are far more stringent than those of spark-ignited automobile gasoline. Any problem in airplane engine due to inappropriate fuel can damage the engine and can cause it to fall down to earth with fatal consequences.

Aviation Gasoline Specifications

Aviation gasoline is the highest performance gasoline product with the most stringent specifications. Aviation fuels must conform to the requirements of ASTM D 910. ASTM D 910 exclude oxygenates. The very stringent specifications, particularly for supercharge octane rating, the MON, the vapor pressure, and the distillation specifications, greatly reduce the options available for blending aviation gasoline without the use of a lead additive. ASTM D 910 set lead additive limit at a 0.53 g/L maximum for grade 100LL. Prior to World War II, a number of avgas grades were manufactured, such as grade 80, grade 91, grade 100, and grade 100 LL, but at present due to very low demand, refineries do not find it economical to manufacture and supply a large number of avgas grades. At present only one avgas grade (100 LL) is manufactured. Table 3-8 lists the specifications of three aviation gasoline grades.

Antiknock Quality

Knock in automobile engines as a result of engine/fuel incompatibility can at worst be annoying noise and generally does not cause engine damage unless it is prolonged and severe. However, knock in aircraft engine is inaudible because of other noises but can be very damaging to the engine. For this reason, the antiknock quality of aviation gasoline is carefully controlled by specifications of different grades. Aviation gasoline antiknock quality is measured at two conditions, and both ratings are part of the specifications. One of the methods measures gasoline octane under conditions comparable to normal cruising and the other under conditions comparable to takeoff or an emergency. For example, an octane rating of 91/98 signifies that it has a cruising rating of 91 octane and a takeoff rating of 98.

Performance Number Scale

Isooctane has the highest octane number of 100. It was discovered[7] that by adding tetraethyl lead to isooctane, it is possible to increase the octane number to greater than 100. With the advances in metallurgy, new engines were developed capable of higher compression ratios and higher power output. These engines required fuel with octane higher than 100.

TABLE 3-8 Aviation Gasoline Specifications as per ASTM D 910-04a

Property		Grade 80	Grade 91	Grade 100	Grade 100 LL	Test method
Knock rating						
Lean mixture octane number	Min.	80.0	91.0	99.9	99.9	ASTM D 2700
Rich octane number	Min.	87.0	98.0			ASTM D 909
Performance number	Min.			130.0	130.0	ASTM D 909
Tetra ethyl lead; mL/L	Max.	0.13	0.53	1.06	0.53	ASTM D 3341
g Pb/L	Max.	0.14	0.56	1.12	0.56	
Dye content						
Blue dye (mg/L)	Max.	0.2	3.1	2.7	2.7	
Yellow dye (mg/L)	Max.	None	None	2.8	None	
Red dye (mg/L)	Max.	2.3	2.7	None	None	
Orange dye	Max.	None	6.0	None	None	
Visual		Clear and bright	Clear and bright	Clear and bright	Clear and bright	
Distillation						
Initial boiling point, °C		Report	Report	Report	Report	ASTM D 86
10 % vol evaporated, °C	Max.	75	75	75	75	
40 % vol evaporated, °C	Min.	75	75	75	75	
50 % vol evaporated, °C	Max.	105	105	105	105	
90 % vol evaporated, °C	Max.	135	135	135	135	
Final boiling point, °C	Max.	170	170	170	170	
Sum of 10% + 50% evaporated	Min.	135	135	135	135	
Recovery vol %	Min.	97	97	97	97	
Residue, vol %	Max.	1.5	1.5	1.5	1.5	
Loss, vol %	Max.	1.5	1.5	1.5	1.5	
Vapor pressure at 38°C, kPa		38.0–49.0	38.0–49.0	38.0–49.0	38.0–49.0	ASTM D 323
Density at 15°C, kg/m^3	Max.	Report	Report	Report	Report	ASTM D 1298, 4052
Freezing point, °C	Max.	–58	–58	–58	–58	ASTM D 2386
Sulfur, Wt %	Max.	0.05	0.05	0.05	0.05	ASTM D 1266, D 2622
Net heat of combustion, MJ/kg	Min.	43.5	43.5	43.5	43.5	ASTM D 4529, D 3338
Corrosion Cu @ 2 h, 100°C	Max.	1	1	1	1	ASTM D 130
Oxidation stability, 5 h						ASTM D 873
Potential gum, mg/100 mL	Max.	6	6	6	6	
Lead precipitate, mg/100 mL	Max.	3	3	3	3	
Water reaction						ASTM D 1094
Volume change, mL	Max.	2	2	2	2	
Conductivity, Ps/m		50–450	50–450	50–450	50–450	ASTM D 2624
Additives						
Anti-cing		Optional	Optional	Optional	Optional	
Antioxidant		Optional	Optional	Optional	Optional	
Corrosion inhibitor		Optional	Optional	Optional	Optional	
Static dissipator		Optional	Optional	Optional	Optional	

TABLE 3-9 Scale for Determination of Octane Number Above 100

Tel mL/gallon isooctane	Octan number	Tel mL/gallon isooctane	Octan number
0.0	100.0	2.1	113.1
0.1	101.3	2.2	113.4
0.2	102.5	2.3	113.7
0.3	103.5	2.4	114.0
0.4	104.4	2.5	114.3
0.5	105.3	2.6	114.5
0.6	106.0	2.7	114.8
0.7	106.7	2.9	115.3
0.8	107.4	3.0	115.5
0.9	108.0	3.2	116.0
1.0	108.6	3.4	116.4
1.1	109.1	3.6	116.8
1.2	109.6	3.8	117.2
1.3	110.1	4.0	117.5
1.4	110.5	4.5	118.3
1.5	111.0	5.0	119.1
1.6	111.4	5.5	119.7
1.7	111.7	6.0	120.3
1.8	112.1		
1.9	112.5		
2.0	112.8		

The need for high-performance gasoline with an octane number greater than 100, the upper limit of the normal octane scale, required am upward extension of the octane scale by equating antiknock quality to isooctane plus x milliliters per gallon of TEL (Table 3-9). To represent octane numbers above 100, a performance number scale was developed. The performance number is the percentage of knock-limited power that can be delivered by an engine on a given fuel compared to what can be obtained with a 100 octane number isooctane fuel. Thus a fuel with a performance number of 120 can provide without knocking 1.2 times the power provided by isooctane fuel. The performance number indicates the maximum power (without knocking) obtainable from a fuel compared with isooctane.

Distillation

The distillation characteristics of avgas are different than those of motor gasoline. Motor gasoline includes heavier petroleum fractions that are less clean burning and promote engine deposits. Avgas has a 90 percent distillation point of 275°F (maximum) and an end point of 338°F (maximum) compared to 290°F and 400°F or higher, respectively, for mogas.

RVP

Motor gasoline normally has a much higher vapor pressure (9 to 15 lb/in^2), which varies seasonally. With a high RVP fuel, the risk of vapor lock during takeoff and climb increases, particularly for aircraft that do not have a gravity-fed fuel system. Avgas RVP is controlled between 5.5 and 7.1 lb/in^2 (38 to 49 kPa).

Heat of Combustion

Net heat of combustion per unit weight of avgas determines how many kilometers an air plane can fly between two refuelings. The higher the net heat of combustion, the larger the distance flown and the less fuel the plane has to carry for a given distance. The carbon-to-hydrogen ratio of the fuel blend components determine the heat of combustion. Aromatics have a lower heat of combustion than that of paraffins. Also, oxygenates have a much lower heat of combustion compared with hydrocarbons and are not allowed in avgas blending. Motor gasolines do not have heat of combustion specifications. Aviation gasoline must yield a minimum of 43.3 MJ/kg net heat on combustion.

Freezing Point

Aviation gasoline is used by small planes that may fly at 30,000 ft or more where the ambient temperatures may be less than –40°F. Aviation gasoline must meet a freezing point requirement of –58°C (–72.4°F). However, aviation gasoline is blended from a low boiling component with a very low freezing point, and the freezing point specification is never limiting.

Additives

Aviation gasoline additives are limited to certain specific additives mandated by international regulations. Use of TEL is prohibited in mogas blends, but its use is necessary in avgas blending to achieve a performance number of 130 or more or the MON specs of 100 required in some grades. TEL concentration varies from grade to grade between 0.13 to 1.06 mL/L or 0.14 to 0.56 g Pb/L.

Aviation Gasoline Blending

Aviation gasoline is the highest performing gasoline product with the most stringent specifications. The important properties are the highest octane number, maximum energy output in terms of net heat of combustion, high hydrogen content, and no olefin content.

Aviation gasoline blends[8] may contain aviation alkylate or 2,2,4 trimethyl pentane (isooctane), a C_7-C_{11} aromatic hydrocarbon, a C_5-C_6 aliphatic hydrocarbon, and a C4 aliphatic hydrocarbon. C_7-C_{11} hydrocarbon is included to provide an appropriate supercharge rating, C_5-C_6 is included to provide the required distillation, and C4 is added to control distillation and RVP.

Aviation gasoline is blended from limited blend components such as the following:

- Aviation alkylate
- Light naphthas
- Toluene or xylene

The most important blend component for avgas blending is aviation alkylate, which may constitute as high as 80 vol % of the blend. Aviation alkylate is identical to the alkylate used for mogas blending, except that it is rerun to remove heavy ends. Straight aviation alkylate has low volatility; therefore components such as light naphtha, isobutane, or isopentane are added. Aromatics such as toluene are added to achieve high octane specially supercharged octane. Supercharge octane of toluene is so high that it cannot be measured directly but is inferred on the basis of supercharge octane of blended product. Toluene/xylenes are not refinery streams and may be imported into the refinery for aviation gasoline blending. The typical 100 LL avgas blend composition[9] is shown in Table 3-10.

TABLE 3-10 Typical Aviation Gasoline Blends* for 100 LL Grade

Blend components	1	2
	Vol %	Vol %
Alkylate	67.0	70.0
Toluene		20.0
Xylene	18.0	
Isopentane	12.0	10.0
Isobutane	3.0	
Lead, mL/L	0.13	0.53

*Reference U.S. Patent 6451075 B1, Sept. 17, 2002.

Additives

Only additives allowed in aviation blends by specifications are added to avgas. These are tetraethyl lead, static dissipater, dye, and antioxidants. TEL is added to avgas blends to achieve high octane/performance number or MON specifications. A static dissipater such as Stadis 450 is added to achieve required conductivity specs of the product. A dye or color is added to identify different avgas grades. The use of an antioxidant is optional to enhance the storage stability of the product.

RACING FUELS

Racing gasoline is different from normal gasoline for street driving in the sense that normal gasolines are blended to address primarily emission considerations rather than performance, whereas racing gasolines are designed mainly for performance and for off-street use. Racing fuels are motor gasoline for high-performance automobiles used in sports such as motor races. Compared with normal automobiles, racing cars have a large high-speed and high-compression engine. Racing cars may typically have 2.4 L naturally aspirated engines in a V8 configuration with an engine speed of 18,000 r/min or more and an engine hp of 750. The compression ratio generally exceeds 14:1. Due to very high compression engines, high octane fuel (more than 100) is required. Unlike street gasoline, use of lead additives such as tetraethyl lead is allowed in racing gasoline blending to achieve a high octane number. Lead levels up to 6 g TEL/gal are not uncommon. Typical properties of some commercial racing gasolines are listed in Tables 3-11and 3-12. Racing gasoline blending requires blend components with high RON and MON and high energy output per unit weight such as alkylate, toluene, xylene, and so on. Unleaded racing gasolines use MTBE as a high octane blend component.

Nonhydrocarbons such as nitrobenzene and methanol have also been used as racing fuels. The advantage of nitromethane is that it yields more power per explosion in the engine because it needs much less air: 1.7 kg air/kg nitromethane compared to 15 kg air/kg gasoline. This implies that many times more nitromethane can be pumped into a cylinder of given volume for complete combustion, increasing engine power output. Nitromethane requires very careful handling, however, because of its explosive properties, and its use is governed by local regulations.

TABLE 3-11 Commercial Racing Gasoline Properties

Property	Units	1	2	3	4	5	6	7	8
RON		114	114	120	114	120	120	110	106
MON		106	106	116	106	116	113	100	96
(RON + MON)/2		110	110	118	110	118	116.5	105	101
Specific gravity		0.728	0.725	0.700	0.725	0.700	0.724	0.725	0.740
RVP	psi	6.5	7.0	6.0	7.0	6.0	4.3	7.0	7.0
Lead content	g/gallon	3.0	4.0	6.0	4.0	6.0	5.0	0.0	0.00

TABLE 3-12 Gasoline Specifications (1995) For Formula -1 Racing

Property	Units	Minimum	Maximum	Test method
RON		92	102	ASTM D 2699
MON		85		ASTM D 2700
Oxygen content	Mass %		3.7	
Nitrogen content	Mass %		0.2	
Benzene	Vol %		5.00	ASTM D 3606
RVP	millibar	350	700	ASTM D 323
Lead Content	g/L		0.005	ASTM D 3237
Density at 15°C	kg/m^3	725	780	ASTM D 4052
Oxidation stability	Min.	360		ASTM D 525
Gums	mg/100 mL		5.0	EN5
Sulfur content	Mass %		0.10	ISO 8754
Cu Strip corrosion			1C	ASTM D 130
Electrical conductivity	pS/m	200		ASTM D 2824
Distillation				ISO 3405
At				
70°C		15	50	
100°C		40	70	
140°C				
180°C		85		
FBP			215	
Residue			2.0	

Note: Maximum nitrogen content of 0.2 wt % excludes the use of nitromethane in the blend.

REFERENCES

1. James H. Garry and Glen E. Handwerk, *Petroleum Refining: Technology and Economics,* (New York, Marcel Dekker, 1993).
2. K. L. Rock, "Cost Effective Solutions for Reduction of Benzene in Gasoline," paper presented at the NPRA Annual Meeting, San Antonio, TX, March 16–18, 1997.
3. L. E. Turpin, "Cut Benzene out of Reformate," *Hydrocarbon Processing,* June (1992): 81–92.
4. G. Yepsen and T. Witoshkin, "Refiners Have Options to Deal with Reformulated Gasolines," *Oil & Gas Journal,* April 8 (1991): 68–71.
5. G. H. Unzelman, "Reformulated Gasolines Will Challenge Product Quality Maintenance," *Oil & Gas Journal,* April 9 (1990): 43–48.
6. A. K. Rhodes, *Oil & Gas Journal,* January 5 (1998): 22–27.
7. David A. Lombardo, "Aircraft Systems."
8. Kurt H. Strauss and Cesar Gonzales, "Future Fuels for General Aviation," ASTM STP 1048, 1989.
9. U.S. Patent 6451075, September 17, 2002, "Low Lead Aviation Gasoline Blends."

CHAPTER 4
KEROSENE

Kerosene is a distillate fraction of crude oil boiling between 300 and 480°F. It is heavier than naphtha and gasoline cut but lighter than diesel cut. The yield of kerosene from a medium gravity crude oil such as light Arabian is approximately 16 vol % (percentage by volume). Kerosene was one of the earliest petroleum products to be produced by refineries. Almost a century ago, before the advent of electricity, kerosene was used in lamps and lanterns for home and street lighting, replacing whale oil. Use of kerosene for lighting declined with the advent of electricity, but kerosene is still used for illumination and cooking by rural communities in many parts of the world where electricity, liquified petroleum gas (LPG), or natural gas are not available and the only other available source of energy is firewood. In India, for example, approximately 12 million tons of kerosene were used in 2006 in rural households for cooking and illumination.

The development of jet engines during the 1950s and the subsequent rapid growth of civil and military aviation during the last half of the century created a huge demand for fuel for civil and military aircraft. Kerosene was preferred to naphtha or lighter petroleum fuels because of the inherent safety in its handling and use. Refining capacities were increased and new product specifications were introduced to meet the quality and volume requirements of jet fuel for the aviation industry. Kerosene used in aircrafts is called "aviation turbine fuel." In 2007, the consumption of aviation turbine fuel worldwide was estimated at 205 million tons, and demand is growing at a faster rate than that of other petroleum products. Small volumes of kerosene are used for other diverse uses such as illumination, cooking, and other uses.

JET ENGINE

In the 1930s, the first turbojet gas turbine engine for aircraft was developed in Germany. In this engine, atmospheric air is compressed by the compressor of the engine, mixed with fuel, and burned. Hot burning gases expand through a turbine and are exhausted to the atmosphere. A greater portion of the energy extracted from the hot gases as they expand through the turbine is used to drive the compressor and engine accessories such as the fuel pump and electrical generator. The remaining small part is available as engine thrust to propel the plane. Another variant of the jet engine is the turboprop engine. In this engine, the energy extracted from hot gases as they expand through the turbine is used to drive a propeller to propel the aircraft apart from driving the compressor and accessories. Jet-powered aircraft had limited use during World War II, but soon after the war, advances in engine design allowed more powerful jet engines to replace slow-moving piston-driven engines almost completely in both civil and military aircrafts. Commercially available kerosene was used in the earlier phase of jet engine development. Kerosene was considered because of its high flash point and lower volatility compared with that of naphtha. A higher flash point allowed safer handling, transportation, and storage of fuel. Kerosene has a very low freezing point, allowing planes to fly at great altitudes. At a height of 36,000 ft above sea level, ambient temperature is estimated to be –70°F.

GRADES AND SPECIFICATIONS

A number of organizations have issued specifications covering aviation turbine fuels. With the exception of Russia and China, jet fuel specifications worldwide are issued by the American Society for Testing and Materials (ASTM) and the British Ministry of Defence. As the aviation fuel supply arrangements became more complex, involving commingling of product from different refineries in joint product pipelines and storage facilities, a number of major fuel suppliers developed common specifications incorporating the requirements of the main international specifications in current use. This document became known as the "Aviation Fuel Quality Requirements for Jointly Operated Systems" (AFQRJOS) checklist. The checklist is recognized by all major aviation fuel suppliers: AGIP, BP, Caltex, Chevron, ELF, ESSO, Gulf, Mobil, Shell, Texaco, and Total. AFQRJOS specifications meets the most stringent requirements of DERD, ASTM 1655 kerosene type Jet A-1, and of IATA Guidance Material–Kerosene type. The International Air Transport Association (IATA) guidance material aims at the availability throughout the world of aviation turbine fuels with similar specifications. Also, the modern jet aircraft fuel system demands a fuel free from undissolved water, dirt, and other foreign contaminants. Part II of the IATA guidance material defines minimum standards to be met by fuel suppliers to assure a clean dry fuel supply to aircraft. To deliver jet fuel in the required conditions requires stringent storage and handling procedures. Part III describes the major international aviation turbine fuel specifications and also a brief description of various additives that may be present in fuel. Two main grades of turbine fuels are in use for civil commercial aircrafts: Jet A-1 and Jet A. Another grade of jet fuel called Jet B is a wide-cut kerosene (a blend of naphtha and kerosene) that is used only in very cold ambient conditions.

Jet A-1

Jet A-1 is a kerosene cut suitable for most turbine engine aircrafts. It has a minimum flash point of 100°F and a maximum freeze point of –47°C (–52.6°F). Jet A-1 meets the specifications of ASTM D 1655 (Jet A-1), British Specifications DEF STAN 91-91 (Jet A-1), IATA Guidance Material (kerosene type), and NATO code F-35. Jet A-1 is the most widely used fuel by all civilian aircraft outside the United States. Table 4-1 lists its specifications.

Jet A

Jet A is identical to Jet A-1 except that it has a higher freeze point maximum of –40°C. It meets ASTM D 1655 (Jet A) specifications. Jet A is used within the United States by domestic and international airlines.

Jet B

Jet B is a wide-cut distillate fuel containing naphtha and kerosene fractions. It can be used as an alternative to Jet A-1, but it has a lower flash point and higher flammability. It is more difficult to handle. It is used in very cold weather operations. It is generally produced to Canadian specifications CAN/CGSB 3.23. Table 4-2 lists the specifications of Jet B.

MILITARY JET FUEL SPECIFICATIONS

The major difference between U.S. military fuels and commercial fuels is the use of certain additives, such as anti-icing, corrosion inhibitors, lubricity improvers, antioxidants, thermal stability improvers, conductivity improvers, and so on. The use of additives in civilian jet fuel is by agreement between user and supplier.

TABLE 4-1 Aviation Turbine Fuels (Jet A-1) Specifications (Joint Fueling System Check List Issue 20)

Property	Units	Min./Max.	Value	Test method
Acidity, total	mg KOH/g	Max.	0.015	ASTM D 3242
Appearance			Clear and bright	Visual
Aromatics	Vol %	Max.	25.0	ASTM D 1319
Total aromatics	Vol %	Max.	26.5	ASTM D 6379
Corrosion, Cu strip: 3 h @100°C		Max.	No. 1	ASTM D 130
Density @ 15°C	kg/m^3	Min.	775	ASTM D 1298
		Max.	840	
Distillation				ASTM D 86
IBP	°C	Max.	Report	
10% recovered	°C	Max.	205	
50% recovered	°C	Max.	Report	
90% recovered	°C	Max.	Report	
End point	°C	Max.	300	
Residue	Vol %	Max.	1.5	
Loss	Vol %	Max.	1.5	
Electrical conductivity	pS/m	Min.	50	ASTM D 2624
		Max.	450	
Existent gum	mg/100 mL	Max.	7	ASTM D 381
Flash point, Abel	°C	Min.	38	IP 170
Freeze point	°C	Max.	–47	ASTM D 2386
Sulfur mercaptan	Wt %	Max.	0.003	ASTM D 3227
or				
Doctor test			Negative	ASTM D 4962
Hydro-processed component	Vol %		Report	
Severely hydro-processed component	Vol %		Report	
Smoke point	mm	Min.	25	ASTM D 1322
or				
Smoke point	mm	Min.	19	
and				
Naphthalenes	Vol %	Max.	3.0	ASTM D 1840
Sulfur	Wt %	Max.	0.3	ASTM D 1266, D 2622
Thermal stability				ASTM D 3241
Filter pressure differential	mm Hg	Max.	25	
Tube deposit rating, visual		Max.	<3	
Viscosity, kinematic @ –20°C	cSt	Max.	8.0	ASTM D 445
Specific energy, net	MJ/kg	Min.	42.8	ASTM D 4529
Water reaction:				ASTM D 1094
Interface rating		Max.	1b	
Separation rating		Max.	2	
Water separation index modified			H	ASTM D 2550 or 3948
Fuel with static dissipastor additive		Min.	70	
Fuel without static dissipastor additive		Min.	85	
Particulate matter	mg/L	Max.	1.0	ASTM D 5452
Lubricity test				
Bocle (wear scar diameter)	mm	Max.	0.85	ASTM D 5001
Additives;				
Antioxidant (Hydrotreated part), mandatory	mL/L		17–24	
in nonhydrotreated part, optional		Max.	24	
Static dissipator (Stadis 450)				
First doping	mL/L	Max.	3.0	
Metal deactivator	mL/L	Max.	5.7	
Corrosion inhibitor			By agreement	
Anti-icing			By agreement	

TABLE 4-2 Aviation Turbine Fuel Jet B Specifications, Canadian Standard General Board (Can/Cgsb 3.22-2002) Wide Cut Kerosene

Property	Units	Min./Max.	Value	Test method
Acidity, total	mg KOH/g	Max.	0.10	ASTM D 3242
Appearance			Clear and bright	Visual
Aromatics	Vol %	Max.	25.0	ASTM D 1319
Corrosion, Cu strip: 3 h @100°C		Max.	No. 1	ASTM D 130
Density @ 15°C	kg/m³	Min.	750	ASTM D 1298/4052
		Max.	801	
Distillation				ASTM D 86
IBP			Report	
10% Recovered			Report	
20% Recovered	°C	Min.	100	
		Max.	145	
50% Recovered	°C	Min.	125	
		Max.	190	
90% Recovered	°C	Max.	Report	
End point	°C	Max.	270	
Residue	Vol %	Max.	1.5	
Loss	Vol %	Max.	1.5	
Electrical conductivity	pS/m	Min.	50	ASTM D 2624
		Max.	500	
Filtration time	Min.	Max.	10	ASTM D 2276
Freeze point	°C	Max.	–51	ASTM D 4305, D 5901, D 5972
Mercaptan sulfur	Wt %	Max.	0.003	ASTM D 3227
or				
Doctor test			Negative	ASTM D 4952
Net heat of combustion	MJ/kg	Min.	42.8	ASTM D 3338, D 4529
Particulate Matter	mg/L	Max.	0.4	ASTM D 2276, D 5452
Smoke point	mm	Min.	25	ASTM D 1322
or				
Smoke point	mm	Min.	20	
and				
Naphthlenes	Vol %	Max.	3.0	
Sulfur	Wt %	Max.	0.4	ASTM D 1266, D 2622
Thermal stability				ASTM D 3241
JFTOT Delta P, 260°C	mm Hg	Max.	25	
Tube deposit rating, visual		Max.	<3	
Vapor pressure, Reid, @ 37.8°C	kPa	Max.	21	ASTM D 5191
Water reaction:				ASTM D 1094
Interface rating		Max.	1b	
Water separation index, modified		Min.	85	ASTM D 2550 Or 3948
Additives	Lbs/1000 BBL		Optional	
Antioxidant antistatic additive, Stadis 450	mL/L		Required	
Corrosion inhibitor			Optional	
Metal deactivator				
Fuel system icing inhibitor	Vol %	Min.	0.10	
		Max.	0.15	

JP-4

JP-4 is a wide-cut jet fuel that was introduced as jet fuel in 1951 in the U.S. Air Force and used extensively in the Korean and Vietnam wars. A major consideration in selecting a wide distillation cut was to maximize the potential availability of fuel during times of war. It was phased out in the United States in the 1990s because of its high volatility and safety concerns; however, it is still used by some other countries although production is limited. JP-4 can be considered a blend of 60 vol % light straight run naphtha, medium straight naphtha, and 40 vol % straight run kerosene. JP-4 has corrosion inhibitor and anti-icing additives. It meets the requirements of U.S. military specifications MIL-DTL-5624U grade JP-4. It also meets requirements of British specifications DEF STN 91-88 AVTAG/FSII. JP-4 can be considered the military equivalent of Jet B. Table 4-3 lists its specifications.

TABLE 4-3 Aviation Turbine Fuel JP-4

Property	Units	Min./Max.	Value	Test method
Acidity, total	mg KOH/g	Max.	0.015	ASTM D 3242
Aniline gravity product		Min.	5250 (1)	ASTM D 1405
API gravity		Min.	45.0	ASTM 1298
		Max.	57.0	
Appearance			Clear and bright	Visual
Aromatics	Vol %	Max.	25.0	ASTM D 1319
Color saybolt			Report	ASTM D 156
Corrosion, Cu strip:				
3 h @100°C		Max.	No. 1	ASTM D 130
Density @ 15°C	kg/L	Min.	0.751	ASTM D 1298/4052
		Max.	0.802	
Distillation				ASTM D 86
20% Recovered	°C	Max.	145	
50% Recovered	°C	Max.	190	
90% Recovered	°C	Max.	245	
End point	°C	Max.	270	
Residue	Vol %	Max.	1.5	
Loss	Vol %	Max.	1.5	
Electrical conductivity	pS/m	Min.	150	ASTM D 2624
		Max.	600	
Existent gum	mg/100 ML	Max.	7	ASTM D 381
Filtration time	Min.	Max.	10	ASTM D 2276
Freeze point	°C	Max.	–58	ASTM D 2386
Hydrogen content	Wt %	Min.	13.5	ASTM D 3343, D 3701
Mercaptan sulfur	Wt %	Max.	0.002	ASTM D 3227
Olefins	Vol %	Max.	5.0	ASTM D 1319
Particulate matter	mg/L	Max.	1.0	ASTM D 2276 , D 5452
Smoke point	mm	Min.	20	ASTM D 1322
Sulfur	Wt %	Max.	0.4	ASTM D 1266
Thermal stability				ASTM D 3241
Filter pressure differential	mm Hg	Max.	25	
Tube deposit rating, visual		Max.	<3	
Reid Vapor pressure, @ 37.8°C	kPa	Min.	14	ASTM D 323
		Max.	21	
Water reaction:				ASTM D 1094
Interface rating		Max.	1b	
Water separation index, modified		Min.	85	ASTM D 2550 or 3948
Additives				
Antioxidant	Lbs/1000 BBL	Min.	6 (2)	
		Max.	8.4	
Antistatic additive, stadis 450	mL/L	Max.	3.0	
Corrosion inhibitor			(3)	
Fuel system icing inhibitor	Vol %	Min.	0.10	
		Max.	0.15	

1. Alternatively, a minimum heat of combustion of 42.8 MJ/kg (18400 Btu/lb) by ASTM D 240/ D 3338 or D 4809.
2. Antioxidant additive is mandatory only for hydrotreated fuels, which is added immediately after hydrotreatment of the product and prior to being sent to storage.
3. Corrosion inhibitor and dosage shall conform to MIL-1-25017-16.

JP-5

JP-5 was introduced in 1952. It is a high flash kerosene meeting the requirements of U.S. military specifications MIL–DTL-5624U grade JP-5. JP-5 also meets the requirements of British specifications DEF STN 91-87 AVTUR /FSII. JP-5 is mainly used by the U.S. Navy for its aircrafts based on aircraft carriers. Its high flash point provides a higher degree of safety in fuel handling. Specifications of JP-5 are presented in Table 4-4.

TABLE 4-4 Aviation Turbine Fuel JP-5

Property	Units	Min./Max.	Value	Test method
Acidity, total	mg KOH/g	Max.	0.015	ASTM D 3242
Aniline gravity product		Min.	4500 (1)	ASTM D 1405
Appearance			Clear and bright	Visual
Aromatics	Vol %	Max.	25.0	ASTM D 1319
Color, Saybolt			Report	ASTM D 156
Corrosion, Cu strip:				
3 h @ 100°C		Max.	No. 1	ASTM D 130
Density @ 15°C	kg/L	Min.	0.788	ASTM D 1298/4052
		Max.	0.845	
Distillation				ASTM D 86
10% recovered	°C	Max.	205	
End point	°C	Max.	300	
Residue	Vol %	Max.	1.5	
Loss	Vol %	Max.	1.5	
Existent gum	mg/100 mL	Max.	7	ASTM D 381
Filtration time	Min.	Max.	15 (2)	ASTM D 2276
Flash point, PMCC	°C	Min.	60	ASTM D93/D 3828
Freeze point	°C	Max.	–46	ASTM D 2386
Hydrogen content	Wt %	Min.	13.4	ASTM D 3343, D 3701
Mercaptan sulfur	Wt %	Max.	0.002	ASTM D 3227
	Btu/lb	Min.	18400	ASTM D 3338/240
Olefins	Vol %	Max.	5.0	ASTM D 1319
Particulate matter	mg/L	Max.	1.0	ASTM D 2276, D 5452
Peroxide number	ppm	Max.	8.0	ASTM D 3703
Smoke point	mm	Min.	19	ASTM D 1322
Sulfur	Wt %	Max.	0.4	ASTM D 1266
Thermal stability				ASTM D 3241
Filter pressure differential	mm Hg	Max.	25	
Tube deposit rating, visual		Max.	<3	
Viscosity, kinematic @ –20°C	cSt	Max.	8.5	ASTM D 445
Water reaction:				ASTM D 1094
Interface rating		Max.	1b	
Water separation index modified		Min.	70	ASTM D 3948
Additives				
Antioxidant	Lbs/1000 BBL	Min.	6 (3)	
		Max.	8.4	
Corrosion inhibitor			(4)	
Fuel system icing inhibitor	Vol %	Min.	0.15	
		Max.	0.20	

1. Alternatively net specific energy, minimum 42.6 MJ/kg (18,300 Btu/lb) by ASTM D 240 is acceptable.
2. A minimum sample size of one gallon shall be filtered.
3. Antioxidant additive is mandatory only for hydrotreated fuels, which is added immediately after hydrotreatment of the product and prior to being sent to storage.
4. Corrosion inhibitor and dosage shall conform to MIL-1-25017-16.

JP-7

JP-7 is a highly refined, high thermal stability fuel developed in the 1960s to meet the high heat sink demand of supersonic air crafts and missiles.[1] It is thermally stable to 550°F. It has high flash, very low aromatic content (maximum 5 percent), a high hydrogen content, and a high heat of combustion. It is blended from kerosene coming from distillate hydrocracker and straight run desulfurized kerosene blend components. A U.S. study showed that a turbine fuel with thermal stability near that of JP-7 can be made from a 50:50 blend of naphthalic straight run kerosene and hydrocracker kerosene.[2] The specifications of JP-7 are listed in Table 4-5.

TABLE 4-5 Aviation Turbine Fuel JP-7 MIL-DTL-38219D (USAF)

Property	Units	Min./Max.	Value	Test method
Aromatics	Vol %	Max.	5.0	ASTM D 1319
Corrosion, Cu strip:				
3 h @100°C		Max.	No. 1b	ASTM D 130
Density @ 15°C	kg/L	Min.	0.779	ASTM D 1298/4052
		Max.	0.806	
Vapor pressure, 149°C	kPa	Max.	20.7	
Vapor pressure, at 260°C	kPa	Max.	331	
		Max.	0.806	
Distillation				ASTM D 86
IBP	°C	Min.	182	
10% recovered	°C	Min.	196	
20% recovered	°C	Min.	206	
50% recovered	°C		Report	
90% recovered	°C	Max.	260	
End point	°C	Max.	288	
Residue	Vol %	Max.	1.5	
Loss	Vol %	Max.	1.5	
Existent gum	mg/100 mL	Max.	5	ASTM D 381
Filtration time	Min.	Max.	15 (1)	ASTM D 2276
Flash point, PMCC	°C	Min.	60	ASTM D 93
Freeze point	°C	Max.	–43.3	ASTM D 2386
Hydrogen content	Wt %	Min.	14.4	ASTM D 3343 or D 3701
Mercaptan sulfur	Wt %	Max.	0.001	ASTM D 3227
Net heat of combustion	MJ/kg	Min.	43.5	ASTM D 2382
	Btu/lb	Min.	18700	ASTM D 3338/240
Particulate matter				ASTM D 2276, D 5452
Origin	mg/L		0.3	
Destination	mg/L		0.5	
Sulfur	Wt %	Max.	0.1	ASTM D 1266
Thermal stability (JFTOT)				ASTM D 3241
Change in pressure drop in 5 h	mm Hg	Max.	25	
Delta TDR spun		Max.	12	
Viscosity, kinematic @ –20°C	cSt	Max.	8.0	ASTM D 445
Water reaction:				ASTM D 1094
Interface rating		Max.	1b	
Separation rating		Max.	2	
Water separation index		Min.	85	ASTM D 2550 or 3948
Fuel system icing inhibitor	Vol %	Min.	0.10	
		Max.	0.15	

Conditions for JFTOT test:
Heater tube max. temperature: 355°C.
Fuel system pressure: 3.45 Mps (500 lb/in^2).
Fuel flow rate: 3 mL/min.
Test duration: 300 min.
Quantity of test fuel: 1 L.

JP-8

JP-8 is the military equivalent of JET A-1, with corrosion inhibitor, anti-icing, antistatic, and other additives. It is mainly used by the U.S. Air Force. It meets the requirements of the British specifications DEF STAN 91-87 AVTUR/FSII. Table 4-6 lists its properties. Table 4-7 lists NATO code and Joint Services designations for U.S. military grades.

TABLE 4-6 Aviation Turbine Fuel JP-8

Property	Units	Min./Max.	Value	Test method
Acidity, total	mg KOH/g	Max.	0.015	ASTM D 3242
API gravity		Min.	37.0	ASTM 1298
		Max.	51.0	
Appearance			Clear and bright	Visual
Aromatics	Vol %	Max.	25.0	ASTM D 1319
Color, Saybolt			Report	ASTM D 156
Corrosion, Cu strip:				
3 h @100°C		Max.	No. 1	ASTM D 130
Density @ 15°C	kg/L	Min.	0.775	ASTM D 1298/4052
		Max.	0.840	
Distillation				ASTM D 86
10% recovered	°C	Max.	205	
End point	°C	Max.	300	
Residue	Vol %	Max.	1.5	
Loss	Vol %	Max.	1.5	
Electrical conductivity	pS/m	Min.	Report	ASTM D 2624
Existent gum	mg/100 mL	Max.	7	ASTM D 381
Filtration time	Min.	Max.	15 (1)	ASTM D 2276
Flash point, PMCC	°C	Min.	38	ASTM D 93/D 3828
Freeze point	°C	Max.	–47	ASTM D 2386
Hydrogen content	Wt %	Min.	13.4	ASTM D 3343, D 3701
Mercaptan sulfur	Wt %	Max.	0.002	ASTM D 3227
Naphthalenes	Vol %	Max.	3.0	ASTM D 1840
Net heat of combustion	MJ/kg	Min.	42.8	ASTM D 4809
	Btu/lb	Min.	18400	ASTM D 3338/240
Olefins	Vol %	Max.	5.0	ASTM D 1319
Particulate matter	mg/L	Max.	1.0	ASTM D 2276, D 5452
Smoke point	mm	Min.	19	ASTM D 1322
Sulfur	Wt %	Max.	0.3	ASTM D 1266
Thermal stability				ASTM D 3241
Filter pressure differential	mm Hg	Max.	25	
Tube deposit rating, visual		Max.	<3	
Viscosity, kinematic @ –20°C	cSt	Max.	8.0	ASTM D 445
Water reaction:				ASTM D 1094
Interface rating		Max.	1b	
Separation rating		Max.	2	
Water separation index		Min.	70	ASTM D 2550 or 3948
Additives				
Antioxidant	mL/L	Min.	17 (2)	
		Max.	24	
Anti-static additive, Stadis 450	mL/L	Max.	3.0	
Corrosion inhibitor			(3)	
Fuel system icing inhibitor	Vol %	Min.	0.10	
		Max.	0.15	

1. A minimum sample size of one gallon shall be filtered.
2. Antioxidant additive is mandatory only for hydrotreated fuels, which is added immediately after hydrotreatment of the product and prior to being sent to storage.
3. Corrosion inhibitor and dosage shall conform to MIL-1-25017-16.

TABLE 4-7 U.S. Military, NATO, and Joint Services Designation of Western Military Aircrafts

U.S. military designation	NATO code	Joint services designation
JP-4	F-40	AVTAG/FSII
JP-5	F-44	AVTCAT/FSII
JP-8	F-34	AVTUR/FSII
JP-8 without FSII	F-35	AVTUR

JET FUEL QUALITY CHARACTERISTICS

Distillation

Most jet fuels are straight kerosene cuts from crude distillation. Typically 300 to 480°F (150 to 250°C) cuts are used for Jet A-1 fuel with a flash point of 100°F. Initial boiling point (IBP) determines the flash point of the fuel. Raising the flash point reduces the kerosene cut yield and increases the naphtha yield. The end point of the cut is decided by the desired freeze point of kerosene. Jet A-1 fuel has a freeze point of –52°F. Raising the freeze point can increase the endpoint temperature of the cut and also the yield of the jet fuel from a given crude oil.

Smoke Point (ASTM D 1322)

This test provides an indication of the smoke-producing properties of jet fuels in diffusion flame. Smoke point is related to the hydrocarbon-type composition of the fuels. Generally the more aromatics in the fuel, the lower is the smoke point and the fuel is smokier. A high smoke point indicates a low aromatic content and a low smoke-producing tendency. The smoke point and luminometer number to which it can be correlated is related to the potential radiant heat transfer from the combustion products of fuel. Smoke point determination is done in a wick-type kerosene lamp. A sample is burned in a closed wick lamp that is calibrated daily against pure hydrocarbon blends with known smoke points. The maximum height of flame that can be achieved with the test fuel without smoking is estimated to the nearest 0.5 mm. The reference fuels used for calibration are isooctane and fuel-grade toluene. Pure isooctane has a reference smoke point of 42.8 mm, whereas 60 vol % isooctane and 40 vol % toluene have a reference smoke point of 14.7 mm.

Aniline Gravity Product (ASTM D 1405/IP 193)

Experimental determination of net heat of combustion is a laborious process. But a fairly accurate estimate of heat of combustion of jet fuels can be made from the aniline point and the API gravity of the fuel. Correlations have been established between the net heat of combustion and product of aniline point and the API gravity of various aviation fuel grades. These relations assume samples to be sulfur free. The test method is intended for use as a guide in cases where an experimental heat of combustion is not available or cannot be made conveniently and an estimate is considered satisfactory instead of an actual determination. In this method, the product of API gravity and aniline point in degrees Fahrenheit is calculated, and the heat of combustion is estimated by a series of empirical equations (Fig. 4-1) or read off tables for various grades.

For kerosene Jet A-1 or Jet-A fuels

$$Q = 41.6796 + 0.00025407^{*} (A^{*}G)$$

For JP-4 fuels

$$Q = 41.8145 + 0.00024563^{*} (A^{*}G)$$

For JP-5 fuels

$$Q = 41.6680 + 0.00024563^{*} (A^{*}G)$$

Where:

Q = Net heat of combustion, MJ/kg on sulfur-free basis

A = Aniline point, °F

G = Gravity, °API

FIGURE 4-1 Net heat of combustion from aniline point and API gravity ASTM D 1405.

Thermal Oxidation Stability (ASTM D 3241)

In subsonic jets, fuel is used to cool the engine components, and its temperature can increase as much as 200°F. In supersonic jets, fuel is used as a heat sink for the engine lubrication system, for cabin air conditioning, and for cooling hydraulic systems. The turbine fuel temperature can increase by 300 to 500°F before they are burned. Because of the high temperature of the fuel, small amounts of solids may form due to oxidation and deposit in heat exchangers, filters, and fuel injectors.

This test method covers the procedure for rating the tendency of aviation turbine fuels to deposit decomposition products within the fuel system. The test results are indicative of fuel performance during aviation turbine operation and can be used to access the level of deposits formed in the engine when liquid fuel contacts a heated surface at a specified temperature. The test method uses the jet fuel thermal oxidation tester (JFTOT) that subjects the test fuel to conditions simulating those in the fuel system of aviation turbines. In a typical JFTOT test, fuel is pumped at a pressure of 500 lb/in^2 and fixed volumetric flow rates through a heater that heats the fuel to 260°C, after which it enters a precision stainless steel filter where fuel degradation products are trapped. The apparatus requires 600 mL of test fuel for a 2.5-h test. The residence time of fuel in the JFTOT apparatus is about 12 s. The essential data derived are the amount of deposits on an aluminum tube and the rate of plugging of a 17-micron nominal porosity filter located just downstream of the heater tube. In actual subsonic engines, the fuel temperature is 100 to 150°C. However, the time period is thousand of hours. High JFTOT test temperature is to simulate[3] the deposits formed over a longer time and larger volume fuel flows through the engine. JFTOT test temperature is higher for fuels of supersonic planes.

Water Reaction (ASTM D 1094)

This test covers the determination of the presence of water miscible components in turbine fuels and the effect of these components on the fuel–water interface. In this method, a sample of fuel is shaken with a phosphate buffer solution at room temperature. The change in the volume of aqueous layer, the appearance of the interface, and the degree of the separation of two phases is taken as the water reaction of the fuel. Water extraction of the aviation fuel reveals the presence of partially soluble contaminants such as surfactants. These contaminants can affect the interface or create emulsions in water or fuel layers, disarm the filter separator quickly, and allow free water and particulates to pass to the engine.

Mercaptan Sulfur (ASTM D 3227)

Mercaptan sulfur if present in the aviation turbine fuel can impart an objectionable odor to the fuel and is corrosive to fuel system components and elastomers. Mercaptans can be almost completely removed by treating processes such as Merox or by hydrodesulfurization processes. Mercaptans in fuel are determined by analytical methods. The fuel sample is dissolved in alcoholic sodium acetate and titrated potentiometrically against a silver nitrate solution. Mercaptan sulfur is precipitated as silver mercaptide. The end point of titration is shown by a large change in the cell potential. From the solution of silver nitrate consumed, mercaptans are estimated.

Acidity (ASTM D 3242)

Aviation turbine fuel may have naturally occurring organic acids or may have traces of acid as a result of a refining treatment. However, even small amount of acids in the fuel are undesirable because they can promote corrosion of equipment and impair water separation characteristics of aviation turbine fuels. In the test method, the sample is dissolved in a mixture of toluene and isopropyl alcohol containing a small amount of water. The resulting solution is titrated against standard alcoholic potassium hydroxide to the end point indicated by the color change of added p-naphtholbenzein indicator solution.

Hydrogen Content (ASTM D 3343)

Aviation fuel composition has a significant effect on soot formation during gas turbine operation. Hydrogen content of aviation fuels is an important parameter that determines the combustion properties of fuels such as the smoke point and luminometer number. Increasing the hydrogen content of aviation fuels not only increases its energy output but it also reduces carbon dioxide emissions from the engine. Hydrogen content can be estimated[4] from fuel API gravity, vol % aromatics, and ASTM distillation 10, 50, and 90 percent points (Fig. 4-2).

$$\%H = 0.06317^{*}G - 0.041089^{*}A + 0.000072135^{*}A^{*}V + 0.00005684^{*}G^{*}V - 0.0004960^{*}G^{*}A + 10.56$$

Or in SI units

$$\%H = (9.1959 + 0.01448^{*}T - 0.07018^{*}A)/S + 0.02644^{*}A + 0.0001298^{*}A^{*}T - 0.01345^{*}T + 2.014$$

Where:

%H	=	Weight % hydrogen
G	=	Gravity, °API
A	=	Volume % aromatics
V	=	Average of 10%, 50%, and 90% distillation data, °F (ASTM D 86)
T	=	Average of 10%, 50%, 90% distillation data, °C
S	=	Specific gravity of fuel, g/mL

FIGURE 4-2 Hydrgen content of aviation fuels.

AVIATION FUEL ADDITIVES

Because additives under certain conditions can exhibit some undesirable side effects, only a limited number are permitted in aviation turbine fuels. The type and concentration of all additives are closely controlled by appropriate fuel specifications. Specifications define the requirements of additives as follows:

Mandatory	Must be present between the defined minimum and maximum limits.
Optional	May be added during manufacture up to the maximum limit.
By agreement	May be added only with agreement of the user/purchaser, within specified limits.

The additives not listed in specifications are not permitted. The following aviation turbine fuel additives are in current use.

Antioxidants

The use of antioxidants is mandatory in fuels that have been hydrotreated or produced by a hydrogen treating process. Straight run fuels often contain natural antioxidants. For this reason straight run fuels have good oxidative stability and do not need antioxidants. However, even the mildest hydrotreating destroys these naturally occurring antioxidants, resulting in a less stable fuel. For this reason kerosene coming from hydrocracker or kerosene that has undergone a hydrodesulfurization process will require antioxidant additives to prevent formation of hydrogen peroxide that can cause rapid deterioration of nitrile rubber components in the fuel system. Antioxidants are added immediately after the hydrogen treatment and prior to the product or component being passed into storage. Antioxidants prevent gum formation and increase storage stability of fuel. Where a finished fuel comprises a blend of hydrotreated and non-hydrogen-treated components, the requirements of a mandatory addition of antioxidants applies only to the portion of blend that has been hydrogen treated. For fuel or a fuel component that has not been hydrogen treated, the antioxidant addition is optional. The permitted concentration of antioxidants in hydrogen-treated jet fuel is a minimum of 17 mg/L and a maximum of 24 mg/L. For fuels that have not been hydrotreated, the maximum limit is 24.0 mg/L.

Metal Deactivator

Metal deactivator (MDA) use is permitted to neutralize the adverse effect of metals on fuel stability. It is normally not needed unless the fuel has undergone a copper sweetening process to remove mercaptans. Metal deactivators are chelating agents that form stable complexes with specific metals. Certain metals such as copper and zinc are very detrimental to fuel stability because they catalyze the oxidation reaction, resulting in fuel degradation. Metal deactivators suppress the catalytic effect of some metals on fuel oxidation and thus fuel storage life and stability. The following material is qualified in most jet fuel specifications; NN–disalicylidene 1, 2- propane diamine in a concentration of active material not exceeding 5.7 mg/L.

Corrosion Inhibitor and Lubricity Improver

Its use is optional to protect storage tanks and pipelines from corrosive rusting and to improve the lubricating properties of the fuel. Use of this additive (typically 9 to 22.5 g/m^3) is mandatory in military jet fuels.

Biocides

Biobor JF additive is used to prevent microbiological growth in hydrocarbon fuels. It is usually used to disinfect aircraft during a period when aircraft is left standing filled or partially filled with doped fuel. The fuel is then used in normal manner. To minimize the deposition of boron compounds in the engine, this treatment is permitted at infrequent intervals. The use of the biocide additive is by agreement between manufacturer and purchaser.

Fuel System Icing Inhibitor

Jet fuel may contain traces of moisture that can freeze at a high altitude and form ice crystals in the fuel. Fuel system icing inhibitors (FSIIs) are additives that prevent ice formation by lowering the freezing point of water. The formation of ice crystals in fuel tanks can block aircraft fuel lines and filters and can cause engine failure. All aviation turbine fuels contain some dissolved water that cannot be easily separated. When an aircraft rises to flight altitude, the fuel cools, and water solubility in the aviation turbine fuel (ATF) decreases. Some of the dissolved water separates out that can remain in a supercooled state or can form ice crystals. When supercooled water meets a rough surface or filter, nucleation and crystallization start quickly, causing line or filter blockages. Anti-icing additives have a limited solubility in jet fuel but are completely soluble in water. When dissolved water separates from fuel, some amount of the additive quickly leaves the fuel and preferentially dissolves in water, which depresses water's freezing point and keeps fuel lines and the fuel filter clear of ice.

FSII is used generally by military aircraft that have no fuel heaters. Use of FSII is mandatory for military grades of jet fuels such as JP-4, JP-5, and JP-8. The only approved grade of FSII is diethylene glycol monomethyl ether (Table 4-8). The typical dosage is 0.10 to 0.15 LV%. Large civilian aircrafts do not require FSII because they are equipped with an electric fuel line heater that keeps the fuel at an appropriate temperature to prevent ice formation. Because of the biocidal nature of these additives, it is also effective in reducing microbiological contamination problems in aircraft tanks and ground handling facilities. Ethylene glycol monoethyl ether is used in Russian TU-1 fuel.

TABLE 4-8 Fuel System Icing Inhibitor

Property	Test method	Ethylene glycol: mono methyl ether	Diethylene glycol: mono methyl ether
Chemical formula		$C_3H_8O_2$	$C_5H_{12}O_3$
Molecular weight		76.1	120.15
Acidity, mg KOH/g	ASTM D 1613	0.09 Max.	0.09 Max.
Specific gravity, 20/20°C	ASTM D 891	0.963–0.967	1.021–1.025
Color, Platinum-Cobalt	ASTM D 1209	15 Max.	15 Max.
Distillation	ASTM D 1078		
IBP, °C		123.5 Min.	191.0 Min.
Dry point, °C		125.5 Max.	198.0 Max.
Ethylene glycol, mass %		0.025 Max.	0.05
Diethylene glycol, mass %			0.5 Max.
pH of 25, vol % solution in water at 25°C	ASTM E 70	5.0–7.0	5.5-7.5
Water Mass % (Karl Fischer)	ASTM D 1364	0.15 Max.	0.15 Max.
Flash Point°C, Min.	ASTM D 93, D 56 or D 3828	43.3	85.0

Static Dissipater Additives

Static electricity is generated by movement of fuel through a high flow rates fuel transfer system. A static dissipated additive (SDA) is added to jet fuel to increase its conductivity and allow dissipation of charge to the ground. That electrostatic charges are dangerous is a well-known fact. The risk can be reduced by correctly handling the product and by observing the well-established safety procedures[5] of earthing and bonding all equipment near or in contact with fuel. However while handling fuels at high loading rates, it becomes impossible to ensure that electrostatic discharges will not occur merely by handling the fuel carefully. The only sure way to prevent buildup of such charges is to raise the conductivity of the fuel to such levels that charge if formed will leak away rapidly.

Conductivity of fuels is usually expressed in pico siemens per meter (1 pS/m) = 1×10^{-12} ohms^{-1} meter^{-1}. Kerosenes have conductivity ranging from 1 to 20 pS/m. For comparison, deionized water may have a conductivity of 10 million pS/m. It has been found that the conductivity of many distilled petroleum products lies between 0.01 and 10 pS/m, and at this level, the products are electrostatically hazardous. Conductivity of hydrocarbons is increased to more than 50 pS/m by the addition of antistatic additives[6] such as "Stadis 450." With conductivity increased to more than 50 pS/m, safe handling of the fuel is possible.

However there are conditions when elevated level of conductivity of fuel can also be detrimental to fuel handling and fuel storage equipment. Excess fuel conductivity has a detrimental effect on certain aviation fuel gauges. These gauges are sensitive to fuel conductivity and can malfunction due to the high conductivity of fuel. The gauges can record an erroneous fuel volume reading, which can result in an in-flight low fuel warning requiring aircraft to make an emergency landing. The instruments utilized to measure volume in fuel tanks actually take into account the dielectric constant of fuel when measuring fuel volumes in tanks. Overconducting fuel can compromise instrument calibration, resulting in erroneous volume readings and possible overfilling of tanks. For this reason, most jet fuel specifications also have an upper limit for conductivity. The conductivity of jet fuel is adjusted to lie between 150 and 600 PS/M at 85°F by adjusting the dosage of antistatic additive.

Thermal Stability Improvers

Some U.S military grades such as JP-8+100 contain an additive package that improves the thermal stability of fuel by 100°F. The additive package contains detergent, dispersant, metal deactivator, and antioxidants. The additive reduces coking and fouling in the engine fuel system. Additive types permitted in various grades are listed in Table 4-9.

TABLE 4-9 Various Additives Inclusion in Jet Fuels

Additive type	Jet A	Jet A-1	JP-4	JP-5	JP-8
	ASTM D 1655	DEF STAN 91–91	MIL-DTL-5624	MIL-DTL-5624	MIL-DTL-83133
Antioxidants	Optional	Mandatory*	Mandatory*	Mandatory*	Mandatory*
Metal deactivators	Optional	Optional	Agreement	Agreement	Agreement
Static dissipator	Optional	Mandatory	Mandatory	Optional	Mandatory
Corrosion inhibitor/ Lubricity improver	Agreement	Optional	Mandatory	Mandatory	Mandatory
Fuel system icing inhibitor	Agreement	Agreement	Mandatory	Mandatory	Mandatory
Biocide	Agreement	Agreement	Not allowed	Not allowed	Not allowed
Thermal stability	Not allowed	Not allowed	Not allowed	Not allowed	Agreement†

*Required in any fuel that has been hydroprocessed; otherwise optional.
†When thermal stability additive is added, fuel is called JP-8+100.

Blending

The jet fuel produced in the refinery is a blend of straight run and hydroprocessed product (Fig. 4-3). Straight run kerosene stream from the crude distillation column may have a high sulfur or mercaptan content. If the mercaptans in the kerosene cut from the crude being processed are higher than jet fuel specifications, these can be removed by a sweetening process such as Merox. If kerosene sulfur is higher than jet fuel specifications, the kerosene stream is desulfurized by processing it in a kerosene desulfurizer. In this unit kerosene feed is treated with hydrogen at an elevated temperature and pressure that removes sulfur in the feed in the form of hydrogen sulfide. If the refinery has a distillate hydrocracker, kerosene stream from the hydrocracker is a high-quality blend component. In cases where the refinery has a delayed coker unit, coker kerosene can be considered as a possible blend component after deep hydrotreating to remove all olefins and other unsaturated compounds. Generally thermally cracked streams containing olefins, nitrogen, or other reactive components are not allowed in jet fuel.

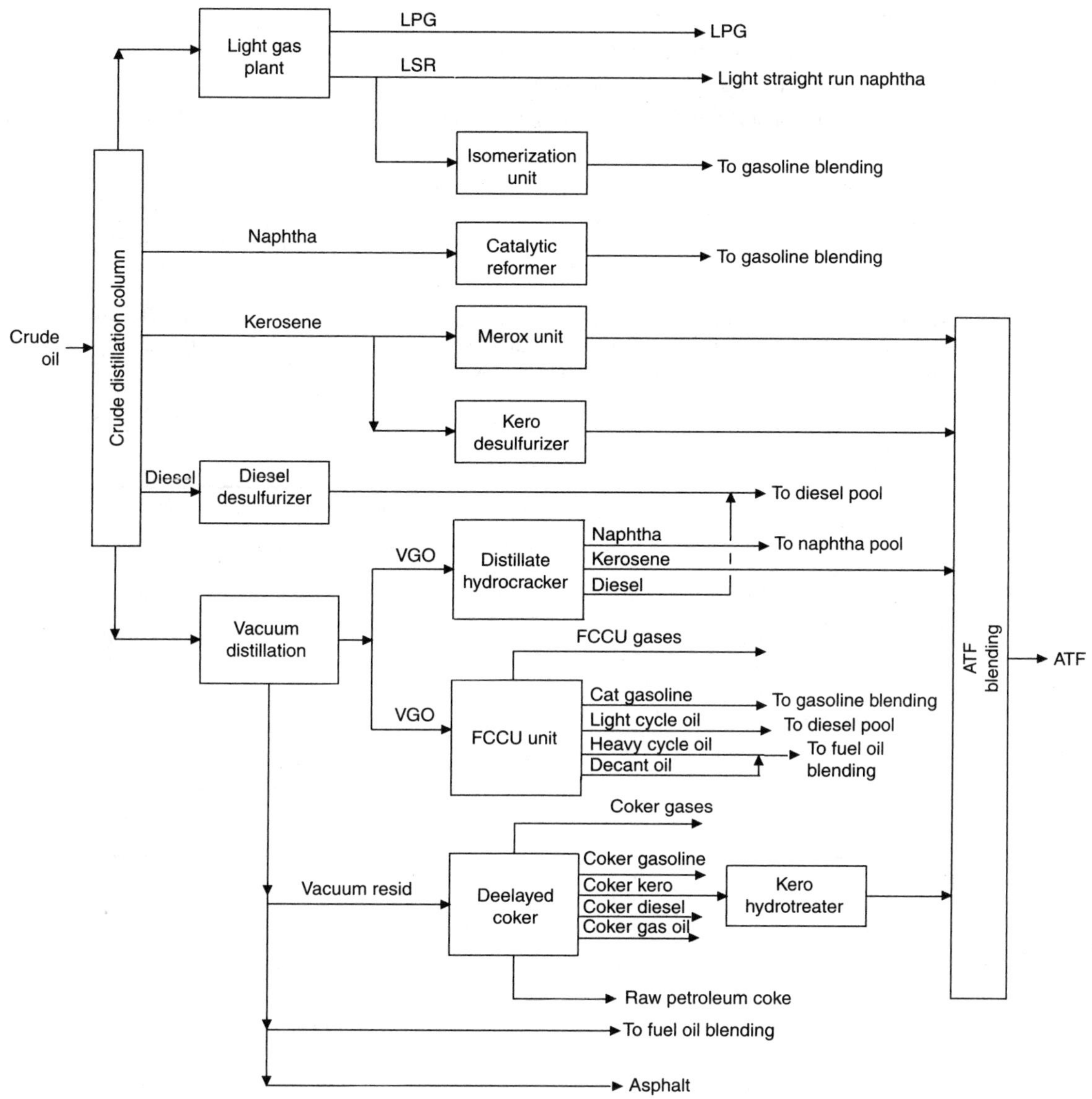

FIGURE 4-3 Aviation turbine fuel blending.

MISCELLANEOUS USES

Apart from its major use as aviation turbine fuel, kerosene is used in a large number of other uses such as the following:

- Illuminant in various lamps
- Fuel in cooking stoves
- Cleaning and degreasing fluid

- Solvent in paints and printing inks
- Low-sulfur fuel in domestic boilers
- Raw material for n-paraffin manufacture
- Manufacture of solvents for insecticides, herbicides, and fungicides to control pests, weeds, and fungi

Kerosene used for these non-ATF uses require minimal processing such as Merox or hydrodesulfurization to remove or reduce sulfur and mercaptans and do not have requirements such as a low freeze point, net heat of combustion, thermal stability tests. Also these kerosenes do not require additives such as antistatic, antioxidants, and so on. Typical specifications of burning kerosenes are listed in Table 4-10. The burning quality of kerosene is tested by IP 10 or ASTM D187. In these tests, kerosene sample is burned for 24 h in a specified lamp under specified conditions. The average rate of burning, the change in the shape of the flame, and the density and color of the chimney deposit are reported. The amount of char formed on the wicks is measured and reported.

TABLE 4-10 Superior Kerosene Oil

Property	Units	Limits	Value	Test method
Acid number strong	mg KOH/g	Max.	Nil	ASTM D 974
Burning test:				
Char value	mg/kg	Max.	20	IP 10
Bloom		Max.	Gray	
Color, Saybolt		Min.	+10	ASTM D 156
Corrosion, Cu strip, 3 h, 50°C		Max.	No. 1	ASTM D 130
Distillation				ASTM D 86
Recovered at 200°C	Vol %	Min.	20	
End point	°C	Max.	300	
Flash point, Abel	°C	Min.	35	IP 170
Relative density	@ 60/60°F		Report	ASTM D 1298
Smoke point	mm	Min.	22	ASTM D 1322
Sulfur	Wt %	Max.	0.20	ASTM D 1266

REFERENCES

1. R. F. Faulkner and J. W. Weber, "Hydrocarbon Scram Jet Propulsion System Development, Demonstration and Application," AIAA-99-492, November 1999.
2. NASA Technical Memorandum 100823, "Availability and Cost Estimates of High Naphthene, Modified Aviation Turbine Fuel," April 1988.
3. P. W. Kirklin and David P. "Aviation Fuels," Thermal Stability Requirements, ASTM STP 1138.
4. J. A. Bert and L. J. Painter, *Method of Calculating Hydrogen Content of Aviation Fuels* (Richmond, CA: Chevron Research Company, January 12, 1973).
5. A. Klinkenberg and Van de Mienne, *Electrostatics in the Petroleum Industry* (Amsterdam: Elsevier, 1958).
6. Manufactured by Octel-Starreon, Newark, NJ.

CHAPTER 5
DIESEL FUELS

The term *diesel* is used for motor vehicle fuel used in compression-ignited engines. Rudolf Diesel invented the compression-ignited engine in 1892 and his name got attached to the compression-ignited engine and the fuel it consumes. In order of economic importance of major oil products in the world, diesel fuel is the most important of all petroleum products. In 2007, petroleum refineries throughout the world processed crude oil at the rate of 85 million barrels per day producing various petroleum products. The production of distillate fuels (480–650°F cut) is estimated at 23 million barrels per day. High-speed diesel engines (1200 r/min and greater) are used to power trucks, buses, tractors, farm machinery, railroad locomotives, passenger cars, yachts, pumps, compressors and small electric generators. Diesel electric locomotives are the dominant type of locomotive used for railroad transportation throughout the world. Diesel electric locomotives use a diesel engine to drive an electric generator. The generator supplies current to traction motors coupled directly to locomotive wheels.

DIESEL ENGINES

Two types of internal combustion engines are used in automotive vehicles; a gasoline engine working on the Otto cycle and a diesel engine. The gasoline engine is mainly used for light-duty vehicles such as passenger cars and is powered by gasoline fuel. The diesel engine is generally used in heavy vehicles such as trucks, buses, locomotives, construction and agricultural machinery. Diesel engines are powered by high-speed diesel fuel. There are some significant differences between diesel engines and gasoline engines. A gasoline engine intakes a mixture of high-volatility gasoline and air and then compresses and ignites the mixture with a spark plug. In the diesel engine, also called the compression-ignited engine, only air is first sucked into a cylinder and then compressed adiabatically, which raises its temperature to diesel fuel ignition temperature. High boiling diesel fuel is next injected into hot compressed air in the cylinder, which starts the ignition. The gasoline engine compresses in the ratio of 8:1 to 12:1 compared with the diesel engine that compresses air in the ratio of 14:1 to 25:1. The higher the compression ratio, the greater is the power generation. Diesel engines operate at a higher temperature and pressure, which makes the engine heavier. Also, diesel engines are noisier and difficult to start in extremely cold weather. Diesel engines belong to a family of lean burn engines. Diesel-powered vehicles have a better fuel economy compared with spark-ignited gasoline-powered engines. Its high-efficiency combustion is due to an overall fuel-to-air ratio below stoichiometric requirements. Diesel engines do not need an electric ignition system, which makes the engine design simpler and more reliable.

SPECIFICATIONS

Diesel fuel specifications are set by the following considerations:

- The performance of the engine (fuel consumption, vehicle behavior, engine life, etc.)
- Fuel handling, safety, and storage stability
- Level of engine emissions affecting air quality standards

In the early years of diesel engine development, increasing engine power and reliability were the principal goals. However, for the past half century, the control and minimization of engine emissions have become the main objectives of the new engine designs. Specifications of diesel fuels are being increasingly formulated by original engine manufacturers to meet the requirements of lower engine emissions and higher engine efficiency. Table 5-1 lists the specifications of some commercial diesels.

Viscosity

Viscosity is a measure of resistance to flow, and it decreases as the temperature increases. A high-viscosity diesel fuel may cause extreme pressure at the injection system and would cause reduced atomization and vaporization of diesel fuel spray. The viscosity of the fuel must be low enough to allow free flow at its lowest operating temperature but high enough to provide lubrication to the moving parts of finely machined injectors. Diesel fuel must be sufficiently viscous to prevent leakage past pump plungers and dribbling at injectors. Viscosity also determines the size of the fuel droplets, atomization, and penetration quality of the fuel injector spray. Diesel fuel viscosity typically lies between 1.5 and 6 cSt at 100°F. A viscosity of 4 cSt at 100°F is considered optimum for atomization, lubrication, and good penetration into the combustion chamber.

Sulfur

Sulfur has a definite effect on the wear of compression-ignited engine components such as piston rings. The fuel sulfur content directly affects the life expectancy of the engine and its components. Fuel sulfur contributes to deposits in the combustion chamber and in the injection system. Sulfur interferes with the working of exhaust gases cleanup devices such as the particulate filter and catalytic converter. At the high exhaust temperatures of heavy-duty engines, sulfur combustion produces sulfuric acid, which increases particulate matter formation. For this reason, sulfur specification was reduced from 1 wt % (percent by weight) a decade ago to 500 to 15 ppm in 2007, in most developed countries of the world.

Cold Flow Properties

The low-temperature operability of diesel fuel is commonly characterized by the cloud point (ASTM D 2500), the cold filter plugging point (CFPP): IP 309 (Institute of Petroleum test method) and the pour point (ASTM D 97). These are defined next.

Cloud Point. The cloud point is the temperature at which wax crystals in diesel begin to settle out and a wax haze or cloud appears in the fuel. This condition exists in cold ambient temperatures. The cloud point is generally 10 to 15°F higher than the pour point.

Cold Filter Plugging Point. Paraffins are an important constituent of all petroleum fuels. When the fuel is cooled, normal paraffin solubility decreases, and some paraffins separate out as wax crystals. The highest temperature at which the fuel when cooled under defined conditions will not flow through a defined wire mesh within a certain time is called the cold filter plugging point (CFPP). CFPP indicates the low-temperature operability of fuel when cooled below cloud point temperature.

Pour Point. The pour point is the temperature at which there is no movement of a chilled sample in a test jar. However, the fuel may still flow under static head at pour point (e.g., from storage tank to pump suction). In a temperate climate, the diesel pour point generally varies between +6 and −6°C.

TABLE 5-1 High Speed Diesel Gas Oil Specifications

Property	Limit	Grades 1	2	3	4	5	6	7	8	9 (1)	10 (2)
Accelrated stability;											
Total insolubles, mg/100 mL	Max.	2.5	2.5		1.5	1.5	2.5				
Acid number, mg KOH/100g;	Max.										
Strong		Nil	Nil	Nil			Nil	Nil	Nil	Nil	
Total		0.5	0.5	0.3	0.3	0.3	0.25	0.2	0.2	0.5	
Appearance visual, 20–25°C								Clear	Clear		Clear and bright
Ash, mass %	Max.	0.01	0.01	0.01	0.005	0.005	0.01	0.01	0.01	0.02	0.01
Color, ASTM	Max.	3.0	3.0	3.5	3.0	3.0	3.0	2.0	2.0		1.5
Carbon residue, ramesbottom;											
On 10 % distillation residue, mass %	Max.	0.20	0.20	0.20	0.20	0.20	0.20	0.20	0.20		0.35
Conradson carbon	Max.									2.0	
Cetane index	Min.			47	46	46	50	48	48		41
Corrosion, Cu, strip, 3 h at 100°C,	Max.	1	1	1	1	1	2	1	1		1B
Classification											
Demulsification Time, Min	Max.				10	10	10				
Density at 15°C, kg/L	Min.			0.820				0.820	0.820	0.850	
	Max.			0.880				0.860	0.860	0.900	0.876
Diesel index	Min.	45	45								
Cloud point, °C	Max.			−4	−1	−12		+1	−7		
CFPP								−11	−19		−12
Distillation, recovered at 350°C Vol %	Min.			85				95	95		95
Distillation, recovered at 366°C Vol %	Min.	90	90								
End Point, °C	Max.				385	385					
Flash point, Penskey-Martein,											
Closed cup, °C	Min.	60	60	61	60	60	60	60	60	60	52
Pour point, °C	Max.	+6	0	−6	−6	−18	−3			3	
Sediment by extraction, mass %	Max.	0.05	0.05							0.1	
Sediment and water by centrifuge, mass %				0.01			0.03	0.01	0.01	0.2	0.05
Sulfur, mass %	Max.	1.0	1.0	1.0	1.0	0.4	1.0	0.19	0.19	1.6	0.0015
Thermal value, gross, MJ/kg	Min.									44.2	
Viscosity, kinematic At 38°C, cSt											
	Min	2.0	2.0	1.7	1.7	1.7	1.5	2.0	2.0	3.0	1.9
	Max.	7.5	7.5	4.3	4.3	4.3	6.0	4.0	4.0	9.0	4.1
Water by distillation	Max.	0.05	0.05				0.05			0.10	
Conductivity, pS/M	Min.									50.00	
Lubricity(3) 60°C, Micron	Max.										520

NOTES
(1) Marine diesel grade.
(2) Ultra low sulfur (15 Ppm) diesel.
(3) Lubricity test by ASTM D 6079.

The pour point, CFPP, and cloud point specifications of diesel are modified in the refinery according to ambient conditions of the areas and the season when diesels are consumed. The options of refineries to adjust cold flow properties of diesel are as follows:

- Adjust cloud point, CFPP, and pour point of diesel fuels by blending with kerosene.
- Use additives that enhance cold flow properties. The additives lower the pour point and the CFPP of diesel.

Cetane Number

The cetane number of diesel fuel is a rating to indicate how quickly a fuel will ignite in a diesel engine. A diesel engine if run on a fuel with a lower cetane number than what it was designed for will be harder to start and noisier, and it will have higher exhaust emissions. The compression ignition quality of diesel fuels is expressed as the cetane number of the fuel. Ignition quality is compared with a mixture of cetane or n-hexadecane ($C_{16}H_{34}$) and alpha methyl naphthalene or 1-methyl naphthalene ($C_{11}H_{10}$). Cetane ignites quickly and is arbitrarily given a cetane number of 100. A substance that is slow to ignite (e.g., alpha methyl naphthalene) is assigned a cetane number of 0. The cetane number of a fuel is defined as the percentage by volume (vol %) of normal cetane in a mixture of normal cetane and alpha methyl naphthalene that has the same ignition characteristics (ignition delay) as the test fuel when combustion is carried out in a standard engine[1] under specified conditions. The test is done in a CFR (Cooperative Research Committee, USA) engine. This is a specially designed engine with adjustable compression. The fuel being tested is injected at 13°(rotation of crankshaft) before the top center (the top center is the moment when the piston has traveled into the cylinder as far as it can go and compression is at its maximum). The engine compression ratio is adjusted until the fuel ignites on the top center. Retaining this compression ratio, the engine is then run on various blends of cetane with 1-methyl naphthalene until a blend is found for which the ignition occurs at the top center. The cetane number is the percentage by volume of cetane in a mixture that has the same performance as the fuel being tested.

In 1962, difficulties in handling alpha methyl naphthalene and its expense led the American Society for Testing and Materials (ASTM) to replace it with 2,2,4,4,6,8,8 hepta methyl nonane (also called isocetane). This component was given a cetane rating of 15, based on an engine rating without affecting the cetane scale. Figure 5-1 presents the properties of all reference fuels used for determination of the cetane number.

Cetane Index

The calculated cetane index is a useful tool for estimating the ASTM cetane number where a test engine is not available for its determination or where the quantity of the sample is too small for use in a test engine. The calculated cetane index is estimated from the density and mid-boiling point temperature by a correlation presented in ASTM D 976 (Fig. 5-2).

The calculated cetane index is particularly applicable to straight run fuels, catalytically cracked stocks, and their blends. However, the calculated cetane index is not applicable to fuels containing additives for raising the cetane number. Also, the correlation is not applicable to pure hydrocarbons, residual fuels, synthetic fuels, or coal tar products.

Diesel Index

The diesel index (DI) (IP 21) is another test for characterizing the burning quality of diesel in the compression-ignited engine. The DI is determined from its aniline point and API gravity (Fig. 5-3). Like the cetane number, a higher DI indicates a higher ignition quality of diesel fuel.

Isocetane

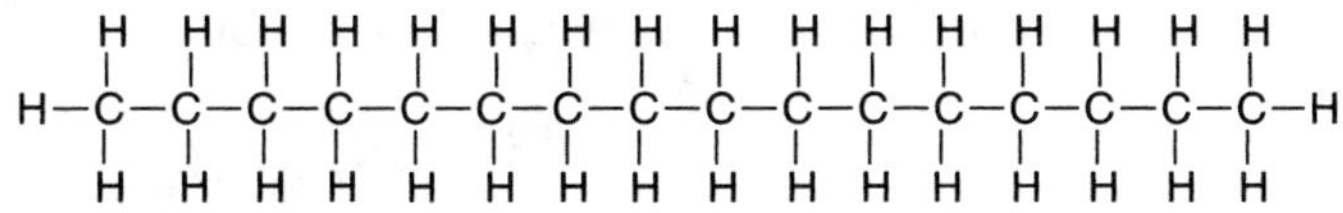

Molecular formula	$C_{16}H_{34}$
Melting point	64°F
Boiling point	549°F
Density, g/CC	0.7733
Flash point	275°F
Cetane number	100

Alpha methyl naphthalene

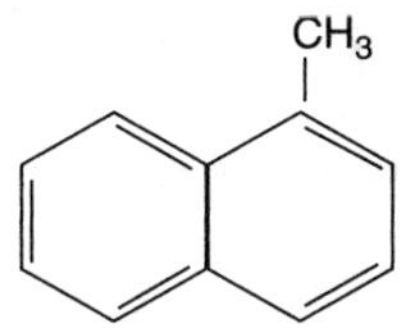

Molecular formula	$C_{11}H_{10}$
Melting point	–30.9°F
Density, g/CC	1.025
Boiling point	472.5°F
Flash point	179.6°F
Cetane number	0

2, 2, 4, 4, 6, 8, 8- Heptamethylnonane

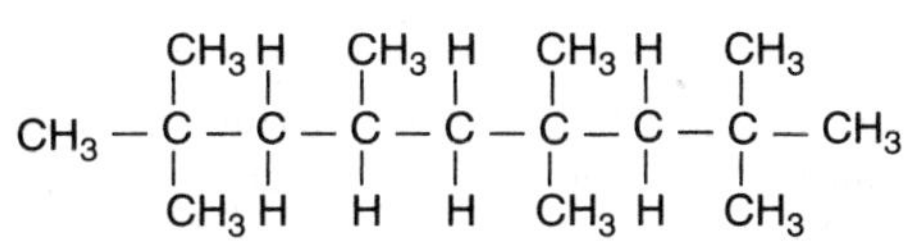

Molecular formula	$C_{16}H_{34}$
Melting point	64°F
Density, g/CC	0.793
Cetane number	15

FIGURE 5-1 Cetane reference fuel properties.

ASTM D 976

$$CI = -420.34 + 0.016{*}G + 0.192{*}G{*}LOG\ M + 65.01{*}LOG(M)^2 - 0.0001809{*}M^2$$

Where:
CI = Calculated cetane index
G = API gravity
M = Mid-boiling point temperature in °F

Or in metric units

$$CI = 454.74 - 1641.416{*}D + 774.74{*}D^2 - 0.554{*}B + 97.803{*}(LOG\ B)^2$$

Where:
D = Density at 15°C
B = Mid-boiling point temperature in °C

FIGURE 5-2 Calculated cetane index.

$$\text{DIESEL INDEX} = \frac{\text{ANILINE POINT (°F)* GRAVITY (°API)}}{100}$$

FIGURE 5-3 Diesel index from aniline point.

The aniline point (ASTM D 611) is the minimum temperature for complete miscibility of an equal volume of aniline and the test sample. The higher the aniline point, the higher the paraffin content or the lower the aromatic content of the diesel.

Back End Volatility

No clear evidence indicates that the 95 percent distillation point of diesel has any effect on the level of particulate emissions. There is, however, a trend to reduce the 95 percent distillation point of diesel to reduce its viscosity and improve cold flow properties.

Water Haze

Water contamination in refinery fuels originates from steam strippers used in refineries to remove light ends from various fuels. Water in fuel is the primary cause of tank bottom corrosion and bacterial growth in storage tanks. Water can cause reduced lubricity in engines causing wear and erosion of moving parts. Certain diesel specifications supplied to military require diesel to be "clear and bright at the ambient temperature," implying thereby that the diesel contains no visible water drops or particulate matter at an ambient temperature or at 77°F, whichever is greater (US MIL-F-16884H, amendment 2). Two forms of water can be present in fuels; dissolved or suspended[2] as tiny droplets of 0.1 to 10 microns. When diesel fuel cools down, dissolved water comes out of solution and forms tiny droplets. If droplets do not settle to the bottom, they form a haze. Heating the fuel can clear up the haze. To meet "clear and bright" or water content specifications, each diesel rundown stream is first cooled to 100°F. The cooled diesel passes through a coalescer and two vessels for dehydration, one containing rock salt and the other containing anhydrous calcium chloride (Fig. 5-4). Water in diesel may be as high as 900 ppm, which may be reduced by the coalescer to 250 ppm. Sodium chloride and calcium chloride drying yield a clear and bright product with water content of less than 50 ppm. The typical haze clear-up temperature of salt-dried diesel is 90 to 100°F. Salt driers can add corrosive chlorides to the fuel. Therefore, after flowing through salt drier, diesel is filtered through a 10-micron filter to remove salt particles and protect downstream

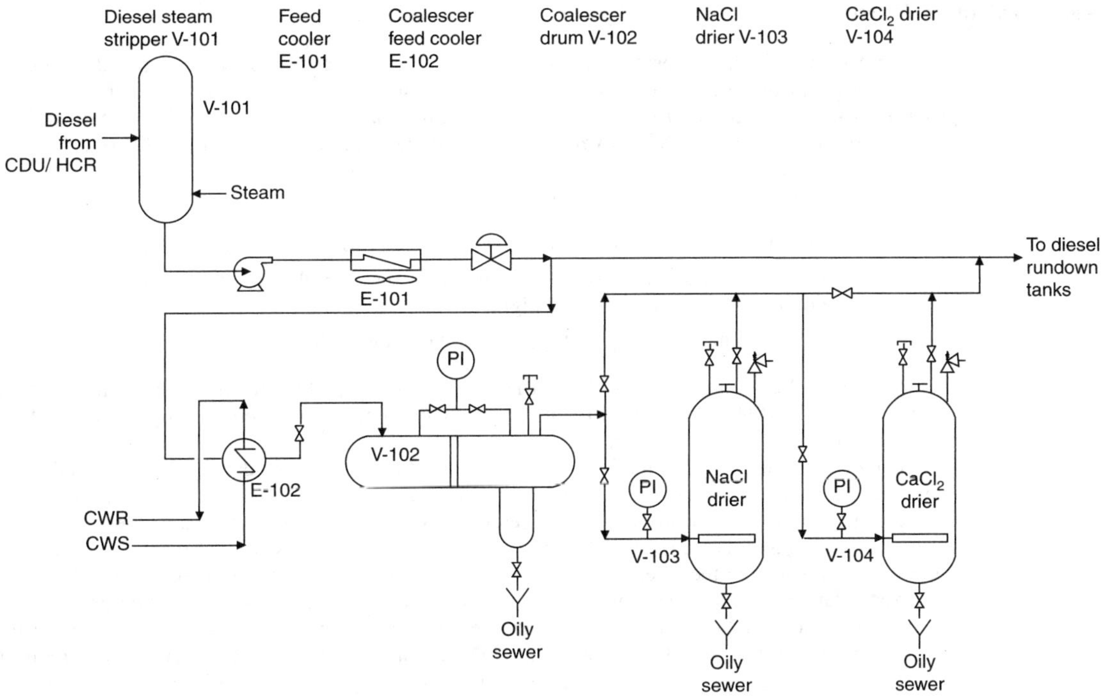

FIGURE 5-4 Process flow diagram for light diesel dehazing.

equipment from salt corrosion. Salt driers require extensive maintenance due to corrosion, bed channeling, salt plugs, and spent salt disposal.

DIESEL FUEL EMISSIONS

Diesel fuel is used in millions of diesel-powered trucks, buses, and private vehicles and stationary diesel engines throughout the world. These engines emit an enormous amount of carbon dioxide, carbon monoxide, nitrogen oxides and particulate matter apart from sulfur oxides. Ambient air quality in many areas of the world with high population densities has deteriorated to a point that it has become a public health problem. Diesel-powered buses, trucks, and other vehicles are responsible for more than 50 percent of diesel engine emissions. Toxic particulate emissions from diesel engines are largely responsible for the health effects of the diesel emissions. The Kyoto protocol[3] of 1992, approved by most countries of the world, calls for reduction of greenhouse gases by 2010 to the 1990 level.

Since 1981, catalytic converters or diesel oxidation devices have been at the core of vehicle emission control systems throughout the world. In this converter diesel exhaust, fumes are passed through a horizontal cylinder containing a noble metal catalyst (platinum or palladium) on an alumina base. As exhaust gas passes over the catalyst, carbon monoxide and unburned hydrocarbons are oxidized to carbon dioxide and water, thereby reducing harmful emissions. Particulate matter emissions are also reduced by 20 to 50 percent. The diesel oxidation device is placed upstream of the diesel particulate filter.

Diesel Particulate Matter

In compression-ignited, or diesel, engines, fuel is burned in an excess of air. Compared with gasoline engines, the lean nature of the diesel-air mixture results in a cooler combustion environment that produces a smaller volume of carbon monoxide and hydrocarbon emissions. Diesel engines, however, do produce a relatively high level of nitrogen oxides and particulate emissions, referred to as diesel particulate matter (DPM).

DPM is a complex mixture of the following

- Very fine (0.04 to 1.0 microns) dry carbon particles and inorganic oxides or sulfates
- A soluble organic fraction that results from the incomplete combustion of diesel fuel and engine lubricating oil that get adsorbed on the high surface area of carbon particles
- A gaseous component consisting mainly of CO, NOx, and sulfur dioxide

Exposure to diesel engine particulate exhaust is known to be human carcinogen, and its abatement is a priority. At least 80 to 90 percent of particulate matter can be removed by the use of diesel particulate filters (DFPs).

The DFP consists of a filter medium encased in a stainless steel cylinder that is positioned on the exhausted stream and designed to collect particulate emissions while allowing exhaust gases to pass through. The DFP filter typically consists of ceramics or silicon carbide to withstand the high temperature of exhaust gases. Once the filter is full of particulate matter consisting of carbon particles, it is burned off to regenerate the filter. Because the diesel engine exhaust temperature is not sufficient to burn particulate matter, a variety of methods are used to achieve the regeneration temperature. For example, a precious metal coating is applied on the filter surface that reduces the ignition temperature or onboard fuel burners are placed upstream of DFP, which can provide a sufficient temperature to burn accumulated carbon. DFP can reduce total particulate matter in exhaust by 85 to 90 percent.

Sulfur in diesel fuel affects the reliability, durability, and emission performance of catalyst-based diesel particulate filters. Sulfur affects the filter performance by inhibiting the performance of catalyst materials upstream or on the filter. Sulfur also competes with chemical reactions intended to reduce pollutant emissions and creates particulate matter through catalytic sulfate formation. In general, the lower the sulfur in diesel fuel, the better the performance of emission cleanup devices on the diesel exhaust. The sulfur content of diesel must not be more than 500 ppm for the DFS device to be effective. Most developed countries in the world are adopting a 15 ppm sulfur limit for this reason. In the past decade, Some diesel fuels contained a high level (1.0 to 1.5 wt %) of sulfur.

Carbon Dioxide

Carbon dioxide is the main cause for concern and the subject of international efforts to try and reduce its output. Carbon dioxide is produced by burning any fossil fuel, such as coal, petroleum products, wood, and so on, and it is believed to cause global warming.

Carbon Monoxide

Carbon monoxide is a poisonous gas produced when petrol or diesel fuel are burned in an engine. Petrol engine exhaust contains significant amounts of carbon monoxide. Diesel engine exhaust virtually does not contain any carbon monoxide because the fuel is burned in excess air.

Nitrogen Oxides

During the combustion of fuel in the internal combustion engine, high temperature and pressure are generated and the nitrogen present in the air combines with oxygen to nitrous oxides, which combine with more oxygen in the atmosphere to produce nitric oxides. The nitrogen oxides (NOx) then combine with ozone to form smog. Unlike hydrocarbons (HCs) and CO particulates (mostly carbon), NOx cannot be removed by oxidation. The NOx removal process calls for the introduction of a reducing stream such as ammonia or urea in the engine exhaust that is passed through a catalyst bed. NOx reacts with ammonia to produce nitrogen and water. The system, called selective catalytic reduction (SCR), has been used on large stationary diesel engines used in power generation.

Hydrocarbons

Internal combustion engine emissions can contain unburned hydrocarbons such as benzene and many other hydrocarbons such as polynuclear aromatic hydrocarbons (PAH), aldehydes, and others, some of which are known carcinogens.

The typical emissions of NOx, CO, SOx, and particulate matter from diesel engines and standards set by the U.S. Environmental Protection Agency (EPA) are listed in Tables 5-2 and 5-3.

TABLE 5-2 Typical Emissions from Stationary Diesel Engines

	Emission factor (g/hp-hr)				
Engine type	NO_X	CO	VOC	SO_X	PM_{10}
Up to 600 hp	14.06	8.5	1.14	0.1645	1.000
Greater than 600 hp	10.86	8.5	1.00	0.1645	0.318

Diesel Storage Stability

Diesel fuel stability is its capacity to resist chemical change caused by oxidation and heat. Good oxidation stability ensures that diesel fuel can be stored for a long period of time without the formation of gum or sludge. Under normal storage conditions, diesel fuel can be expected to stay in a usable condition for 12 months or longer at ambient temperatures of 70°F and 6 to 12 months at 85°F. As diesel get older, a fine sediment and gum start forming, due to the reaction of oxygen in the air with the hydrocarbon, and the rate of these sediment-forming reactions increases exponentially with the temperature. The fine sediment and gum block the fuel filter, leading to fuel starvation of the engine. Frequent filter changes are required to keep the engine running. The gum and sediment do not burn properly in the engine and can cause carbon and soot deposits in injectors and other combustion surfaces.

TABLE 5-3 EPA Standards for Year 2007 Onwards for Heavy-Duty Highway Diesel Engine Emissions

	Emission factor (g/hp-hr)		
Year	NO_X	NMHC	PM
2007	0.2	0.14	0.010

NMHC = Nonmethane hydrocarbons; PM = Particulate matter; Sulfur: Diesel fuel regulations limit sulfur on highway diesel fuel to 15 ppm.

Oxidation Stability Test

The expected life of a diesel fuel is determined by the oxidation stability test (ASTM D 2276). The test measures how much gum and sediment will be deposited after keeping the fuel at 248°F in the presence of oxygen for 16 h. It roughly corresponds[4] to one-year storage at 77°F. A result of less than 20 mg/L of sediment and gum is considered acceptable. The aging process can be accelerated by contact with metal surfaces such as zinc and copper. Metal surfaces can also come from dust and dirt containing trace elements. Fuel composition can also affect aging. Thus diesel blends containing light cycle oils (nonhydrotreated) from fluid catalytic cracking units (FCCUs) can cause rapid aging.

Water in the storage tank can also contribute to aging. Water allows growth of fungi and bacteria, which in turn produces organic acids, making the fuel unstable. The additives described next can prolong the storage life of diesel fuels.

Metal Deactivators

These work by deactivating copper, zinc, and other active metals and stop them from catalyzing the fuel–oxygen oxidation reaction.

Fungicide and Biocides

These work by stopping fungi and bacteria from growing in the fuel and thus prolong the life of the fuel. Fungi and bacteria are normally active at the water–fuel interface. Regularly draining water from a low point in the tank also helps prevent the growth of bacteria and fungi in tanks.

Antioxidants

These work by stopping the oxidation process from taking place. They prevent fuel oxidation and thus help reduce the formation of sediment and gum.

DIESEL FUEL ADDITIVES

Some additives are added to finished diesel fuels to enhance certain properties or for safe handling. These are described next.

Cetane Improvers

The diesel engine is a compression-ignited engine that relies on the fuel to auto ignite and begin combustion. Ignition delay is the time between the injection of fuel into the cylinder and the onset of combustion. If the delay is too long, combustion is violent, noisier, and less efficient with a high level of exhaust emissions. Engines operating on low cetane number fuels are difficult to start, especially in cold weather; produce less power; and consume more fuel. Diesel fuel with a high level of aromatics has a low cetane number. Cetane improver additives are a cost-effective way to achieve higher cetane numbers. Cetane improver additives also allow a higher percentage of cracked components, such as light cycle oils ex FCCU unit to be included in diesel blends.

Cetane improvers are alkyl nitrates such as 2 ethyl hexyl nitrate (Fig. 5-5). These additives break down during combustion to form free radicals that accelerate combustion reactions. The typical dosage is between 100 and 3000 ppm, which increases the cetane number by approximately five to eight numbers.

OCTYL NITRATE

Structure

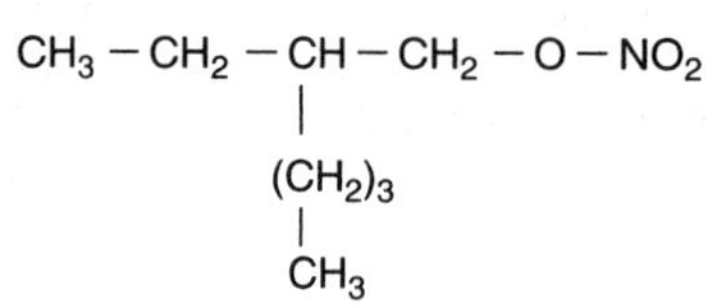

Chemical formula

$C_8H_{17}NO_3$

Molecular weight		175.22
Density	kg/m^3	960–980

FIGURE 5-5 Cetane improver additive.

Pour Point Depressants

Diesel fuels produced from waxy crude oils tend to have high pour point, which can pose significant handling problems during low ambient temperatures due to wax separation in storage tanks and other handling equipment. Refineries sometimes overcome this problem by dumping large volumes of kerosene in diesel to meet pour point specifications. Because kerosene is generally a higher valued product, downgrading any kerosene to diesel represents a loss of revenue to refineries. In certain situations, pour point additives to lower the pour point of diesel may present a cost-effective solution.

Pour point depressants are designed to interfere in the wax crystallization process by modifying the wax crystal structure. The additives are structured such that a part of the molecule is like a paraffin wax crystal that provides nucleation sites and co-crystallization with wax. The other part of the additive molecule is dissimilar to the wax crystal and prevents the extensive growth of wax matrices.

Most pour point depressant additives are polymers and copolymers of methyl acrylates of low molecular weight. The typical additive dosage is between 200 and 1000 ppm.

Antistatic Additives

Ultra-low sulfur diesel, due to removal of sulfur and polar compounds, has very low conductivity. Risks increase with high loading rates, splash top loading, and handling fuels with low conductivity (less than 50 pS/m). Fires and explosions can occur during truck loading or filling a large storage tank due to spark ignition from accumulated electricity. Untreated diesels and gas oils from Middle Eastern crudes typically have conductivity below 10 pS/m at ambient temperatures. Also, a significant decrease in conductivity occurs at low ambient temperatures. The conductivity value of middle distillates can be increased by using antistatic additives such as Stadis 450.[5] The dosage required to raise conductivity to 50 pS/m is typically 1 to 2 ppm.

Lubricity Improvers

The deep hydrodesulfurization of diesel for the production of ultra-low sulfur diesel grades removes most of the sulfur and a significant amount of polar and aromatic compounds that give conventional diesel fuels their adequate lubricity. Loss of lubricity can cause engine problems, mainly the erosion of diesel fuel injection equipment, unless treated with lubricity additives. Lubricity testing of diesels is done on a high-frequency reciprocating rig (ASTM D 6079). However, diesel containing lubricity improver additive cannot be transported through common product pipelines because lubricity improver can contaminate aviation turbine fuel or kerosene. The lubricity improver can be added at the loading terminal.

DIESEL BLENDING

Commercial diesel fuel is a blend of many refinery streams. Blend components may come from straight run diesel and kerosene from the crude distillation column, desulfurized diesels from the hydrodesulfurization unit, diesel stream from the distillate hydrocracker, hydrotreated light cycle oil from the FCCU unit, or hydrotreated coker gas oil. Table 5-4 lists the typical properties of these streams. Blending is done mainly to meet pour point, diesel/cetane index, and sulfur specifications.

TABLE 5-4 Diesel Fuel Blend Components

Property	Units	SRD	Kero	LCO	Hydrocracker diesel	Coker diesel
Specific gravity		0.8495	0.786	0.8825	0.8488	0.879
Sulfur	Wt %	1.29	0.103	0.24	0.036	0.79
Pour point	°F	10.4	−71	−13	40	38.1
Cetane index		50.6			55	
Diesel index		55.3	64	30.2		30
50% ASTM distillation	°F	547	372	541	596	612
95% ASTM distillation	°F	626	433	664	710	711

SRD = Straight run diesel ex crude distillation; Kero = Kerosenes; LCO = Light cycle gas oil ex FCCU unit.

DISTILLATE HEATING OILS

Distillate heating oil refers to distillate product boiling in the kerosene-diesel range. While diesel is designed for combustion in compression-ignited engines, heating oil is intended to be burned in a furnace for space heating. Heating oil properties are very similar to winter grades of high-speed diesel and sometimes can be used as substitutes for each other. In some countries, home heating oil and diesel are taxed differently, and to prevent heating oil misuse as automotive diesel, a dye is sometimes added to it to color it differently for easy identification. Heating oil is used for home heating in cold regions of the world such as North America, Europe, and Asia to keep houses warm and comfortable during the winter months. In the northern states of America, an average single-family home is estimated to consume 17 barrels of heating oil during the winter. The severity of winter weather has a significant impact on demand. Demand for heating oil is seasonal and varies widely according to climatic conditions or geographic locations. In recent years, demand for heating oil is gradually decreasing due to the availability of alternative energy sources such as electric heating from nuclear power plants and the greater use of natural gas and LPG in areas of heating oil consumption. Typical specifications of home heating oils are presented in Table 5-5. Because

TABLE 5-5 Heating Oils Specifications

Property	Limit	Value	Test methods
Acid number, strong, mg KOH/100 g	Max.	Nil	ASTM D 974
Appearance visual, 20–25°C		Clear	
Ash, mass %	Max.	0.01	ASTM D 482
Carbon residue, ramesbottom;			
On 10 % distillation residue, mass %	Max.	0.35	ASTM D 189
Color ASTM	Max.	5.0	ASTM D 1500
Corrosion, copper strip, 3 h, 100°C	Max.	NO. 1	ASTM D 130
Density at 15°C, kg/L	Min.	0.810	ASTM D 1298
	Max.	0.880	
Cloud point, °C	Max.	+2	ASTM D 2500
CFPP		−4	IP 309
Pour point, °C	Max.	−9	ASTM D 97
Distillation, 65 % recovered at °C	Min.	250	ASTM D 86
85 % recovered at °C	Max.	350	
	Max.		
Flash point, Penskey-Martein,			ASTM D 93
Closed cup, °C	Min.	60	
Sediment, mass %	Max.	0.01	ASTM D 473
Sulfur, mass %	Max.	0.3	ASTM D 1552
Viscosity, kinematic at 37.8°C, cSt	Max.	5.5	ASTM D 445
Water, Vol %	Max.	0.05	ASTM D 95

heating oil is burned in a furnace and not in compression-ignited engines, no cetane number specification is necessary, and refineries try to maximize low cetane number material such as FCCU light cycle oil, coker diesel, visbreaker diesel, and so on, in their heating oil blends. However, in some countries, a minor amount of heating oil is also used by remote rural communities for the operation of diesel engines in tractors, agricultural machinery, or in civil construction equipment. For this reason, the cetane specification is included in some home heating oil specifications to make it a dual-use fuel. The typical distillation range is 365 to 705°F. Heating oil is typically blended from kerosene, diesel, and light cycle oil ex FCCU unit. Specifications of heating oils are somewhat similar to winter-grade diesel fuel. The heating oil distillation end point is higher than diesel cuts because particulate emissions are much less critical for heating oils compared to those in diesel fuels. The most significant difference between winter-grade diesels and home heating oils is in their low temperature properties. The cloud point, CFPP, and pour point of home heating oil are fixed and, unlike diesels, they do not change with the season. Thus whereas winter-grade diesel may have a cloud point varying between 17 and 23°F, home heating oil has a cloud point of 36°F. Similarly, the CFPP of winter diesel grades may be −0.4 to 5°F compared with a home heating oil pour point of 26°F. The pour point of winter diesel may vary between −0.4 and −6°F, whereas the pour point of heating oil is 16°F. The density of home heating oil is slightly higher than automotive diesel; thus calorific value of home heating oil on a mass basis is slightly less than that of diesel fuels. Permissible sulfur levels are higher than those for automotive diesel. However, the sulfur level in home heating oils is also likely to fall in future years due to environmental concerns. Home heating oils are stored in small (200- to 400-gal) aboveground tanks and are exposed to very low ambient temperatures during winter months. For this reason, pour and cloud point is kept low to keep it fluid at prevailing ambient temperatures. Oil flows from the storage tank through a strainer to a pump that pressurizes it to 100 to 150 lb/in^2. The pressurized oil flows through an atomization nozzle. The fine oil droplets produced mix with air and are ignited with a high voltage spark. The hot flue gases flow past the heater tubes and exit through a chimney. In a hot water system, water is circulated through the tubes at approximately 180°F for carrying heat to various parts of the house. This heat is transferred to the house by various types of convection or radiation units. Finned tube heaters are most commonly used because of their high efficiency and low cost.

BIODIESELS

Vegetable oils were considered a possible fuel for diesel engines in the early days of their development. In 1893, Rudolf ran his first diesel engine on peanut oil. Earlier results on the use of vegetable oils in diesel engines were not very encouraging due to engine fouling, deterioration of the oil, and viscosity increased with time. It was found that vegetable oils can be transformed into a product that is much more adequate as diesel fuel. Vegetable oils react with an alcohol to form an ester and glycerine in a triesterification reaction (Fig. 5-6). The esters formed have properties that make it suitable for use in diesel engines.

Biodiesel is produced by chemically reacting a fat or vegetable oil with an alcohol in the presence of a catalyst, usually an alkali. The product of reactions is a mixture of methyl esters and glycerol. Glycerol is a valuable by-product.

Biodiesel can be used neat, and when used as pure fuel it is known as B100. However, it is often blended with petroleum-based diesel fuel. When used as a blend, it is known as BXX, where XX denotes the percentage of biodiesel in the blend. Standard international practice is for marketing B5, B20, and B100, with the dominant blend being B5 and B20. There is no or little original engine acceptance of blends other than B5 and B20.

Biodiesel softens and degrades certain types of elastomers and natural rubber compounds in engine over time. Using high-percentage blends can affect fuel system components such as fuel hoses and fuel pump seals that may contain elastomers incompatible with biodiesel.

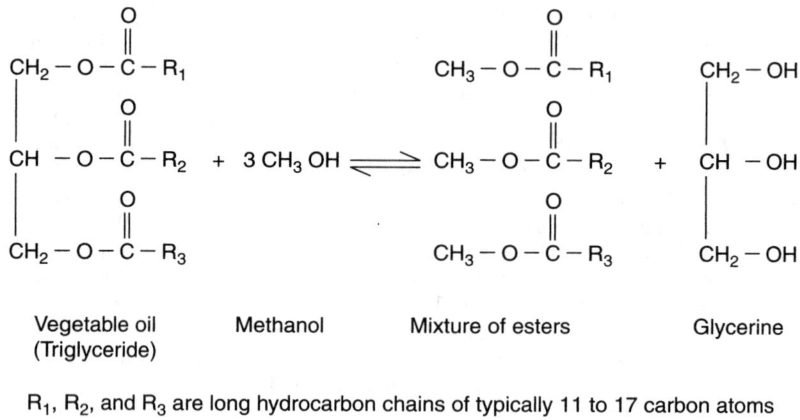

FIGURE 5-6 Biodiesel manufacture.

Manufacturing Process

Vegetable oil used as feed is pretreated to remove gums and neutralize acidity and water. Figure 5-7 shows a block flow diagram of the biodiesel manufacturing process. Vegetable oil, methanol, and catalyst (sodium hydroxide) are pumped to a reactor and agitated for an hour at 140°F. Following the reaction of oil with methanol, ester and glycerol are formed. The glycerol separates out due to the low solubility of glycerol in esters.

After separation of the glycerol, methyl esters enter a neutralization & methanol removal step and separated methanol pass through a methanol stripper, which is a vacuum flash tower or falling film

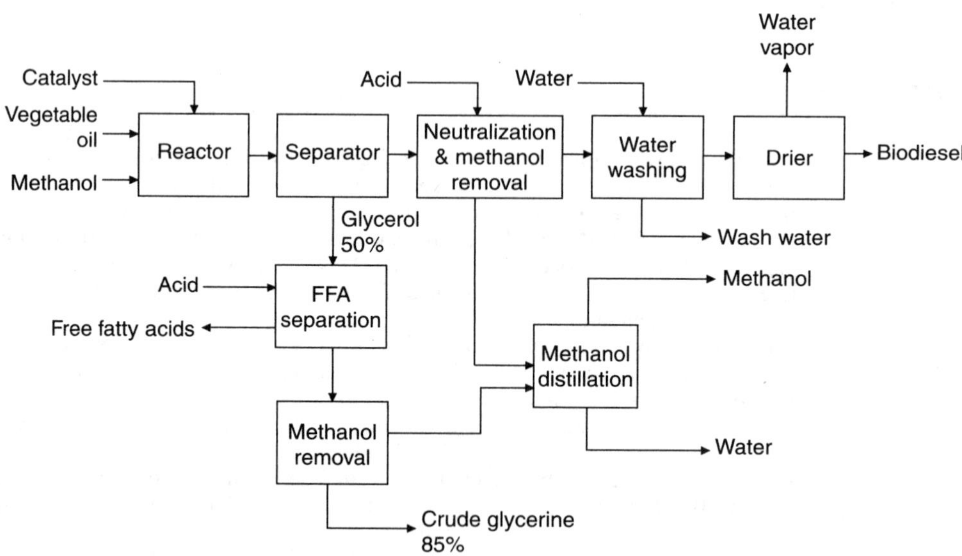

FIGURE 5-7 Block flow diagram biodiesel manufacture.

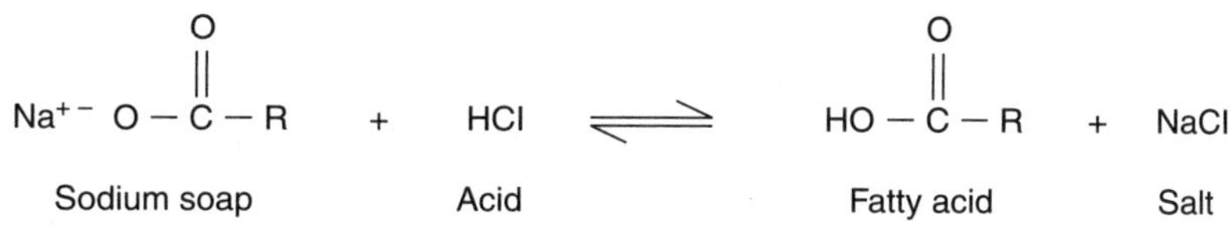

FIGURE 5-8 Sodium soap, acid reaction.

evaporator. Acid is usually added to biodiesel to neutralize any residual catalyst and to split any soap that may have formed during the reaction. Soap will react with acid to form water-soluble salts and free fatty acids (Fig. 5-8).

Salt is removed during the water washing step and fatty acids stay in the biodiesel. The water washing step is intended to remove any remaining catalyst, soaps salts, methanol, and free glycerol from the biodiesel. Neutralization before washing reduces the water required and minimizes the potential for an emulsion to form.

The glycerol stream leaving the separator is about 50 percent glycerol. It contains some excess methanol and most of the catalyst and soap. Acid is added to it to split the soap into free fatty acids and salt. Free fatty acids are not soluble in glycerol and rise to the top from where they can be removed. Salts remain in glycerol, although some may precipitate out. Methanol is recovered from glycerol next by vacuum flash, and glycerol of about 85 percent purity is sold to a glycerol refiner.

An alternative catalyst-free method[6] for transesterification reaction uses a supercritical method at high temperature and pressure in a continuous process. In the supercritical state, oil and methanol are in a single phase, and the reaction occurs spontaneously and rapidly. The process can tolerate water in feedstock; free fatty acids are converted into methyl esters instead of soap. A wide variety of feedstocks can be used. Also, the catalyst removal step is eliminated.

Table 5-6 lists the ASTM specifications of biodiesel. The distinctive features of biodiesel are that it does not have sulfur, it is free of aromatics, it is biodegradable because of the presence of oxygen in its molecule, and it produces significantly less smoke, carbon monoxide, and sulfur emissions.

TABLE 5-6 Biodiesel Specifications (ASTM D 6751-02)

Property	Units	Value	Test method
Flash point, closed cup	°C, Min.	130	ASTM D 93
Water and sediment	Vol %, Max.	0.05	ASTM D 2709
Kinematic viscosity	cSt, 40°C, Min.	1.9	ASTM D 445
	cSt, 40°C, Max.	6.0	
Sulfated ash	Wt % Max.	0.02	ASTM D 874
Total sulfur	Wt % Max.	0.05	ASTM D 5453
Copper strip corrosion	Max.	No. 3	ASTM D 130
Cetane number	Min.	47	ASTM D 613
Cloud point		Report	ASTM D 2500
Carbon residue	Wt % Max.	0.05	ASTM D 4530
Acid number	mg KOH/g	0.8	ASTM D 664
Free glycerin	Wt % Max.	0.02	ASTM D 6584
Total glycerin	Wt % Max.	0.24	ASTM D 6584
Phosphorus	ppm	10	ASTM D 4951
Distillation end point	Max. °C	360	ASTM D 1160

REFERENCES

1. Tentative Method of Test for Ignition Quality of Diesel Fuels by the Cetane Method, ASTM designation D-613.
2. R. L. Brown, Jr., and T. H. Wines, "Improve Suspended Water from Fuel" *Hydrocarbon Processing*, December (1993).
3. United Nation Framework Convention on Climate Change, December 11, 1997.
4. British Petroleum, Australia, *Fuel News*, February 7, 2002.
5. Manufactured by Octel Starreon Company, USA.
6. Bunkyakiat, Kunchana, et al., "Continuous Production of Biodiesel via Transesterification of Vegetable Oils in Supercritical Methanol," in *Energy and Fuels* (American Chemical Society, 2006), pp. 812–817.

CHAPTER 6
RESIDUAL FUEL OILS

In order of economic importance of major oil products in the world, residual fuel oil is the third most important petroleum product after gasolines and gas oils. Nevertheless, residual fuel oils are largely the by-product of refineries that aim at maximizing the production of light and middle distillates. Material remaining after maximizing light and middle distillate production (i.e., atmospheric or vacuum resids) are blended with a minimum amount of distillates to produce salable fuel oils. Refineries generally aim at minimizing the production of fuel oil, the lowest valued petroleum product, to maximize their refining margins. In the early years of petroleum refining, atmospheric distillation residuum, known as long resid, was sold as residual fuel oil without any further processing. As the cost of crude oil increased, it was realized that the atmospheric residue of crude distillation contained valuable heavy distillate that could be further cracked in fluid catalytic cracking units (FCCUs) and hydrocracker units to make more gasoline and middle distillates and thus further decrease fuel oil production. The average refinery production of fuel oil has come down from 20 vol % (percentage by volume) of the crude oil 20 years ago to approximately 14 vol % at present. In 2005, about 80 million barrels per day crude oil was distilled and processed into petroleum products throughout the world, producing 12 million barrels per day or fuel oil. As a result of the increased conversion of heavy distillates into lighter products in refineries, the quality of the fuel oil produced is decreasing, with a higher concentration of sulfur and impurities, higher density and higher molecular weight.

USES OF RESIDUAL FUELS

Residual fuel oil is a bottom-of-the-barrel product blended from the vacuum residue of crude oil distillation, and the residual products from various refining processes, to make cheapest possible fuel.

It is typically a dark viscous liquid of a complex mixture of hydrocarbons, and its composition can vary a great deal depending on the crude processed and the refinery's complexity. Viscosity ranges from 450 cSt to 180 cSt at 120°F. The flash point is typically above 150°F. Since residual fuel oil is thick and viscous, it must be heated before use to reduce its viscosity for easy combustion.

Residual fuels are widely used in industrial applications requiring heat generation. They are also used as fuel for steam generation in power plants and in the boiler "lighting up" facility in every coal-fired power plant where they initiate the combustion process. Among other uses are open hearth furnaces, soaking pits in the iron and steel industry, rotary kilns in cement and lime industry, and for firing heaters for petroleum refining and the manufacture of petrochemicals. A petroleum refinery may consume 8 to 10 percent of its throughput as liquid refinery fuel depending on the complexity of the process. Residual fuel oils are used in industries such as cement, glass, and paper. It is used as a feedstock in fertilizer manufacture where it generates hydrogen by a partial oxidation process.

The marine industry is the largest single consumer of residual fuels (bunker C fuel), consuming almost 33 percent of the world's total residual fuel oil production. Most of the fuel oil used by the world's merchant fleet and other naval ships is residual fuel oil.

DIESEL ENGINES

Sea transport is an energy efficient means of moving freight and carries 90 percent of the global trade (tons-km). The shipping industry generally uses residual fuel oil for economic reasons. The world's shipping fleet is powered mainly by slow and medium-speed diesel engines. A smaller percentage use marine diesel. It is estimated that more than 95 percent of the world's shipping fleet is powered by slow speed (less than 300 r/min) and medium speed (300 to 1000 r/min) diesel engines. Unlike high speed (more than 1000 r/min) diesel engines that use automotive diesel as fuel, low and medium-speed diesel engines use residual fuel oils. The shipping industry uses lower quality residual fuel oil for economic reasons. A large merchant ship powered by diesel engines may consume 150 tons/day heavy fuel oil and may typically carry 3000 to 4000 ton of fuel oil in tanks. Two-stroke low-speed diesel engines are capable of burning the lowest and cheapest quality commercially available residual fuels. The life of diesel engines is inversely proportional to its speed. Reducing the speed of an engine to half is known to increase engine life four times. Slow-speed diesel engines are capable of attaining efficiencies of more than 50 percent.

STEAM BOILERS

Residual fuels are widely used both in stationary and mobile steam boilers of all sizes. No particular problem is encountered in its burning. However, in high-pressure steam boilers that also operate at high temperatures, the fouling of boiler tubes may be encountered due to the presence of vanadium and other metallic compounds in the fuel oil ash. Power generation companies use slow-speed diesel engines in preference to medium-speed diesel engines and steam turbines because of their better efficiencies and longer engine life.

GAS TURBINES

A gas turbine extracts energy from the flow of hot gases produced by the combustion of gas or fuel oil. It has an upstream air compressor coupled to a downstream turbine and a combustion chamber in between. Compressed air is mixed with fuel and ignited in the combustion chamber. The resulting gases pass over the turbine blades, spinning the turbine and mechanically powering the compressor. Finally, gases pass through the nozzle, generating additional thrust, and pass to the atmosphere. Gas turbines are used on high-speed naval boats, ships, locomotives, and in small power plants. Gas turbines have very high power-to-weight ratio compared with reciprocating engines but have a high initial cost. The inherent simplicity of gas turbines makes this type of power plant attractive for many applications. A simple cycle gas turbine for power generation requires a smaller capital investment and actual construction can take as little as a few weeks, compared with years for base power plants. Gas turbines can be turned off and on within minutes, supplying power during peak demand. However, simple cycle gas turbines are less efficient than combined cycle turbines; they are generally used as peaking power plants.

RESIDUAL FUEL OIL SPECIFICATIONS

Marketing specifications have been established by a number of agencies to assure the satisfactory operation of industrial and marine equipment utilizing heavy fuel oils. These specifications include ISO-8217, ASTM D-396, BS 2869, and CIMAC (Conseil International des Machine a Combutiou). Table 6-1 shows ISO-8217 specifications.

TABLE 6-1 Requirement of Marine Residual Fuel Oils (ISO 8217)

Parameter	Units	Limts	RMA 30	RMB 30	RMD 80	RME 80	RMF 80	RMG 380	RMH 380	RMK 380	RMH 700	RMK 700
Density	kg/m^3	Max.	960	975	980	991	991	991	991	1010	991	1010
Kinematic viscosity	cSt, 50°C	Max.	30	30	80	180	180	380	380	380	700	700
Flash point	°C	Min.	60	60	60	60	60	60	60	60	60	60
Pour point												
Winter quality	°C	Max.	0	24	30	30	30	30	30	30	30	30
Summer quality	°C	Max.	6	24	30	30	30	30	30	30	30	30
Carbon residue	Mass %	Max.	10	10	14	15	20	18	22	22	22	22
Ash	Mass %	Max.	0.1	0.1	0.1	0.1	0.15	0.15	0.15	0.15	0.15	0.15
Water	V/V %	Max.	0.5	0.5	0.5	0.5	0.5	0.5	0.5	0.5	0.5	0.5
Sulfur	Mass %	Max.	3.5	3.5	4	4.5	4.5	4.5	4.5	4.5	4.5	4.5
Vanadium	mg/kg	Max.	150	150	350	200	500	300	600	600	600	600
Total sediment	Mass %	Max.	0.1	0.1	0.1	0.1	0.1	0.1	0.1	0.1	0.1	0.1
Aluminum and silicon	mg/kg	Max.	80	80	80	80	80	80	80	80	80	80

Marine Fuel Oils

There are two basic types of fuel oils; distillate and residual. A third type of fuel oil is a mixture of the two basic types, commonly known as "intermediate" (Table 6-2). Distillate fuel oils are made of diesel fractions of crude oil fractionation, whereas residuals are blended from distillation residue such as atmospheric or vacuum bottoms with diesel boiling range cutter stock to reduce their viscosity to specification. In marine industry, distillate fuels are called gas oil or marine gas oil. For example, as per ISO 8217 specifications, marine fuel grades DMX, DMA, DMB, DMC (Table 6-3) are marine gas oils. Residual fuels are called marine fuel oils or residual fuel oils (Table 6-1). Specifications for marine fuels use the first letter "D," signifying "distillate fuel," or "R," signifying "residual fuels." The second letter "M" signifies "marine fuel." Thus DMA is marine distillate fuel A. Residual fuels are designated by letters A to H, K and L, and a number signifying viscosity limit. For example, RMA -30 is "Residual Marine Fuel A" with a maximum viscosity of 30 cSt at 50°C. One important difference between land-based distillate fuels and marine fuels from an environmental perspective is sulfur content. Land-based fuels are generally required by environmental regulations to have lower sulfur levels than equivalent marine fuels.

DMA is a common fuel for tugboats, fishing boats, crew boats, drilling rugs, and ferry boats. Oceangoing ships that take residual fuel oil bunker also take distillate fuels for use in auxiliary engines and sometimes in port areas. The most common residual fuels are grades with a kinematic viscosity of 180 and 380 cSt at 122°F. All marine fuels for safety and insurance reasons are required to have a minimum flash of 140°F (60°C).

TABLE 6-2 Black Diesel for Industrial and Bunkering Uses

Property	Units	Limiting specification
Gravity	API	28.0–33.5
Carbon residue	Wt %, Max.	2.0
Sulfur	Wt %, Max.	1.6
Viscosity	cSt, 100°F	3.3–9.0

1. No visbroken residuum or cracked cutter may be included in the blend.
2. Normal blending components are straight run vacuum resid and light and heavy diesel.
3. Normal blending is to maximum viscosity and after meeting this requirement to the following specifications; (a) maximum carbon residue and (b) maximum SG × sulfur.

TABLE 6-3 Requirements of Marine Distillate Fuels

Parameter	Units	Limits	DMX	DMA	DMB	DMC
Density, 15°C	kg/m^3	Max.		890	900	920
Kinematic viscosity	cSt, 40°C	Max.	5.5	6.0	11.0	14.0
		Min.	1.4	1.5		
Water	V/V %	Max.			0.3	0.3
Sulfur	Mass %	Max.	1.0	1.5	2.0	2.0
Aluminum and silicon	mg/kg	Max.				25
Flash point	°C	Min.	43	60	60	60
Pour point						
Winter quality	°C	Max.		–6	0	0
Summer quality	°C	Max.		0	6	6
Cloud point	°C	Max.	–16			
Cetane index		Min.	45	40	35	

PROPERTIES OF RESIDUAL FUEL OILS

Typical properties of commercial residual fuel oils used in industry are listed in Table 6-4. By definition, residual fuel oils are the product remaining after the distillates have been removed from the crude oil and as such have high concentration of impurities contained in crude oil such as sulfur and metals.

Viscosity

Viscosity is not a measure of residual fuel oil quality, but it is an important specification from a fuel oil handling point of view. Viscosities of residual fuel oils for marine industry range from 80 to 700 cSt at 122°F. Refineries generally produce two grades of residual fuel oils with viscosities of 180 and 380 cSt at 122°F. Heavy fuel oils are heated to approximately 300°F to reach the correct injection viscosity of 10 to 15 cSt in the case of marine diesel engines. Fuel oil heating is also required for direct burning in boilers, in order to get the proper atomization of fuel.

Sulfur

Sulfur content of the residual fuel depends on the vacuum resid sulfur content, which in turn depends on crude processed and refinery processing configuration. Because resid desulfurization processes

TABLE 6-4 Residual Fuel Oil Specifications

Property	Units	1	2	3	4	5	6
Ash	Mass %, Max.	0.1	0.1	0.1	0.1	0.1	0.25
Carbon residue micro	Mass %	15.0			15.0	15.0	20.0
Density, 15°C	kg/L, Max.	0.980	0.980	0.991	0.990	0.985	0.990
Elements, trace	mg/kg, Max.						
Vanadium		55				550	
Aluminum						30	
Sodium		25					
Explosiveness	Vol %, Max.				50		
Flash point, Penskey Martin							
Closed cup	°C, Min.	66	66	66	66	66	60
Pour point	°C, Max.	24	21	–9	20	12	21
Fluidity			Fluid at 0°C				Fluid at 15°C
Pumpability, viscosity, 9°C	Poise, Max.			20			
Sediment by extraction	Mass %, Max.	0.1	0.12		0.12	0.15	
Stability							
ASTM spot test rating	Max.	2	2	2	2	2	2
Compatibility							
ASTM spot test rating	Max.				2		
Sulfur	Mass %, Max.	2.8	3.5	3.5	3.5	3.5	4.00
Thermal stability rating					No. 1 tube		
Thermal value gross	MJ/kg, min.	43.03					
	Btu/lb					18300	
Total sediment	Mass %, Max.	0.15	0.15	0.15	0.15	0.15	
Viscosity, kinematic, 50°C	cSt, Max.	180	80	75	48	180	380
	cSt, Min.			11.8			
Water by distillation	Vol %, Max.	0.5	0.5	0.5	0.5	0.5	
Water and sediment by Centrifuge	Vol %, Max.						0.6

are expensive, any further processing of resid to reduce sulfur would add to the cost of the residual fuel. Sulfur content of most residual fuel oils is between 3.5 and 4.5 wt % (percentage by weight) due to the high sulfur content of vacuum resids, which typically constitutes 60 to 70 percent of blended fuel oil. However, the sulfur in residual fuel oils pose a serious pollution problem. In many land areas where pollution is critical, regulations are progressively prohibiting the use of high-sulfur fuel oils and lowering the fuel oil sulfur level to 1.5 wt % maximum. In industrial boilers, use of high-sulfur fuel oils can lead to excessive corrosion. In steelmaking operations, a portion of the sulfur present in fuel is absorbed in steel, adversely affecting its quality. Flue gas desulfurization (FGD) has been used since the 1930s in shore-based facilities such as power plants to reduce sulfur emissions. In the FGD process, flue gas is scrubbed with an alkaline medium such as a lime solution to neutralize sulfur oxides. But the process produces large amounts of solid waste such as calcium sulfate, which present a disposal problem.

Sulfur emissions from ships' exhaust constitute a significant proportion of total sulfur global sulfur emissions. To control sulfur emissions from ships, MARPOL (the international convention for prevention of pollution from ships) adopted MARPOL[1] annexe VI, which came into effect on May 19, 2005. This convention includes a global upper limit of 4.5 percent sulfur for fuel oils used as fuel in ships. Annex VI contains provisions allowing for special SOx Emission Control Areas (SECAS) such as the Baltic Sea to be established with more stringent controls on sulfur emissions. In these areas the sulfur content of fuel oil used on board ships must not exceed 1.5 wt %. Alternatively, ships must be equipped with an exhaust gas cleaning system to limit SOx emissions.

Seawater scrubbing is emerging as the process of choice for onboard cleanup of sulfur emissions from engine exhaust gases. It can achieve emission reductions equivalent to using 0.1 percent sulfur diesel oil. Seawater scrubbing also removes soot, ash, and unburned oil.

In seawater scrubbing,[2] the flue gases are washed with seawater. Seawater is alkaline (typical pH, 8.3) due to the presence of bicarbonates of various metals. Sulfur oxides present in exhaust gases are thus neutralized and converted to water-soluble sulfites and sulfates. The effluent scrubbing water is highly acidic (pH 3). It is next processed to remove potentially harmful components, diluted with large quantity of sea water, and discharged back to the sea.

Ash Content

Compared with coal, residual fuel oils have a very low ash content and no ash-handling equipment is required. However, certain elements present in fuel oil ash may cause problems in furnaces where very high temperatures are encountered. Sodium and vanadium are particularly troublesome. The vanadium concentration of residual fuels varies quite widely. Vanadium comes from the vacuum resid component of fuel oil. Table 6-5 shows the vanadium and nickel content of some crude oils. Sodium may be present in crude or may come from seawater contamination of crude oil. Because sodium exists mainly as chloride or other water-soluble salts, it can be easily removed in the desalting operation carried out in the refinery. Sodium can be removed by water washing and centrifuging of fuel oil. Vanadium and sodium can cause corrosion and fouling of superheater tubes in high-pressure boilers, with high operating temperatures of 1200°F or above. Similarly, excessive corrosion and fouling of gas turbine blades may be encountered at temperatures of more than 1200°F. At a temperature below the fusion point of ash, there is relatively little corrosion in either boilers or gas

TABLE 6-5 Properties of Vacuum Resids from Various Crude Oils

Property	Units	Arab light	Arab heavy	Kuwait	Aghajari	Gash saran	Taching
Specific gravity		1.020	1.036	1.020	1.007	1.025	0.925
Carbon residue	Wt %	22.4	23.2	17.4	18.0	19.0	7.5
Asphaltene	Wt %	5.5	10.7	5.1	2.9	9.3	0.35
Sulfur	Wt %	4.1	5.1	5.15	3.56	3.3	0.2
Ni + V	mg/kg	87	190	110	228	430	10

turbines. At low operating temperatures, much of the ash passes out of the stack in the form of fly ash with little or no fouling. Thus where fuels with high vanadium must be used, the turbine or boiler operating temperature is lowered below the fusion temperature of ash, typically 1100 to 1200°F. Vanadium salts are converted to vanadium oxide (V_2O_5) during the combustion process, which deposits on furnace tubes and refractory and causes further problems in the furnace operations such as:

- Severe metallurgical attack as temperature is elevated.
- Refractory attack through formation of eutectic (a low melting inorganic compound) on the surface of refractory, which at a critical temperature can destroy refractory.
- Because V_2O_5 is a catalyst for the conversion of SO_2 to SO_3, it can greatly increase the concentration of SO_3 in the stack gases, thus increasing sulfuric corrosion.

Aluminum and Silicon

Residual fuel oils blending use heavy aromatic cutters such as light and heavy cycle oils, decant oil ex fluid catalytic cracker unit. These cutters may contain suspended catalyst fines. Catalysts used in FCCUs and other cracking units are aluminum and silicon zeolites. The measurement of aluminum and silicon content of fuel oil provides an indication of the catalyst fines in fuel oil. In residual fuel oil, catalyst fines can cause significant abrasive wear of engine components such as fuel pumps, injectors, cylinders, and pistons. The centrifuging of marine residual fuel reduces catalyst fines to a safe levels before use in marine engines.

Total Sediment

Before heavy fuel oil blending can be used in a low-speed diesel engine, an essential condition is that it must be free of sediment. Total sediment is one of the key parameters that fuel oil is tested for to make sure it meets the required specification. Too much sediment can result in a gradual buildup of sludge, resulting in blocked filters, pipes, and purifiers. If this occurs, the fuel will not get through the engine and the ship may be disabled on the high seas with a potentially critical situation. Sediment present in residual fuel oils consists of both hydrocarbons and inorganic material. The hydrocarbon material is typically asphaltenes, which are more prevalent in fuels blended from cracked blend components such as visbroken tar. If sediment reaches the engine in a large quantity, it leads to several problems arising from the slow-burning characteristics of asphaltenes, resulting in delayed ignition. Formation of sludge by flocculation of asphaltenes is a time- and temperature-dependent process. Asphaltene separation does not occur in fuels blended from straight run blend components such as straight run vacuum residuum. Heavy fuel oil is centrifuged to reduce sediment to acceptable levels before its use in marine engines.

Density

Residual fuel oils generally have densities in the range of 985 to 991 kg/m^3 at 60°F. However, a small percentage of residual fuel oils may have density exceeding 991 kg/m^3. High density can affect the operation of centrifuge separators installed to separate water and solid particles. Traditional separators can remove water and solids from residual fuels up to 991 kg/m^3 density. In the case of higher density fuel oils, specially designed separators have to be used and separator feed rate and temperature have to be carefully adjusted to ensure the efficient reduction of water, catalyst fines, sodium, and sediment. In case of an accidental oil spill involving high specific gravity (more than 1.0) residual fuel oils, these oils tend to float low in water, making recovery using skimmers difficult. Also, sunken heavy oils may inflict significant damage to seabed resources, fish, and mariculture activities.

Ignition Quality

The operation of all internal combustion engines depends on the ignition quality of the fuel. For spark-ignited engines, the fuel ignition quality is termed as the octane number. For high-speed diesel engines used in automobiles, the fuel used is distillate fuel and the ignition quality is known as the cetane rating. The cetane rating cannot be used for low-speed diesel engines because these engines were not designed for residual fuels. For residual fuel oils used in low-speed diesel engines, an empirical index calculated Carbon aromaticity Index[3] (CCAI) is used. The CCAI is defined below (Fig. 6-1).

$$\text{CCAI} = D - 81 - 141^{*}\ \text{LOG}\left[\text{LOG}\,(V + 0.85)\right] - 483^{*}\ \text{LOG}\left[\frac{T + 273}{323}\right]$$

Where:

CCAI = Calculated carbon aromaticity index

D = Density of residual fuel oil, kg/m^3

V = Viscosity of fuel cSt at temp T.

T = Temperature of viscosity measurement °C

FIGURE 6-1 Calculated carbon aromaticity index (CCAI).

The CCAI normally has a value between 800 and 880. The lower the value, the better the ignition quality. Fuels with a CCAI value of more than 880 can present problems in engines due to poor ignition quality. Modern medium-speed engines tolerate CCAI values up to 870 to 875. Low-speed engines are more tolerant of higher CCAI values. Limits of viscosity and density in international maritime fuel specifications in them provide control of ignition quality for residual fuel grades. For example a 380 cSt (at 122°F) residual fuel at a maximum specification density of 991 kg/m^3 will have a CCAI of 852, whereas a 180 cSt viscosity (122°F) fuel oil with the same density has a CCAI of 861. Ignition quality improves with increasing viscosity and decreasing density.

Ignition quality of heavy fuel oils can vary a lot depending on its blend components. Low ignition quality can cause problems at engine start and during low load operations, ignition delay, and it may also cause fast pressure rise and very high maximum pressure and engine deposits on the piston top, exhaust valves, and, in the case of turbines, on turbine nozzle ring and blades.

Water

The water content of heavy fuel oils varies widely between 0.5 and 1.0 v/v %. Water may come from several different sources. Also, water can be either fresh or more generally saline. It can also be from condensation in the storage tanks. If the water is salt free and well emulsified in heavy fuel oil, the effective energy content decreases with increasing water content, leading to increasing fuel consumption. If heavy fuel is contaminated with seawater, the chlorine in the salt can cause corrosion in fuel handling and fuel injection equipment. The effect of sodium that originates from salt is in its combination with vanadium metal. During combustion, vanadium sodium salts such as sodium vandate are formed that forms deposits on the exhaust valve and in the exhaust gas system and turbocharger.

Heat of Combustion

In almost every application, fuel oil is burned to produce heat. The heat of combustion is a measure of energy available from the burning of a fixed mass of fuel. Calorific value is the most important property of fuel oils and used for comparing the economics of different fuel oils with alternative fuels such as gas or coal. Fuel oil price in term of dollars per million Btu is often used as a reference price for fixing the price of other forms of energy such as natural gas or coal.

The heat of combustion of liquid fuel can be determined in the laboratory by means of a bomb calorimeter (ASTM D-240). In this method, the heat of combustion is determined by burning a weighed sample in an oxygen bomb calorimeter under controlled conditions. The heat of combustion is computed from temperature observations before, during, and after combustion with the proper allowance for thermochemical and heat transfer corrections.

Gross Heat of Combustion

The gross heat of combustion of a liquid or solid hydrocarbon at constant volume is the quantity of heat liberated when a unit mass of fuel is burned in oxygen in an enclosure of constant volume, the product of combustion being gaseous carbon dioxide, nitrogen, sulfur dioxide, and liquid water with the initial temperature of fuel and oxygen and the final temperature of products at 25°C.

Net Heat of Combustion

The net heat of combustion at constant pressure is the quantity of heat liberated when a unit mass of fuel is burned in oxygen at a constant pressure of one atmosphere; the product of combustion is carbon dioxide, nitrogen, sulfur dioxide, and water, all in the gaseous state, with the initial temperature of fuel and oxygen and the final temperature of product of combustion at 25°C.

The laboratory determination of the heat of combustion is a laborious and time-consuming process. It has been found that the heat of combustion of pure hydrocarbons can be estimated fairly accurately from specific gravity and the hydrogen content of the fuel (Table 6-6) based on empirical correlations presented by Gutherie[4] (Fig. 6-2). Because commercial residual fuel oils contain an appreciable amount of water, sediment, and sulfur, it is necessary to correct the estimated values for the effect of these impurities.

TABLE 6-6 Heat of Combustion of Residual Fuel Oils

				Heat of combustion at constant pressure			
Api gravity	Specific gravity	Density lb/gal	% H_2	kcal/kg	Btu/lb	Btu/gal	mm Btu/bbl
0	1.0760	8.980	9.859	9471	17037	152945	6.424
1	1.0679	8.912	9.981	9501	17091	152278	6.396
2	1.0599	8.845	10.101	9531	17145	151610	6.368
3	1.0520	8.779	10.219	9560	17197	150940	6.339
4	1.0443	8.715	10.336	9588	17248	150270	6.311
5	1.0366	8.651	10.451	9616	17297	149599	6.283
6	1.0291	8.588	10.564	9643	17346	148928	6.255
7	1.0217	8.526	10.675	9669	17393	148257	6.227
8	1.0143	8.465	10.785	9695	17440	147587	6.199
9	1.0071	8.404	10.893	9720	17485	146917	6.171
10	1.0000	8.345	11.000	9745	17529	146248	6.142
11	0.9930	8.287	11.105	9769	17573	145580	6.114
12	0.9861	8.229	11.209	9793	17615	144914	6.086
13	0.9792	8.172	11.311	9816	17656	144249	6.058
14	0.9725	8.116	11.412	9838	17697	143586	6.031
15	0.9659	8.060	11.512	9860	17736	142925	6.003
16	0.9593	8.006	11.610	9882	17775	142266	5.975
17	0.9529	7.952	11.707	9903	17813	141609	5.948
18	0.9465	7.899	11.803	9923	17850	140954	5.920
19	0.9402	7.846	11.897	9943	17886	140302	5.893
20	0.9340	7.794	11.990	9963	17922	139652	5.865

$$Q_V = 12400 - 2100^* D^2$$

$$Q_P = Q_V - 50.45^* H$$

$$H = 26 - 15^* D$$

Where:

Q_V = Total heat of combustion at constant volume, KCAL/Kg

Q_P = Total heat of combustion at constant pressure, KCAL/Kg

D = Specific gravity of fuel at 15°/15°C

H = Hydrogen content, Wt % in hydrocarbon

Or

Heat of combustion (gross) in Btu/lb

$$= 57.9^* \text{API} + (17687 - 23\ {}^*\text{sulfur\%})$$

FIGURE 6-2 Heat of combustion of residual fuel oils.

Pour Point (ASTM D 97)

The pour point is the lowest temperature expressed in multiple of 5°F or 3°C at which the movement of oil is observed. In this test, after preliminary heating, the sample is cooled at a specified rate and examined at intervals of 3°C for its flow characteristics. The lowest temperature at which the movement of oil is observed is recorded as the pour point. Most residual fuel oils are blended to have a maximum pour point of 86°F. The pour point requirement of fuel oils is fixed by the ambient condition of storage and handling.

Fluidity

The usefulness of the pour point test in relation to residual fuel oils is open to question, and the tendency to regard pour point as the limiting temperature at which fuel will flow can be misleading. Thus pour point test does not indicate what happens when oil is subjected to considerable pressure such as when flowing out of a storage tank under the gravity head. Failure to flow at pour point is normally attributed to the separation of wax from fuel or due to the effect of viscosity. Also the pour point of residual fuel oil is also influenced by the previous thermal history of oil. The fluidity test is restricted to residual fuel oils. The problem of accurately specifying handling behavior of fuel oils is important because of technical limitations of the pour point test. Various pumpability tests such as ASTM D 3245 have been proposed. The fluidity test[4] covers the determination of the fluidity of a residual fuel oil at a specified temperature. In this test method, the sample as received is cooled at a specified temperature for 30 min in a standard 12.5-mm diameter U tube (Fig. 6-3). The sample is considered fluid if it flows 2 mm in 1 min under a maximum pressure of 152 mm Hg (0.2 atm or 3 lb/in^2). Results are reported as fluid or not fluid at the specified temperature.

Thermal Stability

Most residual fuel oil whether in ships or in industrial applications must be preheated to reduce its viscosity before it enters the burner nozzle or marine engine. Preheating is done in a shell and tube heat exchanger using steam as the heating medium. Preheating may be done to a temperature of 300 to 350°F using medium pressure steam. It has been found that certain fuel oils can cause rapid fouling of the heat exchanger tube, resulting in lowered oil temperatures and necessitating frequent shutdown of the equipment for cleaning. The thermal stability test is designed to predict the fouling characteristics of a given residual fuel oil in preheater equipment.

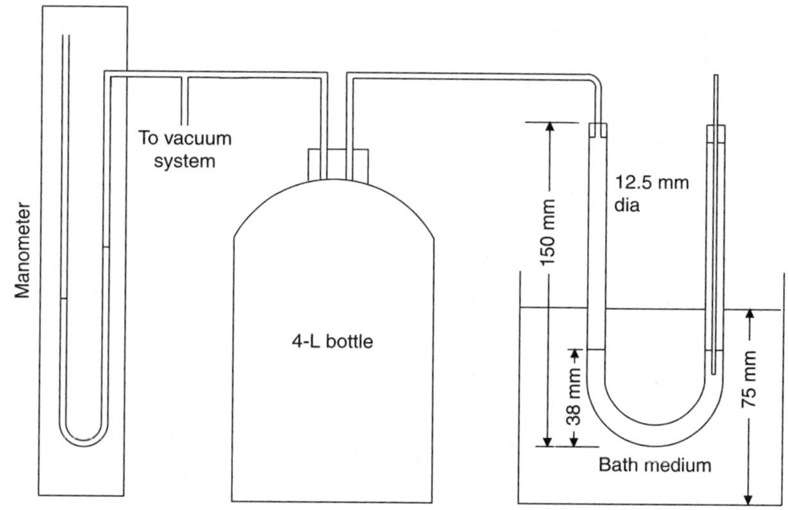

FIGURE 6-3 Fluidity test apparatus.

Thermal Stability Test (ASTM D 1661)

The test apparatus is in the form of a U tube made of heat-resistant glass (Fig. 6-4). A finger air cooler made of glass sealed at one end, about 330 mm long with a 10-mm outside diameter is fitted at one end. A metal thimble consisting of 4-in long seamless mild steel, BWG 16, 0.05 in outside diameter, 1.65-mm wall thickness is reamed to snug fit a heater cartridge and sealed at the other end.

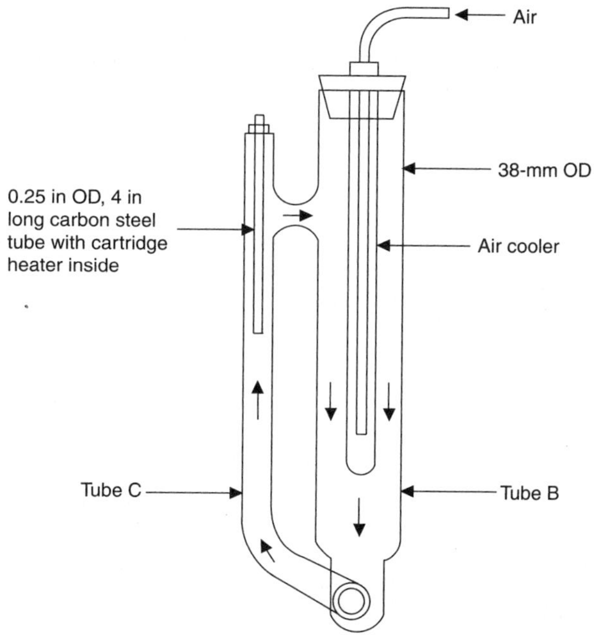

FIGURE 6-4 Apparatus for R.F.O thermal stability test.

The fuel oil (300 cm^3) is preheated to 140 to 160°F and poured into tube B until the oil level is at the top of the short arm connecting tubes B and C. The heater is switched on and power input adjusted so that the thimble temperature is 350°F. Air flow in the cooler is turned on and adjusted to give a temperature of 210°F in tube C.

The fuel oil sample is circulated by thermo-syphon action through the glass apparatus. The fuel oil is in contact with the surface of a steel thimble containing a sheathed heating element. The sample is heated on the hot surface of the steel thimble in the heater chamber, flows up through a riser tube, and then descends in a connecting tube and returns to the heater chamber for recirculation. The temperature of the thimble surface is maintained at 350°F. Heating is continued for 6 h, after which the heater is switched off. At the end of the period, the thimble is removed and examined for sediment formation and discoloration of the surface and compared with reference samples to rate the fuel oil fouling characteristics.

Explosiveness

Residual fuel oils are generally stored in heated tanks at 140 to 170°F. Light hydrocarbon vapor can build up in the head space of the tanks. This can cause flammability and explosion hazards even at temperature below the normal flash point of fuel. Several serious explosions have taken place in fuel oil storage facilities on land and onboard ships. Most explosions occur due to an accumulation of an explosive mixture of gases in the vapor space of fuel oil tanks. The vapor space gases consist predominantly of propane and lighter hydrocarbons with varying proportions of hydrogen sulfide. The light hydrocarbons are believed to be produced because of the delayed decomposition of cracked components such as visbroken tar. Traces of cracking catalysts may aid the process. Hydrogen sulfide may be released from decomposition of high-sulfur stocks. In oil tankers, inert gas blanketing is done to prevent potentially explosive mixtures of hydrocarbon gases and air from forming. Inert gas is obtained by seawater scrubbing of exhaust gases from the tanker boiler plant, which removes all sulfur oxides and particulate matter.

The explosiveness of an atmosphere in tank head space is determined by a device called "Explosimeter." This instrument measures the amount of combustible gas present in a sample drawn from the tank head space. In this instrument, a wire is heated and the sample of gas drawn from the tank is introduced to hot wire, causing combustion. The heat generated increases the resistance of the wire, which is measured by the instrument. The concentration of combustible gas as a percentage of the lower explosive limit is reported as explosiveness[5] of the sample even when the combustible material is unknown.

RESIDUAL FUEL OIL BURNING

The equipment used for burning residual includes storage tanks for storing the fuel oil, pumps for moving the fuel oil to the burner, and a preheater for adjusting the viscosity of the oil fed to the burner. The fuel oil burner is normally a part of the refinery furnace along with the stack and a suitable control system.

Oil Burner

Fuel oil is introduced into furnace through a fuel oil burner. An oil burner is a mechanical device that combines fuel oil with the requisite amount of air before delivering the mixture to the point of combustion. For proper combustion of fuel oil, the oil-air mixture must be well homogenized with a minimum of possible oil droplets. Three types of oil burners are used for residual fuel oil burning

- Mechanical atomizing
- Air atomizing
- Steam atomizing

Mechanical atomizing burners are also known as gun-type burners. Fuel oil is pressurized to a pressure of 100 to 300 lb/in^2 and forced through a nozzle that atomizes it into a fine spray in air and is thus partly evaporated before combustion in the furnace. In rotary nozzle burners, fuel is supplied through the shaft and thrown off a nozzle assembly, rotating at high speed. The fuel is converted into a fine spray. In air-atomized burners, the gravity causes a flow of fuel oil to the burner. Air is supplied through a natural draft or a forced draft using a fan. These types of burners are the most inexpensive and the have lowest operating cost. In steam atomizing systems, atomizing steam is introduced into the burner chamber and comes in contact with the oil stream just before burner tip. The kinetic energy of steam forces the fuel oil into tiny droplets as it leaves the burner. For proper atomization, steam pressure is kept 10 to 15 lb/in^2 higher than fuel oil pressure. Steam consumption is 1.5 to 5 lb/gal oil burned. Dry steam with superheat of 50°F is generally used. Steam consumption is 125 to 200 lb steam per barrel fuel oil.

Fuel Oil System

The purpose of the fucl oil system is to ensure a constant regulated supply of residual fuel oil to the burners of steam boilers and process furnaces. A typical fuel oil system is shown in Fig. 6-5. The system includes facilities for storage, pumping, heating, and distribution of fuel oil at suitable pressure and viscosities so that atomization and burning are possible. Storage is one of the essential needs that must be met with ample provision for availability of at least 5 days' supply at the normal firing rates of furnaces and boilers. If the fuel oil is obtained from more than one source, it may be necessary to blend the stored material, and more than one tank may be required. Oil is delivered to

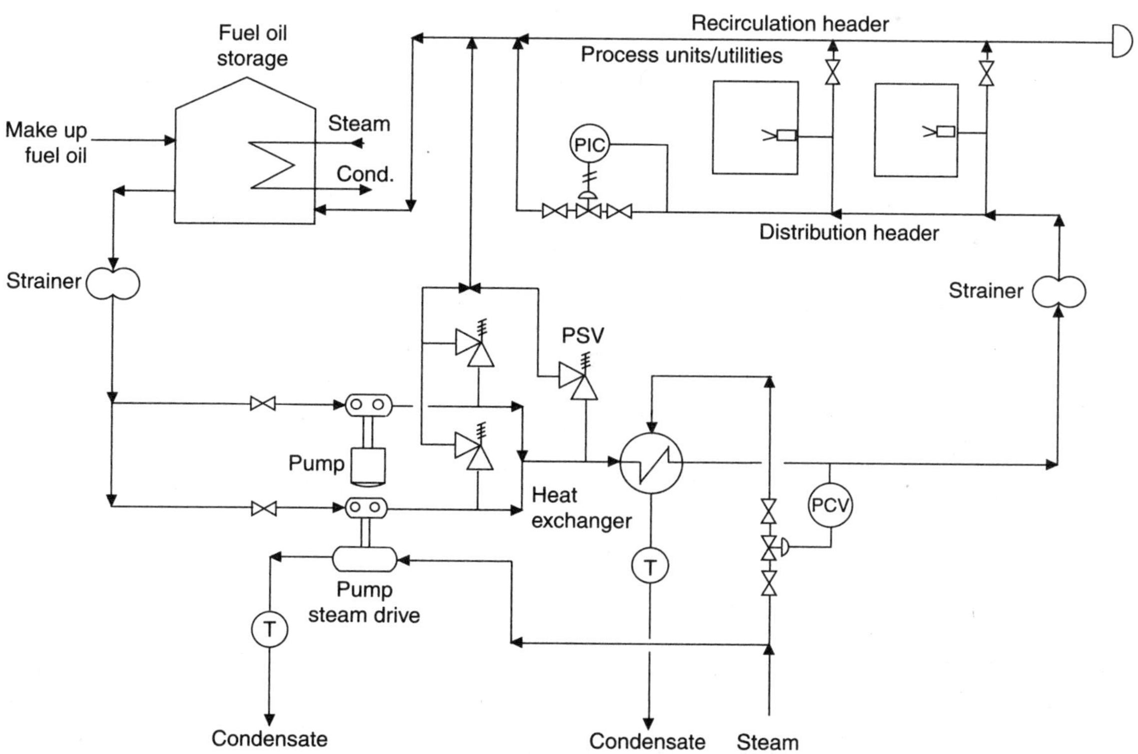

FIGURE 6-5 Residual fuel oil system.

burners at a pressure of approximately 100 lb/in^2 for proper control and atomization. Pump discharge pressure is typically set at 125 to 150 lb/in^2. Pump suction lines from storage tanks are equipped with strainers to intercept dirt and foreign material. Oil is heated in a shell and tube heat exchanger by steam to a temperature sufficient to lower the viscosity of residual fuel oil to 20 to 30 cSt.

Fuel Oil System on Ships

A slightly different fuel oil system is used on ships using residual fuel oil for its propulsion. A typical fuel oil system for use on ships[6] is shown in Fig. 6-6. When using high-viscosity fuels requiring high preheating temperatures, oil from the engine fuel system to the return pipe also has a relatively high temperature. There is a possibility of boiling and foaming in the return pipe in case water is present in the fuel. Also, cavitation in the fuel oil primary suction pump may occur. To circumvent this problem, an additional pump is placed between the service tank and return pipe, which is connected via an automatic deaerating valve to the service tank. With this arrangement, a minimum pressure of 4 bar can be maintained to ensure against boiling.

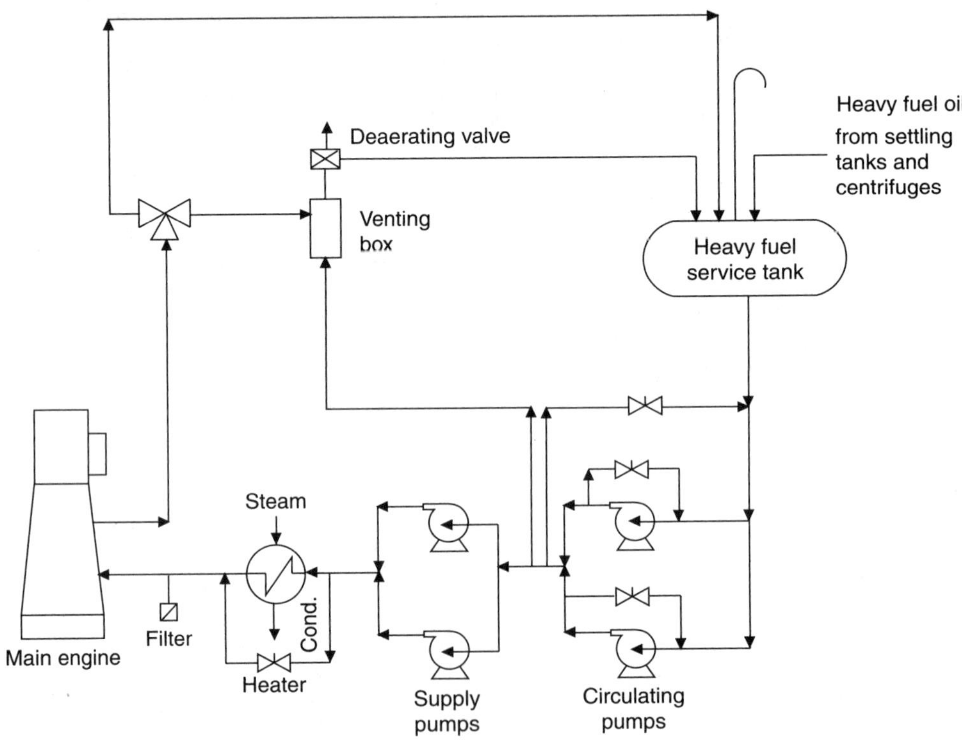

FIGURE 6-6 Merchant ship fuel oil system.

RESIDUAL FUEL OIL BLENDING

Depending on the refinery configuration and process units available, residual fuel oil may be blended from the following refinery streams (Fig. 6-7):

- Atmospheric distillation bottoms (long residue), 650°F+
- Vacuum distillation bottoms (short residue), 1000°F+

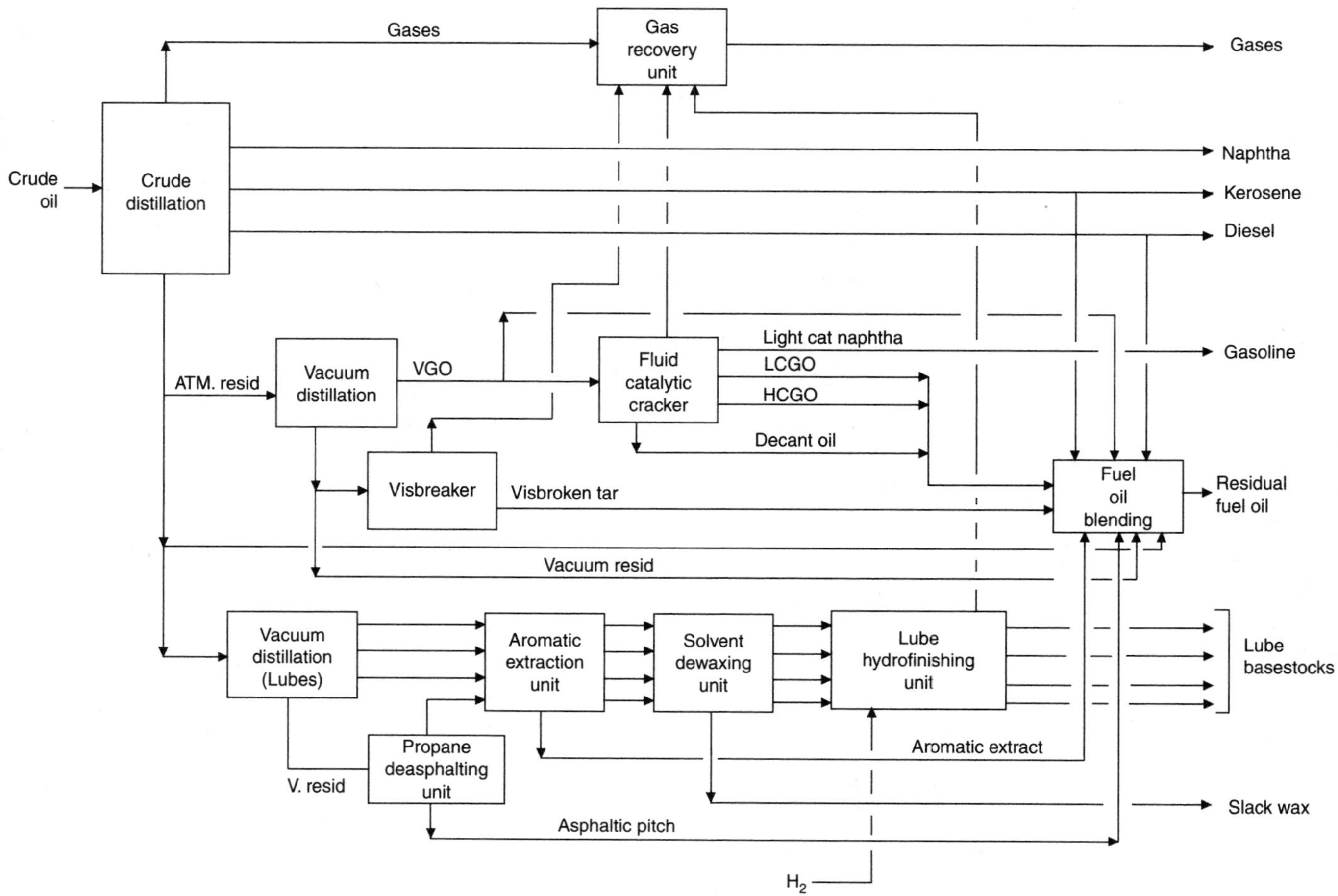

FIGURE 6-7 Refinery residual fuel oil blending.

- Visbroken or thermally cracked long or short residue
- Vacuum residue from resid hydrocracking units such as H oil
- Vacuum resid from resid desulfurization unit
- Cat cracker light cycle oil
- Cat cracker heavy cycle oil
- Cat cracker slurry (clarified) oil
- Gas oil
- Kerosene

Blending is done to make a blend of the available stocks with the least cost to the refinery. Key fuel oil specifications for blending include viscosity, sulfur, and conradson carbon and flash point.

Cracked Fuel Oils

A typical cracked fuel oil may be blended from visbroken vacuum resid and FCCU cutters (light cycle gas oil, medium and heavy cycle gas oils, and decant oils). It is known that fuel oils consisting essentially of thermally cracked hydrocarbons are subject to a stability problem in storage. The quality of the cutter stock is of great importance. Cracked resids have a high concentration of asphaltenes, which must be kept in solution by the use of high aromatic cutter stocks. Use of paraffinic cutter stock can lead to gradual separation of the asphaltenes in fuel oil storage tanks.

Straight Run Fuel Oils

These fuel oils are blended from straight run vacuum resids with straight run diesel and kerosene. Straight run fuel oils do not suffer from the problem of storage stability and do not exhibit the problem of incompatibility if reblended.

Low-Sulfur Fuel Oils

Low sulfur (1.5 to 2.0 percent S) fuel oils can be blended from low-sulfur vacuum resids if crude oil sulfur content is low. However, most export crude has a high sulfur content that produce vacuum resid with 4 to 5 percent sulfur. Fuel oil produced from such resids has a sulfur content of 3.5 to 4 percent. Very few refineries have resid desulfurization or resid hydrocracking units that can produce fuel oils with 1.5 to 2 percent sulfur.

COMPATIBILITY OF RESIDUAL FUEL OILS

One of the major concerns related to compatibility is whether two fuels of different origins can be commingled without having sediment precipitation in the tank. Residual fuel oils blended with thermally cracked residuum and paraffinic distillates may precipitate asphaltenes in storage tanks. Precipitated asphaltenes are also referred to as "hot filtration sediment" or dry sludge. Precipitated asphaltenes contribute to deposits in tankage, to preheater fouling, and to strainer plugging in the fuel-handling system. Asphaltenes can cause overloading in centrifuges installed to remove sediment and particulate matter in marine diesel systems on ships. If asphaltene precipitation occurs during the blending of fuel oil components, the blend components may be termed incompatible. Incompatibility can be determined by making laboratory blends and measuring its stability by the Shell Hot Filtration Test and Spot Test (ASTM D 2781). In this test, equal volumes of residual fuel oil and distillate fuel oils are blended at 149°F. A drop of this blend is allowed to spread on a chromatographic paper of a specified grade at 149°F. The spot thus formed is compared with a series of numbered reference spots. The compatibility of components is rated on the basis of this comparison.

To the extend possible, incompatibility can be prevented by segregating straight run and cracked stocks and avoid putting fuel containing virgin and cracked stocks in the same tank, especially if the straight run fuel is waxy with a high pour point. Usually two cracked fuels, if each is free of sediment, is likely to maintain asphaltene in solution. Also, two fuels of essentially the same viscosity that have the same density can be commingled with a minimum risk of incompatibility.

If the residual component contains only virgin material, it is highly unlikely that intermediates blended from it will be incompatible by themselves. There is no problem in using cracked resids for blending intermediate grades. However, the blend must be tested for stability. Where there is a choice available in the quality of cutter stock to be used in intermediate blends, a cutter with higher density, which also implies higher aromaticity, must be used to improve compatibility.

Whether or not asphaltene precipitation will occur when fuel oil components are blended can be predicted by a procedure developed by Exxon,[7] as follows:

- Estimate the BMCI[8] (Bureau of Mines Correlation Index) of the residual component of blend from its API gravity and mean boiling point (Fig. 6-8). BMCI can also be estimated from API gravity and viscosity. BMCI is indicative of the aromaticity of the stock. Thus the higher the BMCI, the higher the aromaticity. Although the correlation is strictly applicable to distillate stocks, it can be approximated to residual stocks without significant error. BMCI of individual components can be determined by their densities and viscosity or density and mean boiling point. The BMCI of the blend is a linear function of the BMCI of the components.
- Determine the toluene equivalency (TE) of the blend as follows: A sample of residual blend component is dissolved in a mixture of toluene and n-heptane. The mixture is filtered and any undissolved residue is separated and weighed. A number of such mixtures are prepared with increasing toluene concentration until the entire residue is completely dissolved and there is no residue on filtering the solution. The vol % of toluene in the toluene-heptane mixture is the toluene equivalence of the residual blend component. Toluene equivalency (TE) of the blend can be estimated from the TE of the blend component, assuming a component's TE blends in proportion to the amount of asphaltene contributed by each. Thus TE of each component is multiplied by its volume fraction, density, and wt % asphaltenes. The sum of these values is divided by the

Criteria for compatibility

$$\text{BMCI} - \text{TE} \geq \text{K}$$

For compatible blends of cracked resids and conventional cutters, K should lie between 7 and –14.

Where:

BMCI = Bureau of mines correlation index

TE = Toluene equivalency

K = A constant, applicable to a given set of blend components

BMCI of a blend component is defined as follows;

$$\text{BMCI} = \frac{87552}{\text{VABP}} + 473.7 \times \text{SG} - 456.8$$

Where:

VABP = Volume average boiling point in °R

FIGURE 6-8 Residual fuel oil compatibility.

asphaltene content of the blend to give the TE of the blend. TE is a measure of aromaticity needed to maintain all asphaltenes in solution. A higher TE indicates that a higher percentage of toluene in normal heptane mixture is needed to dissolve the asphaltenes. In a blend of resid and distillate cutter stock, TE of the blend will be TE of resid component.

Criteria for compatibility is presented in Fig. 6-8. For most blends of cracked resids and conventional cutter stocks, BMCI must be greater than TE and a delta between BMCI and TE of blend (K value), in the range of 7 to 14 will ensure that the blend is compatible and the sediment as measured by the hot filtration test (HFS) will be lower than 0.2 percent by weight. The equation indicates that if aromaticity of the blend is higher than the aromaticity needed to keep asphaltenes in solution, the blend will be compatible and no asphaltene precipitation will occur.

REFERENCES

1. Krystallon Sea Water Scrubbing, 27th International Bunker Conference, Gothenburg, Sweden, 2006.
2. G. B. G. Nyman and A. Tokerud, "Sea Water Scrubbing Removes SO_3 from Refinery Flue Gases," *Oil and Gas Journal*, 89 (1991): 52–54.
3. Francisco Arvelo Valencia, Isidro Padron Armas, "Ignition Quality of Residual Fuel Oils" *Journal of Maritime Research*, 2 (3), 2005: 77–96.
4. V. B. Gutherie, *Petroleum Products Handbook* (USA: McGraw Hill Book Company, 1960).
5. ASTM D 97, Appendix, "Test for Fluidity of Residual Fuel Oil at Specified Temperature."
6. U.S. military specifications, MIL-F-859E, FTM 1151.1.
7. O. Gradone, "Fuel Quality, On Board Treatment and Responsive Design of Two Strokes, Low Speed Diesel Engines," Marine Fuels, ASTM STP 878 (C. H. Jones, ed.) (Philadelphia: American Society for Testing and Materials, 1985), pp. 111–128.
8. M. G. Griffith and W. C. Siemoud, "Controlling Stability of Residual Fuel Oils," Marine Fuels Symposium, ASTM STP 878 (C. H. Jones, ed.) (Philadelphia: American Society for Testing and Materials, 1985), pp. 227–247.
9. James G. Speight, *Chemistry and Technology of Petroleum* (New York: Marcel Dekker, 1998).

P • A • R • T • 2

PETROLEUM SPECIALTY PRODUCTS

CHAPTER 7
BITUMEN

Bitumen is a black or brown highly viscous liquid or semisolid material that is present in most crude oils and in some natural deposits. Bitumen is a complex mixture of organic compounds containing up to 150 carbon atoms, mainly aromatic, naphthenic, and aliphatic. Bitumen may also contain small amounts of organic acids, bases, heterocyclic compounds nitrogen, oxygen, and sulfur. Some metals are also found in bitumen. Bitumen may typically contain 83 to 86 percent carbon, 9 to 10 percent hydrogen, 1 to 5 percent sulfur, less than 1 percent each of nitrogen and oxygen, and minor amounts of metals such as vanadium and nickel.

Bitumen is one of the oldest engineering materials, in use since the beginning of civilization, about 6000 B.C. Natural bitumen was formed when crude oil from shallow deposits worked its way up through cracks and fissure in the earth. Bitumen was collected as seepage from shallow wells in present-day northern Iraq, a region rich in oil deposits. The first recorded human use of bitumen was by Sumerians in Mesopotamia. They had a thriving shipbuilding industry that used bitumen for caulking and waterproofing ship hulls. In Babylonia, an ancient city of Mesopotamia, bitumen was used for joining bricks in building houses. Statues from that time showed that bitumen was used as a binder for inlaying various shells, precious stones, and pearls. Bitumen was used in that period for the preservation of mummies in Egypt. In 1500 A.D., the Incas of Peru were using a composition similar to modern-day paving bitumen to pave roads and highways.

In United States, the use of asphalt for road paving began in the 1850s. The first asphalt roadway appeared in the early 1870s. In America all bitumen production from the 1870s to 1900 was from Trinidad near the coast of Venezuela, which supplied almost 90 percent of the world demand. Bitumen was obtained from a lake at the depth of about 75 ft. Bitumen obtained from Lake Trinidad was hard and fluxed with a small percentage of heavy petroleum distillate to reduce its viscosity for use in pavements.

Today, almost all over the world, the production of bitumen for paving and other uses is from oil refineries. Approximately 80 percent of bitumen production is used in road building and maintenance as a binder for aggregates. Bitumen for this end use is ever increasing as highways are being built at an accelerated rate in all countries of the world. Bitumen is increasingly being used in building airfields. After road pavement, the next most important use of bitumen is in roofing. Bitumen is used for roofing mainly because of its high resistance to moisture penetration and its good adhesive and cohesive properties. Bitumen deforms slowly and continuously when subjected to shearing forces. It has good weather resistance, good chemical and physical stability, low sensitivity to temperature, a high flash point, low vapor pressure, and good compatibility with other materials used in roofing.

Bitumen is useful anywhere where waterproofing, insulation, or chemical resistance problems are encountered. Bitumen is used in the lining of water reservoirs, canals, and the upstream side of large dams to prevent water seepage and erosion. Bitumen is used in diverse industries such as cable jointing compound, cold sticker compound for sticking, roofing felt over metallic surfaces, timber, battery sealing compound, for making printing inks, in automobiles for undercoating, as a base for paint and lacquers, manufacture of floor covering, and a wide variety of waterproofing and damp-proofing applications. Different end uses of asphalt require different properties and different manufacturing methods. Bitumen or asphalt refer to the same product. In North America, the term "asphalt cement" refers to the residue from crude distillation, generally used in road construction. Elsewhere in the world, this product is known as bitumen.

BITUMEN COMPOSITION

Bitumen is a thermoplastic material: It softens when heated and hardens on cooling. Within a certain temperature range, bitumen is viscoelastic: It exhibits the mechanical characteristics of viscous flow and elastic deformation. Bitumen can be considered a colloidal solution in which asphaltenes are held in suspension in a mixture of oil and resins. Asphaltenes as such are insoluble in oil, but the polar nature of resin molecules keeps asphaltenes in a colloidal suspension. Vacuum residue from crude oil distillation can be split into four different constituents by chromatographic methods: saturates, aromatic, resins, and asphaltenes. The first three together are usually referred to as maltenes. The fractions are classified as asphaltenes or maltenes according to their solubility in n-hexane or n-heptane. Asphaltenes are high-molecular compounds and thus insoluble in these solvents. Maltenes, in contrast, have a much lower molecular weight compared with asphaltenes and are soluble in these solvents. The relative proportion of each fraction varies from crude oil to crude oil. Typical composition is as follows:

Constituent	Wt % (percentage of weight)
Saturates + aromatics	50–60
Resins	25
Asphaltenes	20

Saturates and aromatics are the fraction with the lowest molecular weight, ranging from 300 to 2000. The hydrogen-to-carbon ratio of this fraction is much higher than those of asphaltenes or resins. This fraction acts as the dispersion medium for asphaltenes.

Asphaltenes

The asphaltene molecule[1] has a core of saturated condensed aromatic rings and an aliphatic side chain. Heteroatoms (O, N, S, Ni, V, and others) have been found in rings (Fig. 7-1). The asphaltene molecule is typically composed of 10 or more fused aromatic and naphthenic rings with a significant number of alkyl side chains. The molecular weight of asphaltene is in the range of 2000 to 5000. The higher the asphaltene percentage, the harder the bitumen. Asphaltene constitutes from 5 to 25 wt % of bitumen. When heated, at a low cracking temperature of 750°F, asphaltene decomposition occurs through the cleavage of bridge chain between cluster units. At 800 to 825°F, thermal cracking with coke formation occurs rapidly.

Resins

Resins are composed of heterogeneous polar compounds with a small percentage of oxygen, nitrogen, sulfur, and metals. The molecular weight is typically 800 to 2000. Resins constitute approximately 15 to 25 percent of the weight of asphalt. Resins are aromatic groups with high polarity. Resins have a higher hydrogen-to-carbon ratio compared to those of asphaltenes. Resin fractions are exceedingly adhesive materials and are dispersive agents or peptizers for asphaltenes. Resins can be considered as low molecular weight asphaltenes.

Aromatic Oils

Aromatic oils consist mainly of carbon, hydrogen, and sulfur with a minor amount of nitrogen and sulfur, with a molecular weight of 500 to 900. These oils constitute 45 to 60 percent of the weight of bitumen. These compounds have mainly aromatic ring or naphthenic-aromatic nuclei with a side chain.

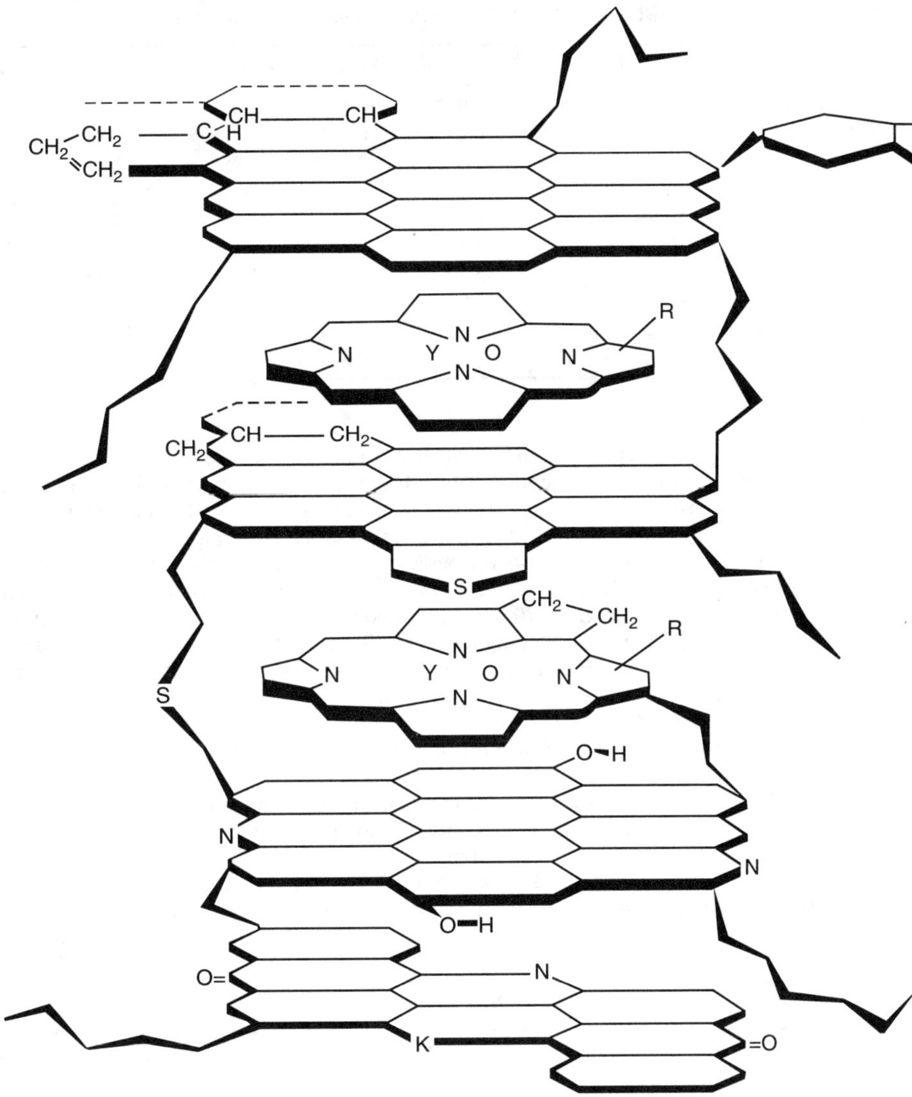

FIGURE 7-1 Hypothetical asphaltene partial molecule.

Saturated Oils

Saturated oils consist mainly of long chain saturated hydrocarbons with some branched chain compounds, cyclic paraffins. Molecular weight is in the range of 500 to 1000. Saturated oils constitute 5 to 20 percent of the weight of bitumen.

BITUMEN FOR PAVEMENT

Bitumen is a by-product of crude oil distillation. In a refinery, crude oil is first distilled under atmospheric pressure and hydrocarbons boiling up to 650°F are removed. The remaining reduced crude is

TABLE 7-1 Distillates and Resid Yield from Various Crude Oils

	Light arab	Bahrain	Heavy arab	Kuwait	Eocene
API	33.9	30.4	28.3	30.5	18.5
SG	0.8555	0.8735	0.8855	0.8735	0.9435
Sulfur	1.86	2.1	2.8	2.5	4.3
Yields LV%					
Gases	2	1.1	2.6	1.5	2.3
Naphthas	18.4	16.5	14.6	26.03	9.8
Kerosene	15.7	14.8	15.8	10.36	6.6
Diesel	21.9	24.2	20.7	18.31	12.1
Vacuum gas oil	23.7	23.9	13.3	11.95	38.2
Vacuum resid	17.3	19.5	33	31.85*	30.8

again distilled under a vacuum to remove most heavy distillates. The cut point between heavy distillates and short residue is approximately 1000°F. The portion of crude remaining after vacuum distillation is termed "short residue." For making paving bitumen, the cut point between heavy distillate and vacuum residue is adjusted to give a residuum of the required viscosity and penetration for use in road paving. Table 7-1 shows the percentage potential yield of bitumen (1000°F + material) from some crude oils. Both the yield and quality of paving bitumen varies widely, ranging from a few percentage points to as high as 70 percent. Some crude oils, generally known as bitumen crudes, yield high-quality paving bitumen simply by distillation, whereas in the case of many other crude oils, either the quality of bitumen produced is unsatisfactory or the bitumen yield is too low to be economical. Table 7-2 shows the properties of vacuum residua from some crude oils. Satisfactory paving bitumen can be manufactured from inferior crude oils alone or in combination with special stocks from other crudes by appropriate processing and blending. Not all crude oils are suitable for making good road bitumen. If crude oil has a high wax content, its viscosity temperature characteristics may not be suitable for making road paving asphalts. Vacuum residue may have a high penetration and too a low viscosity at 275°F. For example, Arab light crude has a moderately high wax and gives vacuum residue, which has poor low-temperature properties. Vacuum residue from this crude has penetration at 77°F of more than 100, which implies a softer bitumen, softer than what is required by road builders in many parts of the world. Whether or not a particular crude is suitable for making bitumen

TABLE 7-2 CDU Vacuum Resid Properties

	Light arab	Bahrain	Heavy arab	Kuwait	Eocene	Aghajari	Gash saran
Density	0.9900	1.0200	1.0191	1.0120	1.0682	1.0070	1.0250
Sulfur	4.26	4.70	5.16	5.36	6.47	3.55	3.30
Nitrogen	0.2900	0.2900		0.3703	0.5057	0.6320	0.8000
Asphaltenes	7.00	7.30	20.60	5.20	18.50	2.90	9.30
Conradson carbon	19.70	21.00		17.70	27.80	18.00	19.00
Vanadium	42		239	92	160	173	310
Nickel	7			18	60	55	120
Iron	1						
Sodium				6			
Viscosity							
122°F	50,000	10,0000					
212°F	694	1160		1500	84,100		2460
Penetration				220			

either directly or by blending or blowing can only be answered by actual laboratory tests. The penetration value of the short residue is variable and depends on the nature of the crude oil and the vacuum distillation operating conditions. The penetration value of vacuum residues from different crude oils can range between 10 and 500.

Lower penetration asphalts can also be produced from vacuum residues feed with a higher penetration by ultrafiltration of the feed through a membrane. The ultrafiltration[2] process is carried out with a pressure differential of 50 to 100 lb/in^2 and with a temperature sufficiently high to keep the vacuum residue at a low viscosity. A temperature of 300 to 400°F can be used for high-temperature stable polymeric membranes; much higher temperatures can be used for ceramics, sintered metals, or glass membranes. If a lower temperature operation is desired, a solvent/cutter stock such as kerosene or diesel may be added to the vacuum residue. Aromatic solvent has a greater solvency for vacuum residue, whereas paraffinic solvent can give a greater rejection of metals. The product is later solvent stripped. The permeate obtained is of a substantially reduced metal content. The retentate is harder than the pitch fraction produced under normal vacuum tower operating conditions. The permeate stream is of a low metal content and can be used as a cat cracker feed. The retentate stream has a higher metal content than vacuum residue feed. The retentate constitutes hard asphalt, which is evident from its lower penetration.

Bitumen Blending

Paving bitumen may be blended from the blend components listed here. Bitumen feedstock for other end uses is also usually a blend of these components (Fig. 7-2):

- Straight run vacuum residuum from select crude oils
- Solvent deasphalter pitch if available
- Heavy aromatic extract
- Blending with additives such as rubbers, and polymers

Blown bitumen has been used as a blend component for road paving in the past, but its use in road paving is declining because of its faster aging characteristics. Table 7-3 shows the typical properties of solvent deasphalter pitch from a solvent deasphalting (SDA) unit.

For paving bitumen, a refinery generally produces a soft bitumen, 250 to 500 penetration stock, and a hard bitumen, 40 to 50 penetration stock. All intermediate penetration grades are prepared by blending the soft and hard grades. Blend properties can be approximated by empirical correlations (Fig. 7-3). Bitumen blending can be done either in-tank or inline. Blending can be accomplished in a number of ways; by a tank mixer, with tank circulation using an external pump-around, or by blowing air or inert gas using a simple spider air mixer placed at the bottom of the tank.

BITUMEN EVALUATION FOR PAVING

The suitability of a given bitumen grade for a paving job can be evaluated only if ambient conditions are available at the site where the bitumen is to be used. Evaluation parameters are discussed next.

Pavement Temperature Index

Data required are average monthly maximum ambient temperature at the site. For a given site, the "pavement temperature index" is the sum of the increment of the average monthly maximum temperature above 75°F. Table 7-4 presents an example of estimating an empirical pavement temperature index (PTI).[3] The input data required is month-wise, average monthly maximum temperature in degrees Fahrenheit. These data are required for an entire year for the site where asphalt is to be used.

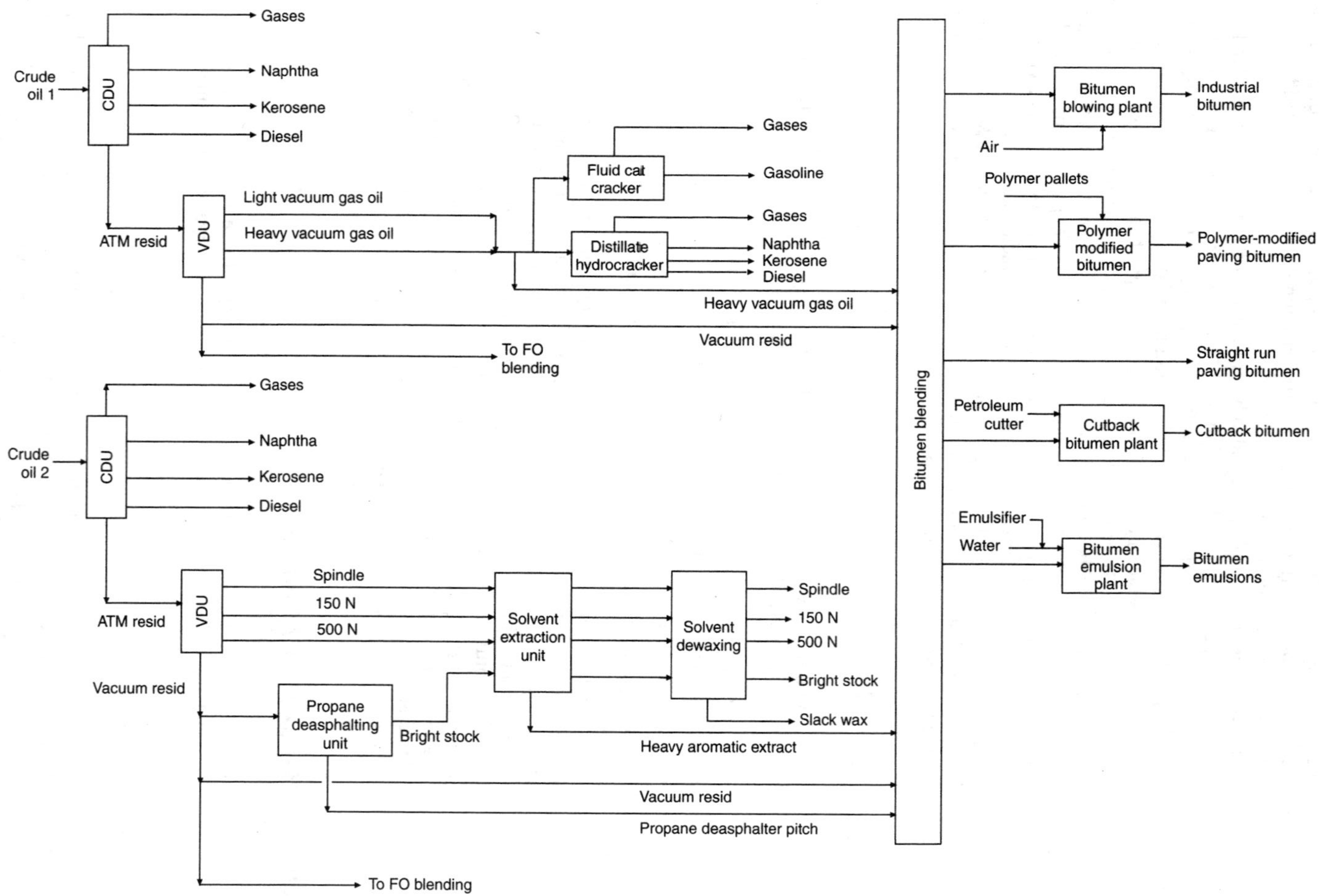

FIGURE 7-2 Refinery bitumen blending.

TABLE 7-3 Properties of Solvent Deasphalted Vacuum Residua (Feed: Vacuum Resids from Mid. Eastern Crudes)

Feed	1	2	3	4	5
Specific gravity	0.9738	1.0030	1.0030	1.0030	0.9738
Viscosity @ 210°F, cSt	375	345	345	345	375
Viscosity @ 450°F, cSt	7.4				
Asphaltene, Wt %		4.2	4.2	4.2	
Penetration, 77°F					800
Solvent deasphalted asphalt					
Asphalt, Wt % feed	28	55	30	14.5	27.4
Specific gravity	1.0839	1.0630	1.1020	1.1400	1.0680
Penetration, 77°F		11	0	0	1
Softening point, °F	200	138	230	311	185
Viscosity 450°F, cSt	100				100
Viscosity 700°F, cSt	7				7
Fraas point, °F		53	>69		

Note: Extraction with C3–C5 solvents in SDA units.

BITUMEN BLENDING

Blend penetration at 25°C

$$\text{LOG } P_m = X^* \text{ LOG } P_a + (1 - X)^* \text{ LOG } P_b$$

Where:

P_a = Penetration of blend component a

P_b = Penetration of blend component b

P_m = Penetration of the blend mix

X = Vol fraction of component a

Blend softening point

$$S_m = X^* S_a + (1 - X)^* S_b$$

Where:

S_a = Softening point of blend component a,°C

S_b = Softening point of blend component b,°C

S_m = Softening point of blend mix

Blend viscosity

$$\text{LOG LOG } (V_m) = X^* \text{ LOG LOG } V_a + (1 - X)^* \text{ LOG LOG } V_b$$

Where:

V_a = Viscosity of blend component a

V_b = Viscosity of blend component b

V_m = Viscosity of blend

X = Volume fraction of component a

FIGURE 7-3 Bitumen blend properties.

TABLE 7-4 Example of Calculation of Pavement Temperature Index (PTI)

1. When a site has average monthly temperature greater than 75°F, PTI is calculated as below; determine delta (temp-75) for every month. All negative deltas are considered = 0. PTI = 208.8.

Month	Average maximum temperature °F	Delta = (T − 75)
January	64.4	0
February	69.3	0
March	78.1	3.1
April	88.7	13.7
May	101.3	26.3
June	110.3	35.3
July	110.5	35.5
August	112.3	37.3
September	107.4	32.4
October	95.5	20.5
November	79.7	4.7
December	67.6	0
	Total	208.8

2. When a site has average monthly temperatures less than 75°F, PTI is defined as the difference between the highest average maximum temperature and 75°F as shown below. Thus PTI is the delta between highest monthly temperature in °F and 75°. PTI = 15.

Month	Average maximum temperature °F	Delta = (T − 75)
January	43	
February	50	
March	60	
April	78	
May	77	
June	85	
July	90	15
August	89	
September	82	
October	72	
November	58	
December	46	

Once the PTI is computed, the project site is classified as "cold," "warm," or "hot" per the criteria listed in Table 7-5. If the project site is classified as warm or hot, the recommended penetration grade as per Table 7-5 is to be used. However if, as per PTI, the site is declared cold, a further calculation is done to classify the cold site into "moderately cold" or "severely cold" by the procedure described next.

TABLE 7-5 Asphalt Selection for Pavements

Pavement temperature index (PTI)	Region classification	Asphalt selection criteria
<30	Cold	Penetration-viscosity method
30–80	Warm	85–100 Penetration grade
>80	Hot	60–70 Penetration grade

Design Freezing Index

The design freezing index (DFI) is an empirical number[4] used to distinguish between shades of cold in cold temperature regions of the world. It is the cumulative number of degree days below 0°C calculated by taking the average of the seasonal freezing index for the three coldest winters in the most recent 30-year period. The DFI is related to pavement temperature at a 2-in depth by the following correlation:

$$T_2 = 318.5 * (DFI)^{-0.03473} - 273$$

Where: T_2 = Pavement temperature at a 2-in depth, in °C
DFI = Design freezing index, in °F days

A DFI of 3000°F days or 1667°C days is used as the boundary between a moderately cold and a severely cold climate. Moderately cold climates have a DFI up to 3000°F days and severely cold climates have a DFI greater than 3000°F days. Soft grades of bitumen cements are used for controlling low-temperature cracking of bitumen in very low temperature zones such as those falling in the arctic regions or at very high altitudes in the mountains. For specifying bitumen quality for such regions, the penetration viscosity number (PVN) is used. A minimum PVN of –0.5 is specified for moderately cold areas where the DFI is less than 3000°F days. A PVN of –0.2 is specified for regions where the DFI is greater than 3000.

PVN

The PVN is an empirical correlation to relate bitumen cement properties (viscosity at 275°F and penetration at 77°F) to its low temperature cracking characteristics.[5] The PVN of a given asphalt can be calculated from its penetration at 77°F and kinematic viscosity at 275°F (Fig. 7-4). Table 7-6 presents

The general PVN equation is:

$$PVN = \frac{(L - X)^* (-1.5)}{(L - M)}$$

Where:

PVN = Penetration viscosity number
L = Logrithm of viscosity in centistokes at 275°F for a PVN of 0 at the given penetration
X = Logrithm of viscosity in centistokes at 275°F of a given asphalt.
M = Logrithm of viscosity in centistokes at 275°F a PVN of –1.5 at the given penetration.

L and M are function of penetration at 77°F

$$L = 4.258 - 0.79674^* \text{LOG (penetration)}$$

$$M = 3.46289 - 0.61094^* \text{LOG (penetration)}$$

PVN can be expressed directly as a function of penetration at 77°F and viscosity in centistokes at 275°F as follows:

$$PVN = -1.5^* \frac{4.258 - 0.7967^*\text{LOG (P)} - \log (V)}{0.795 - 0.1858^*\text{LOG (P)}}$$

Where:
P = Penetration at 25°C (100g.5 s)
V = Kinematic viscosity at 135°C cSt

FIGURE 7-4 Penetration viscosity number (PNV).

TABLE 7-6 Minimum PVN Selection Criteria for Cold Regions

Region	Design freezing index (DFI)	PVN
Moderate cold	<3000°F days	–0.5
Severe cold	>3000°F days	–0.2

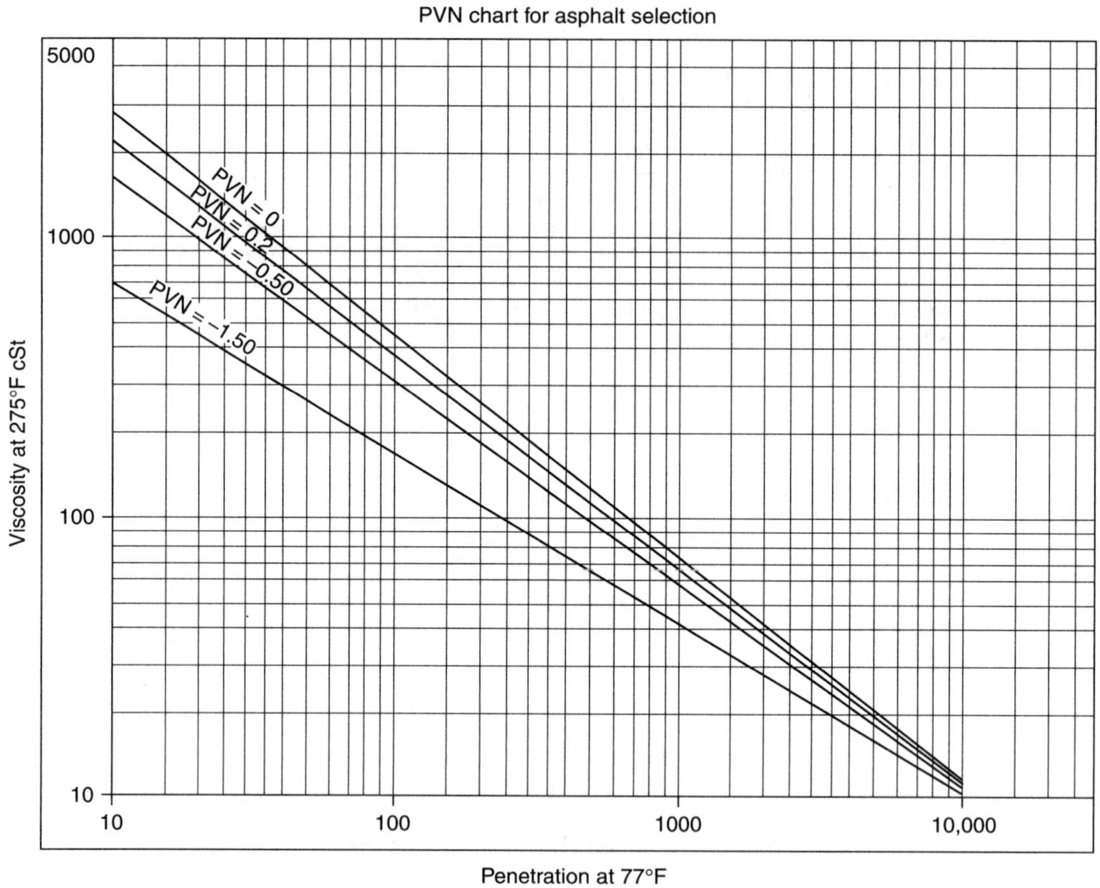

FIGURE 7-5 PVN chart for asphalt selection.

the minimum PVN selection criteria for asphalts in cold climates. Once an optimum PVN for a given climatic condition is estimated, suitable bitumen can be selected based on its viscosity at 275°F and penetration at 77°F (Fig. 7-5).

Penetration Index

The penetration index (PI)[6] and penetration ratios[7] are other empirically defined criteria to evaluate bitumen quality for road paving and other uses. The PI was first defined by Pfeiffer and Doormal[8] in 1936. PI is defined as follows:

$$PI = (20 - 500 * A)/(50 * A + 1)$$

Where: A is the slope of the curve $\log_{10}$ (PEN) vs. temperature

LOG P is a linear function of temperature.

$$\text{LOG P} = \text{A}^*\text{T} + \text{C}$$

Where:

P = Penetration at 25°C (100g5 s)

T = Temperature, °C

A = Slope of log P vs. T graph

C = A constant

Slope of log (pen.) vs. temperature curve (A) can be determined from penetration (25°C) and ring and ball softening point. It is assumed that penetration value for all bitumen is more or less constant at 800 at R & B softening point temperature)

$$A = \frac{\text{Log } 800 - \text{LOG (PEN at 25°C)}}{T_{R48} - 25)}$$

Penetration index is then estimated from the following empirical correlation:

$$PI = \frac{25 - 500^*A}{1 + 50^*A}$$

Penetration ratio (PR)

$$A = \frac{\text{Penetration at 4°C (200g. 60 s)}}{\text{Penetration at 25°C (100g. 5 s)}} \times 100$$

FIGURE 7-6 Penetration and index ratio of bitumen.

The value of A is determined from the plot of log (Penetration) versus temperature. The PI of a bitumen can be estimated from penetration measurements at two temperatures (Fig. 7-6).

All bitumen display thermoplastic properties: Asphalts become softer when heated and harder when cooled. If logarithm of penetration is plotted against temperature, a straight line is obtained such that

$$\text{LOG P} = \text{A}*\text{T} + \text{C}$$

Where: P = Penetration at 77°F

A = temperature susceptibility of the log of penetration or slope of log (penetration) vs. temperature curve

C = constant

The value of A varies approximately between 0.015 and 0.06, showing there is considerable variation in the temperature susceptibility of bitumen from different sources. The value of PI ranges from about –3 for high-temperature-susceptible bitumen to about +7 for highly oxidized low-temperature-susceptible bitumen. Thus PI is a unique function of A. Pfeiffer and Doormaal found that at the ASTM softening point temperature, most bitumen has a penetration of approximately 800. Penetration is measured with a 100-g load for 5 s. Thus the softening point is the temperature at which penetration is 800. The softening point measurement may be made directly with an extra-long penetration needle. The PI of oxidized bitumen[9] may be determined by means of its penetration and softening point. The PI is used to characterize the temperature susceptibility of bitumen at low temperatures. Bitumen with a low PI, less than 0, is more susceptible to temperature changes. Pavement made with these bitumens shows a greater number of transverse cracking incidents caused by thermally induced stresses. Bitumen with a PI greater than 0 is progressively less susceptible to temperature changes. Pavements made with such bitumen experience less transverse cracking and exhibit better low-temperature performance.

Penetration Ratio

The penetration ratio (PR) is another index to estimate the temperature susceptibility of bitumen. It is the ratio of bitumen penetration at 4°C measured with a 200-g load, 60 s to penetration at 25°C measured with a normal 100-g load and 5-s time (Fig. 7-6). The ratio is expressed as a percentage. Paving bitumen must have a PR greater than 25 percent. Any bitumen with a PR less than 25 percent is unlikely to be suitable for paving use.

Paving Bitumen Quality

The most economical way to manufacture road bitumen is to use vacuum residue directly from crude oil. However, not all crude oils yield bitumen of the required hardness, viscosity, and other qualities to meet the requirements of the road paving industry. For direct asphalt manufacture from crude distillation column, the cut point between vacuum residue and heavy vacuum gas oil (HVGO) is the only process variable. Because the cut point is raised by increasing the flash zone temperature, the yield of bitumen falls. Bitumen becomes harder, with lower penetration. Cutpoint can be raised only by a few degrees, however, before the cracking and coking of residue begins. Table 7-7 shows the typical penetration value of bitumen used for pavement in different climates.

TABLE 7-7 Bitumen Grades for Road Paving

Penetration grade	Usage
40–50	Hardest grade: used for hot climate
60–70	For moderate ambient temperatures
85–100	For moderate ambient temperatures
120–150	For moderate ambient temperatures
200–300	Used in extreme cold conditions

Penetration

If vacuum residuum penetration from a given crude is more than 150, the residuum is unlikely to be suitable for use as road paving without blending with other blend components such as residuum from another crude oil, solvent precipitated asphalt, blending with polymer, or blown bitumen.

Viscosity at 275°F

Viscosity at 275°F of a vacuum residuum is an important indicator of asphalt quality. For the 85 to 100 grade, viscosity at 275°F should be 280 cSt or more.

Penetration Index

The PI of bitumen is an indicator of the rate of change of penetration with temperature. For road paving, the penetration index is generally greater than –1.6.

Penetration Ratio

A penetration ratio of less than 25 percent indicates a high-temperature susceptibility of asphalt and thus it is unlikely to be suitable for road paving.

Wax Content

Crude oils with a high wax content are generally unsuitable for making asphalts. The vacuum resids from waxy crude oils have a satisfactory viscosity and penetration relationship but have poor low-temperature properties. Pavement made with high wax asphalt can crack under severe winter conditions.

Asphaltene Content

Asphaltene give hardness to paving asphalts. Higher asphaltene residuum are desirable for making industrial bitumen. The asphaltene level is increased by air blowing, which converts the resins in vacuum residuum to asphaltenes, resulting in a higher softening point and lower penetration values.

BITUMEN GRADING SYSTEMS

Bitumen cement for use in pavement design are classified either according to their penetration or according to their viscosities. Viscosity grading and performance grade (PG) systems are most popular in North America. Elsewhere in the world, a penetration grading system is used. Selecting a grade of bitumen cement for a specific road project depends on several considerations, such as ambient temperatures, rainfall, traffic conditions, and the costs of transporting the bitumen to the site.

Penetration Grading

Based on the depth, a standard needle can penetrate a bitumen binder sample when placed under a 100-g load for 5 s. The test is done at a temperature of 77°F. Penetration grades are listed as a range of penetration units (1 penetration unit = 0.1 mm). Typical asphalt binders used in the paving industry are between 65 and 70 penetration and 85 and 100 penetration (Table 7-8).

Viscosity Grading

The viscosity grading system measures not only measures penetration as in the penetration system but also measures bitumen binder viscosity at 140°F and 275°F. Testing can be done on virgin (AC) or aged (AG) bitumen binders. Grades are listed in viscosity units poises ($cm/g/s^{-1}$) or poises divided by 100. Thus grade AC-5 will have a viscosity of 500 poises at 60°C or 140°F (Tables 7-9 and 7-10).

Superpave Performance Grading

The Superpave program was initiated in 1987 by the U.S. Transportation Research Board (TRB) and the American Association of State Highway and Transportation Officials (AASHTO) as a part of the Strategic Highway Research Program (SHRP). Superpave is a comprehensive method of specifying bitumen for road paving, taking into consideration the specific performance requirements, traffic density, climatic conditions (maximum and minimum ambient temperature, rain and snowfall, etc.), and layer location. It facilitates the selection of the most appropriate asphalt binder, aggregate, and modifier to achieve the required level of road pavement performance. Superpave performance grading is based on the idea that HMA (hot mix asphalt) bitumen binder properties should relate to the conditions where it is used. The system uses the same tests as used in penetration grading and viscosity grading systems. The only difference is that these tests are conducted at specific temperatures dictated by the climatic conditions in the area of use. For example, a bitumen binder used in hot deserts would not be appropriate in a very cold arctic climate. The Superpave system defines the

TABLE 7-8 Penetration Grades of Asphalts (Conforming to AASHTO M20-70/ASTM D 946)

		40–50		60–70		85–100		125–150		200–300	
Test	Test method	Min.	Max.	Min.	Max.	Min.	Max.	Min.	Max.	Min.	Max.
Penetration, 77°F	ASTM D 5-86	40	50	60	70	85	100	120	150	200	300
Flash point, °F	ASTM D 92	450		450		450		425		350	
Ductility, 25°C, CMS	ASTM D 113	100		100		100		100		100	
Solubility in trichloroethylene, %	ASTM D 2042	99		99		99		99		99	
Thin film oven test	AASHTO T 240/ ASTM D 2872										
Loss on heating	ASTM D 6		0.8		0.8		1.0		1.3		1.5
Penetration of residue	ASTM D 5	58		54		50		46		40	
Ductility of residue, 5 cm, per min.cm				50		75		100		100	
Spot test, as and when specified		*		*		*		*		*	
Standard naphtha solvent											
Naphtha-xylene solvent	% Xylene										
Heptane-xylene solvent	% Xylene										

*Negative for all grades.

TABLE 7-9 Viscosity Grades of Asphalts (Conforming to ASTM D 3381/ AASHTO M 226)

Test	Test method	AC–2.5	AC–5	AC–10	AC–20	AC–30	AC–40
Viscosity, 140°F, poise	ASTM D 2171	250	500	1000	2000	3000	4000
Viscosity, 275°F, Min. cSt	ASTM D 2170	125	175	250	300	350	400
Penetration, 77°F	ASTM D 5	220	140	80	60	50	40
Flash point, °F, Min. open cup	ASTM D 92	325	350	425	450	450	450
Ductility, 77°F, CMS	ASTM D 113	100	100	75	50	40	25
Solubility in trichloroethylene, %, Min.	ASTM D 2042	99	99	99	99	99	99
Thin film oven test on residue; Viscosity, 140°F, Max. poise	ASTM D 2170	1250	2500	5000	10,000	15,000	20,000

TABLE 7-10 Viscosity Grades of Asphalts (Conforming to ASTM D 3381/AASHTO M 226)*

Test	Test method	AR-1000	AR-2000	AR-4000	AR-8000	AR-16000
Viscosity, 140°F, poise	ASTM D 2171	1000	2000	4000	8000	16,000
Viscosity, 275°F, Min. cSt	ASTM D 2170	140	200	275	400	550
Penetration, 77°F	ASTM D 5	65	40	25	20	20
Percent of original penetration, Min.			40	45	50	52
Flash point, °F, Min, open cup (original asphalt)	ASTM D 92	400	425	440	450	460
Ductility, 77°F, CMS	ASTM D 113	100	100	75	75	75
Solubility in trichloroethylene, %, Min.	ASTM D 2042	99	99	99	99	99

*Grading based on residue from rolling thin film oven test.

binder selection in relation to climatic conditions and traffic conditions. Superpave performance grading (Table 7-11) is reported using two numbers. The first is the average 7-day maximum pavement temperature in degrees Celsius, and the second is the minimum pavement design temperature likely to be experienced, in degrees Celsius. Thus a PG 64-40 grade is intended for use at a place where the average seven-day maximum pavement temperature is 64°C and the expected minimum pavement temperature is –40°C. The range of high and low pavement temperatures may differ as much as 100°C or more in arctic regions to a much more modest variation (50 to 60°C) in temperate zones. The pavement temperature estimations are used to determine bitumen's suitability for use in that region. In cases where pavement temperature variation between maximum and minimum is more than 90°C, straight run bitumen grades are generally unsuitable, and polymer-modified-type bitumens are used.

TABLE 7-11 Asphalt PG Grades

Low pavement temperature	High pavement temperature °C				
	52	58	64	70	76
–16	52–16	58–16	64–16	70–16	76–16
–22	52–22	58–22	64–22	70–22	76–22
–28	52–28	58–28	64–28	70–28	76–28
–34	52–34	58–34	64–34	70–34	76–34
–40	52–40	58–40	64–40	70–40	76–40
–46	52–46	58–46	64–46	70–46	76–46
–52	52–52	58–52	64–52	70–52	76–52

Correlation between PG temperature limits and conventional bitumen parameters

High temperature

$T_{DSR} = T_{R\&B} + 20$

Where temperature is in °C

$T_{R\&B}$ = Ring & ball softening point

Low temperature

$T_{BBR} = 2^* (T_{FRAAS})$

FIGURE 7-7 Bitumen PG temperature limits.

Note that the numbers in PG grades are pavement temperatures and not air temperature. The pavement temperature is estimated from the air temperature using empirical equations. SHRP high temperature criteria obtained from dynamic shear rheometer (DSR) can be approximately estimated by an empirical relation[10] from the Ring and Ball softening point in degrees Celsius. Similarly, SHRP low-temperature criteria can be approximated from the Frass breaking point of the bitumen (Fig. 7-7).

Bitumen Pavement Structure

An understanding of road structure is necessary to realize the importance of bitumen in road building. Modern road design aims at building flexible road layers, or courses, so that the tensile and compressive stresses imposed by passing traffic are distributed evenly through these layers according to their relative strengths. Therefore neither the base supporting the road nor individual layers are permanently deformed by these concentrated stresses. The individual layers or courses must be made weather resistant and durable. Bitumen plays a major role in meeting these requirements. It binds the aggregates and seals the voids in them. By sealing the voids, bitumen prevents water from seeping into the road base course. For new roads the lowest layer of road is the natural soil of the subgrade. For repairing an earlier road, this layer consists of whatever is left of the old road.

- The subbase is the first layer put down by the road builder and consists of compacted stones, gravel, and sand. Its function is to contribute toward the strength of the road and serve as a base for heavy road rollers.
- The roadbase is the main working layer of the road that gives it its strength and flexibility. It consists of graded mineral aggregates. When the road has to carry heavy traffic, the road base mixture may also contain bitumen, which enhances its performance. The base course is made of a mixture of aggregate and bitumen. It adds to the strength of the road and provides a smooth surface foundation for the top surface layer.
- The wearing course is the top layer, consisting of a more finely textured mixture of aggregate and bitumen. It not only has to provide smooth surface for traffic but it also must be as weatherproof as possible and resist the continued action of abrasive forces imposed by vehicles as they pass along it. The wearing course should also disperse surface water effectively to minimize the danger of skidding.
- The tack coat is a thin layer of emulsified or cutback bitumen that binds a layer of road to the layer beneath.
- The fog seal is a coat of bitumen emulsion applied to an existing surface to seal cracks as maintenance treatment.

- Slurry surfacing is a maintenance treatment in which a mixture of aggregates, fillers, and emulsified bitumen is applied to the road surface for minor shape correction and to improve skid resistance.
- The objective of road recycling is to reclaim the materials originally used in building, bitumen and aggregates, and use them to build the new road. The original layers of road are ripped up, crushed, and reprocessed with a smaller amount of aggregates and fresh bitumen. The recycled material may be processed in situ or transported to another site for reprocessing.

HOT-MIX ASPHALT

Hot-mix asphalt (HMA) is a mixture of hot bitumen and aggregates. Bitumen and aggregates are combined in a mixing facility, where they are first heated, proportioned, and then mixed to produce the desired paving mix. An anti-stripping additive such as hydrated lime, 1 to 1.5 percent by weight of aggregates in the form of powder, may be added to aggregates. The addition of hydrated lime results in reduced rutting, slower aging, and reduced cracking due to fatigue. Hydrated lime is an excellent anti-stripping agent. Lime interaction at the aggregate surface enhances the bitumen-aggregate bond in the presence of moisture, reducing the likelihood of stripping.

There are two types of bitumen mixes; hot mix (HMA) and cold mix (CMA). In cold mix facilities, aggregates are mixed with bitumen at ambient temperature, without heating. Bitumen used is in the form of cutback bitumen or in emulsion form. Bitumen binds crushed stones and gravel, commonly known as aggregate, into firm, hard surfaces for roads, streets, and runways. Cold mix bitumen is used for light load secondary roads.

HMA Plant

HMA is used for heavy traffic density roads. The important components of an HMA plant are a rotary dryer, burner-blower, exhaust fan, dust collection system. The HMA plant also has bitumen heating and storage facility, pug mill, reclaimed bitumen paving (RAP) area, bitumen storage tanks, and bitumen concrete storage bins. HMA facilities can be classified into three categories: batch mix, continuous mix, and drum type. The most popular design is the batch mix and drum mix. Aggregates from the storage bin feed into the rotary dryer where they are dried by hot burner exhaust gases. The dryer temperature is 275 to 325°F. In the batch facility, bitumen is added to the aggregate after drying. The mixing takes place in a pug mill type mixer. A batch mix facility has a plant tower that includes vibrating screens, bucket elevators, storage bins, and a weigh hopper and mixer. As the hot aggregate leaves the dryer, it drops into a bucket elevator and is transferred into a set of vibrating screens that drop the aggregates into individual hot bins according to their mesh sizes. To control the aggregate size distribution in the final batch mix, aggregates from various bins storing different size aggregates are mixed until the desired mix and weight is obtained. RAP (reclaimed bitumen paving) may also be added. Concurrent with the aggregate weighing, liquid bitumen is pumped from a storage tank into a bitumen bucket, where it is weighed to achieve the desired aggregate-to-bitumen ratio in the mix. Aggregate from the weigh hopper is first dropped into the pug mill and dry mixed for 5 to 10 s. The liquid bitumen is next dropped into a pug mill where it is wet mixed until homogeneous. The hot mix is conveyed to a hot storage silo or dropped directly into a truck and transported to the road paving site and spread in a layer with a paving machine. While still hot, the paving mixture is compacted with heavy road rollers to produce a smooth paving surface. The bitumen binder content is generally between 5 and 8 percent of the total weight of HMA. In the drum mix facility, the aggregate is not only dried and heated within the drum but also mixed with bitumen.

The drum mix facility, unlike the batch mix facility, does not incorporate a grading screen unit; therefore the feed aggregates must be presized and mixed into appropriate proportions before entering the mixing drum. Aggregates are fed to the drum at the burner end. As the drum rotates, the aggregate as well as combustion products move forward toward the other end of the drum. The aggregates and combustion products move in concurrently, giving them a longer time in the drum. Bitumen is introduced into the drum toward the last third of the time. Higher VOC emissions are released in a drum-type operation. Exhaust gases from the drum after passing through a knockout drum are cooled by water spray, next through a cyclone to stack. Most of the pollutants from the hot mix are particulates, sulfur dioxide, and other gases coming from the burner-blower, bitumen storage, and loading and transportation. Particulate emissions are controlled by secondary devices such as water sprays, cyclones, bag house, and so on.

BITUMEN TEST METHODS

Penetration (AASHTO T-49, ASTM D 5)

The penetration test is a measure of the consistency or relative hardness of bitumen. In this test, a sample of bitumen cement is placed in a suitable container and its temperature brought to 77°F. A weighted needle is then brought to bear on the surface of bitumen for a given period of time. The load on the needle is 100 g, and the time interval is 5 s. The penetration of the needle into the bitumen in units of 1/10 mm is termed the penetration of bitumen.

Softening Point (AASHTO T-53, ASTM D 36)

The softening point is defined as a temperature at which a bitumen sample can no longer support the weight of a 3.5-g steel ball. Two samples of bitumen are confined in a brass ring suspended in a beaker of water. As water is heated at a given rate, the bitumen softens and eventually drops along with the ball through the ring. The softening point is defined as a mean of the temperatures at which two disks soften enough to allow each ball enveloped in bitumen to fall a distance of 25 mm or 1 in. For bitumen having a softening point below 175°F, a water bath can be used. For bitumen having a softening point above 175°F, glycerin is used. The softening point is particularly significant for roofing bitumen.

Ductility (ASTM D 113, AASHTO T-51)

The ductility test measures bitumen binder ductility by stretching a standard-sized briquette of bitumen binder to its breaking point. The stretched distance in centimeters at the breaking point is then reported as ductility. The test is conducted at 77°F. A briquette of bitumen is cast under standard conditions and of specified dimensions. It is then brought to standard test temperature. It is next pulled or extended at a specified speed until the thread of bitumen connecting the two ends breaks. The length of pull in centimeters at which the thread of material breaks is designated as its ductility, which is an important property in many applications. However, the presence or absence of ductility is more important than its numerical value. Bitumen with good ductility has better binding properties compared to those lacking it. But bitumen with high ductility is more temperature susceptible.

Fraass Breaking Point (IP-80, NFT 66026, DIN 52012)

The Fraass breaking point is the temperature in degrees Celsius at which a film of bitumen binder of specified and uniform thickness will break under defined loading conditions. A sample of bitumen binder is applied to a metal plate at an even thickness. The plate is next submitted to a constant cooling rate and flexed repeatedly until the binder layer breaks. The temperature at which the first crack appears is reported as the Fraass breaking point.

Spot Test (AASHTO T-102)

The spot test is in the form of paper chromatography. This test is used to determine whether or not bitumen cement has been damaged during processing due to overheating. The damage, called thermal cracking, can occur if bitumen is heated and held at a temperature greater than 850°F for a half hour or so. The actual temperature and induction period may vary depending on the nature of material; its asphaltene, resin, and trace metals content. Cracked bitumen tends to be less ductile and more susceptible to aging effects. A drop of asphalt bitumen is dropped on a filter paper. If the spot formed is uniformly brown, the test is negative. If the spot formed is brown with a black center, the test is positive.

Oliensis Test (ASTM D 1370)

This test is mostly used for roofing bitumen. In this test a layer of bitumen is placed on felt substrate that has been saturated with bitumen. If there is a gradient of oil concentration, oil will migrate from one type of bitumen to another causing embrittlement in oil-deficit bitumen. In this test, a 1-mm layer is spread in a small pan and allowed to cool. The surface is dusted with a roofing grade of talc, and small drops of bitumen saturant are put on top. The specimen is put into an oven at 60°C for about 72 h. If there is a significant migration of oil from the bitumen, a saturant drop will develop a dark oily ring. The width of the ring must not exceed 0.5 mm.

Aging Tests

Typical aging simulation tests are as follows:

- Thin film oven test
- Rolling thin film oven test
- Pressure aging test

Thin Film Oven Test (ASTM D-1754, AASHTO T-179)

The thin film oven (TFO) test simulates short-term aging by heating a film of asphalt binder in an oven at 325°F (163°C) for 5 h. The effect of heat and air are measured from changes that occur in physical properties measured before and after heating.

Rolling Thin Film Oven Test (ASTM D-2872, AASHTO T 240)

The rolling thin film oven (RTFO) test simulates short-term aging by heating a moving film of bitumen binder in an oven for 85 mins at 325°F (163°C). The effects of heat and air are determined from changes in physical properties measured before and after the test. The moving thin film is created by placing the asphalt binder sample in a small jar and then placing the jar in a circular metal carriage that rotates within the oven. This test is considered superior to the TFO test because it can be carried out in a shorter time and with asphalt binder modifiers. It also allows the continuous exposure of the fresh bitumen binder to heat and air flow.

Pressure Aging Vessel Test (AASHTO PP1)

The pressure aging vessel (PAV) test simulates the effect of long-term bitumen binder aging that occurs to HMA pavement over a period of 5 to 10 years. PAV is an oven/pressure vessel combination that takes RTFO-aged samples and exposes them to high-temperature and high-pressure air for a period of 20 h. The test conditions are as follows:

Air pressure	300 lb/in^2 (2070 kPa)
Temperature	195°F (90°C), 212°F (100°C) or 230°F (110°C) (depending on expected climatic condition)

Chewing Test

Originally the degree of asphalt binder softening was determined by chewing.[11] A sample of asphalt binder was literally chewed to subjectively determine its softness. This method is no longer in use.

Absolute Dynamic Viscosity (AASHTO T 51, ASTM D 113)

This test is carried out at 140°F (60°C) to approximate the maximum HMA pavement surface temperature during the summer months. This test measures the time taken by a fixed volume of bitumen binder at the specified temperature to be drawn up through a capillary tube with a fixed closely controlled vacuum.

Rotational (or Brookfield) Viscometer Viscosity (AASHTO TP 48, ASTM D 4402)

Rotational viscometer (RV) viscosity is used to measure high-temperature viscosity at 275°F (135°C). This test measures the torque required to maintain a constant rotational speed (20 r/min) of a cylindrical spindle submerged in bitumen binder at a constant temperature. This torque is converted into a viscosity reading and displayed on the viscometer. RV high-temperature measurements simulate binder workability at mixing and HMA lay-down temperature. For testing polymer-modified asphalt binders, RV is more suitable than a capillary viscometer because rubber particles sometimes clog the capillary and cause a faulty reading.

Dynamic Shear Rheometer (AASHTO TP 5)

A dynamic shear viscometer (DSR) is used in the Superpave system for testing binder viscosities between 115°F (60°C) and 180°F (90°C), which is the actual maximum pavement anticipated temperature in the area where asphalt binder is to be used. The basic DSR test uses a thin asphalt binder sample sandwiched between two plates. The lower plate is fixed while the upper plate oscillates back and forth across the sample at 1.59 Hz (10 radians/s) to create a shearing action. These oscillations are meant to simulate a shearing action corresponding to traffic speed of 90 km/h. The test is used to predict rutting tendencies and fatigue resistance of asphalt binders.

Bending Beam Rheometer (AASHTO–TP-1)

The bending beam rheometer (BBR) is used in a Superpave system to test asphalt binder at low temperatures where failure is usually due to thermal cracking. In this test a bitumen beam is subjected to a small load (100 g) for 240 s. Using basic beam theory, BBR calculates beam stiffness S_t and the rate of change of that stiffness "m" as load is applied (Fig. 7-8). The BBR test is used to simulate bitumen binder stiffness after 2 h of loading at minimum HMA design temperature. Creep stiffness S_t is related to thermal stresses in HMA pavement due to shrinking; the "m" value is related to the ability of bitumen to relieve these stresses. Superpave binding specifications

Bending beam rheometer test

$$S_t = \frac{P^* L^3}{4^* b^* h^{3*} \Delta_t}$$

Where:

S_t = Creep stiffness at time t = 60 s.

P = Applied constant load (100 g or 0.98 newton or 980 mN)

L = Distance between beam supports, 102 mm

b = Beam width, 12.5 mm

h = Beam thickness, 6.25 mm

Δ_t = Deflection at time t = 60 s

FIGURE 7-8 Bending beam rheometer test for bitumen.

require a maximum limit on creep stiffness and a minimum limit on "m" values or the ability to relieve thermal stresses without cracking.

Direct Tension Tester (AASHTO TP-3)

The direct tension tester (DTT) test basically loads a small sample of asphalt in tension until it breaks. The failure strain is then calculated from the following equation:

$$S = L_i/L_s$$

Where: S = failure strain
L_i = change in length corresponding to specimen maximum loading
L_s = effective length

If a bitumen binder has a high BBR creep stiffness, indicating high thermal stress, it must have minimum failure strain, indicating it will stretch rather than crack to meet Superpave binder specifications. A direct tension tester is used in a Superpave system to complement the BBR test in testing bitumen binder at low temperatures. DTT is used because creep stiffness as measured by BBR is not sufficient to predict thermal cracking in some asphalt binders that exhibit high creep stiffness (more than 300 MPa). High creep stiffness values implies that the bitumen binder will possess high thermal stresses in cold weather as a result of shrinkage. Under such conditions, asphalt binder becomes more susceptible to cracking in cold weather due to high thermal stresses. However, some asphalt binders, especially modified asphalt binders, may be able to stretch enough without breaking so they can absorb high thermal stresses without cracking. DTT identifies such binders. DTT is only used for testing bitumen binders with a high BBR creep stiffness (300 to 600 MPa). Bitumen binders with creep stiffness lower than 300 are deemed satisfactory, and DTT is not needed.

Bitumen Purity (AASHTO T-44, ASTM D 2042)

Bitumen cement for various binding applications should consist of pure bitumen. Any impurity will be detrimental to bitumen cement performance. Mineral/inorganic impurities can be quantified by dissolving a sample of bitumen in trichloroethylene or 1,1,1 trichloroethane and filtering through a filter paper. Any remaining undissolved residue is an impurity. Water if present can be quantified through distillation.

TYPES OF BITUMEN

Apart from straight run bitumen, the following other types of bitumen are produced and used in the industry:

- Cutback bitumen
- Bitumen emulsion
- Polymer-modified bitumen
- Oxidized or blown bitumen

Cutback Bitumen

Cutback bitumens are penetration grades blended with a petroleum solvent that evaporates when applied. Unlike HMA, cutbacks do not require heating before spreading on the pavement. Cutbacks are used in many applications because the solvent reduces asphalt viscosity for use at a lower temperature. Cutbacks are used in road applications for priming road surfaces because they can penetrate and bind the surface layers and thus create waterproofing. Other road building uses are tack coats, fog seals, and slurry seals in roads and highways. After cutback asphalt is applied, the petroleum solvent evaporates, leaving behind bitumen cement. A cutback bitumen is said to have cured when petroleum solvent has evaporated.

Classification. The classifications of cutback bitumen is based on the rate of evaporation of distillate in the cutback, as follows:

- Rapid cure (RC)
- Medium cure (MC)
- Slow cure (SC)

RC cutbacks are blended with naphtha or gasoline that is highly volatile. MC cutbacks are blended with kerosene, a medium volatility solvent. SC cutbacks are blended with diesel or vacuum gas oils with the lowest evaporation rate. Composition of some cutback bitumen are shown in Table 7-12.

Each cutback grade may also be further classified according to its viscosity. The number assigned to each grade corresponds to lower limit of kinematic viscosity at 60°C (Tables 7-13 and 7-14). The upper limit of viscosity is twice the lower limit viscosity. For example, MC 250 indicates a medium

TABLE 7-12 Typical Composition of Cutback Bitumen

	Grade	30	70	250	800	3000
	Viscosity, 60°C, cSt	30–60	70–140	250–500	800–1600	3000–6000
Rapid curing						
	Bitumen, Vol %	—	65	75	83	87
	Naphtha, Vol %	—	35	25	17	13
Medium curing						
	Bitumen, Vol %	54	64	74	82	86
	Kerosene, Vol %	46	36	26	18	14
Slow curing						
	Bitumen, Vol %	—	50	60	70	80
	Vacuum gas oil, Vol %	—	50	40	30	20

Basis normal base bitumen 80/100 grade.

TABLE 7-13 Cutback Bitumen Properties (Medium Curing Cutback)

Property		Test method	MC-30	MC-70	MC-250
Kinematic viscosity	60°C, cSt	ASTM D 2170	30–60	70–140	250–500
Flash point	°C, Min.	ASTM D 3143	66	66	66
Distillation, Vol % of total		ASTM D 402			
Distillate to 360°C					
190°C			0–15	—	—
225°C			15–60	10–50	10
260°C			50–85	40–80	15–55
316°C			80–100	80–95	60–87
Residue from distillation	Vol %				
Residue properties					
Ductility, cm	25°C, Min., cm	ASTM D 13	25	55	100
Penetration	25°C	ASTM D 5	90–180	90–180	120–250
Solubility in trichloroethylene	Min. %	ASTM D 2042			99
Water content	Vol % Max.	ASTM D 95			0.2

TABLE 7-14 Cutback Bitumen Properties (Slow Curing Cutback)

Property		Test method	SC-70	SC-250	SC-800	SC-3000
Kinematic viscosity	60°C, cSt	ASTM D 2170	70–140	250–500	800–1600	3000–6000
Flash point, COC	°C, Min.	ASTM D 92	66	80	90	105
Distillation test	Vol %	ASTM D 402	10–30	4–20	2–12	0-5
Total distillate to 360°C						
Distillation residue						
Kinematic viscosity	60°C, cSt	ASTM D-2170	400–7000	800–10,000	2000–16,000	4000–35,000
Asphalt residue						
Residue to 100 penetration	Vol % Min.	ASTM D-243	50	60	70	80
Residue properties						
Ductility, cm	25°C, Min. cm	ASTM D 113	100	100	100	100
Water content	Vol % Max.	ASTM D 95	0.5	0.5	0.5	0.5

cure cutback bitumen grade with a viscosity range of 250 to 500 cSt at 60°C. Road oil is a SC-type cutback asphalt. SC cutbacks can be produced directly from a crude vacuum distillation column either by leaving some heavy vacuum gas oil in the bottom asphalt product or blending asphalt product with diesel or vacuum gas oil. The viscosity of road oil depends on the amount of oil left and its quality. Road oils are used as a dust palliative to settle dust and bind gravel on roads and highways.

Classification by Consistency. In this system, each of the three types (RC, MC, and SC) is produced in six standard grades. The prefix of each type denotes the type; SC for slow curing, MC for medium curing, and RC for rapid curing. The suffix or grade number denotes the consistency range of material. In each type, grade 0 is the most fluid; grade 5, the most viscous. At room temperature, the consistency of grade 0 material resembles that of heavy cream; grade 5 has the consistency[12] of heavy molasses.

Cutback Bitumen Manufacture. In the batch process, mixing of bitumen and diluent (kerosene, diesel, or other) is done in a tank. Bitumen is heated to 280 to 285°F after which metered amounts of diluent are introduced below the level of the bitumen. Mixing is done by recirculation through a pump until uniform density is achieved and there is no change in density or viscosity with time. In a continuous process, the mixing of bitumen and diluent is done in a colloid mill that produces a very well-dispersed cutback. The most used cutbacks are MC. This type has kerosene as a diluent. The use of diesel in place of kerosene is preferable because of its lower volatility and higher flash point. RC cutbacks contain naphtha and pose high fire risks. For this reason, use of RC cutbacks is avoided if MC can be used instead. For road priming, SC cutbacks give the best results and are used wherever feasible.

The use of cutback asphalt is decreasing for the following reasons:

- Environmental regulations: Cutback asphalts contain petroleum distillates that evaporate into the atmosphere during their application, adding to pollution levels. In contrast, emulsified asphalts only lets water vapor into the atmosphere during their usage.
- Loss of high-energy products: The petroleum solvents used require higher amount of energy to manufacture compared with water and emulsifying chemicals used in emulsified asphalts.

Bitumen Emulsions

Emulsified bitumen is simply a suspension of small bitumen cement globules in water assisted by an emulsifying agent such as soap. Emulsifying agents assist by imparting an electrical charge to the surface of bitumen cement globules so they do not coalesce. Bitumen emulsions can be used in almost any application where cutback bitumens are used. The advantage of bitumen emulsions includes the ability to handle asphalt with minimal or no heating and the absence or significant reduction of cutter in the binder. Standard grades of emulsions contain approximately 60 percent bitumen by weight and 40 percent water. Grades containing 70 percent or more bitumen are becoming popular for a sprayed seal operation. Bitumen may contain additives such as petroleum cutters and polymers. Disadvantages are their slow initial curing rates and higher cost. Emulsions effectively reduce asphalt viscosity for use at lower temperatures. Bitumen emulsions are used for various applications. For road building, emulsions are used for tack coats, fog seals, slurry seals, and bituminous surface treatment. Types of bitumen emulsions are shown in Table 7-15. Bitumen emulsion properties are shown in Table 7-16.

TABLE 7-15 Types of Asphalt Emulsions According to Their Electric Charge

Kind	Type	Viscosity grade	Application
Anionic	RS	RS-1, RS-2	Spraying
	MS	MS-2	Mixing and spraying
	SS	SS-1.SS-1h	Mixing and spraying
Cationic	RS-C	RS-2C, RS-3C	Spraying
	MS-C	SM-C	Mixing (sand) and spraying
	SS-C	SS-C,SSCh	Mixing and spraying

RS = Rapid setting; MS = Medium setting; SS = Slow setting; C = Cationic emulsion; h = Lower penetration/harder asphalt cement.

TABLE 7-16 Bitumen Emulsion Properties

Property	Test method	Cationic slow setting	Cationic rapid setting*
Binder content, Wt %	ASTM D-244	60–63	70–72
Saybolt furol viscosity, SFS	ASTM D-244	20–50	>80
Residue, g/cc	Ip-91	0.05–0.25	0.05–0.30
Elastic recovery, 10°C, %	Ductilometer		52

*With polymer.

Anionic Emulsions. Anionic emulsions have a negative charge. Because mineral rock also has a negative charge, there is no electrostatic attraction between the emulsion and mineral rock. For an anionic emulsion to break requires particles to get so close to each other that the repulsion forces are overcome by the gravitational forces. Emulsion breakup is a two-step process of flocculation and coalescence.

Cationic Emulsions. Cationic emulsions have a positive charge and hence facilitate a direct and very rapid reaction between the emulsion and an aggregate or pavement. The size of the charge or zeta potential affects stability. The other form of breakup is evaporation of water, leaving bitumen to flocculate and coalesce.

Generally, all emulsions appear as a viscous brown liquid when initially applied. When bitumen cement starts to adhere to surrounding material such as aggregates, existing surface, subgrade, the color changes from brown to black, and emulsion is said to have broken. As water evaporates, emulsion begins to behave more and more like pure asphalt cement. Once all the water has evaporated, the emulsion is said to have set. The time required to break and set depends on the type of emulsion, the application rate, the temperature of the surface onto which it is applied, and environmental conditions. Under most circumstances, emulsion sets in about 1 to 2 h. Bitumen is not soluble in water and to keep it in colloidal form dispersed in water is not easy. Emulsifiers of both the cationic and anionic type are based on salts of long chain fatty acids found in oils and fats.

Anionic Emulsifiers. These are based on a fatty acid reaction with an alkali such as caustic potash or caustic soda (KOH/NaOH). This salt or soap is an active emulsifier. The nonpolar tail of this molecule is hydrophobic and hence aligns itself to asphalt molecules. The polar end containing K^+ or Na^+ aligns itself with water phase and provides solubility in water. Sodium or potassium ions impart a charge to water, which will be balanced by the opposite charge on the surface of asphalt. The emulsifiers generally used are a mixture of fatty acids, and these impart different level of charge, or "zeta potential."

Cationic Emulsifiers. These are based on acid salts of amines prepared from fatty acids. They may be fatty diamines, fatty quaternary compounds, or reaction products of polyamines and carboxylic acids. The type of emulsifier determines the number of charges it imparts to bitumen surface, or, in other words, these determine the zeta potential. The mineral composition of the aggregates partially dictates whether anionic or cationic emulsion should be used.

Zeta Potential. Zeta potential is the electric potential between the surface of bitumen particles and the bulk solution. Zeta potential is determined by the amount of emulsifier adsorbed on the surface of asphalt. A double layer of ions and counterions exists in solution surrounding each particle of bitumen. The form of the double layer depends on the concentration and ionic density of the emulsifier and pH. A large zeta potential indicates a greater double layer, faster movement, and a greater repulsion between particles and more stable emulsion. The pH affects the way in which bitumen adsorbs emulsifier and is thus critical to the double layer and zeta potential.

Advantages of Emulsions. Emulsions are water based. They have no flash point. They are not flammable or explosive. Drums of emulsion kept in the sun will not expand or burst. As water-based emulsions, they do not pose a health risk to workers or cause burns. Spillage are of no environmental importance. If a spill occur on a waterway, emulsion breaks and settles to the bottom as a harmless organic mass. Grass and other plants can grow through it. A spill on the side of the road actually encourages plant growth by keeping moisture and warmth in the ground. Plants can grow and break through it. An emulsion viscosity is low; hence it can be used even at low temperature where it can wet the aggregate even at –10°C. At this temperature, a cutback bitumen with 20 to 30 percent kerosene will be required. The drawback of a higher cutter content is that the residual bitumen is even softer, which can cause bleeding later on. If bitumen is used alone, a much higher application temperature is required. For hot asphalt mix, asphalt will reach road temperature in about 2.5 min. The rapid increase in asphalt viscosity can prevent stone wetting, stripping, or patch disintegration.

Compared to cutbacks, the curing rate of asphalt emulsions is much faster A lower viscosity emulsion that has good adhesion will not only improve adhesion but will coat stone better than hot asphalt or even a cutback. Generally 5 to 10 percent less asphalt in the form of emulsion is required for the same job. In penetration patching the emulsion will find voids more easily and will not get hung up in the top layers. This can create patches that are not soft and will not bleed. In sealing applications, the emulsion goes further up the sides of the stone, creating a better bond. In residential streets, this can save 10 percent on binder requirements.

Bitumen Emulsion Manufacture. Referring to the process flow diagram in Fig. 7-9, the process described here is for cationic emulsions. For anionic emulsions, different emulsifying chemicals and additives are used.

Heated bitumen from storage at 265°F is pumped by positive displacement metering pump to static mixer where it is mixed with 2 to 2.5 % vol kerosene. Kerosene is pumped from a storage tank via metering pump P-102. Fluxed bitumen flows to the colloid mill or homogenizer, where a hot aqueous solution of emulsifiers, additives such as calcium chloride, latex, and so on, depending on the type of emulsion desired, joins fluxed bitumen. In the high-speed homogenizer, bitumen is sheared to micron-sized particles that remain in suspension. Emulsion is next cooled in plate-type heat exchanger by cooling water.

For preparing an aqueous chemical solution of emulsifiers, water is heated in plate-type heat exchanger by heat exchange with a heat transfer fluid, which in turn is heated in a diesel-fired heater. Water and emulsifier chemicals, acid, and calcium chloride are mixed in static line mixer. Metered amounts of these chemicals flow from their individual storage tanks to the static line mixer via diaphragm-type dozing pumps. Latex, if required for emulsion thickening, is pumped directly to a colloid mill via a dozing pump. Similarly, a small amount of emulsifier (dope) is pumped direct to the bitumen feed pump suction.

Process Variables. The main process variables are particle size, pH, the exit temperature of emulsion, and delta between bitumen and soap temperature. These variables affect the particle size and particle size distribution. Particle size is the most important variable and is determined by the mill's operating parameters, such as rotation speed, gap dimensions in the grinding elements. There is an optimum exit temperature and bitumen temperature/soap temperature delta that must be determined experimentally for every system. Typically emulsion exit temperature may be 185 to 195°F and delta between asphalt temperature and soap temperature around 175 to 190°F.

Bitumen Feed Quality. The base bitumen grade used for emulsion making depends on the ambient conditions where the product is to be used. Generally it is 60/70 penetration for warmer climates with temperature over 100°F to 150/200 penetration for moderately cold climates with temperatures of 32°F or lower.

Bitumen Density. Specific gravity of all bitumen is more than 1, in the range of 1.000 to 1.050. An emulsion based on such bitumen will have a tendency to settle down in water. For this reason, both the density of bitumen and water is adjusted. Bitumen density is reduced by blending it with a small amount of petroleum cutter stock (0 to 2 vol %). Water density is increased by dissolving calcium chloride, which increases density to more than 1.

Cutter. Cutter gives a temporary softening to bitumen. Bitumen will have then improved coating ability to aggregate. The cutter also helps a prime coat emulsion to penetrate a base course. Also the cutter has to be sufficiently volatile not to leave any residue in the bitumen, which can give a lowered viscosity. Kerosene is the most widely used cutter. Approximately 2 to 2.5 vol % kerosene is added to reduce bitumen density and viscosity and increase penetration to the desired range.

Dope. In emulsion manufacture, some surfactant is sometimes added directly to bitumen. This is termed "doping," which facilitates the emulsification of difficult bitumen and improves its stability without affecting its breaking characteristics. The temperature at which bitumen phase is introduced

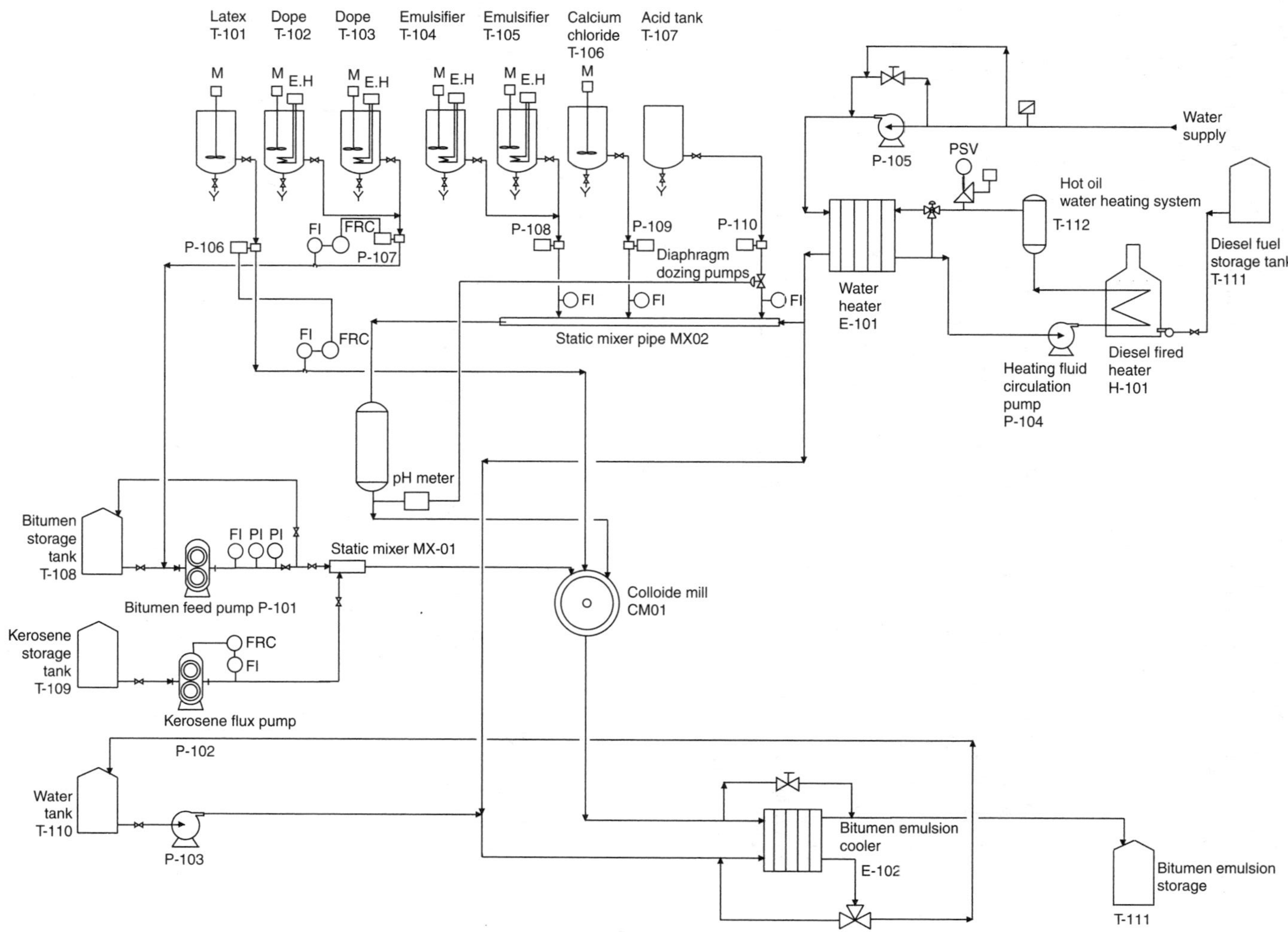

FIGURE 7-9 Bitumen emulsion plant.

in the mixer is sufficient to maintain the bitumen in a liquid phase. A temperature above 210°F is usually necessary. The temperature of the water phase is so adjusted that emulsion should not boil when it leaves the mixer. If necessary, the mixing operation is carried out under pressure. Water phase temperature is maintained below 160°F. Surfactant concentration is generally less than 2 wt % in solution. When the surfactant is cationic, surfactant concentration in the solution is approximately 1 percent by weight. In case of acidic bitumen containing naphthenic acids with emulsifying activity, the addition of emulsifying surfactant may not be required. The addition of an inorganic or organic base to aqueous phase is required.

Mixing. Breakdown in storage of an emulsion is by flocculation and coalescence. For this reason asphalt is processed through a colloid mill to reduce asphalt to a very fine particle size, 0.2 to 2 microns, which increases emulsion stability. Aggregates can be classified by their mineralogical type; when moisture is present, however, all aggregates have a net negative charge. If cationic emulsion is used, the breaking mechanism is by physiochemical reaction with stone in which emulsified particles electrostatically plate out the surface of the aggregate. The greater the charge on the aggregate, the faster the reaction. As the water evaporates, asphalt particles come together by the mechanism of flocculation and coalescence and become asphalt.

Storage. Emulsion storage is done between 160 and 195°F. Freezing or boiling of the emulsion can break the emulsion. Higher temperatures can be used, provided storage pressure is increased to avoid boiling. It is possible to modify certain characteristics of emulsions, such as stability, rate of breaking, and viscosity, by incorporating more than one surfactant in the formulation, provided the surfactants are mutually compatible. Other additives used in emulsions are as follows:

- Stabilizers
- Antifreeze
- Thickeners (natural or synthetic lattices added to aqueous phase)

When there is high content of salt in bitumen, water is pulled into bitumen droplets caused by osmosis. In cationic emulsions, calcium chloride is added to the emulsion to increase the stability and to prevent swelling of bitumen droplets. In anionic emulsion manufacture, trisodium polyphosphate is used for the same purpose.

Clay Emulsions. Clay emulsions are structured differently from other bitumen emulsions. The emulsifying agent is colloidal clay instead of soap or chemical emulsifier. The clays used are bentonite, kaolin, and so on. The clay carries a negative charge and swells in water The application of clay emulsion includes protective coating, adhesives, paper sizing, flooring mastic, and insulation mastic. Clay imparts special properties to bitumen, which allow special uses such as tree grafting, water tank repairing, coating buildings, and so on.

In the manufacture[13] of clay emulsion, the emulsifier solution is replaced with a clay solution. A slurry of colloidal clay such as Bentonite (Bentonite is a colloidal clay with a particle size of 0.2 to 2 microns in diameter) and water is prepared by mixing approximately 5 percent clay in a high-speed mixer. The water and colloidal clay are mixed for 10 to 15 min until the clay is fully hydrated and the clay cannot absorb more water. Mixing is carried out at the ambient temperature and the viscosity of slurry is 100,000 to 180,000 cP at 77°F. The pH of the clay emulsion is maintained between 5 and 9.5. Control of pH during emulsification and later storage is critical to the stability of emulsion and the particle size. Suitable agents are acidic salts such as aluminum phosphate, sodium phosphate, and potassium dichromate.

Bitumen Binder Modifiers

Some bitumen binders require modifications to meet specifications. In recent years, traffic volumes, loads, and tire pressures have increased, which can cause more rutting and cracking. Modifiers can increase bitumen binder stiffness at normal service temperature while decreasing its stiffness at low temperatures to improve resistance to thermal cracking. There are many bitumen binder modifiers available, and the selection is done with the following objectives in mind:

- Lower viscosity at high temperature associated with construction.
- Higher viscosity at high temperature to reduce rutting and shoving.
- Lower stiffness and faster relaxation to reduce thermal cracking.
- Increased adhesion between the asphalt binder and the aggregate in the presence of moisture. This is achieved by an anti-stripping additive (0.5 percent by weight). Hydrated lime is the best anti-stripping agent. Lime interaction at the aggregate surface enhances bitumen aggregate bond and substantially reduces moisture damage. Lime has many other benefits, such as reduced rutting, slower aging rate, and reduced cracking due to fatigue.

Polymer-Modified Bitumen

Polymer-modified bitumens are penetration-grade binders used in roads, airfields, and other high-stress applications such as road intersections with heavy traffic density and busy highways with heavy truck movement. Modified bitumen binders have been used for extreme climatic conditions such as hot deserts or regions with an arctic climate where aging of normal bitumen binders is rapid and pavement distress frequent. There are three main categories of polymers that have been used for modifying bitumen binders. Most widely used for road pavements is thermoplastic elastomers. They confer elasticity to the bitumen binder to allow it to recover past deformations. Examples of this type of polymers are as follows:

- Styrene-butadiene-styrene (SBS)
- Styrene-isoprene-styrene (SIS)
- Styrene-butadiene
- Polybutadiene in the form of crumb rubber from used tires

When mixed with asphalt, SBS is swollen by the oily fraction of the asphalt but maintains its microstructure and confers elastomeric properties to bitumen binder without losing processibility at high temperatures. There is of course alteration in the colloidal equilibrium of asphalt because of the removal of part of the maltenic fraction, but the overall effect is of increased elasticity. The solubility of SBS in bitumen is only partial, which leads to another problem: macroscopic separation from asphalt during long-term high temperature storage. The problem is accentuated as the percentage of polymer increases. Crucial to good balance between solubility and good mechanical properties of polymer-modified asphalts (PMA) is the polymer blending step in bitumen.

Polymer-modified binder typically contains 2 to 8 percent by weight of polymers. Polymer modification can greatly improve the performance properties of the bitumen binder. At usual summer temperatures of pavement of 140°F, under traffic load, bitumen is not able to maintain the original shape of the pavement, leading to a permanent deformation of the pavement known as rutting. In low temperatures, bitumen gets brittle and tends to crack. Both these distresses are not avoidable for neat bitumen, thus resulting in shorter pavement life. Bitumen composition suitable for road application with good low-temperature performance and good high-temperature rutting resistance have been obtained[14] by mixing oxidized bitumen with a penetration index of 0 to 2 with 2 to 3 percent thermoplastic rubber of the polystyrene or butadiene type. However, the addition of polymer can significantly increase the cost of the bitumen binder.

Polymer-Modified Bitumen Manufacture

Referring to the process flow diagram in Fig. 7-10, a metered quantity of bitumen (50 to 70 penetration) from bitumen storage at 300°F is pumped by positive displacement steam-jacketed pump P-101 to

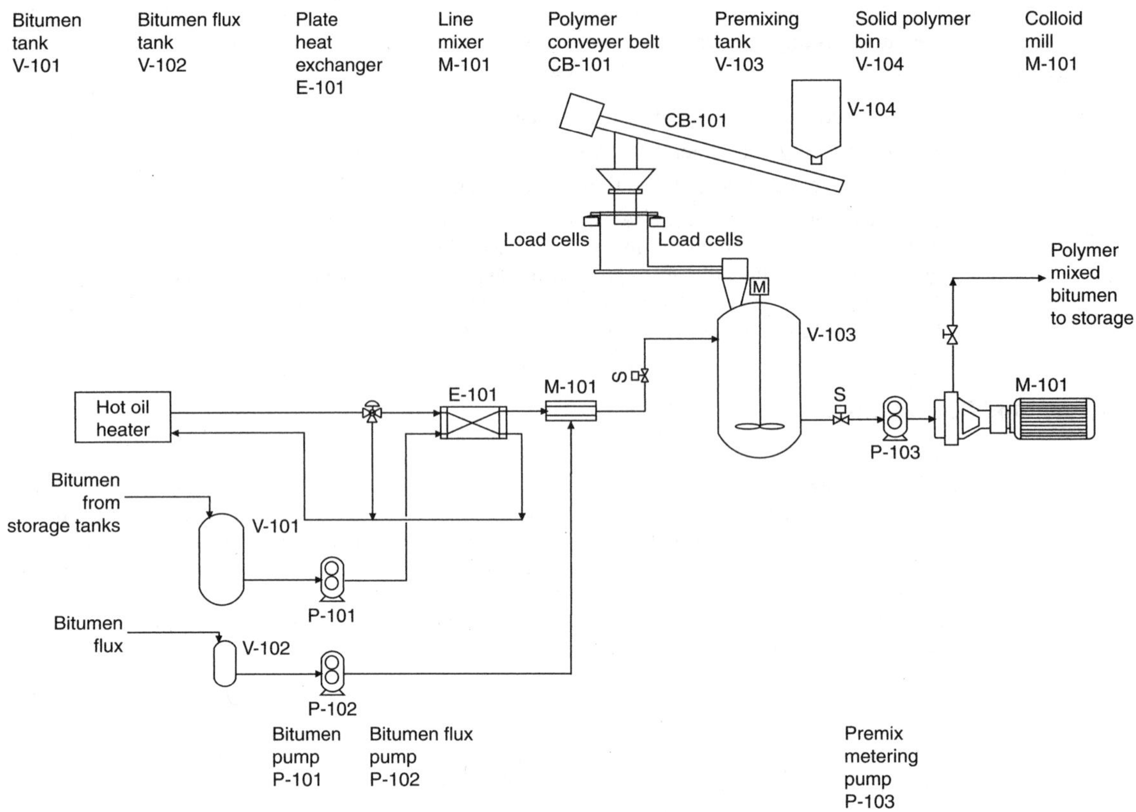

FIGURE 7-10 Polymer bitumen manufacture.

plate-type heat exchanger E-101. In the heat exchanger, bitumen is heated to 365°F by circulation of heat transfer oil. After heating, bitumen is mixed with 2 to 4 vol % of flux oil, normally kerosene or diesel, in a line mixer. Flux oil from storage tank V-102 is pumped to line mixer M-101 via a metering pump P-102. A flow controller controls the flow of flux oil at 2 to 4 percent of bitumen flow. Fluxed bitumen from the line mixer flows to premixing tank V-103 where solid polymer pallets are mixed with fluxed bitumen. Mixing is done with a powerful high-speed mixer. The quantity of the flow of the solid polymer pallets to the mixing tank is controlled at about 8 wt % of bitumen by load cells. Solid polymer pallets stored in polymer bins move to the mixing tank via conveyer belts. The liquid level in the tank is controlled by a level controller. The bitumen-polymer mix from the mixing tank flows to positive displacement pump P-103 suction, which in turn delivers this mix to colloid mill or homogenizer M-101 where mixing and particle-size reduction of bitumen and polymer take place. The effluent from the homogenizer is sent to product storage tanks. Among the various polymers used in bitumen modification, are natural rubbers and crumb rubbers from used tires have been successfully used for many years. These modified binders are known to have good rutting and fatigue resistance. The blending process is similar to one described for other polymers. Rubber is shredded and ground to a fine powder, typically 15 to 30 mesh size, and then mixed with bitumen at a high temperature, 330 to 340°F. For good mixing, bitumen is fluxed with approximately 2 wt % heavy vacuum gas oil. Rubber content is approximately

8 wt % of bitumen. The rubber or crumb rubber binder can also be added to aggregate directly, in an HMA plant, instead of mixing with bitumen.

Oxidized Bitumen

The objective of the bitumen blowing is to increase the softening point, viscosity, and lower penetration of asphalt. The bitumen blowing process is used for the manufacture of industrial bitumen with a high softening point and low penetration. These grades are designated by their softening point in degrees Celsius and penetration at 25°C. For example, grade 95/25 indicates an average softening point of 95°C and a penetration of 25. The physical properties of air blown bitumen depends on feedstock composition and bitumen blowing unit operating conditions. Air blowing increases the amount of asphaltenes and decreases the levels of saturates, aromatics, and resins. The weight percent of asphaltene typically ranges around 7 to 8 percent in straight run vacuum residues. In air-blown product, asphaltenes range around 33 percent by weight.

The bitumen blowing process is used mainly for the production of industrial bitumen grades. Blown or oxidized bitumen is used for the roofing industry and other industrial applications such as paints, cable insulation, protective coatings of pipes, fishnet coating against fungus growth, soundproofing panels manufacture, making printing inks, production of bitumen felts for roofing, waterproofing of buildings.

Until recently, the bitumen blowing process was extensively used by refineries for the production of paving asphalts. However, bitumen specifications for paving are becoming more stringent, requiring road asphalts with better low-temperature performance and a longer pavement life. Blown bitumen is not considered a good blending component of paving bitumen. Air-blown bitumen is chemically less stable because of a combination of the increased asphaltene level and the decreased aromatic content. This decrease in stability leads to faster aging and the deterioration of bitumen blends containing air-blown asphalt component. Road paving asphalts containing air-blown asphalts blend components are known to age many times faster. In one study[15] an asphalt blend containing 39 percent air-blown asphalt and 61 percent flux was found to age 5.4 times faster compared with samples prepared without air-blown blend component. The aging tests on samples were done at 190°F and 1 atm as per ASTM aging test methods.

AIR BLOWING PROCESS

Referring to process flow diagram Fig. 7-11, in the bitumen blowing process, hot bitumen from a vacuum distillation unit at 500°F is introduced into bitumen blowing reactor C-101 from the bottom. Air is filtered and compressed in compressor CP-101. Compressed air at approximately 100 lb/in^2 is introduced into the reactor through a sparger and bubbles through the heated bitumen. Bitumen oxidation reactions take place and reactor temperatures increase. Reactions are exothermic and can go out of control at temperatures greater than 525°F. To control reactor temperature and foaming, steam is introduced into the reactor. Gases produced in the reactor are routed to incinerator F-101 through knockout drum C-104. Pressure in the blowing reactor is controlled to approximately 2 bars by safety valves that open to asphalt relief sumps. The product is drawn from the bottom of the bitumen reactor into blown bitumen receiver C-103, which also controls liquid level in the reactor. Oxidized bitumen product is cooled to 250°F in heat exchanger E-101 by exchanging heat with a diesel stream in a closed loop. Heated diesel is cooled back to its original temperature by air-cooled exchanger E-102. Sometimes small amounts of additives (termed catalysts), such as ferric chloride or phosphorus pentaoxide in 0.1 to 0.5 percent concentration by weight, are added to accelerate oxidation reactions and thus reduce residence time. Many proprietary bitumen blower reactor designs are available for more efficient air-bitumen contacting. "Biturox" technology is one such example. Improved feedstock/air contact due to use of high-performance mixers reduces the amount of air used for oxidation and considerably reduces undesirable atmospheric emissions (Fig. 7-12).

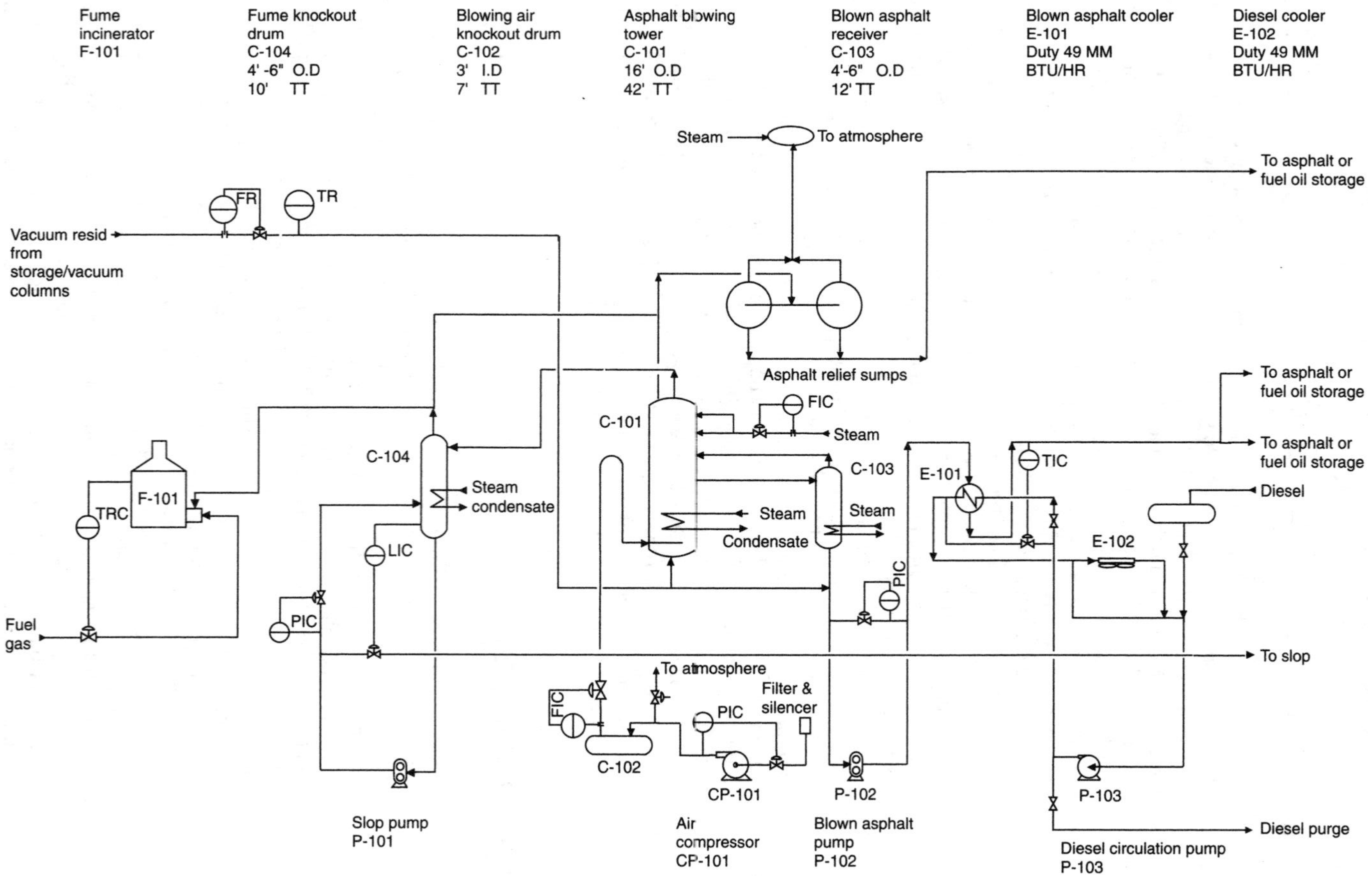

FIGURE 7-11 Bitumen blowing plant.

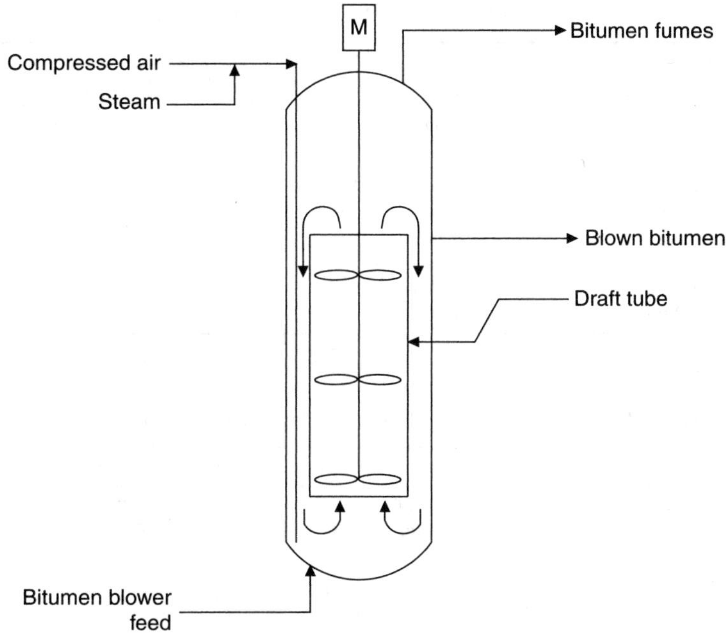

FIGURE 7-12 Bitumen blowing reactor.

Operating Variables

Operating variables for bitumen blowing are as follows:

- Composition of feedstock
- Reactor temperature
- Residence time
- Air-to-feed ratio
- Air feed contacting efficiency
- Use of catalyst

For a given feed, the desired softening point and penetration of the product can be achieved by adjusting these operating variables in laboratory tests. The properties of air-blown asphalts strongly depend on air blowing conditions and feed properties. The time to reach the desired consistency is greatly reduced at higher operating temperatures. Feed and product properties and operating conditions of a refinery bitumen blowing unit are shown in Tables 7-17, 7-18, and 7-19. Operating temperatures are generally kept at 450 to 600°F. It is known that hardening and oxidation rates are higher for bitumen blown at higher temperature.[16] Saturates in bitumen are preferentially converted to asphaltenes on blowing compared with aromatic oils. The contents of asphaltenes increases linearly with the amount of air blown. This is consistent with the proposed mechanism that suggest saturates and nonpolar aromatics are first converted into resins and then resins are converted into asphaltenes.[17]Air blowing increases the molecular weight of asphaltene molecules. Gel permeation chromatography (GPC) can be used to identify air-blown material in asphalt binders.

TABLE 7-17 Bitumen Blowing Unit Feed and Product Properties

Feed	Vacuum residue Ex Kuwait crude						
	Weight fraction	SG	Viscosity, 210°F cSt	Sulfur Wt %	Pour point °F	Penetration 77°F	Softening point, °C
Feed							
Vacuum resid	1.0000	1.016	1500	5.1	120	220	
Heavy vacuum distillate	0.0000	0.936	30	3.5	106		
Total feed	1.0000	1.016					
Product							
Bitumen	0.9799	1.028	6400			60/70	45–56
Losses	0.0201						

TABLE 7-18 Bitumen Blowing Unit (Batch Operation) Feed and Product Properties

Feed	Vacuum residue Ex Kuwait crude						
	Weight fraction	SG	Viscosity, 210°F cSt	Sulfur Wt %	Pour point °F	Penetration 77°F	Softening point, °C
Feed							
Vacuum resid	0.9500	1.016	1500	5.1	120	220	
Heavy vacuum distillate	0.0500	0.936	30	3.5	106		
Total feed	1.0000	1.012	800–1100				
Product							
Bitumen	0.9700	1.028	3000–5000*			10–20	85–95
Losses	0.0300						

*Viscosity at 300°F.

TABLE 7-19 Bitumen Blowing Unit Operating Conditions*

Operating condition	Units	
Feed temperature	°F	390–410
Bed temperature	°F	500–520
Air blowing rate*	SCF/BBL	167
Residence time	H	4.00

*For 60/70 penetration asphalts.

INDUSTRIAL USES OF BITUMEN

Roofing

Next to paving, the most important usage of bitumen is in roofing. There are two basic types of bitumen roofs; a built up roof and a prepared roof.

Built-Up Roofing. These are constructed in place by building up layers of saturated felts or other substrates (fiber glass, polyester, etc.) protected by bitumen. Mineral aggregates or another coating is applied at the top surface. These are generally used for commercial buildings with large area and low slope (less than 1/8 pitch) roofs where roofing must be continuous to ensure water runoff and waterproofing. The material may be factory-made impregnated felt that is glued to the roof with hot bitumen. The roof may be built up of one or several layers of felt and bitumen. The hot system uses saturated felt and mopping bitumen applied hot for adhesion. A coating of bitumen with a cover of mineral particles is applied to the top layer to insulate the roof from extreme ambient temperatures both cold and hot.

Prepared Roof. These are factory-made roofs. They include membranes in rolls and with shingles on the surface. The substrate may be felt, metal, or mineral. Prepared roofs are attached by heating or with cold bitumen emulsions. Polymer-modified bitumen is increasingly being used for roofing.

Bitumen Quality

The properties of bitumen for roofing are shown in Tables 7-20 and 7-21. The main requirements are that it be flexible enough to resist cracking at low ambient temperatures and stiff enough to withstand deformation due to roof movement at the highest service temperature. Blown bitumen grades 90/18 or 85/25 are widely used. Polymer-modified bitumens are now being increasingly used for roofing. Bitumen used as a felt saturant must have a viscosity low enough to soak the felt evenly. It must have a penetration greater than 40. The softening point should be 120 to 160°F. It should not have any volatile material and have a high flash point. Penetration grades 100 to 200 are ideal. Additives such as fungicides, herbicides, and other mildew retardants are added to bitumen.

TABLE 7-20 Properties of Roofing Bitumens

Properties	Units	1	2	3
Roof slope	Min.	0	>1/12	>3/12
	Max.	1/12	<3/12	
Softening point	°C	60–65	75–80	85–950
Penetration	1/10 mm, 100 g 5 s	25–40	20–30	15–25
Flash point	°C	232	232	232
Ductility	Cm	10	3	

TABLE 7-21 Properties of Oxidized Grade Butumens

	Test method	Grade 75/30	Grade 85/25	Grade 85/40	Grade 95/25	Grade 105/35	Grade 115/15
Softening point °C;	BS2000/IP 58						
Min.		70	80	80	90	100	110
Max.		80	90	90	100	110	120
Penetration, 25°C	BS2000/ASTM D86	30	25	40	25	35	15
Loss on heating, 5 h 163°C, Max.	BS2000/IP45	0.2	0.2	0.5	0.2	0.5	0.2
Solubility in trichloroethylene		99.5	99.5	99.5	99.5	99.5	99.5

Bitumen Felts

Substrates used are sheets composed of organic or inorganic fibers. The most common substrates are synthetic, cellulose-type fibers, fiberglass, polyester, and nylon. These are manufactured on machines, similar to those used in paper manufacturing. The substrate passes through a sump bath containing molten soft bitumen and is nest rolled and cooled before winding on to rolls. The felt is saturated to 140 to 225 percent by weight with bitumen. Saturated felt is the basic unit or pile.

Mopping bitumen is used to adhere it to the other piles and the roof surface and to provide the base of the final roof coating.

Pipe Coating

Metal pipes are subject to corrosion, particularly when they are laid buried in the ground or under water. These pipes are protected from corrosion by a protective coating of bitumen. This coating may consist of bitumen mixed with mineral aggregates (crushed rock, gravel, or sand), finely divided mineral fillers (limestone dust, portland cement, or talc or kaolin), and an organic fiber filler (polyester, polyamide, polyfluorocarbon, or polyarylene sulfide). Inclusion of a small weight percent of organic fiber has a strong influence on the cracking properties of the coating. Table 7-22 shows the composition of one such formulation.[18] The preferred bitumen for this application is air-blown bitumen with a penetration of 5 to 25. The coating is applied at a temperature of 350 to 500°F. The thickness of the mastic coating on the pipe depends on pipe diameter, which may vary from 2 to 48 in. Coating thickness may vary accordingly from 0.5 to 40 mm.

A low viscosity primer is first applied that allows good wetting and penetration into rough surface pits. The pipe is next coated with a full-strength coat.

TABLE 7-22 Buried/Underwater Pipe Coating Composition*

Ingredient	Type	Wt %	Properties
Mineral aggregates	Crushed rock/gravel/sand	55–65	30 Mesh size
Mineral fillers	Lime stone dust/portland cement/ talc/kaolin	20–30	100–200 Mesh
Bitumen	Air blown	10–15	Penetration: 5–25
Synthetic organic fiber	Polyester/polyamide/polyfluorocarbon or poly (aryleyene sulfide)	0.1–0.5	Fiber length 15 mm Fiber weight 1–16 Denier

*Mixing and application temperature = 350–500°F.

Paints

Paints based on solvents and industrial bitumen have wide uses. These may contain vegetable drying oil, mineral fillers, pigments, or aluminum paste. Composition of such materials are up to 60 percent grade 90/18 or harder (oxidized) bitumen in a volatile solvent. Aluminum paints for metals may contain approximately 17 to 30 percent bitumen, 37 to 58 percent solvent, and 20 to 28 percent aluminum paste (aluminum powder plus solvent).

Flooring

Bitumen is blended with aggregates, sand, and limestone dust to produce a flooring mastic. These are used in floor making as a water barrier or for filling cracks to prevent water seepage. Mastic can be applied by spreading. The bitumen used is industrial grade with a softening point of 156°F. Table 7-23 presents the composition of a bitumen mastic for floor making.[19]

TABLE 7-23 Bitumen Floor Mastic Composition

Component	Wt %
Bitumen binder	9.5
Aggregates	26.2
Sand	37.1
Lime dust	27.2

Hard Grades

Hard bitumen grades are produced using processes similar to those used for penetration grades, but they have lower penetration values and a higher softening point. These are hard and more brittle and used primarily in the manufacture of bitumen paints and enamels. Typically these grades have a penetration less than 12 and are designated by prefix H and softening point limits in degrees Celsius. Table 7-24 shows the specifications of some harder grades of bitumen used in the industry.

TABLE 7-24 Properties of Hard Grade Butumens

	Test method	Grade 80/90	Grade 100/120
Softening point °C;	BS2000/IP 58		
Min.		80	100
Max.		90	120
Penetration, 25°C	BS2000 PART 49/	6	2
	ASTM D86 IP 49	12	10
Loss on heating, 5 h 163°C, Max.	BS2000 PART 45/IP45	0.2	0.2
Solubility in trichloroethylene	BS 2000 PART 47	99.5	99.5
	IP 47		
% By mass, Min.		99.5	99.5

STORAGE AND HANDLING OF BITUMEN

Bitumen is stored and maintained at elevated temperature during production and transportation to the end user. A number of aspects related to quality maintenance and safety must be kept in mind. When handled properly, bitumen can be reheated or maintained at elevated temperatures without any adverse effect on its properties. However, if bitumen is subjected to excessive heating at elevated temperatures in storage or in application, it can harden and produce poor performance in service. Storage and pumping temperatures of various bitumen types and grades are shown in Table 7-25.

Storage Tank Heating Systems

Bitumen tanks may be heated by one of the following ways:

Direct-Fired Heating. In this method of heating, oil or gas is burned in a burner inside a tube and hot exhaust gases are circulated through a coil immersed in bitumen. Cooled exhaust gases are next vented to the atmosphere through a high stack. The heat input and temperature increase in tank is regulated by controlling the fuel input to the burner. The difficulty of controlling the heat input and heating coil skin temperatures, which may exceed 570°F, make this method suitable only for small bitumen tanks or for mobile bitumen tanks.

Heat Transfer Oils. Bitumen is heated in this method by circulating a heat transfer oil, heated in a gas- or oil-fired furnace. Hot heat transfer fluid is circulated by a pump through heating coils in bitumen and back to the external furnace. The heat input and hot oil temperature can be precisely controlled. There is no danger of heating coil skin temperatures reaching any dangerous levels. Such a system is suitable for any size bitumen tank. Steam can be used in place of hot oil but is not preferred because of the following reasons. To heat the tank to 400°F requires the use of high-pressure steam. In the event of any leak in the steam pipe, the direct contact of water with hot asphalt can cause explosion. Water can also cause corrosion in tanks.

TABLE 7-25 Asphalt Storage and Handling Temperatures*

Bitumen grades	Pumpimg temperature °F Min	Storage and handling temperature °F Max
Penetration		
450	194	374
350	203	374
200	212	374
100	221	392
70	230	392
50	239	392
40	257	392
35	257	392
25	275	428
15	284	428
Hard		
H80/90	320	446
H100/120	374	446
Oxidized		
75/30	302	446
85/25	329	446
85/40	329	446
95/25	347	446
105/35	374	446
115/15	401	446
Polymer modified	239	338
Cutback†		
50 S	149	320
100 S	158	338
200 S	176	356

Note: Normal storage temperature should not exceed 338°F. Typical storage temperatures are between 320 and 338°F for all grades excluding cutbacks.

*Air pollution control (Northern Ireland) Guidance for processes, process guidance note GNB 6/1 Version 1, March 1998.

†STV (standard tar viscometer viscosities), IP-72.

Electrical Coils. The electrical heating of bitumen tanks is the cheapest and simplest way and is used for smaller tanks. The advantage of the system is that heating can be controlled precisely by electric controllers.

Bitumen, other than cutback grades, has a flash point in excess of 482°F, which is well in excess of the maximum recommended handling temperatures. A flammable atmosphere can occur in storage tanks when temperatures exceed the flash point of bitumen. There are two dangers related to this temperature increase. The atmosphere above the product may become flammable or explosive. An accidental contact with a spark from a hot metal surface such as an exposed burner tube flame could trigger an explosion. Even without an external source of ignition, by the mechanism of auto ignition, ignition of bitumen can occur when exposed to air at as low temperatures as 482°F. To prevent such situations, some safety measures are necessary. The heating system is designed taking into consideration the following factors:

- Minimize the risk of overheating. Overheating can cause a loss of volatile matter and the homogeneity of bitumen.
- Minimize the risk of oxidation, which can significantly affect the quality of asphalt.

- Minimize heat losses from the tank by providing adequate insulation. Vertical cone roofed tanks are preferred over horizontal tanks because of their lower surface area from where heat losses occur.

  Overheating is avoided by the following measures:

- A maximum working temperature is specified for each tank.
- Minimum and maximum liquid levels for safe heating are marked on all dipsticks.
- Minimum and maximum levels for safe heating are clearly marked on all bitumen tanks.
- No attempt is made to reheat bitumen in a tank when the bitumen level is below the safe heating level.
- During heating, the tank is vented to the atmosphere. Tank contents are circulated to assist heat transfer and prevent the formation of hot spots.

REFERENCES

1. H. C. Chen and A. G. Bridge, Modern Engineering & Technology Seminar, Taiwan, July 3–21, 1978.
2. U.S. Patent 5173172, "Production of Hard Asphalts by Ultrafiltration of Vacuum Residua," issued August 19, 1991.
3. Department of Army and the Air Force, USA, "Bitumen Pavement Standard Practice," TM 5-822-8/AFM 88-6, Chapter 9.
4. Department of Army and the Air Force, USA, ETL 110-1-139, June 2, 1990.
5. N. W. McLeod, "A Four-Year Survey of Low Temperature Transverse Pavement Cracking on Three Ontario Test Roads," *Proceedings of the Association of Asphalt Paving Technologists*, 41 (1972).
6. "Design Techniques to Minimize Low Temperature Asphalt Pavement Transverse Cracking," Research Report 81-1, Asphalt Institute, December 1981.
7. "Paving Bitumen Specifications," Indian Standards, IS 73-1992.
8. J. P. H. Pfeiffer and P. M. Van Doormal, *Journal of the Institute of Petroleum Technologists* USA 22 (1936): 414; J. P. H. Pfeiffer, *Properties of Asphaltic Bitumen* (Elsevier, 1950), pp. 166–170.
9. *Shell Bitumen Handbook* (Shell Bitumen Company, 1991): 74–75.
10. A. A. Yousefi, "Rubber Modified Bitumen," *Iranian Polymer Journal* 11 (2002).
11. M. G. Lays, *Handbook of Road Technologies* (Gordon & Breach Science Publication, USA 1990).
12. V. B. Guthrie, *Petroleum Product Handbook* (New York: McGraw-Hill, 1960), pp. 13–15.
13. U.S. Patent 6776833, "Emulsion of Bitumen in Colloidal Clay and Water Slurry," issued August 17, 2004.
14. U.S. Patent 5863971, "Preparing Bitumen Composition," issued May 1990.
15. J. R. Hardee, "Physical and Chemical Characteristics of Superpave Binders Containing Air Blown Asphalts," Project MBTC 2049, Henderson State University, Arkadelphia, AR.
16. N. Y. Vassiliev, R. R. Davison, S. A. Williamson, and C. J. Glover, *Industrial and Engineering Chemistry Research* 40 (2001):1773–1780.
17. M. A. Quddus, S. N. Sarwar, and F. Khan, *Fuel* 74 (1995):684–689.
18. U.S. Patent 4052219, "Pipe Cost Composition," issued April 10, 1977.
19. U.S. Patent 4371400, "Mastic Asphalt Composition," issued February 1, 1981.

CHAPTER 8
PETROLEUM COKE

Petroleum coke is a black carbonaceous solid material produced as a by-product of delayed coking or fluid coking units in refineries. There is a large world market for petroleum coke as fuel because of its high calorific value, low ash, and discount pricing relative to coal. This ensures a ready market for petroleum coke in power utility stations and cement plants. The largest single nonfuel use of petroleum coke is in the manufacture of carbon anodes for the aluminum smelting industry which accounts for almost 80 percent of all petroleum coke produced. Other important uses of petroleum coke are in the manufacture of graphite anodes for electric arc furnaces (used in steel plants, phosphoric acid, and calcium carbide manufacture) and in titanium dioxide (TiO_2) manufacture. Each end use requires a different quality of petroleum coke. Both physical and chemical characteristics of petroleum coke determine its suitability for a specific use. Coking is a thermal cracking process used in refineries to maximize residuum conversion to distillates and thus minimize low-value fuel oil production. Petroleum coke is a by-product of coker units. If the coke produced has high sulfur and metals, it is sold as fuel for power generation or cement plants. If the by-product coke is of a low sulfur and low metal content, it can be further upgraded in value by the coke calcining process. Calcining reduces the moisture and volatile hydrocarbon content of raw petroleum coke. Calcined coke is used in the manufacture of anodes for aluminum smelting, for graphite electrodes for electric arc furnaces, and many other end uses.

MANUFACTURING PROCESSES

Raw petroleum coke is a by-product of coking units of refineries. There are two coking processes: the fluid coking process and the delayed coking process. Due to their lower capital cost, most of the coker units built in refineries use the delayed coking process. Also the properties of the coke produced in the fluid coker units are much inferior to that produced in delayed coker units. For these reasons, most of the petroleum coke produced in the refineries is from delayed coker units. Coking is a noncatalytic thermal cracking process based on the concept of carbon rejection. The heaviest hydrogen-deficient portion of feed (asphaltenes, resins, etc.) are rejected as coke, which contain essentially all the feed metals and ash and a substantial portion of feed sulfur and nitrogen. Feed sulfur removal from liquid product ranges from 40 to 75 percent, depending upon feed conradson carbon, and nitrogen removal from liquid product is greater than 50 percent of that of feed.

DELAYED COKING PROCESS

Delayed coking is a thermal cracking process in which feed (vacuum residue or heavy distillates) is rapidly heated in a furnace to approximately 800 to 840°F and then confined in a reaction zone (coke drum) at the proper temperature and pressure until the unvaporized part of the furnace effluent is converted to vapor and coke. Coking reactions are very complex. Three distinct steps take place:

- Partial vaporization and mild cracking (visbreaking) of feed as it passes through the furnace coil
- Cracking of vapor as it passes through the coke drum
- Successive cracking and polymerization of liquid trapped in the drum until it converts into vapor and coke

Products from delayed coking process are gases, gasoline, gas oil, and coke. Coking is an endothermic reaction, with a furnace supplying the necessary heat to complete the coking reactions. Coke produced from the vacuum residue feed is called green coke and used in the manufacture of carbon anodes after calcining. The properties of the coke produced are a function of coker feed properties, which in turn depend on the crude processed in the refinery. For example, if crude processed in the refinery has a high sulfur and high metal content, the vacuum residuum produced will also have high sulfur and metals and the coke produced will have high sulfur and high metals. For producing petroleum coke with low metals and low sulfur content suitable for anode making for aluminum smelters, crude oil processed in the refinery must have a low sulfur and low metal content. The feed to coker units is generally vacuum residue from crude distillation units.

For the production of premium grades of petroleum coke (needle coke) used in making graphite electrodes, the feed to the delayed coker is highly aromatic heavy distillates such as desulfurized decant oil ex refinery fluid catalytic cracking unit (FCCU) or thermal tars produced from naphtha cracker units. Naphtha cracker thermally cracks naphtha or heavier feed in the presence of steam to ethylene, propylene, and C4 olefins. Thermal tars are the unconverted residue, highly reactive, highly aromatic, used as a feed component for needle coke manufacture.

Raw petroleum coke as produced in a refinery coker unit (green coke) has a high volatile matter content (8 to 15 wt % [percent by weight]) and a low density. It is further processed in a coke calciner unit, where it is heated to high temperatures to drive out the volatile matter. The calcining process improves coke density, its strength, and its electric conductivity. Calcined coke is suitable for anode making for the aluminum industry and for other uses such as graphite electrodes manufacture, TiO_2 manufacture, and so on.

Delayed Coking Process Description

Referring to the process flow diagram in Fig. 8-1, vacuum residuum fresh feed enters the coker fractionator V-104 bottom surge zone. Fresh feed is mixed with the recycle condensed in the bottom section of the fractionator and is pumped by heater charge pump P-101 through coker heater H-101. In the heater the charge is rapidly heated to the desired temperature for coke formation. Steam is injected into each of the heater coils to maintain the required minimum velocity and residence time and to suppress coke formation in heater tubes. The vapor liquid mixture leaving the furnace enters coke drums V-101 or V-102 where the trapped liquid is converted to coke and light hydrocarbon vapor. Total vapor leaves the drum to enter the coke fractionator. A minimum of two drums are required. One drum receives furnace effluent while the other drum is being decoked. The coker drum overhead vapor flows to coker fractionator tower V-104 and enters below the shed section. Coke drum effluent vapor is first quenched with fresh feed and then washed with hot gas oil pumpback in the trayed wash section. These operations clean and cool the effluent product vapor and condense a recycle stream at the same time. This recycle stream, together with fresh feed, is pumped from the coker fractionator to the coking furnace. The washed vapor passes to the rectifying section of the tower. A circulating heavy gas oil pumparound stream withdrawn from the pumparound pan is used to remove heat from the tower, condensing the major portion of heavy gas oil and cooling the ascending vapor. The hot pumparound stream of coker gas oil withdrawn from the fractionator is used to reboil the towers either in the vapor recovery plant or for steam generation. The coker gas oil product is air cooled in E-109 to storage temperature. The fractionator overhead vapor is partially condensed in overhead condenser E-101 and E-102 before flowing to fractionator overhead drum V-105. The vapor is separated from liquid in this drum. Vapor flow under pressure controls the suction of the gas compressor in the gas recovery section. The top of the fractionator is refluxed with a part of the condensed liquid in the overhead drum via reflux pump P-103. Sour water is withdrawn from the overhead drum and routed to the sour water treating facility.

Vapor Recovery

Referring to the process flow diagram in Fig. 8-2, coker gas comes from the fractionator overhead drum, after passing through knockout drum V-210, and flows to multistage compressor K-201.

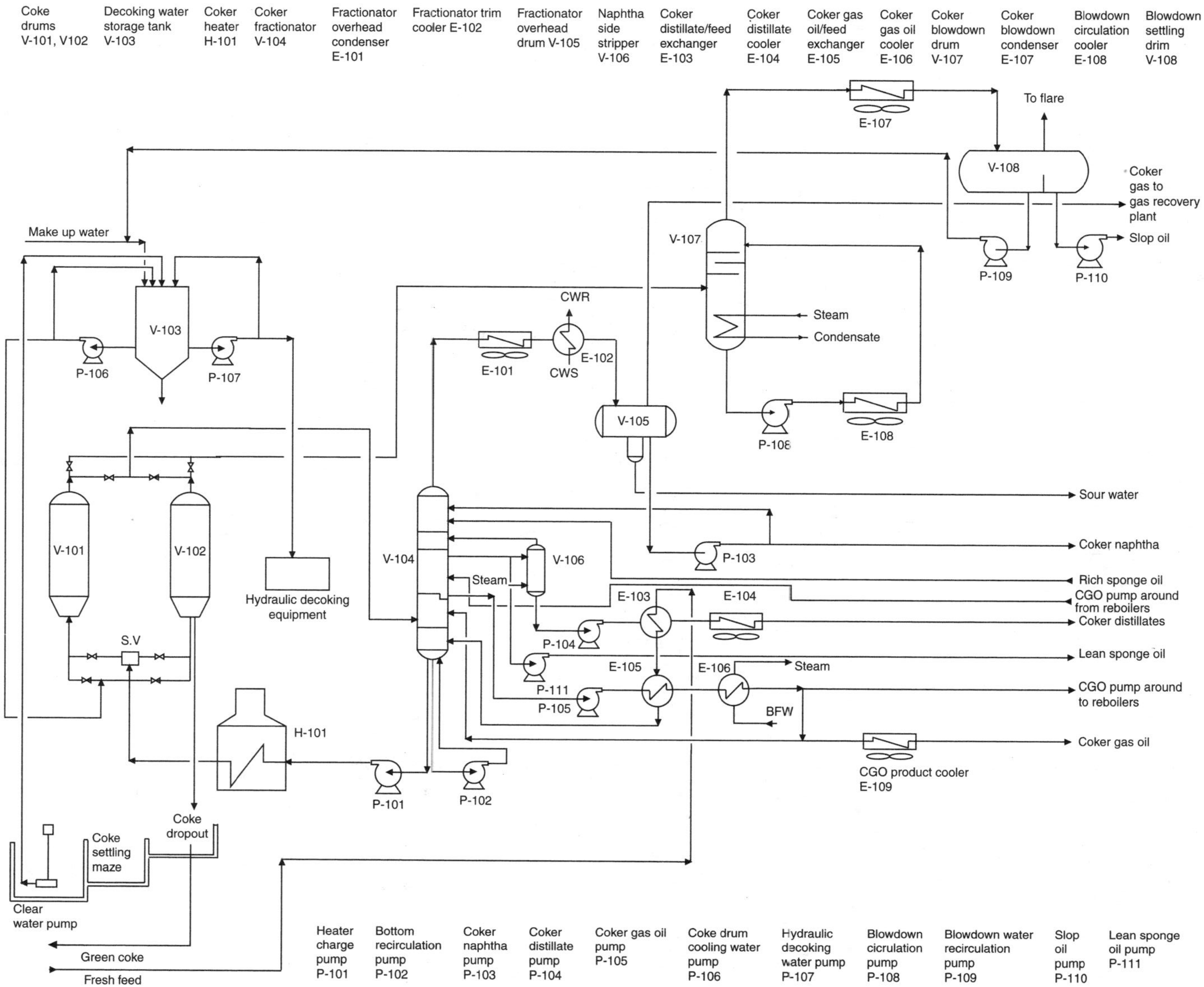

FIGURE 8-1 Delayed coker unit coking section.

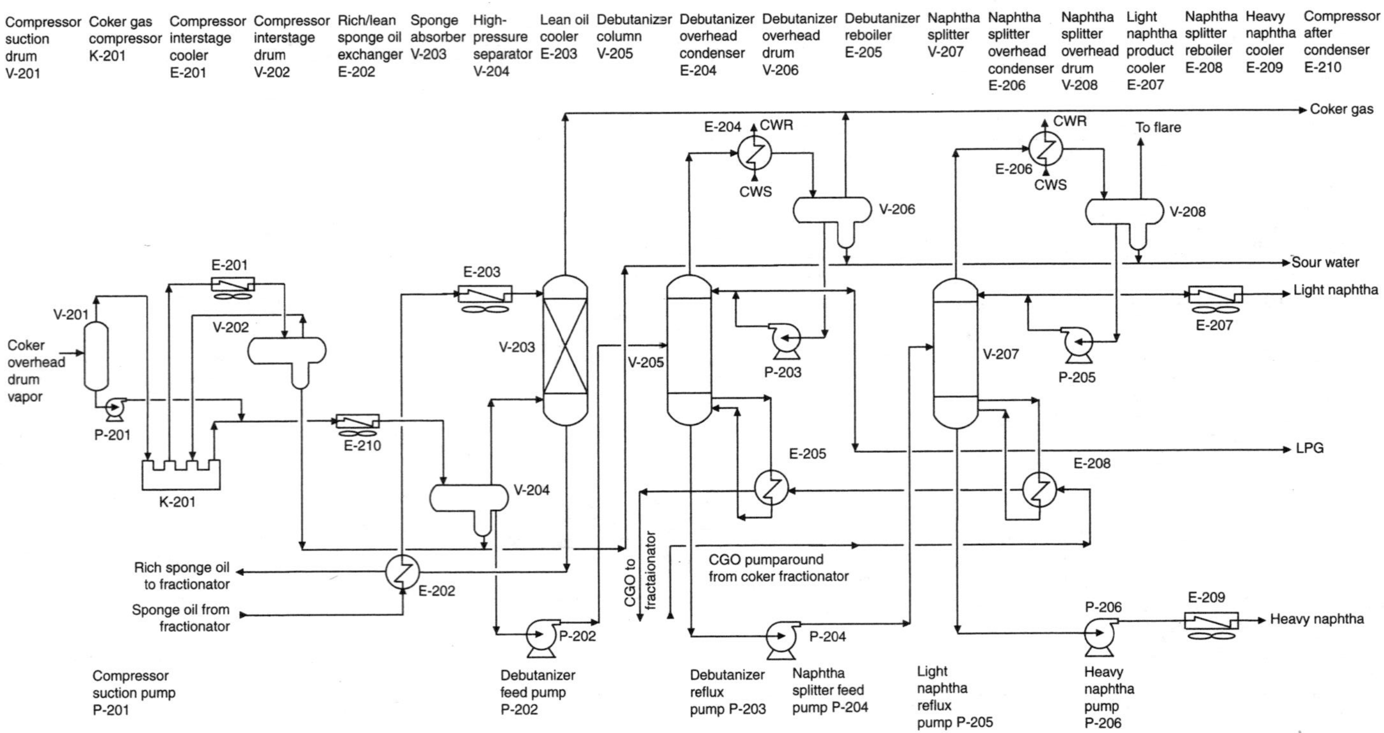

FIGURE 8-2 Delayed coker vapor recovery section.

The compressed gas is cooled and any condensed liquid is separated in high-pressure separator V-204. Vapor from the high-pressure separator enters sponge absorber V-203 near the bottom. In the sponge absorber, a liquid coker distillate stream (sponge oil) withdrawn from the coker fractionator is used to absorb the heavier hydrocarbons from the gas. The rich sponge oil is returned to the fractionator column after heat exchange with the incoming lean sponge oil. Coker gas leaving the sponge absorber goes to a gas processing plant or refinery fuel system. Liquid from high-pressure separator drum V-204 is pumped via pump P-202 to a debutanizer column V-205 where butane and lighter gases are removed as top product while the bottom product flows to naphtha splitter column V-207. In the naphtha splitter column, the feed is split into a light naphtha cut as overhead product and heavy naphtha cut as bottom product.

Coke Drum Decoking

Delayed coking is a batch process, using at least two coke drums. One drum at a time is in operation, receiving hot residuum feed from heater coils, while the other drum is disconnected and is under the decoking cycle.

The decoking operation consists of the following steps:

1. *Steaming*. The full coke drum is steamed out to remove any residual oil liquid. The mixture of steam and hydrocarbons is first sent to the fractionator and next to the coker blow-down system where hydrocarbons (wax tailings) are recovered.
2. *Cooling*. The coke drum is water filled and allowed to cool below 200°F. Steam generated during cooling is condensed in the blow-down system.
3. *Draining*. The cooling water is drained from the drum and recovered for reuse.
4. *Unheading*. The top and bottom heads are removed in preparation for coke removal.
5. *Decoking*. Hydraulic decoking is the most common decoking method. High-pressure water jets are used to cut the coke from the coke drum. Water is separated from the coke fines and reused.
6. *Heading and testing*. After the heads are replaced, the drum is tightened, purged, and pressure tested.
7. *Heating up*. Steam and vapor from hot coke drum are used to heat up the cold coke drum. Condensed water is sent to the blow-down drum. Condensed hydrocarbons are sent to either the coker fractionator or the blow-down drum.
8. *Coking*. The heated coke drum is placed on stream and the cycle is repeated.

OPERATING CONDITIONS

Table 8-1 presents the operating conditions of a delayed coker unit processing vacuum residuum feed. The delayed coker process yields are shown in Table 8-2. Table 8-3 presents the typical properties of feed and the products of a delayed coker unit.

Needle Coke Manufacture

Generally, the feedstock for delayed coker units is vacuum residuum that produces sponge coke. Needle coke is produced in delayed coker units from heavy distillate feedstocks.

There is a great demand in industry for low-sulfur needle coke, which can be processed into graphite electrodes. Graphite electrodes are required to have a low coefficient of thermal expansion (CTE) to withstand severe conditions of thermal shock and the stresses encountered in the operation of electric

TABLE 8-1 Delayed Coker Operating Conditions

Operating parameter	Units	Value
Heater coil outlet temperature	°F	927
Drum outlet temperature	°F	802
Drum pressure	lb/in^2	24.5
Flash zone temperature	°F	748
Unit combined feed ratio (CFR)		1.24
Tower top temperature	°F	208
Accumulator pressure	lb/in^2	10.7

TABLE 8-2 Delayed Coker Process Yields

Stream	Wt fraction
Feed products	1.0000
Fuel gases and H_2S	0.1190
Light naphtha	0.0140
Heavy naphtha	0.0530
Kerosene	0.1720
Diesel	0.3000
Coker gas oil	0.1490
Coke	0.1930

TABLE 8-3 Delayed Coker Feed and Product Properties

Property	Units	Coker feed	Light naphtha	Heavy naphtha	Kerosene	Diesel	Gas oil	Coke
API			80.9	63.9	44.6	29.7	19.9	
Density	g/cc	0.9738	0.6661	0.7242	0.8035	0.8778	0.9346	
Bulk density	Lbs/ft^3							47
Sulfur	Wt %	1.44	0.14		0.21	0.74	1.1	3.2
Flash point	°F							
Freeze point	°F				–54			
Pour point	°F					35.1		
RVP	lb/in^2		12.2					
Conradson carbon	Wt %	12.1						
RON			81.8					
Distillation	ASTM D 1160	°F						
	5%	924						
	10%	973		186	335	550	733	
	20%	1031						
	30%	1071						
	50%			214	390	613	833	
	90%			252	463	699	958	
Nitrogen	Wt %						0.23	
Viscosity	cSt, 122°F						25.6	

steel melting furnaces. The performance of these electrodes depends to a great extent on the characteristics of the needle coke raw material used to fabricate these electrodes. Needle coke for graphite electrodes must have a very low CTE ($1–2 * 10^{-7}$ in/in/°C). Unfortunately, there is a scarcity of high-aromatic, low-sulfur feedstocks required for this premium coke segment. Table 8-4 lists the desirable feedstock properties for needle coke production.

The Bureau of Mines Correlation Index (BMCI) is a function of the volume average boiling point and API gravity (Fig. 8-3). BMCI is an indicator of the aromaticity of the feed. The CTE of the coke produced is a function of the feed BMCI. As the BMCI increases, the CTE decreases, passes through a minimum value (Fig. 8-4), and increases again.[1] Optimum CTE is between 1 to 4×10^{-7} in/in/°C. BMCI is thus a convenient specification for screening feedstocks for needle coke production. Sulfur must be below 0.6 percent by weight and ash less than 0.3 percent by weight. Higher sulfur can cause puffing of electrodes. High ash increases resistivity of electrodes and decreases its mechanical strength. Calcined coke

TABLE 8-4 Needle Coke Feed Characteristics

Aromatics	Wt % Min.	75
Sulfur	Wt % Max.	0.5
Asphaltene	Wt % Max.	1.0
Boiling point	°F	>800
BMCI		95–130

$$BMCI = \frac{87552}{VABP} + \frac{67029}{133.3 + G} - 456.8$$

Or

$$BMCI = \frac{87552}{VABP} + 473.7^{*} SG - 456.8$$

Where:

BMCI = Bureau of Mines Correlation Index
VABP = Volumetric boiling point, °R
G = API gravity
SG = Specific gravity

FIGURE 8-3 Bureau of mines correlation index.

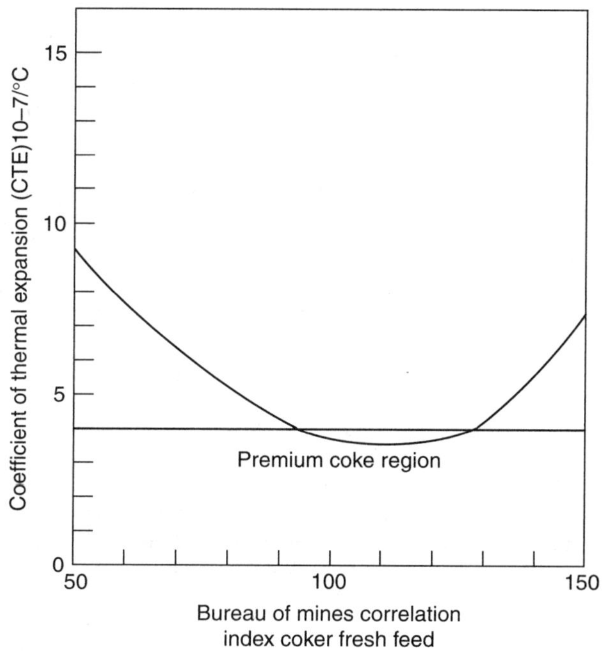

FIGURE 8-4 CTE as a function of BMCI.

with high sulfur or ash is not considered suitable even if other key properties meet the quality standards of needle-grade coke. Thus quality and price of needle-grade coke depends highly on the properties of feedstock used for making needle coke.

Feedstocks

Delayed coker feedstocks for needle coke production are high aromatic, heavy refinery streams such as the following:

- Decant or slurry oil from fluid catalytic cracking process
- Aromatic extract from lube oil units
- Ethylene cracker pyrolysis tars

An important feedstock for needle coke manufacture is decant, or slurry oil, from the FCCU of the refinery. In refinery FCCUs, vacuum gas oil, either as such or after desulfurization, is cracked in the presence of a catalyst to yield mainly gases, gasoline, light cycle gas oil, heavy cycle gas oil, and decant oil. Decant oil is the heaviest component produced as a result of cat cracking. The yield of decant oil is typically 8 to 10 percent by weight of the FCCU unit feed. Properties of decant oil from untreated vacuum gas oil and desulfurized vacuum gas oil (VGO) are shown in Table 8-5. Decant oil also contains a significant amount of suspended catalyst fines, which make any further processing in a catalytic unit (hydrosulfurization, hydrocracking, etc.) difficult. These fines deposit on the downstream processing unit catalyst and deactivate it in a very short time.

TABLE 8-5 Properties of Decant Oils

Property	Units	1	2
Specific gravity		1.0187	0.9339
Distillation	°F		
10%		604	616
30%		685	662
50%		732	708
90%		936	800
Sulfur	Wt %	0.37	1.15
Conradson carbon	Wt %	2.63	
Pour point	°F	8.6	84
Bromine number		6.47	
Viscosity	100°F, cSt	69.42	
	210°F, cSt	5.65	5.6
Heptane insoluble	Wt %	0.01	

Feed Blending

Referring to the process flow diagram in Fig. 8-5, decant oil is split into two cuts, a light cut containing about 85 to 90 percent by weight of feed and a heavy cut containing 10 to 15 percent feed by vacuum flashing or fractionation. All the catalyst fines in the feed are concentrated in the bottom cut. The top cut can now be desulfurized to less than 0.5 percent sulfur. After fractionation, the light cut is desulfurized at moderate operating conditions without significantly saturating aromatic hydrocarbons, polycyclic hydrocarbons that are the most desirable coke precursors. It is essential that the light cut has low conradson carbon (less than 3 percent), by adjusting the cut point. Feed with high conradson carbon requires high hydrogen pressure to prevent catalyst deactivation, which also results in saturating aromatics.

After hydrotreating the light cut, may be blended back with heavy cut to prepare coker feed for needle coke. Desulfurized decant oil may also be blended with steam cracker pyrolysis tar (ex ethylene plant). The typical properties of pyrolysis tar are listed in Table 8-6. Composition is adjusted so as to give an asphaltene content of less than 1 percent and a hydrogen content of less than 8 percent by weight. The percentage of pyrolysis tar is 25 to 50 percent by weight. Such feedstock can produce a coke suitable for graphite electrode making, with a CTE of less than 0.5×10^{-6}/°C.[2] Optionally, light cut may be thermally cracked in a thermal cracker unit. The thermal cracker converts a part of the feed into lighter products and a heavier tar fraction with more aromatic content. This heavy polymeric fraction (40 to 50 percent by weight of feed) has a hydrogen content of approximately 8 percent by weight, and it is an excellent coker feedstock.

Needle Coke Operation

Table 8-7 presents the delayed coker operating conditions for needle coke production. Heavy coker gas oil with an initial boiling point (IBP) of 550°F is recycled back to the coker. Pyrolysis tars as

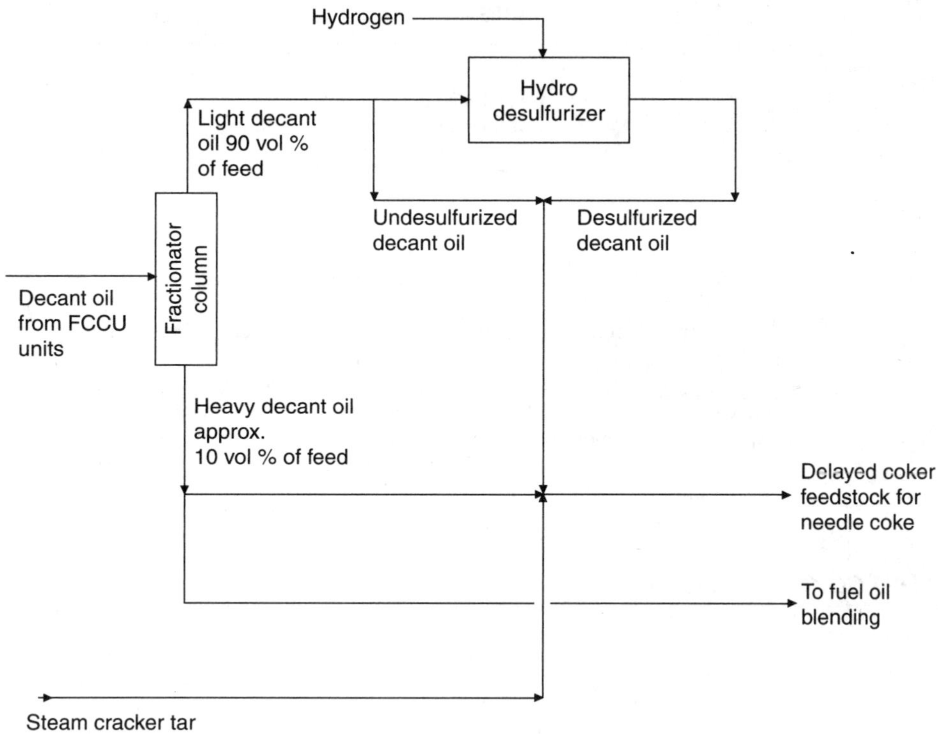

FIGURE 8-5 Needle coke feed blending.

TABLE 8-6 Steam Cracker Tar Properties

Property	Units	1	2	3
Specific gravity		1.1238	1.0835	1.1313
Distillation				
IBP		446	340	311
10%		554	480	327
30%		636.8	550	376
50%		734	700	450
60%		842		514
70%			826	
Recovery	Vol %	68	70	55
Mol wt		319	398	
Carbon	Wt %	91.7	90.97	93.1
Hydrogen	Wt %	6.78	7.62	6.8
Nitrogen	Wt %	0.21		
Sulfur	Wt %	1.2	0.5	0.22
Convadson carbon	Wt %	23.1	16	25.1
Pour point	°F	66	37	
Bromine number		15.44	17.98	12.88
Viscosity	100°F, cSt	4079	1154	14456
	210°F	206	27	80
Heptane insoluble	Wt %	21.1	19.2	22.8

TABLE 8-7 Needle Coke Operating Conditions

Operating condition	Units	
Coke drum temperature	°F	880–950
Coker drum pressure	lb/in^2	20–135
Combined recycle ratio		0.2–0.8

such are unsuitable for the commercial production of premium coke by delayed coking process because the tar converts to coke in the coils of the delayed coker furnace after a very short operating period, requiring frequent coil decoking and an uneconomical run length. Coker feed is generally a blend of desulfurized decant oil and pyrolysis tar.

The drum temperature, pressure, and feed recycle ratio are higher than those of conventional sponge coke operations. A high drum temperature lowers the coke CTE. A furnace fouls more rapidly with higher temperatures. Coke becomes harder and coke removal times increase dramatically with higher temperatures. A 15°F increase in drum temperature decreased the CTE by 30 percent but tripled the coke cutting time.

FLUID COKING PROCESS

Figure 8-6 shows a simplified process flow diagram of a fluid coker unit, which is very similar to the FCCU of the refinery except that coke is circulated between the reactor and the burner instead of catalyst. Unlike delayed coker, fluid coking is a continuous process that uses a fluidized bed reactor to thermally crack petroleum residuum to produce more valuable lighter products such as gas, naphtha, kerosene, diesel, and gas oils. The residue is sprayed into a fluidized bed of hot coke. The use of a fluidized bed permits the coking reactions to be conducted at a higher temperature and a shorter contact time than that with delayed coking. These conditions result in a lower yield of coke. The process uses two vessels: a reactor and a burner. Coke particles are circulated between these two to transfer heat from the burner to the reactor. The heat is generated by burning a part of the coke. The reactor contains a bed of coke particles that is kept fluidized by the introduction of steam. The residuum feed is injected directly into the reactor and distributed uniformly over the surface of coke particles where cracking reactions take place, resulting in the formation of lighter products that vaporize. No preheat is necessary because circulating coke particles supply the heat required for coking reactions and the control of reactor temperature. Vapor products leave the bed and pass through the cyclones, which removes most of the entrained coke. The vapor is then discharged into the bottom of a scrubber where the remaining coke is removed and the product cooled to condense into a heavy tar. The resulting slurry is recycled back to the coking reactor. The overhead product from the scrubber is sent to a fractionator for the separation of gases, naphtha, and light and heavy gas oils. Coke particles in the reactor flow down into a stripping zone at the base where stripping steam removes any product vapor from between the coke particles. The coke then flows down a stand pipe and into a riser, which leads to a burner. Steam is used to induce the upward flow of coke into a burner. The average bed temperature is maintained by regulating air flow as required to burn a part of particulate coke. Operating conditions of a typical fluid coker are shown in Table 8-8. Flue gases from a burner bed pass through cyclones and are discharged to a stack through a control valve to control burner pressure. Hot coke is returned to the reactor through another stand pipe and riser assembly. Because coke is one of the products, it must be withdrawn continuously from the burner bed to maintain the coke inventory in the burner at a constant level. To prevent circulating coke from becoming too coarse, large coke particulate is removed through a quench elutriator and replaced with smaller seed particles. Normal coke circulation from the coke burner to the reactor is 5 to 10 lb per pound of feed. Cooled coke removed from the system is 20 to 200 mesh. Typical yields from a fluid coker unit are listed

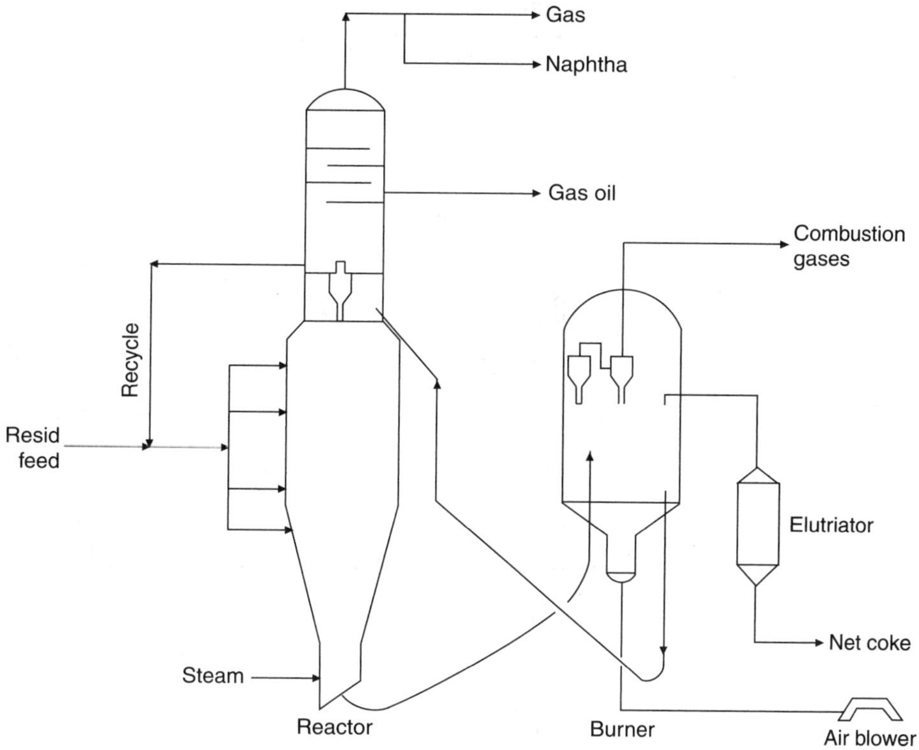

FIGURE 8-6 Fluid coking process.

TABLE 8-8 Fluid Coker Operating Conditions

Operating parameter	Units	Reactor	Burner
Operating temperature	°F	950	1025
Operating pressure	lb/in^2	10.2	10.9
Fluid velocity	ft/s	1–3	2–3
Bed depth	ft	30–50	10–15

in Table 8-9. Coke produced by the fluid coking process finds a limited application in the cathodic protection of pipelines and packing media for anode baking furnaces.

Coke Calcining

Petroleum coke as produced in the delayed coker is a low-valued product that can be used only as fuel for power generation, in the cement industry, and in metallurgical operations. Green coke is processed in another processing unit called the coke calcination unit. In the calcining process, raw petroleum coke is heated to a high temperature in a gas-fired rotary kiln to remove moisture and volatile matter. The calcination process produces coke that has a higher density, higher physical strength, and increased electrical conductivity. Most petroleum coke is calcined in rotary-type kilns. Calcined coke product is used in the aluminum smelting industry for carbon anode manufacture, for graphite electrode manufacture, for TiO_2 production, and also to increase the carbon content of iron and steel products. Green coke specifications suitable for calcining are shown in Table 8-10.

TABLE 8-9 Fluid Coker Unit Yields

Coker Operation; Recycle Cut Point(RCP) = 975°F

Property	Units	
Feed	Heavy Arabian	
Cut point		700°F +
Gravity	° API	11.5
Conradson carbon	Wt %	14.4
Sulfur	Wt %	4.5
Metals		
NI + V	ppm	134
Product yield		
Feed	wt fraction	1.0000
Products	wt fraction	
H_2S		0.0080
C3–		0.0710
C4		0.0150
C5+		0.7300
Gross coke		0.1760
Coke burned		0.067
Net coke production		0.109

TABLE 8-10 Green Coke Specifications Suitable for Calcining

Property		Value
Sulfur	Wt % Max.	3.3
Vanadium	ppmw, Max.	200
Nickel	ppmw, Max.	300
Iron	ppmw, Max.	400
Volatiles	Wt % Max.	11

Process Description

Referring to the process flow diagram in Fig. 8-7, raw petroleum coke (green coke) lumps, 50 to 100 mm in size, are fed from feed hopper V-101 to a weigh feeder and then to rotary kiln R-101. A typical industrial kiln may be 150 to 250 ft long and 5 to 15 ft in diameter. The kiln is lined inside with refractory bricks and made of steel. It is rotated with a train of gears. The kiln is sloped down from the feed side to the discharge side at a slope of 0.50 to 0.75 in/ft. The coke progresses down the kiln countercurrent to the hot combustion gases. As its temperature increases to maximum, the feed moisture is driven off and also volatile matter is stripped from the coke. Devolatilization occurs at 900 to 1800°F. Further down, in the calcining zone (2200 to 2550°F), dehydrogenation and some desulfurization reactions occur and coke density increases. The rotary kiln heating zones are shown in Fig. 8-8. At the discharge end, the rotary kiln has a burner in the firing crown. The burner is used to preheat the refractory before start-up and supply some of the heat for calcining. Most burners are fired by natural gas. The temperature in the kiln is monitored by optical pyrometers. The temperature is controlled by the amount of gas, excess combustion air, kiln rotation speed, and raw coke feed rate. The degree of calcination is estimated from the density and electrical resistivity of the calcined coke. Water vapor originating from the moisture in feed coke, hot combustion gases, unburned gases, tar from volatile matter in feed coke, and coke fines flow out from the feed end of the kiln into "After Burner" R-102 where excess fuel and coke fines are burned with air. Both the incinerator and kiln are operated under slightly negative pressure using an induced draft fan or a tall stack. Hot gases are routed through waste heat boiler H-101 or through a bag filter to trap particulate matter before

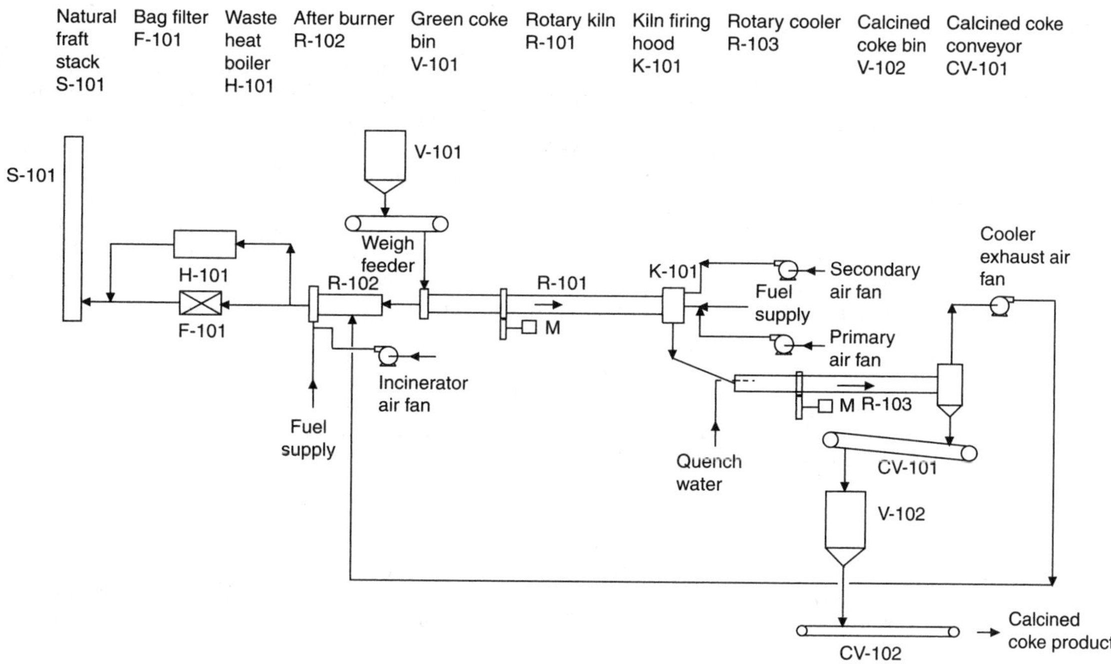

FIGURE 8-7 Coke Calcining unit.

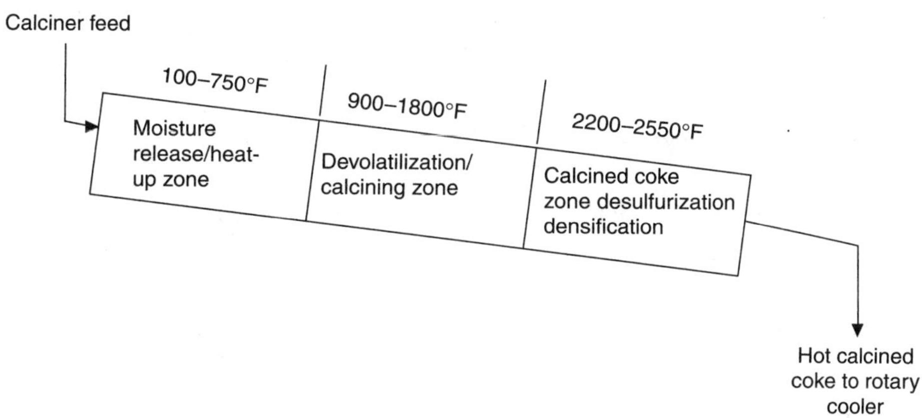

FIGURE 8-8 Rotary kiln heating zones.

discharging to stack. The coke residence time in the kiln is 40 to 60 mins, and it drops off the discharge end into a refractory-lined chute and then into rotary cooler R-103. The cooler is also a cylindrical pipe with a smaller diameter of shorter length and rotated at higher revolutions per minute than the kiln. Water is sprayed into the front end to contact the hot coke, taking advantage of very high heat of vaporization of water for cooling. The coke residence time in the cooler is approximately 20 min after which it is discharged onto a high-temperature conveyer belt or into screw feeders. The water spray rate is adjusted to give a product outlet temperature of 250 to 350°F, in order to keep the calcined product dry. Calcined coke product from the cooler is routed to storage silos through a system of conveyer belts and elevators.

Operating Variables

Coke in the kiln must be heated at a very slow rate during the initial devolatilization process (930 to 1120°F) so that the mesophase or liquid crystal part does not bloat or distort (popcorn) during the evolution of the volatile matter. Petroleum coke with anisotropic (needle structure) and/or with high volatile matter must be calcined with slowed-up heat rates to produce high-density and low-porosity coke. The quality of calcined coke is a function of the quality of the green coke feed to the calciner unit. Coke from different crude oils produces green coke with different physical properties: particle size, density, porosity, structure, volatile matter, and ash content. Feed particle size must be uniform for good control of the residence time of the coke in the kiln. Without proper sizing and feeding of the rotary kiln, slides can occur, dumping most of the material out of the kiln. Table 8-11 lists the properties of the industrial calcined coke samples.

TABLE 8-11 Properties of Commercial Calcined Petroleum Cokes

Property	Units	1	2	3
Moisture	Mass %	0.15	0.15	0.15
Ash	Mass %	0.3–0.5	0.3–0.5	0.3–0.5
Fixed carbon	Mass %, Min.	99.00	99.00	99.00
Sulfur	Mass %	1.3–1.7	1.0–1.3	0.5–0.6
Real density	g/cc	2.03–2.06	2.03–2.06	2.03–2.06
Vibrated bulk density	g/cc (8–14 MESH)	0.84–0.87	0.97–1.00	0.77–0.80
Metals				
Nickel	ppmw	100–200	400–500	250–350
Iron	ppmw	150–250	200–300	150–250
Vanadium	ppmw	50–100	400–500	250–350
Aluminum	ppmw	100–200	200–300	100–200
Calcium	ppmw	200–300	150–250	200–300

PETROLEUM COKE TYPES

Different physical forms of coke are produced in the delayed coker:

- Sponge coke
- Needle coke
- Shot coke

Sponge Coke

Sponge coke is dull black with an amorphous appearance. It is produced from vacuum resids of low to moderate asphaltene concentration. Straight run vacuum residues tends to produce a large percentage of isotropic or amorphous cokes. These are visibly very porous and are called "sponge coke." They may have some percentage of embedded shot coke. A low shot coke percentage is specified for anode grades of sponge coke. A highly desirable quality of petroleum coke to be used in the production of carbon anodes is the high grindability index. Grindability is a relative measure of the ease with which coke may be pulverized. Shot coke in sponge coke samples can cause excessive power consumption and is thus an undesirable impurity. The addition of 0.5 to 10 percent by weight of an oxygen-containing carbonaceous material (sawdust, lignite, low rank coals, etc.) which decompose under coking conditions promote the formation of sponge coke.[3]

Coke produced from a delayed coker is known as green sponge coke. It is mainly composed of carbon but also contains 10 to 15 percent of volatile hydrocarbons together with impurities such as sulfur, vanadium, and nitrogen. If green coke has a sufficiently low level of sulfur and metals, it may be suitable for calcining for making feedstock for aluminum smelter anodes. This higher quality green sponge coke is often described as "anode quality." Petroleum coke with a high sulfur and high metal content that is not suitable for anode making is used as a fuel in various applications and known as "fuel grade." Its use is frequently limited by its high sulfur content, which restricts its use in power generation. A major consumer is the cement industry where impurities present in coke are absorbed in the cement product and not released to the atmosphere. The net price a refiner realizes for fuel-grade petroleum coke is a function of its sulfur content and coke hardness. Coke hardness is measured by the Hardgrove Grindability Index (HGI). Petroleum coke with a HGI lower than 30 is more difficult to market; petroleum cokes with a HGI greater than 45 are relatively easy to market. The HGI of petroleum coke is determined by the coker unit feed quality and operating conditions.

Needle Coke

Thermal cracking of aromatic fractions produces graphite crystals in the form of needlelike ocicular structures. This is called "needle coke." It has a higher density, a very low coefficient of thermal expansion, and high electrical and thermal conductivity.

Needle coke is a premium-grade petroleum coke. Its hardness is due to its dense mass formed with a structure of carbon threads or needles oriented in a single direction. Needle coke is highly crystalline. It can withstand temperatures as high as 5000°F. Compared with calcined petroleum coke used for making aluminum anodes, needle coke has a lower sulfur and ash content and a higher density. Needle coke is mainly used for making large-diameter graphite electrodes used in electric arc furnaces for steel production. Needle coke CTE lies between 0 to 4×10^{-7}; that of sponge coke is 8 to 18×10^{-7}. Shot coke has a CTE of more than 20×10^{-7}. Table 8-12 presents the typical coke specifications for sponge and needle coke for various end uses.

TABLE 8-12 Typical Coke Specifications

		Sponge	Needle
Green coke			
Sulfur	Wt %	<3.0	<1.5
Metals			
V	ppm	<350	
Ni	ppm	<300	
Si	ppm	<150	
Fe	ppm	<270	
Volatile matter	Wt %	<12	<6
Calcined coke			
Moisture	Wt %	<0.5	<0.5
Volatile matter	Wt %	<0.5	
Ash	Wt %	0.5	0.5
Sulfur	Wt %	<3.0	<1.5
Metals			
V	ppm	<350	
Ni	ppm	<300	
Si	ppm	<150	
Fe	ppm	<270	
Density	g/cc	2.04–2.08	>2.12
VBD		>0.8	
CTE	$1/\,°C^* \, 10^{-7}$		<4.0

Shot Coke

Shot coke appears as hard spheres and is produced from high asphaltene and low API feedstock. Heavier coker feedstocks produce shot coke when the delayed coker is operated for maximum liquid products and minimum coke. The shot coke concentration is influenced by the concentration of polar aromatics above a certain molecular weight. Shot coke formation can be suppressed by increasing delayed coker pressure and/or the recycle ratio. Blending aromatic material in feed and/or increasing the recycle ratio tends to reduce shot coke formation. Shot coke is a very hard material. It is preferred in certain petroleum coke uses such as TiO_2 manufacture.

PROPERTIES OF CALCINED COKE

Density

The real density (RD) of raw green coke, 1.3 to 1.4 g/cc, is measured on –200 mesh particles using a helium pyknometer. After calcinations, anode grade coke density increased from 2.05 to 2.08 g/cm^3. Calcined needle coke for graphite electrodes can reach an RD of 2.13 g/cm^3. In low-sulfur coke, RD increases as temperature increases, so this property is used to control the calcining process. For higher sulfur coke, increasing the temperature beyond 2280 to 2370°F actually decreases RD due to desulfurization, which creates small pores in coke.[4,5]

Electrical Resistivity

The electrical resistivity of calcined coke decreases with increasing calcination temperatures. Petroleum coke actually changes from an insulating material to an electrical conductor upon calcination.[6]

Mercury Apparent Density

The Hg (mercury) apparent density (AD) is a measure of the porosity and density of calcined coke. Anisotropic needle-type coke produces a higher AD upon calcining than isotropic sponge coke or shot coke. Calcination above 2280°F decreases AD for all coke types with the exception of low-sulfur cokes. Low-sulfur coke decreases in AD only beyond 2460°F. AD decreases very rapidly in high-sulfur cokes with the onset of desulfurization.[7]

Vibrated Bulk Density

Vibrated bulk density (VBD) of calcined petroleum coke is of great importance for the aluminum industry. VBD is measured on screened particle fraction. About 100 g of screened 20 to 48 mesh calcined coke is vibrated in a graduated cylinder and its volume measured. Calcined coke for most prebaked anodes must have a VBD of 0.84 to 0.86 g/cc or higher. Prebaked anode density correlates with the VBD of calcined coke. Monitoring the quality of green coke is necessary for controlling calcined coke VBD. Low heat-up rates in calciners can improve VBD. The porosity of calcined coke increases and VBD decreases if the calciner heat-up rate is greater than 90°F/min.[8] Volatile matter (VM) of raw coke and the CTE can be correlated with VBD. Coke is calcined, ground to flour, mixed with coal tar pitch, extruded into small rods, baked and graphitized to 5200°F. CTE is measured next. With the CTE as a parameter, VBD can be estimated as a function of VM of raw coke by empirical correlations.

Hardgrove Grindability Index

The HGI (Hardgrove Grindability Index) is a measure of the hardness of coke and can be measured on both raw/green coke and calcined coke. The property is most useful for green coke and is important for fuel-grade cokes that need to be crushed before burning in a power plant. The test uses a sized fraction of 14 to 28 mesh that is placed in a grinding apparatus using two rotating disks with a groove that contains coke and steel balls. Weight is placed on the disks and the apparatus is rotated for 60 revolutions. The coke is then retrieved and the amount of –200 mesh present correlates with an HGI value. The HGI value also provides a rough indication of the volatile matter in the coke. Higher HGI values indicate that the coke is soft and more grindable and also has high volatile matter. HGI is also influenced by coke structure. Raw shot coke has a low HGI of 28 to 50, and raw needle coke has an HGI as high as 70 to 100. Raw coke HGI has been correlated with the VBD of calcined coke. Raw coke HGI of 75 to 85 is required to produce good VBD calcined coke.

Air and Carbon Dioxide Reactivity

Reactivities of calcined coke in air at two different heat-up rates and in carbon dioxide (CO_2) are determined to provide information as to how an anode will behave in a smelting pot. For air reactivity, weighed samples are placed under a purge of air and heated until the ignition of the sample takes place. The ignition temperature is converted to a reactivity value expressed as percent per minute. For CO_2 reactivity, the sample is weighed and put in a purge of CO_2 while the sample is heated automatically to 1832°F (1000°C). After cooling, the sample is weighed to determine the percentage of sample that has reacted. Calcining temperatures and the resulting coke densities (RD, AD, and VBD) and resistivity affect the air and CO_2 reactivity of the calcined coke. Metals such as Na, Ca, and V catalyze both air and CO_2 reactivity, whereas sulfur tends to inhibit the CO_2 reactivity.

Shot Coke Content

Shot coke cannot be used for making anodes in the aluminum smelting industry. Shot coke balls are made of two layers of materials with different CTE values. These small balls (2 to 4 mm in diameter) fracture at the interface of the layers when calcined, due to the difference in their CTEs. Layers of shot particles lead to cracks in the anode because the pitch binder adheres only to the outer layer of particles.[9] A visual inspection is done on anode-grade calcined product samples for shot coke content. Small quantities of shot coke are used after calcinations in nonanode applications such as TiO_2 manufacture.

Screen Sizing

Aluminum smelters require strict specifications on the amount of different size fractions of the calcined coke. To make carbon anodes, calcined coke is first screened in the calciner plant to separate out different size fractions. These are then recombined with pitch and recycled anode butts in the prebaked anode manufacturing plant. Quantities of various size fractions are adjusted to produce optimum anode density. Factors that influence the size of the calcined coke particulates are the size of raw coke feed to calciner and the handling steps required to deliver the product to the anode fabrication plant. Delayed coker coke cutting and handling methods also influence the size of green coke feed to the calciner.

Chemical Properties

Volatile Matter. Volatile matter (VM) is a weight loss on heating of coke. The test is done on a 60-mesh sample. The coke is placed in a covered platinum crucible and heated to 1740°F at a controlled rate in a furnace. The weight loss of the sample as the percentage of feed is the volatile matter of the coke. Typically the VM of green coke is between 8.5 and 12.5 percent by weight. This is reduced to approximately 0.4 percent after calcining at 1650 to 1830°F. The VM of the raw coke is correlated with the VBD. Some VM is burned during calcinations, which accounts for some of the calcining yield loss. In the calcining process, the devolatilization of coke starts at approximately 900°F and is complete at 1650 to 1850°F.

Hydrogen Content. Hydrogen content is determined by combustion in oxygen. This property provides a measure of the calcination of the product. During calcination, most of the hydrogen is evolved before a temperature of 1800°F is reached. Certain end uses of calcined petroleum coke specify low hydrogen content product, for example for use in polycarbonate plastics where low hydrogen in carbon is needed to prevent the formation of hydrochloric acid during the chlorination of carbon monoxide.

Moisture and Ash. Moisture is determined by oven drying. Ash is determined by muffle furnace ashing of the coke sample. Calcined coke must be dry to avoid problems with screening and the fabrication of carbon anodes. The ash content of the calcined coke is in the range of 0.1 to 0.3 percent by weight.

Sulfur and Metals (V, Ni, Si,Fe, Ca, Na). The sulfur content is determined by the X-ray fluorescence (XRF) technique. Metals are determined by inductively coupled plasma (ICP) spectroscopy. In aluminum smelting, any excess metal in the coke migrates to the aluminum metal because coke is consumed during the process. Some grades of aluminum metal require very low values of certain metals depending on the end use of aluminum. Sulfur and some metals also affect the air and CO_2 reactivity of calcined coke. The concentration of metal in coke increases upon calcining due to the loss of volatile matter, sulfur, hydrogen, nitrogen, and moisture. During calcination up to 2460°F, some coke desulfurization, 10 to 15 percent weight loss, may also take place.

Nitrogen. Sulfur and nitrogen in calcined needle coke cause problems of "puffing," or swelling, of the coke in green electrodes during the graphitization process. Sulfur and nitrogen are evolved during calcinations at 2550 to 2900°F.

USES OF PETROLEUM COKE

Carbon Anodes

Aluminum reduction cell in modern smelters use two types of carbon anodes, "prebaked" and "Soderberg." Carbon anodes must be dense, strong, electrically conductive, and of high carbon purity. Anode properties depend on the quality of the calcined petroleum coke and the pitch used for binding anodes. The uniformity of the coke is important to permit suitable anode fabrication. Deficiencies in anode quality can affect cell performance, efficiency, and metal purity.[10]

Prebaked Anodes

Prebaked anodes are widely used in the electrolytic smelting of aluminum as well in arc melting and the holding furnaces of steel mills. The first step in the preparation of both types of anodes includes the preparation of a paste (petroleum coke mixed with pitch binder). The paste preparation includes crushing, grinding, and screening of petroleum coke and clean spent anodes (butts) and blending the coke with pitch binder in a steam-jacketed mixer. The binder used is generally coal tar pitch. Coke and tar pitch are mixed in ratio of 7 to 1-3. The paste is then formed into rectangular blocks using a press. Anodes at this stage are known as green anodes. Typical dimensions of anode for aluminum smelting are approximately 1250 mm in length, 700 mm in width, and 500 mm high. The green anodes are baked in a direct-fired ring furnace at 1800 to 2200°F. A direct-fired ring furnace uses pitch and tar to isolate and seal off green anode blocks from the atmosphere during a 28-day baking cycle. The baking process carbonizes the pitch, removes volatile matter in the pitch, and increases conductivity. The baked anode is then fitted onto a rod that can be aluminum or iron for the electrical connection. Prebaked anodes require a long baking time at a high temperature often with uneven shrinkage.

Soderberg Anodes

The Soderberg anode is formed continuously from a paste of petroleum coke and coal tar pitch added to the top of a rectangular steel casing typically 6 to 8 m long, 2 m wide, and 1 m high. While passing through the casing, the paste bakes, forming carbon to replace the anode being consumed. The baked portion extends into molten electrolyte. The top of a steel box is connected to a hopper in which is fed a paste mixture of milled petroleum coke and pitch that feeds in a downward direction to replace the carbon being consumed by the process. The heat of the molten bath bakes the carbon-pitch mixture, thus forming baked anode. The Soderberg anode is widely used in the manufacture of aluminum and also in submerged arc furnaces for the production of silicon, phosphorus, ferroalloys such as ferrosilicon, and so on. Soderberg anodes, although less popular than prebaked anodes, are nevertheless widely used. These anodes suffer from some disadvantages. The use of Soderberg anodes results in the emission of a large volume of fumes, poor uniformity of conductivity, high resistivity for electrical current, and a low yield of carbon from the pitch component.

The specifications of anodes and the process of its manufacture can substantially affect the performance and profitability of aluminum smelter. Anode contributes significantly to a smelter operating cost. Factors determining net consumption of anodes are shown in Table 8-13.

TABLE 8-13 Factors Determining Net Consumption of Anodes in Aluminum Smelters

Property	Units	
Net consumption	kg carbon /ton Al	400–500
Cell factor		270–310
Current efficiency		0.82–0.95
Bath temperature	°F	1730–1800
CO_2 reactivity residue		75–90
Air permeability	npm	0.5–5.0
Thermal conductivity	$W/(m \cdot k)$	3.0–6.0
Air reactivity residue	%	60–90

Coal Tar Pitch

Coal tar pitch is a mixture of high molecular weight polycyclic aromatic compounds obtained as a residue from the distillation of coal tar. It is used as a binder in the production of anodes for aluminum and for graphite electrode manufacture. Coal tar pitch is distinctly superior to petroleum-derived pitch for this service because coal tar pitch has a higher specific gravity, yields a higher coking value, and has a higher baked carbon density.[11] The quality of pitch is determined by its softening point (SP), toluene insoluble (TI) content, quinoline insoluble content, coking value, beta resins content, ash, density, and viscosity. The binding property of the hard pitch is due to the presence of high molecular weight resins that determine its softening point and coking value. The SP of the pitch should permit easy mixing with petroleum coke. The TI content of pitch indicates its high molecular weight component (more than 1000). The TI content of pitch should be in the range of 25 to 32 percent by weight. Quinoline as a solvent dissolves about 60 to 70 percent of benzene insoluble (BI). The insoluble part is termed quinoline insoluble (QI). The QI content of pitch should be in the range of 10 to 15 percent by weight. QI is a component with a molecular weight more than 5000, and this helps in improving the

TABLE 8-14 Coal Tar Pitch Properties

Property	Units	1	2	3
Softening point	°F	228–235	203–217	230
Benzene insoluble	Wt %	30–37	28–29	
Toluene insoluble	Wt %			27.2
Quinoline insoluble	Wt %	10–12	9–11	13.2
Beta resions	Wt %	>25	20	14.00
Coking value	Wt %	>58	>55	57.10
Ash	Wt %	<0.2	<0.3	0.19
Density	g/cc	1.21		1.33
Sulfur	Wt %			0.61

QUINOLINE

Molecular formula C_9H_7N

Molecular weight 129.16

Density 1.093 g/cc

Boiling point 460°F

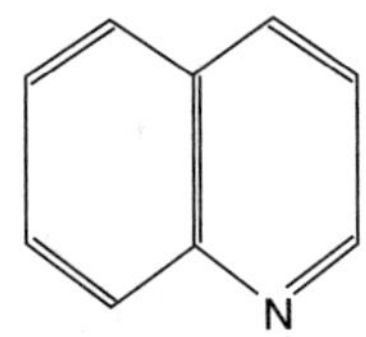

FIGURE 8-9 Structure of quinoline.

mechanical properties of electrodes. The soluble part, the difference between TI and QI, is termed beta resins, having a molecular weight between 1700 and 2100, and this component determines the binding strength of the pitch. Table 8-14 shows the typical properties of tar pitch used in anode making.[12]

Quinoline (C_9H_7N) is naturally found in coal tar. Its molecular weight is 129.16; density, 1.093; and boiling point, 460.4°F. The structure of the quinoline molecule is shown in Fig. 8-9.

ALUMINUM SMELTING

In aluminum smelting, aluminum metal is obtained by the electrolytic reduction of aluminum oxide as per this equation:

$$2\ Al_2O_3 + 3\ C = 4\ Al + 3\ CO_2$$

Electrolysis is done by passing a direct current through a molten electrolyte bath containing dissolved alumina. The electrolyte is primarily cryolite (sodium aluminum fluoride, potassium fluoride, and sodium aluminum tetrafluoride). The electrolyte bath is maintained at 1740 to 1780°F. Carbon is the only material that can withstand and remain inert to the hostile environment of molten electrolyte and aluminum. Carbon anode is consumed by the electrolytic oxidation. Oxygen from the dissolved alumina is released at the carbon anode forming CO_2 gas. At the cathode, elemental aluminum deposits as a pool at the bottom of the reduction cell. Actual carbon consumption is 0.38 to 0.4 tons per ton of aluminum produced. Some of the reactions that lead to excess carbon consumption are as follows:

$$C + O_2 = CO_2$$

$$C + CO_2 = 2\ CO$$

The first reaction is due to air reactivity of the carbon anode with atmosphere at the top (exposed portion of anode). The second reaction shows a CO_2 reaction that can occur at the bottom of the anode. Both these reactions contribute to additional carbon consumption.

As a prebaked anode is consumed, it is individually lowered further in the bath, keeping anode and cathode distance constant. When the anode become too small, it is removed from the pot and sent for cleaning, breaking, and stub cleaning. The cleaned used anode pieces called butts are reused as a coarse carbon component of the green anode mix. The rods/stubs are also cleaned and returned to be rejoined with newly baked anodes ready for service.

TITANIUM DIOXIDE PRODUCTION

Titanium is an important metal used in aerospace. However the most important use of titanium is in the manufacture of TiO_2, which is used extensively in the manufacture of paint, paper, plastic, and inks. TiO_2 is the most widely used white pigment because of its brightness and very high refractive index. It is an effective opacifier in powder form where it is employed as pigment to provide whiteness and opacity to products such as paints, coatings, plastics, papers, inks, foods, and medicines (pills and tablets). In cosmetics and skin care products, TiO_2 is used both as pigments and as a thickener. This pigment is used extensively in plastic and other applications for its ultraviolet (UV)-resistant properties where it acts as a UV blocker and reflector. TiO_2 is used in most sunscreens because of its high refractive index and resistance to discoloration under UV light. Over 90 percent of titanium production is used in the manufacture of TiO_2. It has replaced lead additives in paint due to its better environmental properties and superior "hiding power" compared with those of alternative materials.

Two process are used to produce TiO_2: the sulfate process and the chloride process. The feedstock for the more popular chloride process is a mineral rutile or processed ore containing over 90 percent TiO_2. The ore is mixed with calcined petroleum coke, and the two are reacted in a fluidized bed with chlorine at 1650°F as follows:

$$TiO_2 + 2Cl_2 + 2C = TiCl_4 + 2\,CO$$

$$TiCl_4 + O_2 = TiO_2 + 2Cl_2$$

The reaction yields $TiCl_4$ and chlorides of all impurities present. Carbon is generally provided by calcined coke. Petroleum coke specifications for TiO_2 manufacture are presented in Table 8-15. Although the stoichiometry of the reaction requires only 200 kg coke per ton of TiO_2, the actual coke consumption is approximately 220 kg per ton TiO_2. The reaction is exothermic, and accurate temperature control is essential. The mixed chlorides are cooled and low volatile impurities (e.g., iron, manganese, and chromium chlorides) are separated by condensation and removed from the gas stream with any unreacted solid starting material. $TiCl_4$ vapor is condensed to a liquid followed by fractional distillation to produce an extremely pure colorless liquid with a freezing point of –11.2°F and a boiling point of 276.8°F. Being a vapor phase distillation process ensures that potentially discoloring trace contaminants can virtually be eliminated with subsequent benefit to pigment color.

TABLE 8-15 Petroleum Coke Specifications for Titanium Dioxide (Tio_2) Manufacture

Property		Value
Fixed carbon	Wt %, Min.	99
Sulfur	Wt %, Max.	3.5
Iron	ppmw, Max.	700
Vanadium	ppmw, Max.	200
Volatile matter	Wt %, Max.	0.5
Moisture	Wt %, Max.	0.1
Ash	Wt %, Max.	0.5
Granulometery		
+6 mesh	Max. % Taylor	100
–60 mesh	Max. % Taylor	8

The next processing step is oxidation of $TiCl_4$ to TiO_2. The feed is reacted with oxygen in an exothermic reaction to form TiO_2 and liberate chlorine, which is recycled to a chlorination step. The high reactor temperature ensures that only the rutile crystal form is produced. After cooling, the gas stream passes through a separator to collect pigment particles, which are next treated to remove any adsorbed chlorine gas. The design and operating condition of the reactor are critical to high-quality pigment production.

Calcined coke for this application must be sized (smaller than 5 mm) for use in fluidized chlorinator. The coke is reacted with TiO_2 in a fluidized bed reactor in the presence of chlorine gas. For this application, low-moisture (less than 0.3 percent by weight), low-hydrogen (less than 0.2 percent by weight), hard calcined coke with a minimum amount of fines (less than 0.5 percent on 60 mesh) is desired. Any moisture in the coke tends to form hydrochloric acid, and coke fines have a tendency to be blown out of the fluidized bed. Metals form slag in the chlorinator, and sulfur in the coke should be low: 2.5 to 3.5 percent by weight maximum to minimize pollution. About 1 ton of chlorine is required for 5 to 6 tons of TiO_2 pigment depending on the iron content of the rutile. Chlorine is largely consumed as ferric chloride. Ferric chloride is further converted to ferric hydroxide by reacting with lime. Resultant calcium chloride and iron oxides are safely disposed off. About a third of the chlorine ends up as hydrogen chloride.

STEEL PRODUCTION

Carbon Raiser/Recarburizer: Charge Carbon

Several different types of calcined cokes are used to increase the carbon level of steel. During the purification and melting of steel (scrap iron), oxygen is injected along with lime, which reacts with impurities and forms a slag on the top of molten metal. Oxygen burns the carbon in the charge, producing additional heat in the melt. Initial charge coke added with scrap steel must have high carbon purity. The sulfur content of the coke must be low (1 to 3 percent by weight). As the steel is further refined, lower and lower sulfur content (less than 0.1 percent by weight sulfur) petroleum coke must be used to raise the carbon content of the steel. Sulfur is undesirable in steel because it causes brittleness in the metal.

Injector Carbon

Injector carbon is normally sized low-sulfur petroleum coke. Sizing is necessary (less than 5 mm) to prevent plugging of the injector system. The injector is used to put carbon into melted steel below the foamy slag. The amount of sulfur is critical, and the amount of nitrogen in the coke can cause problems with the heat treatment of cast steel parts. Very high temperatures during coke calcining decrease the nitrogen content of calcined petroleum coke (CPC).

Ladle Additives

Ladle additives require carbon material with very low sulfur (less than 0.1 percent by weight), low nitrogen, and low hydrogen content. Low hydrogen in CPC is desired to prevent the hydrogen embitterment of steel. Graphite scrap or crushed old graphite electrodes are commonly used. Coke made from ethylene tar is very low in sulfur and nitrogen for use as a ladle additive.

Blast Furnace Lining

Low-sulfur CPC is used for making special refractory bricks for blast furnace linings. Bricks are made using CPC with a pitch-type binder and baked in a special pit-type furnace. The coke used must be low in sulfur to prevent sulfur contamination of steel.

Chemical Processing

Sized CPC with a low hydrogen content is used in the gasification process in a Stauffer-type carbon monoxide (CO) generator for making CO. Further synthesis produces phosgene gas ($COCl_2$) used along with bisphenol for production of high-strength polycarbonate plastics. Low hydrogen content CPC is needed to prevent the formation of hydrochloric acid during the chlorination of CO.

GRAPHITE ELECTRODES

Needle coke is mainly used for the production of graphite electrodes for the steel industry. This use accounts for almost 10 percent of the total petroleum coke produced. An electric arc furnace (EAF) is used in smaller scale steel production units using scrap steel as feed. EAFs employed in these mills use heat generated by an electric arc to melt raw material, primarily scrap metal, and reprocess it into steel. Graphite electrodes are a critical input into EAF steel production. Electrode consumption is of the order of 4 lb per ton steel. Graphite electrodes are also used in ladle furnaces for refining steel and for remelting steel in foundries. Graphite electrodes are a large column of virtually pure graphite. EAF steel production requires temperatures virtually as high as 5000°F to melt scrap metal. Heat is generated as electricity passes through graphite electrodes and creates an electric arc. An EAF typically uses three columns of electrodes at one time. Each column typically consists of three electrodes joined together. Graphite electrodes range in diameter from 3 to 32 in and 8 to 9 ft in length. The trend in EAF operations is to design an EAF with a higher electric energy input that requires larger diameter electrodes. The larger the diameter of electrode, the more electricity it is able to conduct. A majority of electrodes sold are 28 in in diameter with a gradual shift toward larger diameters up to 32 in. Smaller ladle furnaces used for keeping steel molten use smaller diameter graphite electrodes. The graphite electrodes must possess a low value for the CTE because of severe thermal shocks that occur in electrical smelting. During the manufacture of these graphite electrodes, the electrodes are heated to 3600 to 5500°F to provide energy to convert coke to crystalline graphite. If the carbon feed contains any sulfur, it will decompose, causing rapid irreversible expansion of the carbon body, thus mechanically weakening the electrode. The phenomenon is called puffing. Thus the carbon used must have a low sulfur content. Another important property that an EAF needs in graphite electrodes is low electrode consumption, less than 4.5 lb per ton steel produced. The electrode is manufactured using sized calcined needle coke with approximately 27 percent coal tar pitch. The paste so formed is extruded so the needles in the coke are in the direction of the extrusion. Green electrodes are baked and impregnated several times with petroleum pitch before lengthwise graphitization in a special electrically heated furnace at 5200°F. Specifications of needle coke for the manufacture of large-diameter graphite electrodes are shown in Table 8-16.

TABLE 8-16 Graphite Electrodes, Typical Petcoke Specifications

Specification	Units	
Ash	Wt %	0.2
Apparent density	g/cm^3	1.6–1.72
Porosity	%	22–28
Transverse strength	lb/in^2	1300–2000
Young modulus	$\times 10^6$ lb/in	0.9–1.4
Electrical resistance	$\times 10^5$ omega in	18–33
CTE	$\times 10^{-6}$ / °F	0.2–1.8

Semigraphite Electrodes

Semigraphite electrodes are produced from calcined sponge coke for specialty electric furnaces such as manufacturing phosphorus and TiO_2 (sulfate process). Low sulfur (less than 1.0 percent) calcined coke is mixed with coal tar pitch, extruded into electrodes, and then partially graphitized.

REFERENCES

1. U.S. Patent 4043898, issued August 23, 1977.
2. U.S. Patent 4740293, "Premium Coke from a Blend of Pyrolysis Tar and Hydrotreated Decant Oil," issued April 26, 1988.
3. U.S. Patent 4096097, "Method of Producing High Quality Sponge Coke or Not to Make Shot Coke," issued June 20, 1978.
4. P. Rahdey, "Structural Changes in Petroleum Coke During Calcinations," *Light Metals* (1967).
5. E. E. Hardin, E. C. L. Beilharz, and P. J. McCoy Lester, "A Comprehensive Review of the Effect of Calcinations at Various Temperatures on Coke Structure and Properties When Calcined at Various Temperatures," *Light Metals* (1993): 501–508.

6. E. E. Hardin and R. E. Gehlbach, "Calcined Petroleum Coke for Aluminum Industry," Great Lake Carbon Corp. information booklet, March 1992.
7. E. E. Hardin, E. C. L. Beilharz, and P. J. McCoy Lester, "A Comprehensive Review of the Effect of Calcinations at Various Temperatures on Coke Structure and Properties When Calcined at Various Temperatures" (Part II), *Light Metals* (1994): 571–581.
8. P. Rhedey, "Structural Changes of Petroleum Coke During Calcinations," *Light Metals* (1967).
9. P. J. Elis and J. D. Bacha, "Shot Coke," *Light Metals* (1996): 477–484.
10. E. E. Hardin and R. E. Gehlbach, "Calcined Petroleum Coke for Aluminum Industry," Great Lake Corporation information booklet, March 1992.
11. E. R. McHenry and Kopper Industries, Inc., "Proceeding of the Fifth Australasian Aluminum Smelter Technology Workshop," Sydney, Australia, October 22–27, 1995.
12. A. J. Chaudry and P. L. Singh Harihar, "Process Development for Production of Hard Pitch from Coal Tar for Electrode Industry," *ISIJ International* 39(1999): 10–14.

CHAPTER 9
CARBON BLACK

Carbon black products are direct descendants of earlier "lamp blacks," a black pigment produced and used by various civilizations to create rock paintings more than 3000 years ago. For a long time, black pigment was obtained by charring organic materials such as wood or bone or collecting soot from burning oils in wick lamps. These early blacks were not very pure and differed greatly in their chemical composition from present-day carbon blacks, which are almost pure carbon. The blacks produced were used as black pigment for making inks, eye cosmetics, and other such uses.

Modern carbon black is virtually pure elemental carbon in the form of colloidal particles that are produced by the incomplete combustion or thermal decomposition of gaseous or liquid hydrocarbons. Its physical appearance is in the form of pellet or black fluffy powder. It is made up of carbon atoms arranged in plane layers of aromatic rings. The layers are very much unlike the crystalline carbon form of graphite where layers are connected in a three-dimensional network. Electrons can easily move along the layers; therefore carbon black conducts electricity. Electron microscopic examination indicates that primary particles are spherical or near spherical in shape with a diameter of 10 to 100 nanometers (nm).

Carbon blacks do not exist as primary particles. Primary particles fuse to form aggregates (50 to 500 nm), which may contain a large number of primary particles. The shape and degree of branching of the aggregates is referred to as the structure. The greater the number of particles, the higher the structure. Carbon black aggregates in turn may form large agglomerates of 5000 nm or larger that are held together by van der Waals forces. The bond between aggregate particles is strong enough to resist mechanical working, whereas the bond between aggregates in agglomerates is relatively weak and is easily destroyed during mixing or pelletization operations.

An estimated 9 million metric tons of carbon blacks are manufactured worldwide at present, and almost 70 percent of it is consumed by the automobile tire industry. Other rubber products such as carpets, belts, hose, shoe soles, and so on, account for 20 percent of carbon black consumption. The remaining 10 percent is used in a large number of applications such as black pigments, printing inks, battery cells, additives in plastics, and cosmetics. Carbon black is used in plastics, paper, pigment, and other diverse uses. An important use of carbon black is as black pigment for coating, for plastics to make them ultraviolet (UV) resistant or conducting, and for printing inks for newspapers and magazines. Carbon blacks are used in the production of cosmetics, such as eye shadow, eyeliner, and mascara and also in lipsticks and nail enamels. Also smaller volumes of more crystalline carbon black, such as acetylene black, are used in dry cell batteries.

The main use of carbon black in the tire and rubber industry is as a reinforcement and filler. The addition of carbon black to rubber improves the wear resistance of tires and other properties. The importance of carbon black in this industry is demonstrated by the fact that before 1920 when no carbon black filler was used, the tread life of a tire was 3000 miles. Today the tire tread containing some 30 percent carbon black lasts 40,000 miles or more. The stability against UV radiation is also enhanced because black pigment hinders light. The content of carbon black is the reason why so many rubber products are black. Carbon black in rubbers and plastics not only makes the substance black, but it also make them UV stable and heat conducting: taking heat away from the tread and thus producing less static. Practically all rubber products where tensile and abrasion wear properties are important use carbon black as a reinforcing material. Rubber such as styrene butadiene rubber (SBR) may be blended with up to 50 percent by weight (wt %) of carbon black to improve its tensile strength and wear resistance.

MANUFACTURING PROCESSES

The production of some form of carbon black for use as a pigment was practiced for more than 2000 years in ancient civilizations. It originally involved burning of oils and fats in open pans in a restricted supply of air and collecting the soot on a cooled surface. The modern-day lamp black manufacturing process is an improved version of that centuries-old practice of producing carbon black by burning oil in a wick lamp and collecting soot. Present-day carbon black manufacturing started in the United States in the 1860s. Carbon black was produced by burning natural gas in a restricted supply of air and collecting the soot. These blacks became the principal pigment in the manufacture of printing ink. By the 1920s, the rubber vulcanization process was discovered and carbon black became the most important rubber reinforcement ingredient. Its demand increased exponentially.

Today, a variety of processes are used to produce carbon black. Most processes involve partial combustion in a restricted supply of air or thermal decomposition of hydrocarbons such as oil or natural gas. The characteristics of carbon blacks produced vary, depending more on manufacturing process employed and less on feedstock. Therefore carbon blacks are classified by their manufacturing methods. Major processes used to manufacture carbon blacks are as follows:

- Channel process
- Gas black process
- Thermal black process
- Acetylene black process
- Lamp black process
- Furnace black process

Furnace black process is the most important one for the production of carbon black and allows production of nearly all types of carbon blacks required for rubber, plastics, paint, and pigment manufacture. Lamp and gas blacks are important alternatives to furnace blacks. These processes yield carbon blacks having properties that partially overlap with those obtainable from furnace black process (Table 9-1). The acetylene black process produces high-conductivity carbon blacks specifically for the dry cell industry.

TABLE 9-1 Pigment Black Properties by Various Processes

Property	Units	Furnace blacks	Gas blacks	Lamp blacks
Particle size	nm	10–80	10–30	110–120
Jetness	M_y (DIN 55979)	210–270	230–300	200–220
Tinting strength	(IRB3 = 100), %	60–130	90–130	25–35
Oil absorption	g/100 g	200–500	220–1100	250–400
Volatile matter	Wt %	0.5–6.0	4–24	1–2.5
DBP absorption	cc/100 g	40–200		100–120
pH value		6–10	4–6	6–9

DIN = German standards.
IRB = Industry reference black.
M_y = Jetness is a measure of blackness, expressed as a value termed "M_y" value.

CHANNEL BLACK PROCESS

During the first half of the 20th century, carbon blacks were produced predominantly in the United States and exported worldwide because the feedstock, natural gas was cheap and available in sufficient quantities. The channel process for carbon black manufacture uses natural gas as raw material. A large numbers of small flames fed by natural gas from ceramic burners impinge on the underside of a water-cooled steel channel, depositing a layer of carbon black that is periodically removed by scrapping devices. These iron channels, which moved slowly back and forth over the flames, have since lent their name to this manufacturing process. The deposited carbon black is scrapped off into a funnel-shaped trough and transferred by screw conveyers to storage silos. Not all carbon is deposited on iron channels, and because no filters were used at that time, a large part of carbon black was released into the atmosphere together with waste gases. The yield of carbon black was only 5 percent. The channel process produced carbon black with a particle size of 10 to 30 nm. It was possible to vary carbon black particle size by altering the distance of the burner tip from the channel and changing the natural gas and air flow rates. The channel process produced several grades of fine carbon blacks that had maximum color and were fully reinforcing in rubber. The channel process was completely discontinued in 1970s as more economical raw materials and more efficient and less polluting manufacturing methods evolved.

GAS BLACK PROCESS

The gas black process was developed in Europe in 1935. The properties of this black were similar to that of U.S. channel black and quickly gained popularity as the reinforcing component of tires. The gas black process uses vaporized oil instead of natural gas as a feedstock. Coal tar distillates are the preferred feedstock. These oils are heated in a vaporizer, and the resultant vapors are carried by a hydrogen-rich gas to the burner. The hydrogen-rich gas is called producer gas ($CO + H_2$). The flame is allowed to impinge on water-cooled rollers. Most of the carbon black formed is deposited on rollers, and the remainder is collected in filters. Both the streams are combined and processed further into pellets for export. Like the channel process, the gas black process is based on the principle of thermal oxidative decomposition and operates as an open system where carbon black is formed in a diffusion flame. Air has free access to the system, resulting in surface oxidation and thus producing acidic blacks. The process of charging carrier gas with vaporized oil provides a means of controlling particle size or specific surface area. Carbon black particle size is in the range of 10 to 30 nm.

THERMAL BLACK PROCESS

In this process, natural gas is thermally decomposed into carbon and hydrogen at elevated temperatures of 2500 to 3000°F in a cyclic operation. The reaction is conducted in a checker brick furnace. In the make cycle, natural gas feed is passed through the heated refractory brickwork in a furnace. Because the natural gas decomposition reaction is endothermic, the furnace cools off. Natural gas decomposes into carbon and hydrogen. The furnace is reheated by combusting natural gas and hydrogen evolved by the decomposition reaction, and the make cycle is repeated. The cycle is repeated every 5 to 8 min. Reactor effluent is quenched by a water spray and carbon black is separated from hydrogen in cyclones and bag filters. Hydrogen gas is recycled to process fuel for heating the furnace.

Because the thermal black process is carried out at comparatively lower temperature and long residence times, thermal black differs from other carbon blacks; carbon black particles grow slowly to very large sizes: 200 to 500 nm. The typical N_2 surface area is 6 to 15 m^2/g; oil absorption is 30 to 50 cc/100 g. Percent volatile content is low: 0.1 to 0.5 percent by weight. Thermal blacks are used for mechanical rubber goods with high filler content or for specialty applications in which their unique compounding properties cannot be matched by carbon blacks produced by other processes. Thus medium thermal (MT) black is often used at high loading (200 parts per 100 part rubber) in the manufacture of wire insulation,[1] matting, and mechanical goods.

ACETYLENE BLACK PROCESS

Acetylene black is used for producing dry cells, mainly to improve the depolarizing properties of manganese dioxide. Acetylene black is also used as an additive to antistatic and electrically conductive rubbers and plastic materials. Acetylene black is produced by the continuous thermal decomposition of acetylene gas in the absence of oxygen. The reactor is heated initially by burning acetylene to a temperature of 1112°F. The air flow is stopped when the required reactor temperature is reached while acetylene flow continues. Acetylene rapidly decomposes into carbon and hydrogen. The heat of reaction is dissipated by cooling coils through which cooling water is circulated. The reaction continues until the flow of feedstock is disrupted. The carbon black formed is in the form of a black-colored aerosol. It enters a settling chamber and is periodically removed. Hydrogen is vented to the atmosphere. Due to reaction conditions and unique feedstock properties, acetylene black differs from other carbon black grades. Homogeneous hydrocarbon feedstock, long residence time, and high reactor temperatures yield a very pure carbon black that exhibits a greater degree of crystallization than the blacks obtained by other processes. The shape of particles is not spherical and it is difficult to densify or pelletize acetylene blacks. Table 9-2 shows the properties of acetylene black.

TABLE 9-2 Typical Properties of Acetylene Blacks

Property	Units	Value
Average particle size	nm	35
Nitrogen surface area	m^2/g	69
Bulk density*	g/mL	0.26
Acetone absorption	mL/5 g	38
Electric resistivity	ohm-cm	0.19

*For 50% compressed sample.

LAMP BLACK PROCESS

Lamp black has been in use since prehistoric times and is probably the oldest pigment known to humankind. For use as a watercolor, carbon black was mixed with glue and prepared in sticks and sold as India ink. Currently, lamp black is used as black pigment in printing inks, crayons, shoe polishes, carbon paper, ceramics, and cements. Lamp black is a deep black pigment consisting of amorphous carbon in a fine state of division. It is obtained by the imperfect combustion of highly carbonaceous substances such as resins, resinous wood, fatty oils and fats, paraffin oils, and coal tar oils are burned with an insufficient supply of air. A considerable part of carbon may be deposited in the form of soot. This soot, however, is not pure carbon but retains variable proportions of tarry products of imperfect combustion. These tarry products impart to lamp black a more or less pronounced warm brownish hue except in those cases in which it has been prepared by processes especially designed to remove the tarry products.

Today, lamp blacks are prepared by the imperfect combustion of coal tar distillates. Generally, coal tar residues such as naphthalene and anthracene oils are used to enrich water gas ($CO + H_2$), and the mixture is then burned from a slotted pipe. The smoky flame impinges on cold rollers, and the black is removed by mechanical scrapers.

There are a variety of lamp blacks. The typical properties of lamp black are shown in Table 9-1. They are generally of large particle size (110 to 120 nm) and display a high degree of structure. Lamp black possesses little reinforcing ability in rubber. Selection of pigment blacks for industrial applications is mostly done on the basis of color criteria, with the primary particle size playing the most significant role. The finer the particles of pigment black, the deeper the color. The better the stabilization of pigment to binder, the more bluish the undertone appears.

FURNACE BLACK PROCESS

Most carbon black production throughout the world is now from furnace black. The furnace black process allows for the production of nearly all the types of carbon blacks required for rubber, plastics, paints, inks, and so on.

Furnace black process uses a heavy aromatic oil as feedstock. The primary feedstock is introduced into a hot gas stream generated by burning a secondary feedstock such as natural gas or oil. Primary feed vaporizes and then pyrolyses in the vapor phase to form carbon black and gases.

REACTOR

The furnace process reactor (Fig. 9-1) is a refractory-lined tubular vessel. A liquid or gaseous fuel such as natural gas or liquid fuel is reacted with an oxidant (air) in the combustion zone to form hot combustion gases. After the combustion zone, the reactor has a zone of converging diameter and a feedstock injection zone. Here carbon black feedstock is injected at one or more points into the path of the hot combustion gases. The hot combustion products and feedstock pass downstream in the reactor to a primary reaction zone. Following the feedstock introduction, feedstock is mixed, atomized, and vaporized into the combustion gas stream. The mixture of combustion gases and vaporized feedstock than enters the reaction zone where feedstocks undergo exothermic pyrolysis reactions. The temperature may reach 2400 to 3200°F, and feed oil undergoes complex reactions forming carbon, hydrogen, carbon monoxide, carbon dioxide, and trace amounts of other hydrocarbons. The residence time of carbon formed in the reactor until it reaches quench point is very short, in the order of 15 to 20 ms.[2] Carbon is in the form of carbon black particles and aggregates. The reaction zone may be considered made up of two zones, the primary reaction zone and a secondary reaction zone. The primary reaction zone extends from the feed introduction zone to approximately 3.8 times the reaction zone diameter. The remaining length of the reaction zone extending up to the quench introduction is the secondary reaction zone. Carbon black is formed in the primary reaction zone. In the secondary reaction zone, a controlled amount of air is injected into the reactor that raises the temperature of the effluent stream by approximately 200 to 400°F.

Secondary air accelerates the rate of surface-modifying reactions in the secondary reaction zone without interfering with reactions that normally occur in the primary reaction zone. Pyrolysis reactions are next stopped by the introduction of a water quench. Cooled gases leave the reactor for heat recovery and carbon black separation.

Three variables are decisive for the operation of the carbon black reactor:

- Carbon black oil feed rate
- Fuel gas rate
- Combustion air rate

These variables are not really independent. The volume of combustion air determines the volume of exhaust gas produced and thus fixes the plant throughput. The fuel gas required for energy production is mostly employed in such volumes that the volume of oxygen introduced with the combustion air is present in deficiency. For a given fuel gas and combustion air rate, the carbon black

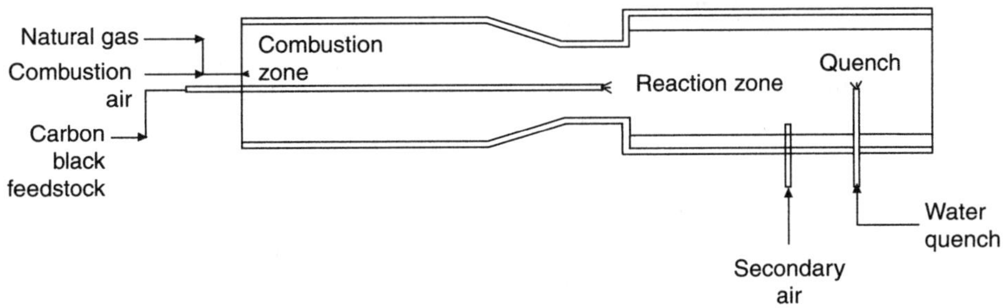

FIGURE 9-1 Carbon black furnace reactor.

feed rate is also fixed. Increasing the carbon black oil feed rate leads to coarser carbon black particles (larger primary particle size) with a higher yield. On the contrary, lower carbon black oil feed rates lead to correspondingly finer carbon black (smaller primary particle size) with a lower yield. The structure is controlled[3] by recycling a part of tail gas or by the introduction of a potassium chloride additive, 4 to 8 g/kg oil feed.

Process Description

Referring to the process flow diagram in Fig. 9-2, controlled quantities of process air and natural gas or other hydrocarbon fuel (secondary feedstock) are introduced into refractory-lined reactor V-101 in the combustion zone. Combustion air is supplied by a centrifugal blower and is preheated from 900 to 1200°F before introduction into the reactor. Carbon black feedstock is pumped from storage tanks T-101/102 through preheater E-102 where it is heated to 260°F and enters the reactor's feed injection zone. Carbon black and tail gas are formed in the primary reaction zone. Optionally more air may be introduced downstream of the primary reaction zone to modify the surface properties of the carbon black. The pyrolysis reactions are stopped by an injection of quenching water in the quench zone of the reactor. Effluent from the reactor is essentially carbon-laden flue gases containing some combustibles. Tail gas may consist of almost 50 percent water vapor, 35 percent nitrogen, and the rest is made up of CO, CO_2, and H_2, with trace amounts of hydrocarbons. The imperfect combustion of carbon black feedstock converts approximately 50 percent of the feed into carbon and another 50 percent into tail gas. These quenched gases are at a temperature of 1300 to 1500°F. The reactor effluent exchanges heat first with combustion air and next with carbon black feedstock in an oil preheater. Air heater E-101 and oil preheater E-102 are heat exchangers through which combustion air or oil flow countercurrent to tail gas. The higher combustion air temperature results in a better yield of carbon black. Tail gas after heat exchange in air preheater and oil pass through cyclones and silicon-coated bag filter F-101 where almost 100 percent of the particulate matter is removed. Inlet temperature of the flue to bag filter is controlled at 535°F. Clean gases containing combustibles from the bag filter are routed to carbon black pallet dryer V-103. Carbon black is recovered from the bag filter by reverse flow of the flue gases, collects in hoppers, and is conveyed to pellitizer K-102 through micro pulverizer K-101. Pelletization is done by mixing carbon black with a near equal quantity of water and a small quantity of pellet binder additives. Indirectly fired rotary drum dryer V-103 is used to dry the wet pellets. Water vapor from the dryer, drawn by an exhaust blower, exits to the atmosphere. The dry carbon black is lifted by a bucket elevator and passed through a screen to separate lump from the product, and a magnetic separator separates any accidental contamination of the product before discharging it into a product storage tank by product screw conveyers. Lumps separated from the product are returned for reprocessing.

Feedstock

For better yield, carbon black manufacture requires feedstock with a high content of aromatic hydrocarbons. Aromatics containing a number of condensed rings are most suitable because the carbon to hydrogen mass ratio increases with the number of rings. In classifying carbon black feedstock, density is an important parameter because density increases with increasing aromaticity. Also the Bureau of Mines Correlation Index (BMCI) is used for more detailed classification. Feedstock for furnace process are carbon black oils such as the following:

- Decant oil (ex FCCU)
- Aromatic extracts from lubricating oil manufacturing units
- Pyrolysis tar ex naphtha cracking units for ethylene plants
- Coal tar and creosote oil ex coal tar distillation units

The properties of carbon black feedstocks are shown in Table 9-3. These high aromatic feedstocks may be mixed to obtain feed with optimum feed density and BMCI.

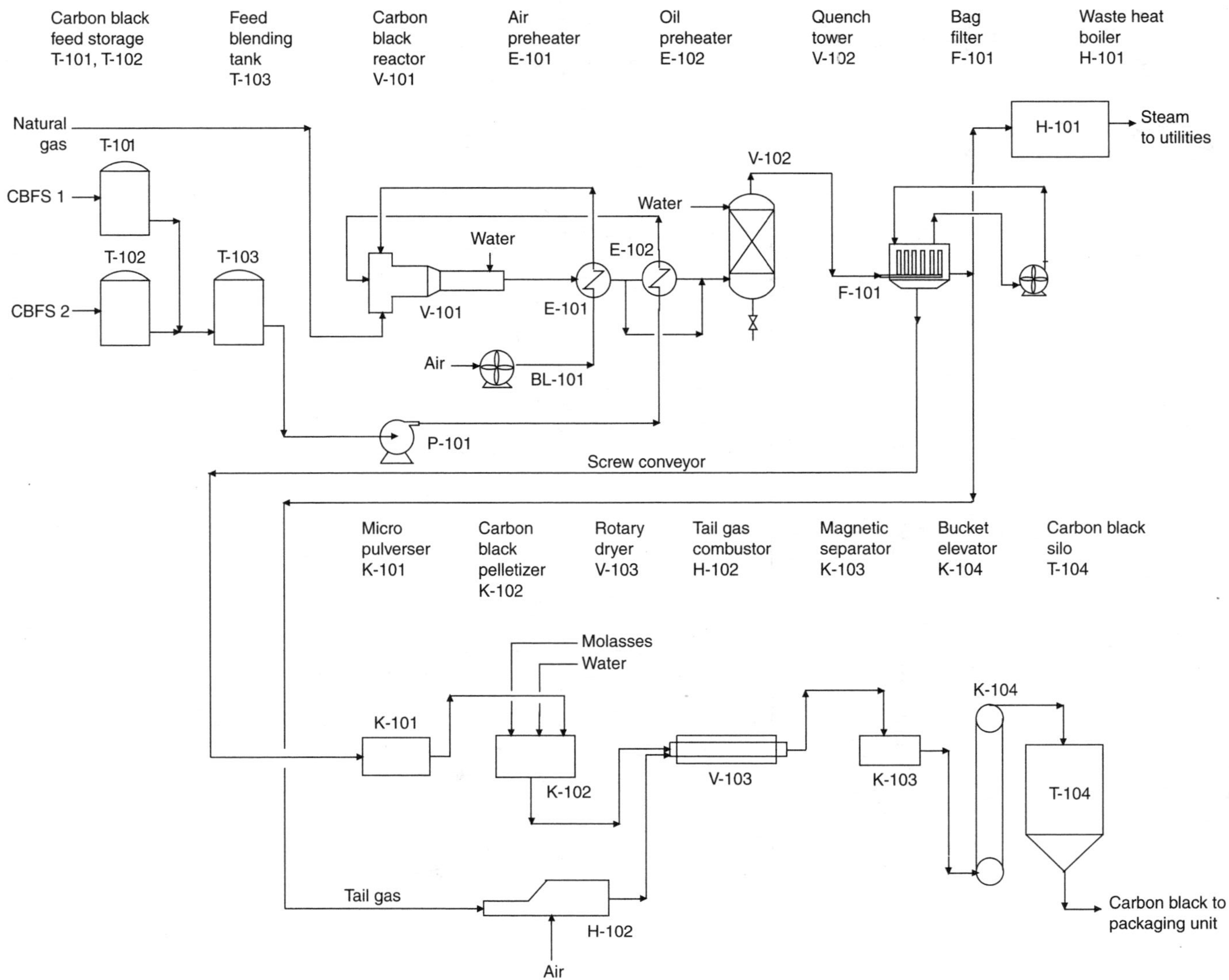

FIGURE 9-2 Carbon black manufacture furnace process.

TABLE 9-3 Carbon Black Feedstocks

Property	Units	FCCU decant oil	Ethylene plant tar naphtha cracking	Ethylene plant tar go cracking	Extract solvent refining decant oil	Coal tar
		1	2	3	4	5
Gravity	°API	–0.2	4.1	–3.8	8.7	–7.8
Specific gravity		1.0777	1.0435	1.1081	1.0093	1.1439
Distillation						
	5%	610	517	450	577	480
	50%	763	665	818	703	704
	90%	961			809	804
BMCI		127	114	135	97	160
Viscosity	SUS 210°F	46	60	130		35
Ash	Wt %	0.012	0.002	0.007	0.002	0.007
Toluene insoluble	Wt %	0.024	0.007	0.18	0.02	0.22
Pentane insoluble	Wt %	1.2	15.1	24.2	0.38	1.7
Sulfur	Wt %	1.9	0.05	3.1	3	0.6
Na	ppmw	1.3	1.1	14	<0.1	7.4
K	ppmw	0.3	<0.1	0.2	<0.1	0.3

SUS = Saybolt Universal Second.

Tail Gas

The composition of tail gas from the furnace black process is shown in Table 9-4. Although acetylene and higher hydrocarbons make up to less than 1 percent of tail gas by volume, they account for nearly 10 percent of the calorific value of tail gas. The relative amount of hydrogen and carbon monoxide can vary with the feedstock used; the calorific value of tail gas is in the range of 50 to 65 Btu/standard cubic feet. The tail gas contains almost 50 percent by volume (vol %) of water vapor. Tail gas is dried by cooling from 535 to 110°F with an air cooler and separating the condensed water. Water removal increases the calorific value of the tail gas to 90 to 110 Btu/standard cubic feet, which can be used in special designed burners to generate power or used as a fuel in the carbon black manufacturing unit.

TABLE 9-4 Carbon Black Tail Gas Composition

Property	Units	Raw tail gas	Tail gas after drying
Composition			
Hydrogen	Vol %	6.5–12	6.5–12
Carbon monoxide		6.5–12	6.5–12
Nitrogen		30–62	30–62
Water		30–45	5–10
Acetylene/C2+		<1.0	<1.0
Calorific value	Btu/SCF	50–65	90–110

Carbon Black Quality

The furnace process is a very versatile and can produce carbon blacks for all grades of tire and rubber products and also pigment-grade carbon blacks by varying the operating conditions of the process. The carbon black structure may be reduced in the presence of alkali metal ions in reaction

TABLE 9-5 Types of Carbon Black Used in Tires

ASTM designation	Name	Abbreviation	Particle size, nm	Tensile strength, mPa	Relative laboratory abrasion	Relative roadwear abrasion
N110	Super abrasion furnace	SAF	20–25	25.2	1.35	1.25
N220	Intermediate SAF	ISAF	24–33	23.1	1.25	1.15
N330	High-abrasion furnace	HAF	28–36	22.4	1.00	1.00
N300	Easy processing channel	EPC	30–35	21.7	0.8	0.9
N550	Fast extruding furnace	FEF	39–55	18.2	0.64	0.72
N683	High modulus furnace	HMF	49–73	16.1	0.56	0.66
N770	Semireinforcing furnace	SRF	70–96	14.7	0.48	0.6
N880	Fine thermal	FT	180–200	12.6	0.22	
N990	Medium thermal	MT	250–350	9.8	0.18	

zones. Other additives such as alkaline earth metal compounds increase the specific surface area. Because the furnace process typically operates in an oxygen-deficit closed system, volatile matter content is low: typically 0.5 to 6 percent and with a high pH of 6 to 10.

The furnace process can produce carbon blacks grades from N-100 to N-700 with different particle sizes and physical properties (Table 9-5). Carbon blacks for compounding into tread rubber typically have a nitrogen surface area of 120 to 165 m^2/g and a dibutyl phthalate absorption (DBPA) value of at least 120 mL/100 g. For pigment-grade carbon blacks, the operating conditions are adjusted to give a finer carbon black product with a t area greater than 400 m^2/g and a DBPA value of 80 to 120 mL/100 g. The "t" area, also known as the statistical thickness surface area (STSA), is the external area of carbon black and is equal to the nitrogen surface area minus the porosity.

OXIDIZED CARBON BLACKS

The objective of oxidation, also called "aftertreatment," is to increase the volatile content of carbon blacks for various end uses. In the case of rubber processing, oxidation of carbon black results in a considerable reduction in modulus and a significant increase in scorch time if carbon black has been previously oxidized. The specifications of carbon black for use in rubber industry contain two properties that are significantly affected by addition of the carbon black to rubber: 300 percent modulus and scorch time.

Modulus 300 percent is the tensile stress necessary to elongate a specimen to 300 percent of its original length. Scorch time is a measure of the curing time of rubber. Generally, a low scorch time is related to fast-curing rubber and a high scorch time is related to slow-curing rubber. During carbon black manufacture, modulus and scorch time are carefully monitored. In case rubber has too high a modulus or too low a scorch time, compared with specifications, carbon black is considered off specification for use in the rubber industry.

To decrease modulus and/or to increase scorch time, carbon black is treated with dilute nitric acid, 0.1 to 1.0 percent. A decrease in modulus is a function of the acid strength and treatment time. For a given treatment time (0.5 to 1.0 h), the modulus decreases as the acid strength increases. The scorch time rises as a percentage of the original increase as nitric acid strength is increased. After treatment with nitric acid, carbon black pellets are dried at 400 to 550°F to obtain dry carbon black pellets. A significant oxidation of carbon black is achieved.[4]

The volatile content of carbon blacks is an important specification for the use of carbon blacks in pigments. Carbon blacks with acid surface groups are particularly suitable for lacquer binders. The basic requirements for any color black are high jetness, easy dispersion in an appropriate vehicle, and good viscosity stability in the resultant formulation. The properties of black responsible for these characteristics are its degree of jetness on a nigrometer scale, the particle size, the volatile content, and to some extent the physical nature of the surface of the particles and the chemical nature of the volatile matter adsorbed therein.

The surface groups of carbon blacks interact with the molecules of the binder, which enables them to disperse in the binder of the lacquer and affect the color of the lacquer. The starting feed is high- or medium-color carbon blacks produced by a channel or gas black process under an oxidizing atmosphere. Acidic oxides may be generated on the surface of furnace black or increased on the surface of channel/gas blacks whenever these carbon blacks are subjected to an oxidative aftertreatment. The oxidation treatment may be carried out as nitric acid in a liquid phase or in a gaseous phase with air or a NO_2 and air mixture. Carbon black is oxidized in a fluidized bed with a mixture of NO_2 and air at a maximum temperature of 570°F, which must not be exceeded. The degree of oxidation depends on the residence time and concentration of reactants. For use as pigment, process conditions are adjusted so that approximately 15 percent volatiles are obtained. It is possible to produce carbon blacks with volatile content more than 15 percent, but this is not required. The reactions are exothermic, and feed rates of carbon and oxidants are adjusted to control the temperature below 570°F. Table 9-6 shows the properties of carbon black before and after the oxidation treatment.[5] After oxidation, carbon black is treated with hot air at a maximum of 535°F to remove any nitric oxide adsorbed on the surface of the carbon black.

TABLE 9-6 Oxidation of Carbon Black

Property	Units	Feed	Oxidized carbon black 1*	Oxidized carbon black 2†
			1	2
Mean primary particle size	nm	17	17	17
N_2 Surface area	m^2/g	215	370	340
Nigrometer index		71	70	71
Oil absorption	g/100 g	680	480	500
Volatile matter	Wt %	5.8	17.00	14
pH value		4.0	2.5	3.2

*Oxidized with a mixture of 9.5 vol % nitric oxide and air at 437°F; low space velocity.
†Oxidized with a mixture of 10.0 vol % nitric oxide and air at 446°F; high space velocity.

CARBON BLACK PROPERTIES

The electron microscope image of carbon blacks shows they have a complicated structure with some spherical particles fused together. The size of the spherical particles is termed "particle size," and the size of the particle chain is called "structure." Various functional groups such as hydroxyl and carboxyl groups are found on the surface of the carbon black, and their composition and amount affects many properties of carbon black. The most important properties of carbon blacks are as follows:

- Particle size
- Structure
- Surface activity

These properties have a large effect on some important physical properties of carbon blacks for use in rubber, pigments, and plastics.

Particle Size

The typical carbon black particle size ranges from 8 nm for furnace blacks to 300 nm for thermal blacks. In rubber processing, finer particles increase reinforcement and abrasion resistance and improve tensile strength. Surface area is used by the industry to define carbon black fineness. The larger the surface area (m^2/g), the finer the carbon black and the more difficult it is to mix in rubber and other media. In pigment and ink manufacture, the diameter of the spherical particles is the key property that directly affects blackness and the dispersibility of carbon black in resins or other vehicles. In general, the smaller the particle size, the higher the blackness of carbon black. Dispersion, however, becomes difficult due to the increase in interaggregate attractive forces.

Structures

During the manufacture of carbon black, primary carbon particles fuse to form aggregates (Fig. 9-3). The shape and degree of aggregate branching is referred to as structure. A high-structure carbon black is one in which aggregates are composed of many particles clustered together with considerable branching and chaining. Increasing the structure typically increases the modulus, hardness, electrical conductivity, and compound viscosity. In pigment manufacture, structure also affects blackness and the dispersibility of carbon black. Generally, an increase in structure size improves dispersibility but lowers blackness. Carbon black with large structure shows excellent electric conductivity. DBPA absorption is a measure of structure, with a high number representing a higher structure.

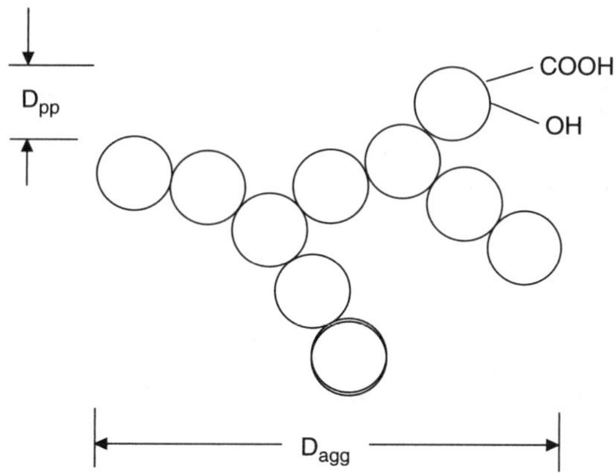

D_{agg} = Aggregate diameter

D_{pp} = Particle diameter

D_{agg} = 50–400 nm

D_{pp} = 10–75 nm

FIGURE 9-3 Carbon black structure.

Surface Activity

All carbon blacks have chemisorbed oxygen complexes (carboxylic, quinonic, lactonic, or phenolic groups) on their surfaces to varying degrees depending on the condition of manufacture. These surface groups are called volatile content and expressed as weight loss in percent after heating a dried sample of carbon black to 1742°F. Some carbon blacks have their volatile content increased by oxidation or "aftertreatment" by various oxidizing agents such as nitric acid, nitric oxide, air, and so on.

In rubber processing, such groups can have a significant effect on the vulcanization process and on the carbon black reinforcement of rubbers. Surface activity can affect abrasion resistance, tensile strength, hysteresis, and modulus. In pigment manufacture, a higher volatile content of carbon black increases dispersibility and dispersion stability. The affinity of carbon black with ink, paint, or varnish depends on the type and amount of functional groups. Carbon blacks with the large number of hydroxyl groups given with oxidation treatment greatly enhance their affinity for printing inks or varnishes.

Porosity

Porosity is the fundamental property of carbon black that can be controlled during the manufacturing process. Increasing the porosity reduces the density of the aggregate. This allows increased carbon black loading, which in turn leads to increased modulus and electrical conductivity for a fixed loading.

The porosity of carbon black is the difference between the nitrogen surface area and the electron microscope surface area. The electron microscope surface area is calculated from the observed average electron microscope particle diameter. The difference is due to the pores on the surface of the carbon black.

SECONDARY PROPERTIES

Density

Carbon black as it is formed is a powdery material with a very low density. It has poor conveying properties and is dusty, which makes handling very difficult. Densification reduces the extent of these drawbacks but degrades the dispersibility. The densification process involves the removal of occluded air, either dry or in a water slurry. Sometimes a small amount of binder is added during pelletization to improve pellet integrity for optimum bulk handling.

Color

Carbon black has always been the most widely used black pigment because of its coloring ability, cost effectiveness, and performance. Carbon blacks are available in a wide range of jetness levels to meet the coloring requirements of all types of plastics. Simple coloring requires only enough carbon black to achieve opacity; usually in the range of 0.5 to 3.5 percent loading. Grade selection depends largely on the degree of jetness required by the end product. Carbon black appears black because of its natural ability to absorb energy from all wavelengths of light. Smaller particle blacks are better absorbers of light than the larger particle blacks because of the greater surface area available for light absorption. Therefore, smaller particle blacks have a higher jet appearance. In general, to obtain the maximum light absorption needed for optimal jetness and tint strength, a carbon black with a higher surface area is required. In addition to jetness, there are other important appearance considerations for plastics that can be controlled by the selection of the appropriate carbon black grade such as undertone, tint strength, and gloss. Undertone is a subtle secondary color underlying jetness. Undertone exhibited by carbon black ranges from warm brown to cool blue. Tint strength is the relative ability of carbon black to darken a colored resin. Gloss level can range from a very shiny to a matte finish.

Conductivity

Carbon black is far less resistant to a flow of electricity than the plastic in which it is dispersed. Therefore, carbon black can be used to lower the resistivity of plastics imparting antistatic, semiconductive, or fully conducting properties. The loading level of carbon black is an important factor in the conductivity of the composite. Higher loading increases conductivity; however, carbon blacks can vary significantly in their inherent conductivities. The conducting properties of carbon black increase with decreasing particle size, increasing structure, and decreasing volatile content. Some grades require higher loading than others to achieve the same level of conductivity in the resultant composite. The end use of conductive plastics ranges from electronic packaging to business machines, hospital equipment requiring antistatic properties.

CARBON BLACK TEST METHODS

Electron Microscope Surface Area

Measurement of particle size or related property of surface area is one of the principal procedures used in evaluating and controlling carbon black quality. Particle size can be precisely determined by electron microscopy and surface area estimated. Specific surface area[6] is derived directly from surface mean diameter (d_{sm}) using the correlation in Fig. 9-4. Note that the electron microscope surface area (EMSA) determination assumes unattached spheroidal particles. These values are useful for comparing different carbon blacks because they are based on full particle size distribution. However, this method is not a true surface area measurement and will give a low value when applied to highly porous samples. Typical rubber grades of carbon blacks have a mean diameter range from 17 nm (N101) to 107 nm (N762). N 900 is considerable larger (285 nm). The specific surface area of rubber grades based on the particle size (EMSA) varies from 9 to 138 m^2/g. After rubber, the next most important use of carbon black is as black pigment in coatings, inks, cosmetics, and so on. Carbon blacks for these uses are graded according to their level of blackness or "color"; for example, high color furnace (HCF), medium color furnace (MCF), and low color furnace (LCF). Carbon black grades used as black pigment (HCF) are of much smaller diameter particle size (9 to 16 nm) in comparison with rubber grades.

$$S = \frac{6000}{\rho^* d_{sm}}$$

Where:

ρ = Density of carbon black, g/cm^3

d_{sm} = Mean particle diameter, nm (nanometer)

S = Surface area, m^2/g

FIGURE 9-4 Carbon black surface area (EMSA).

Nitrogen Surface Area (BET Area) (ASTM D-3037)

This method[7] measures the total surface area of carbon black particle because nitrogen gas is able to cover the full surface area including pores without any interference from surface organic groups. In this method, a nitrogen/helium mixture is passed through carbon black in a sample cell immersed in liquid nitrogen. After the adsorption is complete, the liquid nitrogen reservoir is removed and the cell is placed in a heating mantle. The adsorbed nitrogen is desorbed, which is detected using a thermal conductivity sensor on a gas chromatograph. The resulting peak area yields the volume of nitrogen desorbed. Once the volume or weight of desorbed nitrogen is adsorbed as a monolayer on their surface, the surface area can be calculated as shown in Fig. 9-5.

$$S = \frac{W_m{}^* N^* A}{M}$$

Where:

S = Surface area of the sample, m^2

W_m = Weight of monolayer, g

N = Avogodro's number, 6.024×10^{23}/mole

A = Cross-sectional area of adsorbate

M = Molecular weight of adsorbate

Specific surface area (m^2/g) can be calculated by dividing surface area by actual weight of sample.

FIGURE 9-5 Nitrogen surface area.

Iodine Number (ASTM D-1510)

The adsorption of iodine from an aqueous solution of iodine and potassium iodide provides a rapid and convenient procedure for surface area measurement. The iodine number[8] is the weight in grams of iodine absorbed per kilogram of carbon black. The Iodine number is an indicator of surface area of carbon black: the smaller the particle size, the greater the iodine number. However, this method is not suitable for blacks with a high volatile content or a high amount of extractable matter. For such blacks, adsorption of cetyl trimethyl ammonium bromide (CTAB), from an aqueous solution of black, is preferred. Any residual feed contained in carbon black interferes with the wetting of the surface of the carbon black and produces a lower than normal value for a given particle size. Such residual material results from a very short reaction time during the manufacturing process.

For carbon black grades that have surfaces that are relatively clean regarding residual stock and have not been exposed to oxidation during their manufacture, the nitrogen surface area is nearly identical with the iodine surface area. For those grades that contain significant feedstock residual, the nitrogen surface area is the best indicator of the relative average particle size and hence reinforcement potential.

When the surface of the carbon particle is oxidized in the manufacturing process, surface pitting occurs, and the relationship between the particle size and nitrogen surface area is disturbed. The interior surface of the pits or micropores can significantly increase the nitrogen surface area, leaving the particle size unchanged.

CTAB Surface Area (ASTM D-3765)

The CTAB surface area test is designed to overcome both the problems mentioned in the iodine and nitrogen surface area tests. In this test, molecules to be adsorbed (cetyl trimethyl ammonium bromide) are far larger than the nitrogen molecule and also very effective wetting agents. The ratio of N_2 surface area to the CTAB surface area is a measure of carbon black surface porosity. The larger this ratio, the greater the surface porosity. The ratio of CTAB to iodine number is a measure of the surface chemical activity. The higher the ratio, the greater the reinforcing potential of rubber and the lower the hysteresis losses.

Dibutyl Phthalate Absorption Number (ASTM D-2414)

Carbon black structure is an important criterion in printing inks and other industries. It affects the viscosity, flow characteristics, and color density. The measure of the structure of furnace carbon black is DBP absorption (dibutyl phthalate absorption). In this test DBP is titrated into a given mass of carbon black as it is worked in a specially designed working chamber. As the test proceeds, the mixture begin to change into a very viscous paste. During transition, the machine torque required for mixing increases very rapidly. The end point is defined as the attainment of a certain torque value.

The carbon black structure as indicated by DBP absorption inhibits the elasticity of the compound. The increased structure manifests itself in many ways. Viscosity, modulus stiffness hardness, conductivity, and dispersibility are all increased.

Compressed DBP (CDBP) (ASTM D-3439)

It has been observed that the DBP value decreases in nearly all cases if mechanical work is applied to carbon black samples. To approximate the structure remaining after mixing more closely, a test is devised in which carbon black is compressed four times at 24,000 lb/in^2 prior to the normal DPB test procedure. The CDBP procedure is known to better predict rubber compounding properties.

Tint Strength (ASTM D-3265)

A gray paste is made by mixing the carbon black with zinc oxide and a reference oil. The reflectance of gray paste is measured and reported as percent tint strength relative to the industry standard. Because carbon black was first used as a pigment, rather than a rubber reinforcing agent, it was observed that certain carbon blacks produce a deeper black tint than others. Tint strength is related inversely to aggregate diameter. Tint strength is indicative of the ultimate reinforcement potential of carbon black.

Toluene Discoloration Test (ASTM D 1618)

Hydrocarbon extractable in toluene from carbon black can be used to estimate toluene extractable matter in carbon black. About 2 g of carbon black is dissolved in 20 cm^3 toluene. The filtrate is tested for light transmittance in a spectrophotometer at 425 nm. The result is expressed as percent light transmission, the values starting from 100 percent (zero extractable) to a lower value as extractables increase (minimum 80 to 85 percent). Toluene extractables is of value to the carbon black manufacturing process, which must adjust operating conditions to limit toluene extractables feedstock residual matter at a specified value.

Volatile Content

The volatile content of carbon black is an indication of the amount of chemisorbed oxygen and/or other gases on its surface and is determined by measuring the loss of weight on calcining a dried carbon black sample at 1782°F.

APPLICATION AND USES

Rubber Reinforcing Agent

Carbon black is the predominant reinforcing filler used in rubber compounds. A typical passenger car or truck tire may contain approximately 30 percent by weight carbon black. The improvement in rubber properties is a function of the physical and chemical properties of carbon black. The most important physical and chemical properties of carbon black that affect rubber properties are aggregate size, shape (structure), particle size, surface activity, and porosity. The other nonfundamental properties include the physical form and residual matter. The physical form of carbon black (beads or powder) can affect its handling and mixing characteristics and thus rubber properties. The ultimate degree of dispersion is also a function of the mixing procedure and equipment used.

The ASTM classification currently used for rubber grades of carbon blacks was introduced in 1968. In this classification, the grade number consisted of a letter followed by a three-digit number. The prefix, either "N" or "S," indicates whether the grade is normal or slow curing. Regular furnace blacks are denoted by "N," indicating normal vulcanization rate; posttreated furnace and channel blacks are denoted by "S," slow vulcanization rate. When the system was developed, channel blacks were still popular and were slow curing. With the growing popularity of the furnace black process, slow curing grades mostly disappeared, and at present all rubber grades of carbon blacks carry the prefix "N." The first of the three digits is used to provide a measure of mean particle diameter as measured by an electron microscope, expressed in nanometers (10^{-9} meters). The possible particle diameters were grouped as shown in Table 9-7. The second and third digit refer to other aspects of behavior in rubber, including iodine absorption and structures. These are assigned by ASTM D-24 to new products as various manufacturers develop them. In general, lower structure blacks are assigned lower numbers and higher structure blacks are assigned higher numbers.

TABLE 9-7 ASTM D 24 Designation of Carbon Black Grades

First digit	Typical average particle size, nm
0	1–10
1	11–19
2	20–25
3	26–30
4	31–39
5	40–48
6	49–60
7	61–100
8	101–200
9	201–500

N100, N200, and N300 series blacks are generally used in tire treads for their abrasion resistance, whereas N300 and N500 are used in the carcass of radial tires. N600 and N700 series are also used in carcass and less critical performance applications and slow-speed tires. Rubber compounds for tires should have improved tread wear and traction, reduced rolling resistance, and keep heat generation at minimum level, whereas in nontire rubber applications, they should conform to properties such as higher resilience and dimension stability in extruded goods.

Prior to the introduction of ASTM, carbon blacks for rubber and tire manufacture were named according to advancements in furnace process technology. Carbon blacks were named according to the manufacturing process and their end use and denoted by various letters representing different qualities of carbon blacks. For example, SRF, or semireinforcing furnace, was developed in 1928. It was so named because its rubber reinforcing properties and particle size (70 nm) were intermediate between that of channel black and thermal black. Some old nomenclature for carbon blacks and their applications are shown in Table 9-8.

TABLE 9-8 Carbon Black Types Used in Tires

Industry name	Abbreviation	ASTM designation	Particle size, nm
Super abrasion furnace	SAF	N110	20–25
Intermediate SAF	ISAF	N220	24–33
High-abrasion resistance	HAF	N330	28–36
Easy processing channel	EPC	N300	30–35
Fast extruding furnace	FEF	N550	39–55
High modulus furnace	HMF	N683	49–73
Semi reinforcing furnace	SRF	N770	70–96
Fine thermal	FT	N880	180–200
Medium thermal	MT	N990	250–350

Effect of Particle Size

Particle size is the fundamental property that has a significant effect on rubber properties. Finer particles lead to increased reinforcement, increased abrasion resistance, and improved tensile strength. For example, pure gum vulcanite of SBR rubber has a tensile strength of 2.5 mPa and an almost nonexistent abrasion resistance. Mixing it with 50 percent N110 carbon back increases the tensile strength 10-fold. Typical particle size ranges from 20 to 25 nm for N110 (SAF) to 250 to 300 nm for

N990 (MT). Because reinforcement is a function of particle size or surface area, the logical starting point in choosing a carbon black grade is to access the level of reinforcement required for the rubber. Viscosity of the compound will increase while dispersibility will decrease as a result of smaller particle size. Also dispersing finer particle size requires increased mixing time and energy.

Effect of Structure

Carbon blacks do not exist as primary particles. Primary particles fuse to form aggregates that may contain a large number of particles. The shape and degree of branching is referred to as its structure. The structure level of carbon black ultimately determines its effect on several important rubber properties. Increasing carbon black structure increases modulus, hardness, and electrical conductivity and improves dispersibility but increases compound viscosity. Carbon black structure as indicated by DBP absorption inhibits the elasticity of the compound much in the same way as reinforcing steel inhibits the elasticity of reinforced concrete. The stiffening effect is more pronounced with structure than with particle size. Increased structure manifests itself in many ways: Viscosity, modulus, stiffness, hardness, conductivity, and dispersibility are all increased, and extrusion surface imperfections, swell, and tear strength are all decreased.

Effect of Surface Activity

Surface activity of carbon black is a function of the manufacturing process and the heat history of carbon black. Surface activity is difficult to measure directly but is indirectly indicated by its volatile content and the pH of the carbon black sample. Surface activity has its effect on rubber properties such as abrasion resistance, tensile strength, hysteresis, and modulus. The effect of surface activity on cure characteristics very much depends on the cure system in use. All carbon blacks have chemisorbed oxygen complexes (carboxylic, quinonic, lactonic, or phenolic groups) on their surfaces to a varying degree depending on the condition of manufacture. In the carbon black industry, these surface groups are called "volatile content." Some carbon blacks have their surfaces that have been purposely oxidized by aftertreatment with various oxidizing agents to increase their volatile content, thus allowing maintenance of satisfactory rheological properties at higher loading in liquid systems. A desirable effect of higher volatile content is the enhancement of dispersibility. Chemisorbed oxygen complexes tend to act as surfactants in improving the wetting characteristics of carbon black.

Effect on Conductivity

Various grades of carbon blacks can be compounded into rubber to provide rubber with high electrical resistance or good conductivity. For example, cable manufacturers employ high loading of medium thermal black to produce rubber with resistivity as high as 10^{12} ohm-cm. For conducting or antistatic properties, acetylene blacks with resistance as low as 0.2 ohm-cm can be used. Conducting rubber finds applications in antistatic tire treads, hospital flooring, etc.

Carbon Black Pigments

The use of carbon black as a pigment and ink was known to ancient civilizations as early as 3500 B.C. A form of carbon black (soot) was used for making inks, for making eye cosmetics and kohl in the Middle East, North Africa, sub-Saharan Africa, and India.

Carbon blacks today are used in many strategic industries; for making black coatings, printing inks, and as a pigment for plastics and to impart antistatic properties to rubbers and plastics. Carbon black is used in making eye cosmetics such as mascara, eye shadows, and nail enamels. Compared to their usage in rubber and tire industry, the use of carbon black for nonrubber uses is very low, less than 10 percent of total carbon black production.

TABLE 9-9 Pigment Grade Carbon Black Properties

Property	Units	HCC channel black*	HCC gas black	HCF furnace black
Average particle diameter	nm	14	13	13
Volatiles	Wt %	10	16	9.5
N_2 Specific surface area	m^2/g	695	430	560
Nigrometer index		65	63	64

*For the purpose of classification of pigment grades, gas blacks are considered channel black class.

Carbon blacks for use as pigment are classified according to an internationally recognized nomenclature system that signifies the manufacturing process and the carbon black jetness. The first two letters indicate the strength of pigment; high color (HC), medium color (MC), regular color (RC), and low color (LC). The final letter indicates the manufacturing process: for example, high color channel (HCC), regular color furnace (RCF), and so on. Carbon blacks with a nigrometer scale value of 70 or below are considered "high color"; medium color blacks have a nigrometer scale value of 70 to 78. Regular color blacks have a nigrometer value[9] of 80 to 90. The volatile matter content of carbon black is another important property for color blacks. Volatile content affects the tone and color properties and also determines the flow properties of lacquers or enamels. "Regular color blacks" have a volatile content of less than 5 percent by weight. "Medium color blacks" have a volatile content of 5 to 10 percent by weight. "High color blacks" have a volatile content typically of 10 to 18 percent. Table 9-9 shows some properties of high color blacks.[10] The channel process produces better quality high color pigments because of the higher surface area and higher volatile matter compared with that from the furnace process.

The basic requirements for any color black are as follows:

- High jetness
- Easy dispersion in appropriate vehicle
- Good viscosity stability in formulation

The nigrometer scale of carbon black is a measure of intensity of the blackness or jetness and is closely related to particle size. In general, the lower the particle size of carbon black, the lower the scale value and the more intense the blackness.

The nigrometer index is determined as follows: Carbon black (0.1 g) is carefully ground together with linseed oil varnish until a homogeneous paste is obtained. The paint is quickly painted on to a slide, and light transmission is measured through the slide with a nigrometer. The recording that takes place is known as the nigrometer index. A small number means a high degree of depth of color and a large number, a low degree.

High jet black lacquers are used in the automobile industry, for optical instruments, for solar collectors, and so on. In these black lacquers, carbon black is the preferred pigment over all others because it allows coloring with the highest degree of blackness and has practically unlimited resistance to light and heat. For such lacquer systems demanding the highest blackness and gloss, carbon blacks of HCC area (high color channel blacks) or HCF (high color furnace black) are preferred.

In the case of the channel process, which uses a natural gas feed, carbon black is produced in a large number of small flames that burn against cooled iron rail (channels) on which carbon black is deposited. Carbon black is formed in a gaseous atmosphere that has excess atmospheric oxygen. As a result, oxygen-containing groups with an acid character and with a pH of approximately 3 develop on the surface of carbon blacks. However, because of the poor economics of the channel process (3 to 6 percent carbon black yield) and adverse environmental impact, the process is no longer used.

The gas black process operates essentially in a similar way to the channel process. Here, instead of natural gas, a coke oven gas ($CO + H_2$) loaded with coal tar/oil vapor is conveyed to the burner. The flame strikes against cooled rotating rollers on which carbon black deposits. In this process too, formation of carbon black takes place in an oxidizing atmosphere, and therefore the carbon produced displays an acidic reaction in water. The yield of carbon black from this process is 10 times higher than that from the channel process, and the gas black process competes with the furnace process.

Because of the great similarity of this process to the channel process, *carbon blacks produced by gas process are also classified with channel blacks.* The furnace process operates in a closed system. The volume of atmospheric oxygen used in the combustion of fuel gas or oil is much less and the process operates in an oxygen-deficient atmosphere. For this reason furnace blacks have much less surface oxides and show neutral to weakly alkaline reactions (pH 7 to 10).

Carbon Black Coatings

Coating compositions are used for decorative, protective, and functional treatment of many kinds of surfaces. These surfaces include coils, metals, appliances, furniture, hardboard, lumber, plywood, marine, automobiles, cans, and paperboard. Some coatings, such as those on undersea pipelines, are for protective purposes. Others, such as those on automobiles, are for both protective and decorative purposes. Some coatings resist the fouling of ship and boat bottoms from corrosion. Coatings are used in silicon chips, printed circuit panels, coatings on wave guide fiber for signal transmission, magnetic coatings on video tapes and computer disks, and more.

Surface coating compositions are generally more or less viscous liquids with three base components:

- A film-forming substance or a combination of substances called a binder
- A pigment or combination of pigments
- A volatile liquid

The combination of binder and volatile liquid is called the vehicle. Vehicles may be in solution form or the dispersion of binder particles in a nonsolvent. Pigments are finely divided insoluble solid particles dispersed in coating vehicles and are distributed throughout the binder in the final film. Surfactants may also be added and are typically used as pigment dispersants.

Pigments in coating composition provide opacity and color. The amount and type of pigment control such properties as gloss of the final film and can have important effect on mechanical properties. Some pigments inhibit corrosion. Further pigment affects the viscosity. Carbon blacks are common pigments used in coating applications.

An important variable determining the performance of carbon black in coating composition is its surface area. The higher the surface area of carbon black, the better the color properties of the resulting coating. Surface area is inversely proportional to the particle size. Particle size affects such properties as jetness, gloss, and blue tone. There are several different measures of surface area, that is, nitrogen surface area (BET method), iodine surface area, and CTAB surface area. For coating usage, the external surface area excluding pores should be between 350 and 400 m^2/g.

To provide a coating with overall color properties, smaller particle pigments are desirable. However, smaller particle sizes also result in an increase in the viscosity of the coating. Also, particle size and surface area also adversely affect its dispersibility into coating composition. In manufacturing coatings, it is desirable to disperse pigment in such a way as to achieve a stable dispersion where most of it is separated into individual particles. The mechanism of dispersion of pigment involves wetting, separation, and stabilization. It is well known that the higher the surface area of the pigment, the more difficult the pigment is to wet and disperse in the vehicle. Poor pigment dispersion may lead to the deterioration of pigment properties. Dispersion stability may also suffer. For these reasons, commercially available pigments for high color coatings applications are a compromise between surface area and dispersion quality and stability.

Another important variable is the structure of the carbon black. Structure is measured by the dibutyl phthalate absorption (DBP) value. DBP is a measure of the structure or branching of the carbon product. In general, the greater the structure, the better the dispersibility of carbon black. In contrast, the greater the structure, the higher the viscosity of the coating composition. Also, higher structure results in a generally poor color performance; lower gloss and jetness. For coatings usage, DBP values should be in the range of 80 to 120 cc/100 g.

For coating usage, the preferred carbon blacks are those that have been oxidized to increase the oxygen functionality on the surface prepared by a reaction with nitric acid or ozone. The increase in functionality on the surface gives rise to a decrease in pH.

The use of modified carbon black provides a significant advantage and cost savings by reducing the viscosity of the formulation, which allow higher loading of carbon in formulation. The milling time is reduced. Modified carbons provide improved jetness, blue tone, and gloss.[11]

Conductive Coatings

Electrically conductive liquid coatings containing carbon blacks are essential in many applications, including primers for automotive plastic parts and coatings for electrostatic dissipation control. Electrostatic conductive coatings are increasingly being used in applications such as magnetic tape coatings, aircraft paints, spark-resisting equipment enamels, primers for electrostatic spraying, coatings for resistive heating, and inks used in printed circuits. Most of these end uses are satisfied by coatings utilizing a conductive filler at various loading levels. Carbon black is the preferred conductive filler for several important reasons because of its cost effectiveness and because it is readily dispersed and stabilized in a liquid system.

A highly conducting carbon black is the combined effect of these four properties:

- Surface area
- Structure
- Porosity
- Volatile content

For high conductivity, carbon black must have a high surface area or fine particle size, high structure, high porosity, and low volatile content. As the carbon black loading in a conductive coating is increased, the surface resistivity (reciprocal of surface conductivity) decreases. Several percent weight of carbon black in coating is necessary before a significant drop in surface resistivity is noted. The amount of carbon black required to obtain a given surface resistivity is resin dependent, typically in the range of 12 to 34 percent by weight. The thickness of the film has an influence on the surface resistivity. From a 1- to 3-mil wet thickness, resistivity decreases gradually. Above 3 mils, the change is negligible.

Plastics

Carbon blacks are added to plastics for the following objectives:

- Coloring and opacifying
- Stabilization against UV light
- Reinforcing of mechanical properties
- Modifying electrical conductivity

Coloring

Usually carbon blacks are first converted into a plastic concentrate. The concentration of carbon black in the master batch ranges from 20 to 60 percent, which allows easy dispersion in plastics.

Carbon blacks are available in wide range of jetness levels to meet the coloring requirements of all types of plastics. Simple coloring requires only enough carbon black to achieve opacity, usually in the range of 0.5 to 3.5 percent loading. Grade selection depends largely on the degree of jetness required by the end product. In addition to jetness, there is another important appearance consideration for plastics that can be controlled by the selection of appropriate carbon black grades such as undertone, tint strength, and gloss. Undertone is a subtle secondary color underlying jetness. Undertone exhibited by carbon black ranges from warm brown to cool blue. Tint strength is the relative ability of carbon black to darken a color resin. Gloss level can range from very shiny to a matte finish. Carbon blacks appear black because of their natural ability to absorb energy from all wavelengths of light. Small-particle blacks are better absorbers than large-particle blacks because of the greater surface area available for light absorption. Therefore, small-particle blacks have a higher jet appearance. In general, to obtain the maximum light absorption needed for optimal jetness and tint strength, a carbon black grade with a high surface area should be chosen. Low-structure carbon blacks exhibit more of a jetness than higher structure blacks.

UV Stabilization

Stabilization against UV light is important for plastics exposed to direct sunlight. The continuous exposure of plastics to sunlight can result in photo oxidation and consequent failure. Photo oxidation can be prevented by incorporating 2 to 3 percent carbon black with a 15- to 20-nm particle size.

Reinforcement

As a reinforcement filler of plastics, the effect of carbon black is similar to that on rubber, with smaller particle sizes providing maximum reinforcement. Thus N110 provides the highest potential reinforcement but is difficult to disperse. N220 and N330 are preferred for this use.

Electrical Conductivity

Carbon black is far less resistant to the flow of electricity than the plastics in which they are dispersed. Therefore, carbon black can be used to lower the resistivities of plastics to impart plastic antistatic, semiconductive, or conducting properties. The end uses of conductive plastics range from electronic packaging to business machines, hospital equipment requiring antistatic properties, semiconducting shields, shielded power cables, and so on. In general, higher loading increases conductivity; however, carbon blacks can vary significantly in their inherent conductivity, depending on particle size, structure, and volatile matter. Smaller particle size increases conductivity. Also higher structure increases conductivity. Volatile or oily matter decreases conductivity.

Many molded and extruded plastic compounds are required to conduct away electricity accumulating on the surface. Antistatic properties are incorporated[12] by 20 to 25 percent loading with N472-type carbon black.

PRINTING INKS

An important application of carbon black is as a pigment in lithographic and printing inks and as toner in photocopiers. Printing inks are manufactured from these four components:

- Pigment and/or dyes that constitute the coloring agent
- A resin that serves to carry the pigment and bond it to the substrate
- Additives to control such performance characteristics as drying, adhesion, and rub resistance
- Oil and solvent to dissolve the resins

Printing inks can be categorized into following types:

1. *Water-based inks.* Includes all flexographic, gravure, and screen inks that have water as the solvent or diluting component.
2. *Oil-based inks.* Includes all lithographic and letterpress inks that have oil as the primary solvent/diluting component.
3. *UV inks.* Includes all flexographic, gravure, screen, and lithographic inks that are cured by polymerization on exposure to UV or electron beam energy. UV-cured inks are becoming more popular in the printing industry. Pigment black effectively absorbs light both in the visible and UV band. The blacker the ink, the longer the drying time.
4. *Solvent-based inks.* Includes all flexographic, gravure, and screen inks that have an organic solvent as the primary solvent/dilution component.

Carbon black is the coloring agent in printing inks and its quality is governed by the following quality criteria:

- High color density (color depth)
- Blue tint
- Brilliance
- Yield (mileage)
- Correct rheology

The requirements listed represent a compromise because high color density implies small particulate size pigment blacks that are not prone to a blue tint in printing inks. The color tone of pigment blacks depend mainly on particle size (medium particle size) and the dispersion rate achieved. The finer the particles of pigment black, the deeper the color and the browner the pigment will appear. The larger the particle size, the lower the color depth and the more pigment will appear blue. Brilliance or gloss is not particularly affected by particle size. Structure, in contrast, plays an important role in achievable gloss. Color density, tone, and gloss are a function of the degree of coverage and dispersion achieved.

Pigment blacks are used 100 percent to color black printing inks. They are also used to tint or tone gray and other colors such as brown, olive green, and so on. Decreasing particle size increases color depth and brown tint. Increasing particle size reduces color depth and produces blue tint. Fine pigment blacks have a relatively high surface area, which requires a higher binding agent demand, resulting in a higher viscosity.

Structure

Pigment black structure is an important criterion in printing ink manufacture. A number of quality factors depend on structure. These are as follows:

- Dispersion properties
- Viscosity
- Flow characteristics
- Color density
- Brilliance
- Permanence
- Conductivity

As mentioned earlier, the measure for the structure of pigment blacks is dibutyl phthalate absorption (DPB) absorption. DBP values for printing ink blacks are in the range of 40 to 125 mL/100 g pigment black. Pigment black with DBP values lower than 90 mL/100 g fall into the category of "low structures." DBP values greater than 90 indicate "high structure." A higher structure pigment black disperses easily in ink formulation. However, higher structures also cause a significant viscosity increase, whereas lower structure blacks has a less pronounced thickening effect and result in better flow characteristics for the printing ink. High-structure pigment blacks cause low gloss and rub resistance. When gas blacks are used, the criteria is no longer DBP but oil requirements, which gives a good indication of thickening properties. Gas blacks come with high structure and with good flow characteristics.

Newspaper Inks

Newspaper inks are made with medium- to high-structure pigment blacks. One of the less desirable characteristics of conventional newspaper ink is its tendency to smudge. Pigment black structure is a factor in determining smudge resistance. A low structure improves the resistance to smudging. A high structure makes smudging easier. Beaded blacks are used in place of powder. The beaded version does not produce any dust and greatly facilitates handling in an automated process. Oil-beaded blacks are becoming more popular on account of their trouble-free service.

COSMETICS USAGE

Carbon black is used in coloring cosmetics, particularly those intended for use in the eye area. The color is generally blended with other permitted color additives to give a wide variety of possible shades. The amount of color used depends on the desired color effects. Minor amounts of carbon blacks are also used in soaps, temporary hair color rinses, creams, and lotions. Generally, carbon blacks manufactured by the furnace process or channel process are permitted. Carbon blacks are known to contain a variety of manufacturing impurities, most notably polynuclear aromatic hydrocarbons. Depending on the nature of feedstock quenching and aftertreatment procedures, other impurities may also be present. The poly nuclear aromatics (PNA) present in carbon black intended for cosmetic usage must not exceed 5 ppm, and the benzopyrene level should not exceed 20 ppb. PNAs are known to be strongly bound to carbon blacks. For this reason, bound PNAs are not available for absorption from cosmetic products.

Carbon blacks used in cosmetics are manufactured by the channel or furnace process with a very low level of impurities. The lead level in the product must not exceed 20 ppm and arsenic not more than 2 ppm. Low PNA N550-6 or low PNA N774-6 developed for pharmaceutical applications are generally used. An estimation of human exposure to carbon black in various cosmetics[13] is listed in Table 9-10.

TABLE 9-10 Carbon Black Exposure from Cosmetic Products

Cosmetic product	Product used per day, mg	Color concentration, % Wt	Color used per day, mg	Color available for absorption, %	Color exposure, mg/day
Ingested cosmetics					
Lipstick	20	1.00	0.20	50	0.1
External cosmetics					
Eye liner	10	30.00	3.00	50	1.5
Brush-on-brow	5.5	30.00	1.65	50	0.83
Eye shadow	7.9	30.00	2.37	50	1.19
Mascara	25	10.00	2.50	50	1.25
Blushers and rouge	14	2.00	0.28	50	0.14
Makeup, foundation	265	5.00	13.25	50	6.625
Nail enamel	56	2.00	1.12	0	0

Minor amounts of pharmaceutical-grade carbon blacks are used in the confectionary as a tinting agent. Small quantities of carbon blacks are used for a large number of applications such as the tinting of synthetic fibers, as an anticaking agent in cement and fertilizer, and a source of resistance in ceramics resistors, and so on.

REFERENCES

1. V. B. Guthrie, *Petroleum Product Handbook* (New York: McGraw-Hill, 1960), pp. 15–12.
2. U.S. Patent 5100637, "Carbon black for compounding into tire tread rubber" issued March 31, 1992.
3. U.S. Patent 4636375, "U.S. Patent 4636375, issued January 13, 1987" issued January 13, 1987.
4. U.S. Patent 4075140, "Oxidizing Carbon Blacks and Use of It in Elastomers," issued February 2, 1978.
5. U.S. Patent 4366138; "Carbon Black Useful for Pigment and for Black Lacquers," issued October 22, 1979.
6. J. B. Donnet, R. C. Bansal, and M. T. Wang, *Carbon Black,* 2nd ed (New York: Marcel Dekker, 1993).
7. P. H. Emmett and T. Dewitt, *Industrial and Engineering Chemistry Analytical Edition* 13 (1941): 28.
8. W. R. Smith, F. S. Thornhill, and R. I. Bray, "Surface area and properties of Carbon Black," *Industrial & Engineering Chemistry* 33 (1941): 28.
9. U.S. Patent 3620792, "Process for Treating Carbon Black," issued November 16, 1971.
10. U.S. Patent 4366138, "Carbon Blacks Useful for Pigment for Black Lacquers," issued December 28, 1982.
11. U.S. Patent 5895522, "Modified Carbon Products with Leaving Groups for Inks and Coatings," issued April 2, 2003.
12. H. S. Katz and J. W. Milewski, *Handbook of Fillers for Plastics* (New York: Van Nostrand Reinhold, 1987).
13. U.S. Cosmetics, Toiletry and Fragrance Association (CTFA) submission to the Department of Health and Human Services, "Carbon Blacks for Cosmetics," November 19, 1986.

CHAPTER 10
LUBE BASE STOCKS

Lubricants are required in machines to reduce friction and wear between moving parts. Lubricant base stocks make up a large portion of finished lubricants, from about 75 to 80 percent in automotive engine oils to 90 percent or more in some industrial oils. Thus, base stocks contribute significantly to the finished product properties. Base stock has a major impact on the viscosity, volatility, low temperature fluidity, solvency for additives and contaminants, demulsibility, air release, foam, oxidation, and thermal stability of the finished product. Petroleum lubricating base stocks are made of a higher boiling portion of crude oil that remains after the removal of lighter hydrocarbons. Starting material for their manufacture is usually atmospheric residue boiling above 650°F. However, atmospheric resid from every crude is not suitable for lubricating oil manufacture. Careful selection of a base stock is key to formulating a quality finished lubricant. Base stock properties are related to base stock composition. Base stocks contain three types of hydrocarbon; paraffins, naphthenes, and aromatics. In the paraffin group, isoparaffins are the preferred type because they exhibit excellent oxidation stability, low volatility, and good viscosity characteristics. Normal paraffins, however, are not a desirable component because of their poor cold flow properties such as pour point, cold filter plug point (CFPP). Aromatics are good for the solvency of additives and contaminants but generally have poor oxidation stability and high volatility. Naphthenes also have good low temperature fluidity and oxidation stability. Sulfur and nitrogen are often present in combination with hydrocarbons in a base stock, particularly the aromatics. Sulfur can improve oxidation stability but may also contribute to deposit formation and color instability. Nitrogen promotes oxidation and deposit formation. The preferred lubricant base stocks for a wide range of product application contain predominantly isoparaffins and naphthenes with a balance of aromatics and sulfur for proper solvency and oxidation stability.

The manufacture of lube base stocks from crude oil involves a series of steps aimed at the removal of certain undesirable components resulting in a base oil that meets the performance requirements of lubricating oils. There are two basic routes for making lube base stocks; the conventional process, consisting of solvent extraction, solvent dewaxing, and hydrofinishing, and the hydroprocessing route, consisting of lube hydrocracking, hydrodewaxing, and deep hydrotreating. The hydrotreating route produces higher viscosity index (VI) lubes with superior quality but cannot produce high-viscosity lube base stocks.

CONVENTIONAL PROCESS

The conventional lube base stock manufacturing process consists of the following steps:

- Vacuum distillation of atmospheric resid to yield several distillate cuts and vacuum resid
- Propane deasphalting of vacuum residuum to yield bright stock and asphaltic pitch
- Solvent extraction of vacuum distillates and bright stock to remove aromatics and improve the viscosity index of lubricating oil base stock
- Solvent dewaxing of distillate cuts to yield slack wax and various lube cuts, which improves the cold flow properties such as pour point and, cloud point of the lube base stock
- Hydrofinishing or clay treatment to improve color, oxidation stability, and thermal stability of lubricating oils

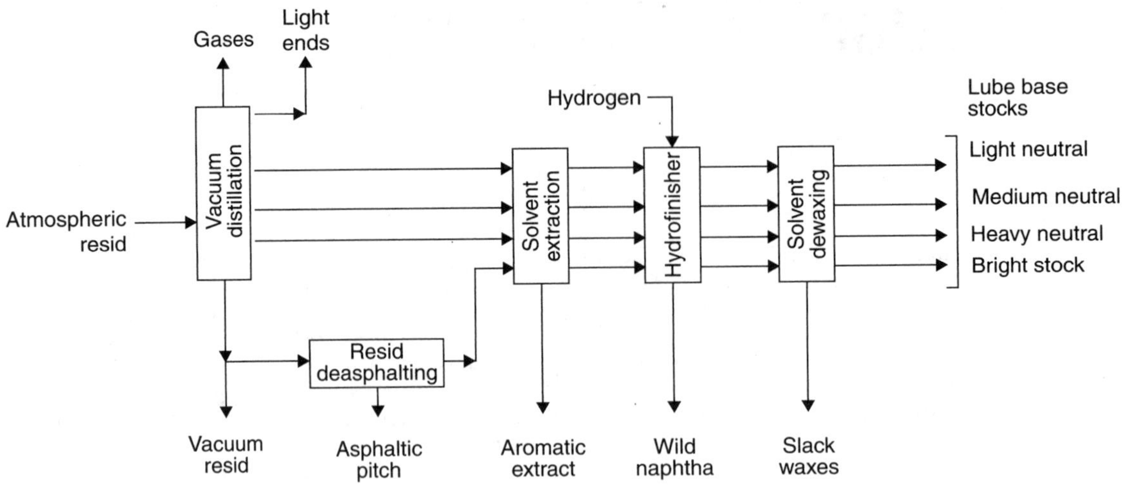

FIGURE 10-1 Conventional lube base stocks manufacture.

Figure 10-1 shows the processing scheme for a conventional lube plant comprising vacuum distillation, furfural extraction, MEK dewaxing, and propane deasphalter. Tables 10-1 and 10-2 show the material balance and stream qualities for a plant processing Arabian Light crude.

TABLE 10-1 Lube Oil Plant Overall Material Balance*

Stream	Product	Wt fraction on ATM resid	Cut point, °F	Specific gravity	Viscosity, cSt, 122°F	Sulfur, Wt %
1	Atmospheric resid	1.0000	633.3	0.9710	830	4.20
2	Vacuum resid	0.4700	919.4	1.0320	131,500	5.40
3	Feed to PDA	0.1900				
4	Vacuum overhead oil	0.0160	365	0.8640	3	1.50
5	Vacuum top excess	0.0820	494.6	0.8800	4	1.50
6	Asphalt	0.1368		1.0740	*	6.40
7	Light extract	0.0576		1.0011	39	6.10
	Medium extract	0.0678		1.0136	140	5.80
	Heavy extract	0.0539		1.0088	1786	5.50
	Bright stock extract	0.0161				
8	Wild naphtha	0.0093				
9	Light slack wax	0.0205		0.8100	8	0.25
	Medium slack wax	0.0204		0.8330	15	0.07
	Heavy slack wax	0.0134		0.8500	28	0.10
	Bright stock slack wax	0.0066		0.8700	65	0.31
10	Light neutral	0.0721		0.8613		
	Medium neutral	0.0759		0.8729		
	Heavy neutral	0.0419		0.8824		
11	Bright stock	0.0297		0.8984		
12	Net vacuum resid	0.2800				

*Comprising vacuum distillation, propane deasphalter, solvent extraction, and *MEK* Dewaxer.
Basis: Light Arabian crude processing

TABLE 10-2 Lubricating Oil Cut Properties from Lube Plant

	Specific gravity	Viscosity, cSt 104°F	Sulfur, Wt %	VI	Oil content, Wt %
Solvent extracts					
Light	1.016	68.6	3.00		
Medium	1.012	565	5.00		
Heavy	1.015	2640	5.00		
Bright stock	0.983	—	—		
Slack waxes					
Light	0.820	0.02			6.4
Medium	0.869	0.03			9.9
Heavy	0.890	0.08			30.5
Bright stock	0.905	0.16			15.1
Base oils					
Light	0.860	19.10	0.10	95	
Medium	0.880	60.20	0.20	96	
Heavy	0.885	131.50	0.15	96	
Bright stock	0.897	473.00	0.35	97	

Vacuum Distillation

Vacuum distillation separates the atmospheric residue into a series of fractions representing different viscosities/molecular weight ranges. The objective is to isolate hydrocarbons with the proper boiling range and viscosity characteristics suitable for lubricant manufacture. These fractions, by convention, are identified by "neutral numbers," which are, in fact, their Saybolt Universal second (SUS) (measure of kinematic viscosity) at 100°F. Typically the neutral number ranges from 90 to 600 for various straight run fractions (Table 10-3). For example, "90 neutral" indicates a straight run lube cut with a viscosity of 90 SUS (approximately 17.5 cSt) at 100°F. A process flow diagram of a refinery vacuum distillation column for lube manufacture is shown in Fig. 10-2. Operating conditions of a refinery vacuum distillation unit (VDU) column are shown in Table 10-4.

TABLE 10-3 Viscosities of Raw Lube Cuts from Vacuum Distillation and Propane Deasphalter Unit

Neutral number	Viscosity, 100°F SUS	Viscosity, 100°F cSt
50	50	7.38
60	60	10.35
70	70	13.08
80	80	15.66
90	90	18.13
100	100	20.53
150	150	31.90
175	175	37.45
200	200	42.90
250	250	59.25
300	300	64.65
350	350	75.46
500	500	107.90
650	650	140.32

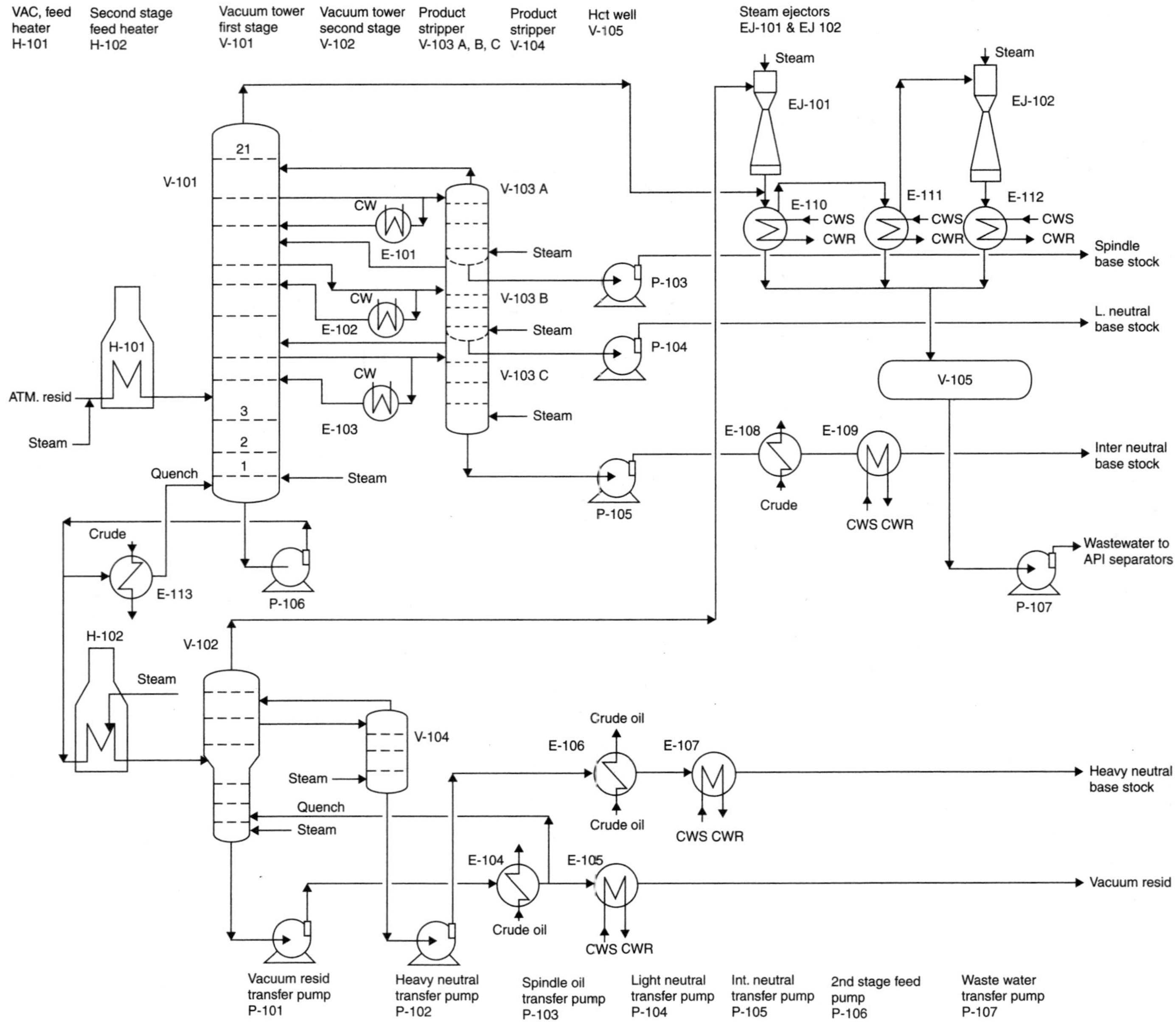

FIGURE 10-2 Vacuum distillation for lube base stocks manufacture.

TABLE 10-4 Operating Conditions for Vacuum Column Lube Operation

	Operating conditions	
	1	2
Pressure		
Top, mm Hg	97	66
Flash zone, mm Hg	156	106
Temperature		
Top, °F	293	185
Flash zone, °F	765	740
Fractionation efficiency		
Streams	Overlap 95/5, °F	
Vacuum gas oil–100N	62	
100N–195N	88	
195N–340N	104	
340N–650N	128	
650N–residuum	139	

Propane Deasphalting Unit

Vacuum residue contains heavier base oils that are separated from asphaltenes and resins by a solvent deasphalting process. Propane is the preferred solvent. The oil separated by solvent deasphalting of vacuum residue is known as Bright stock, which is much more viscous and heavier than straight run vacuum distillates. Solvent deasphalted base stocks are identified by their SUS viscosity at 210°F. Thus, "150 Bright Stock" implies a nominal viscosity of 150 SUS at 210°F. Table 10-5 shows the typical operating conditions for a PDA unit.

TABLE 10-5 Propane Deasphalting Unit Operating Conditions
Feed: Vacuum residue from Aghajari crude

	Viscosity 1350 cSt at 210°F
Propane-to-feed ratio, V/V	5.30
Extractor pressure, lb/in^2	455
Extractor top temperature, °F	154
Extractor bottom temperature, °F	126
Furnace coil outlet temperature, °F	410
Deasphalted oil viscosity, cSt, 210°F	38
DAO conradson carbon, Wt %	2.1
PDA asphalt penetration, ASTM D5	5

Solvent Extraction

The objective of solvent extraction of lube base stocks is to improve the viscosity index (VI) of lube base stocks. The feed to the unit are vacuum distillate cuts such as spindle, light neutral, intermediate neutral, heavy neutrals, and bright stock (produced from propane deasphalting of vacuum resid). The operation of the unit is same for all the grades except the operating conditions such as the solvent-to-feed ratios and feed temperatures, which are adjusted as required. All the feeds are processed in a solvent extraction unit in a blocked-out operation. Solvent extraction process removes undesirable components such as aromatics (low VI) and compounds containing heteroatoms such as oxygen, nitrogen, and sulfur, present in vacuum distillates and residual stocks.

The separation of low lubricating oil quality components is controlled by the quantity of the solvent employed and by the solvent temperature. The solvents that have been used in the past are benzene, phenol, nitrobenzene, liquid sulfur dioxide, furfural, and many others. Except for furfural, these solvents are no longer used because of their high toxicity and low biodegradability. Furfural is by far the most popular solvent because it is nontoxic and biodegradable. The molecular structure of furfural is shown in Fig. 10-3. The physical properties of furfural are shown in Table 10-6. In recent years, another solvent, normal methyl pyrrolidine (NMP), has claimed popularity. Figure 10-4 shows the molecular structure of NMP. Many new solvent extraction plants are being built using NMP solvent, and the physical properties of furfural and NMP are shown in Tables 10-6 and 10-7.

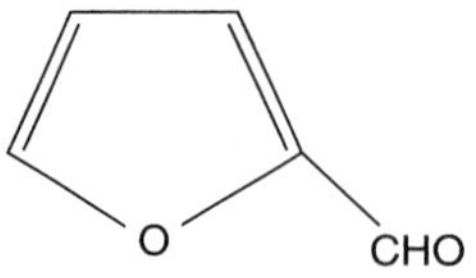

FIGURE 10-3 Furfural structure.

The following advantages are claimed for NMP solvent over furfural in lube oil extraction plants:

- NMP requires a much smaller solvent-to-feed ratio, resulting in smaller plant size.
- The stability of NMP is better than that of furfural; therefore solvent consumption is much lower than that for furfural solvent.
- NMP is noncoking and has lower toxicity.

The solvent extraction process produces high-quality lubricating oils characterized by a high VI, good thermal and oxidation stability, light color, and a good additive response. The by-product extract phase, rich in aromatics, is used as a carbon black feed stock, rubber extender oils, and many other nonlube uses. Figure 10-5 shows a process flow diagram of a refinery solvent extraction unit. The lube feed stock (distillate or deasphalted residual stock) is contacted with solvent in an extraction tower of a rotary disk contactor where both the solvent and feed are fed continuously and the raffinate phase, consisting mainly of paraffins and some solvent, and the extract phase, consisting of mainly aromatics and some solvent, are separated. The raffinate and extract phase are then taken to separate solvent recovery trains, consisting mainly of heating furnaces and distillation columns. Both raffinate and

TABLE 10-6 Properties of Furfural Solvent 2-Furfuraldehyde*

Formula	$C_5H_4O_2$
Formula weight	96.082
Specific gravity	1.161
Melting point, °F	–33.7
Boiling point, °F, 760 mm	323.1
Viscosity, CP, 77°F	1.494
Refractive index	1.5262
Flash point, °F, CC	138.2
Ignition temperature, °F	599

*At 1 atm, furfural forms an azeotrope with 65.0% water, boiling at 207.5°F.

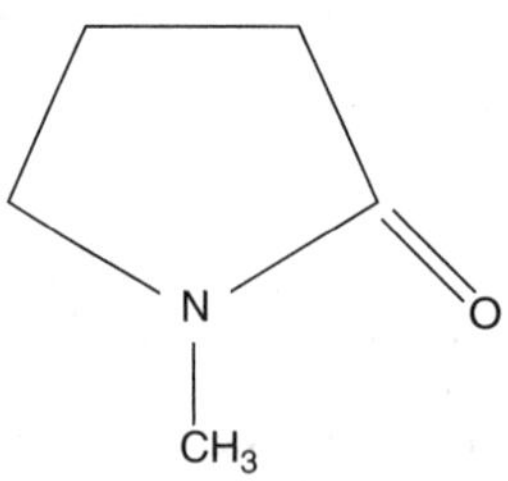

FIGURE 10-4 n-Methyl-2-Pyrrolidinone structure.

TABLE 10-7 Properties of n-Methyl-2-Pyrrolidinone Solvent

Formula	C_5H_9NO
Formula weight	99.13
Density	1.0279
Melting point, °F	–11.92
Boiling point, °F	395.6
Viscosity, CP, 100°F	1.65
Refractive index	1.468
Flash point, °F	196
Surface tension, dynes/cm	40
Evaporation rate (NBAC = 1)	0.03
Kauri butanol value	350

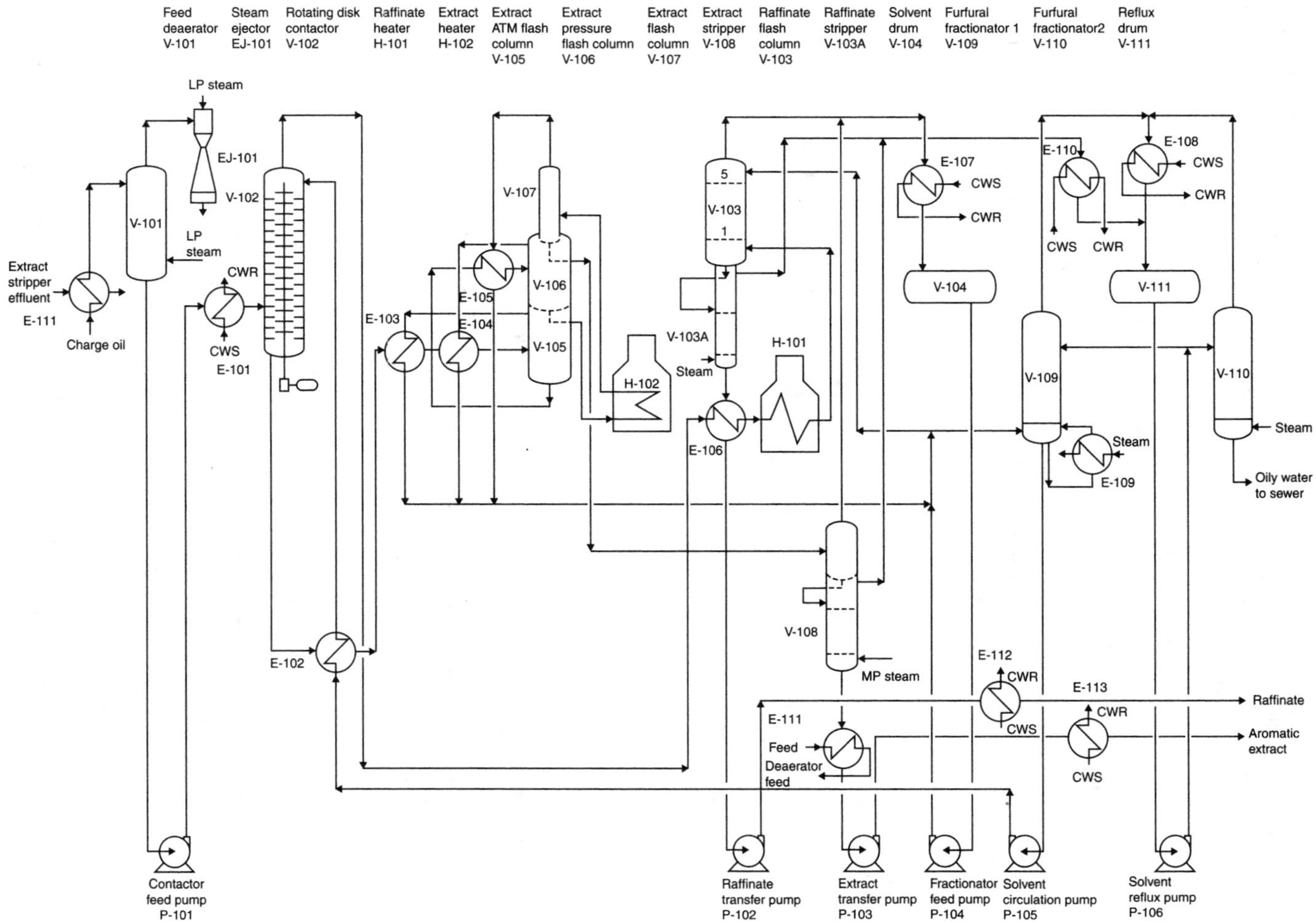

FIGURE 10-5 Furfural extraction plans.

extract streams are steam stripped to remove traces of solvent from these streams. Feed and solvent flow rates are also controlled, and these are contacted in a countercurrent manner in the tower. Raffinate stream exits from the top of the tower and is routed to a solvent recovery section for separation of solvent from this stream. The extract stream containing the bulk of the solvent exits the bottom of the extraction tower and is routed to the recovery section for removal of solvent from this stream. Solvent is separated from the extract phase by multiple effect evaporation at various pressures followed by vacuum flashing and steam stripping under vacuum. Solvent is separated from the primary raffinate by vacuum flashing and steam stripping under vacuum. The overhead vapors from steam strippers are condensed and combined with solvent condensate from recovery sections and are distilled at low pressure to remove water from solvent. Furfural forms an azeotrope with water and requires two fractionators. One fractionator separates furfural from azeotrope, and the second separates water from azeotrope. Water is drained to a oily water sewer. Solvent is cooled and recycled to extraction tower. Typical operating conditions, yields, and stream properties for a solvent extraction unit processing Middle Eastern crude are shown in Tables 10-8, 10-9, and 10-10.

TABLE 10-8 Furfural Extraction for Lube Manufacture Lube Base Stocks Yield
Feed: Vacuum distillates, propane deasphalted oil from Aghajari (Iranian) crude

	Spindle oil	Intermediate oil	Heavy oil	Vacuum residue
Yield on crude, Wt %	12.1	5.5	4.5	20
Deasphalted oil yield, Wt %				40
Products	Spindle oil	Intermediate oil	Heavy oil	Bright stock*
Raffinate yield, Wt %	56	57	54	63
Extract yield, Wt %	44	43	46	37
Properties (Raffinate)				
Viscosity index	95	95	90	90
Viscosity,† 102°F, cSt	10.4	62.8	143.5	507.5
Viscosity,† 210°F, cSt	2.7	10.2	13.7	31

*Bright stock yield on vacuum residue.
†Viscosity after dewaxing.

TABLE 10-9 Solvent Extraction of Lube Feedstocks Operating Conditions
Basis: Feed vacuum distillate and bright stock feed ex 34.5° API light Arabian crude
Furfural extraction unit

Operation*,†	100 N	300 N	650 N	150 BS
Feedstock nominal viscosity, cSt, 100°F	20.53	64.65	140.32	530.3
cSt, 210°F				31.9
Solvent-to-feed ratio, V/V	1.5	1.8	2.8	2.9
Furfural miscibility temperature at ratio, °F	244.4	267.8	276.8	311
Extracted raffinate VI	95	96	96	97

*Term "100 N," or 100 neutral, indicates that the nominal viscosity of feedstock at 100°F in SUS units is 100. "300 N" and "650 N" operations are similarly defined.
†Term "150 BS," or "150 bright stock," indicates that the nominal viscosity of feedstock at 210°F is 150 SUS.

TABLE 10-10 Furfural Extraction of Bright Neutral Operating Conditions
Feed: Bright neutral from light Iranian (Aghajari crude)

	Operating conditions	
	1	2
Solvent-to-feed ratio, V/V	1.470	1.720
Extractor temperature		
Top, °F	269.6	266
Bottom, °F	197.6	176
Raffinate density, gm/cc	0.9095	0.9068
RI, 140°F	1.48	1.483
Furfural content		
Raffinate, ppm	50	20
Extract, ppm	40	50

Solvent Dewaxing

The next step in conventional lube oil manufacture is removal of wax to improve the flow characteristics at low temperatures. Feed, the waxy oil, is mixed with a solvent (such as methyl ethyl ketone or methyl isobutyl ketone and the mixture is cooled from 10 to 20°F (6 to 12°C) below the desired pour point of lube oil. The wax crystals that are formed are removed by filtration. A process flow diagram of a refinery solvent dewaxing unit is shown in Figs. 10-6 and 10-7, and the operating conditions for a solvent dewaxing unit are shown in Tables 10-11 and 10-12. Properties of some dewaxing solvents such as methyl ethyl ketone (MEK), methyl isobutyl ketone (MIBK), benzene, and toluene are shown in Table 10-13.

Lube Hydrofinishing

Some base stocks produced by conventional processing scheme of solvent extraction and dewaxing require a finishing step, such as hydrofinishing or clay treatment, to improve color, oxidation stability, and thermal stability lubricating oil base stock produced. Hydrofinishing consists of passing the heated oil and hydrogen through a bed of catalyst. The process removes some color bodies and some unstable compounds containing nitrogen and sulfur. It saturates residual olefins to form paraffins. This process stabilizes base stock color and improves demulsibility and air release characteristics. A slight improvement in oxidation resistance may also result. An alternative process is clay treatment, which removes dark-colored and unstable molecules. Hydrofinishing represents a relatively mild operation at relatively low temperatures and pressures. A process flow diagram for a lube hydrofinishing unit is shown in Fig.10-8. Table 10-14 presents the overall lube yield from a conventional lube plant. Operation conditions for lube hydrofinishing unit are shown in Tables 10-15 and Table 10-16.

Hydroprocessing Route

Lube Hydrocracking. An alternative method to solvent extraction is hydrocracking to reduce aromatics in the base stocks. In this process, the aromatics are converted to naphthenes, paraffins, and fuel components by breaking carbon to carbon bonds at a high temperature and high hydrogen pressure, in the presence of a catalyst, rather than their physical removal. Depending on the severity,

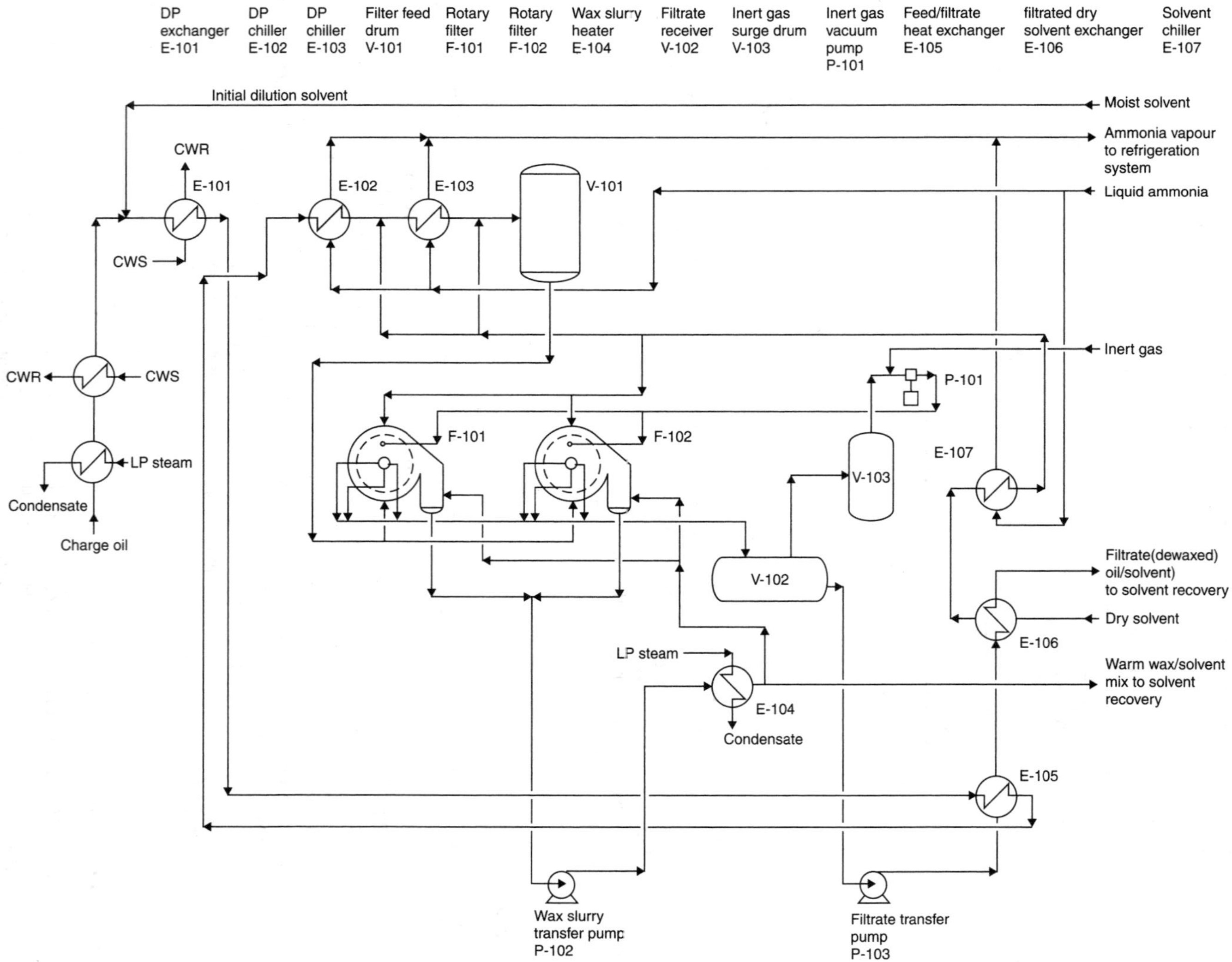

FIGURE 10-6 Solvent dewaxing of lube oils.

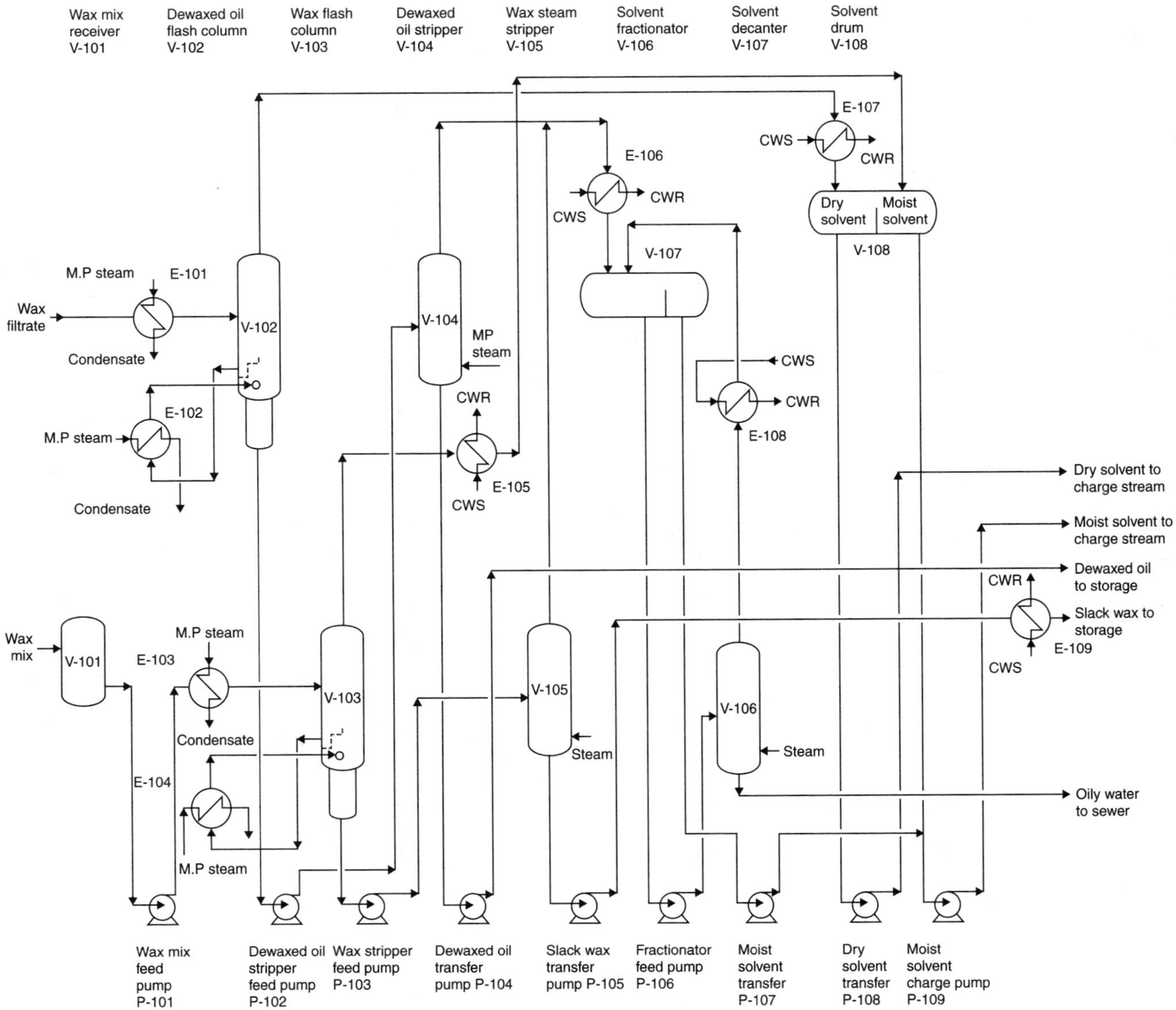

FIGURE 10-7 Lube dewaxing (solvent recovery section).

TABLE 10-11 Solvent Dewaxing of Lube Base Stocks Operating Conditions
Feed: Ex waxy asian crude

Cut, °F	840–915			
Density	0.8585			
Viscosity, cSt, 210°F	5.81			
Pour point, °F	134.6			
Wax content, Wt %	69			
Solvents		MEK + BZ + TOL	MIBK + BZ + TOL	MIBK
Solvent composition:	Vol %			
Methyl ethyl ketone		40		
Benzene		30	30	
Toluene		30	30	
MIBK			40	100
Solvent-to-feed ratio		5/1	5/1	5/1
Solvent addition temperatures	°F			
	158	1.0	1.0	1.0
	96.8	1.0	1.0	1.0
	32			
	–7.6	1.5	1.5	1.5
	–13	1.5	1.5	1.5
Total solvent-to-feed ratio		5.0	5.0	5.0
Average chilling rate				
Between 158 and 132°F	°F/min	2.5	2.5	3
Below 32°F	°F/min	1	1	<0.5
Dewaxing temperature	°F	–13	–13	–13
Wash solvent-to-feed ratio	V/V	1.5/1	1.5/1	1.5/1
Dewaxed oil				
Yield	Wt %	32.2	23.7	27.4
Pour point	°F	24.8	26.6	6.8
Kinematic viscosity	cSt, 210°F	11	11.4	12.1
Viscosity index		61	67	58
Slack wax				
Yield	Wt %	67.8	73.9	72.6
Conjgealing temperature	°F	144.5	138.56	135.5
Delta T (dewaxing temp-pour point)	°F	37.8	39.6	19.8

BZ = benzene; MEK = methyl ethyl ketone; MIBK = methyl isobutyl ketone; TOL = toluene.

TABLE 10-12 Solvent Dewaxing Operating Conditions
Feed: Bright neutral Aghajari crude

Solvent-to-feed ratio, V/V	4.00
SDA dosing, ppm	200
Chiller outlet temperature, °F	6.8
Filter feed drum temperature, °F	8.6
Cold wash-to-feed ratio	1.00
Cold wash temperature, °F	–0.4
Solvent composition	
Methyl ethyl ketone, Vol %	50
Toluene, Vol %	50
Viscosity of dewaxed oil, cSt, 210°F	35
VI of dewaxed oil	93
Pour point of dewaxed oil, °F	23
Flash point of wax, °F	424

TABLE 10-13 Properties of Lube Dewaxing Solvents

		MIBK	MEK	Benzene	Toluene
Molecular weight		100.1	72.11	78.1	92.1
Boiling point, °F		240.62	175.28	176.18	231.08
Specific gravity		0.802	0.805	0.879	0.867
Viscosity, CP, 68°F		0.585	0.425	0.649	0.587
Water solubility, 68°F					
	Wt % water	2.41	11.95		
	Wt % ketone	2.04	27.33		
Azeotropic data					
	Wt % water	24.3	11	8.33	13.5
	Boiling point, °F	189.86	156.02	156.65	183.38
Flash point	°F, PMCC*	60	19	12	40
Freezing point	°F	–112.47	–124.42	41.95	–138.98
Specific heat (liquid)	68°F, cal/g	0.46	0.498	0.411	0.392
Heat of vaporization	1 atm, cal/g	86.5	106	94.1	86.8

*PMCC = Pensky–Martens Closed Cup.

essentially all sulfur and nitrogen is eliminated from the base stock. This process changes the structure of many molecules in the feedstock. Aromatics are converted into naphthenes, many naphthene rings are broken open, and many paraffinic molecules are rearranged. Lube hydrocracking generally results in base stocks with improved oxidation and color stability. However, with very low aromatic levels, additives and contaminant solubility are a concern for the finished product formulation. The lube hydrocracking process is similar to distillate hydrocracking, a process used for the production of distillate, that is, naphtha, kerosene, diesel, and so on, from vacuum gas oil feed. However, processing conditions are less severe, which minimizes cracking. The process produces molecules that have a high VI and greater oxidation stability. The hydrocracking process allows a great deal of flexibility relative to crude source for the production of high-quality lube base stocks from inferior feedstocks. Although hydrocracking is less dependent on feed stocks than solvent refining, feed stocks still can have a significant impact on the product properties. Hydrocracking vacuum distillates usually produces base stocks in the 95 to 105 VI range. Higher operating severity can increase this to 115+ VI but with loss of yield. Use of a high wax (paraffin) content feed will result in even higher VI stocks. However, the hydrocracking produces predominantly lower viscosity base stocks (approximately 15 to 110 cSt at 100°F) due to cracking of larger, heavier molecules into smaller, lighter molecules. Thus hydrocracked base stocks cannot be used in many heavy industrial and engine oil products for which these must be blended with solvent-refined base stocks or other thickening agents. As described earlier, hydroprocessing results in lower aromatics, sulfur, and nitrogen than most solvent-refined base stocks. In some lubricant applications, hydroprocessed base stocks do offer desirable benefits for formulating finished products, but, they have lower additive and contaminant solubility, which needs to be addressed when formulating many lubricants. Additives response of hydroprocessed lube base stocks is different from those of solvent-refined base stocks. However, they do offer significant benefits in selected lubricant applications. This is particularly true when oxidation stability is critical and where additive content is low, such as in turbine oils. They also provide lower volatility lubes, more economically. Hydrocracked base stocks can be viewed as an important component for lube formulation that is used to develop finished products with superior performance but not a complete replacement of solvent extracted lubes base stocks.

The process flow diagram for a hydrocracker unit for producing lubricating oils is shown in Fig. 10-9. Operating conditions and product qualities are presented in Tables 10-17 and 10-18.

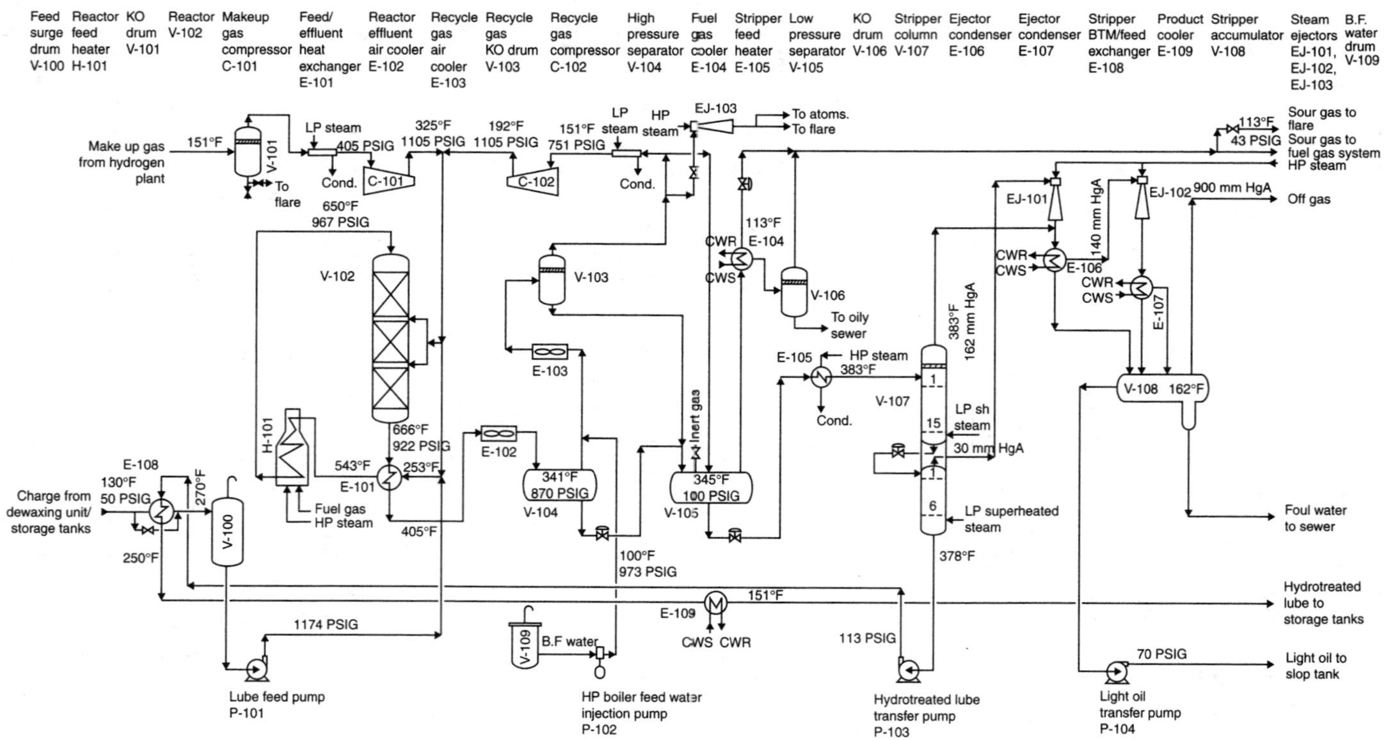

FIGURE 10-8 Lube hydrofinishing unit.

TABLE 10-14A Material Balance, Lube Plant
Feed: Aghajari crude

Distillation	Spindle	Light neutral	Intermediate neutral	Heavy neutral	Short resid			
Yield on crude, Wt %	2.3	9.8	5.5	4.5	20			
Deasphalted oil yield on short resid, Wt %					40			
Furfural extraction neutral	**Spindle**	**Spindle**	**Intermediate neutral**	**Intermediate neutral**	**Heavy neutral**	**Heavy neutral**	**Bright stock**	**Bright stock**
	95 VI	60 VI	95 VI	70 VI	90 VI	70 VI	90 VI	60 V1
Raffinate yield, Wt %	56	—	57	78	54	75	63	
MEK dewaxing								
Raffinate yield, Wt %	70	82	79	80	77	80	74	77
Hydrofinishing								
Yield, Wt %	99	98	99	98	98	98	98	97
Overall yield	39	80	44	61	41	59	46	75

TABLE 10-14B Product Properties

	Spindle	Spindle	Intermediate neutral	Intermediate neutral	Heavy neutral	Heavy neutral	Bright stock	Bright stock
	LVI	HVI	LVI	HVI	LVI	HVI	LVI	HVI
Viscosity, 120°F, cSt	2.7	2.7	10.2	8.1	17.3	13.7	39.5	31
Viscosity index		95	—	95		90		90
Pour point, °C	−12.2	−12.2	−6.7	−6.7	−6.7	−6.7	−6.7	−6.7
Color ASTM, max.	1–2	0–1/2	3–4	1–1.5				

TABLE 10-15 Lube Hydrofinishing Unit Operating Conditions and Yields
Feed: 851–972°F Aghajari cut undewaxed and unextracted lube cut
Catalyst: Co-Mo on Alumina

Properties		Operating conditions 1	Operating conditions 2
Pressure, lb/in^2		924	924
Temperature, °F		707	662
LHSV		1	0.5
Gas-to-oil ratio, SCF/BBL		2807	2807
Feed and product properties	**Feed**	**Product 1***	**Product 2†**
Density	0.9095	0.8878	0.89
Viscosity, 100°F, cSt	52	41.46	48.73
Color, ASTM	8+	2	2–
Total sulfur, Wt %	1.4	0.14	0.17
Conradson carbon, Wt %	0.014	0.014	0.018

*Product 1 as per operating conditions 1.
†Product 2 as per operating conditions 2.

TABLE 10-16 Lube Hydrofinishing Operating Conditions
Feed: SAE-30 Stock, 27.7 API

Reactor		
Reactor inlet pressure, lb/in^2		981
Reactor outlet pressure, lb/in^2		936
Feed inlet temperature, °F		649
Feed outlet temperature, °F		666
Space velocity, LHSV	Hr^{-1}	1.0
Flash drums		
HP flash drum pressure, lb/in^2		885
HP flash drum temperature, °F		351
LP flash drum pressure, lb/in^2		114
LP flash drum temperature, °F		345
Recycle gas rate	NSCF/BBL	2416
Makeup gas (95% purity)	NSCF/BBL	197
Stripper column		
Feed temperature, °F		383
Column top temperature, °F		383
Number of plates		15
Flash column		
Pressure, mm Hg A		30
Temperature, °F		378
Recycle compressor		
Recycle gas inlet pressure, lb/in^2		765
Recycle gas inlet temperature, °F		151
Recycle gas outlet pressure, lb/in^2		1119
Recycle gas outlet temperature, °F		192

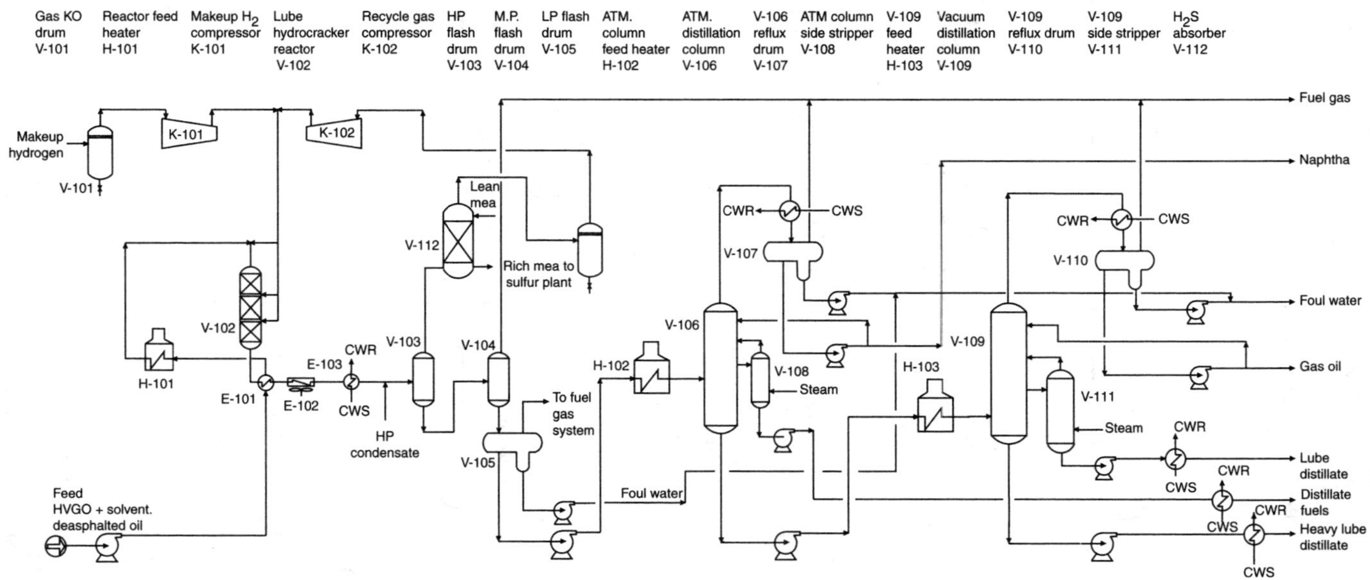

FIGURE 10-9 Lube hydrocracking unit.

TABLE 10-17 Lube Hydrocracking Operating Conditions

Operating parameter	Units	
Catalyst		Ni-Mo/Silica-alumina base
Reactor inlet temperature	°F	750
Reactor pressure	lb/in^2	>2500
Hydrogen partial pressure at reactor outlet	lb/in^2	2000
Space velocity	LHSV	0.5–1.0
Hydrogen rate	SCFB	9000
Hydrogen consumed	SCF/BBL	800–1000
Catalyst life	YR	1–3

TABLE 10-18 Lube Hydrocracking Feed and Product Properties

Property	Units	Feed	Product
Feed			
Deasphalted oil			
API gravity		21.5	
Viscosity	SUS, 210°F	60.1	
Viscosity	cSt, 210°F	10.4	
Pour point	°F	95	
Sulfur	Wt %	1.3	
Carbon residue	Wt %		
Product			
650°F + Lube cut after dewaxing			
VI			109
Viscosity	SUS, 210°F		43.3
	cSt, 210°F		5.3
Sulfur	Wt %		0.03
Carbon residue	Wt %		0.05
Yield	Vol % feed		42

CATALYTIC DEWAXING

In conventional lube making processes, wax (n-paraffin) is physically separated from the feedstock by solvent dewaxing to improve the cold flow properties of lube oil. In catalytic dewaxing processes, n-paraffins are isomerized to isoparaffins in the presence of a catalyst. Catalytic dewaxing is an alternative to solvent dewaxing, particularly in regions with a low demand for waxes. Feed to the process can be slack paraffin wax, microcrystalline wax, or Fischer-Tropsch wax. The pour point may range from –9 to –15°C. Feed stock is mixed with hydrogen, heated to the desired reactor temperature, and charged to a fixed bed hydrodewaxing reactor. Catalyst in the reactor selectively cracks or isomerizes the feed paraffin using a zeolite shape selective catalyst. Catalytic dewaxing can be applied to either solvent extracted or hydrocracked base stocks to reduce the pour point. Isoparaffins have high VI, low pour points, and a good resistance to oxidation. Because the process converts rather than remove paraffin, the process yield of lube base oil is higher compared to that from the solvent dewaxing process.

In the catalyst dewaxing process,[1] feed is first passed over an isomerizing catalyst. The catalysts employed have an acidic component and a hydrogenation component. The acidic component comprises an intermediate pore size silico-alumino-phosphate[2] (SAPO) molecular sieve. The hydrogenation function is provided by platinum and/or palladium metals. By intermediate pore size is meant an effective pore structure in the range of about 5.3 to 6.5 Å when the molecular sieve is in the calcined form. Intermediate pore size molecular sieves typically admit molecules having kinetic diameters of 5.3 to 6 Å with little resistance. The most preferred silico-alumino-phosphate is SAPO-11, although SAPO-31 and SAPO-41 are also used. The most preferred metal component is platinum. Some isomerization of n-paraffins takes place, thus reducing the pour point of the feed. The effluent from the first reactor passes over a second catalyst. This catalyst comprises an aluminosilicate zeolite having a pore size that admits straight chain normal paraffins either alone or with slightly branched chain paraffins but which excludes more highly branched hydrocarbons such as naphthenes and aromatics. Zeolites such as ZSM-5, ZSM-11, ZSM-12, ZSM-23, with platinum and/or palladium hydrogenation function are typically used. In this reactor bed, the waxy paraffins undergo mild cracking reactions to yield nonwaxy products.

The combination of two catalysts produce a greater yield of lube oil than achieved with either catalyst alone. Pour point is reduced both by isomerization reactions and cracking of normal paraffins. Viscosity is also reduced due to cracking reactions. Due to selectivity of the catalyst employed, gas and light ends yield is minimal. Feed nitrogen must be less than 10 ppm for a reasonable catalyst cycle length.

The catalytic operating conditions employed depends on the feed properties and pour point desired. Operating conditions are presented in Table 10-19. The intermediate pore size aluminosilicate zeolite (dewaxing catalyst) may be used in the same reactor as the silico-alumino-phosphate molecular sieve (isomerization catalyst) or it may be used in a separate reactor. When both catalysts are placed in the same reactor, the isomerization catalyst is layered on the top of the dewaxing catalyst. The effluent from the dewaxing reactor is sent to a hydrofinishing unit to provide a more stable lubricating oil. Catalysts used for hydrofinishing are conventional Co-Mo or noble metals such as Pt or palladium on an alumina base.

TABLE 10-19 Catalytic Isomerization Dewaxing Operating Conditions
Feed: Lube hydrocracker unconverted bottoms

Property	Units	Value
Reactor temperature	°F	480–550
Reactor pressure	lb/in^2	1100
Space velocity	LHSV	1.00
Hydrogen-to-oil ratio	SCF/B	2000
Hydrogen consumption	SCF/B	300–500

Deep Hydrotreating

Changes in finished lubricating oil specifications toward higher quality lubes is causing many lube plants to incorporate hydroprocessing technology into their existing conventional lube plant. These plants maintain the solvent extraction section. Effluent from the solvent extraction plant is fed to a hydrotreating unit that operates under moderate operating conditions to saturate the remaining aromatics and produce high VI and highly saturated lube. If wax production is important, the effluent from the hydrotreating unit is sent next to the solvent dewaxing unit. If wax is less important, the hydrotreated stream can be catalytically dewaxed[1] to further increase VI and yield. Hydrotreating in this case represents a more severe set of operating conditions than hydrofinishing but a less severe condition than that for lube hydrocracking. At higher pressures and with selected catalysts, aromatics rings become saturated to become naphthenes. In addition to converting naphthenes, hydrotreating can remove most of the sulfur and nitrogen. The operating conditions are so chosen that allow for retention of selected aromatic compounds, which in turn has a positive effect on oxidation stability and deposit control. This process is particularly suitable to manufacture very high-quality stocks for turbine applications.

Various processing steps (lube hydrocracking, hydrodewaxing, or deep hydrotreating) can be combined in a number of configurations to make group I, group II, or group III base stocks. A conventional lube-making scheme comprising solvent extraction, hydrofinishing, and solvent dewaxing can make only group I lube base stocks (Fig. 10-1). Figure 10-10 shows lube hydrocracking followed by hydrodewaxing to give Groups II and III lube base stocks. Figure 10-11 shows solvent extraction

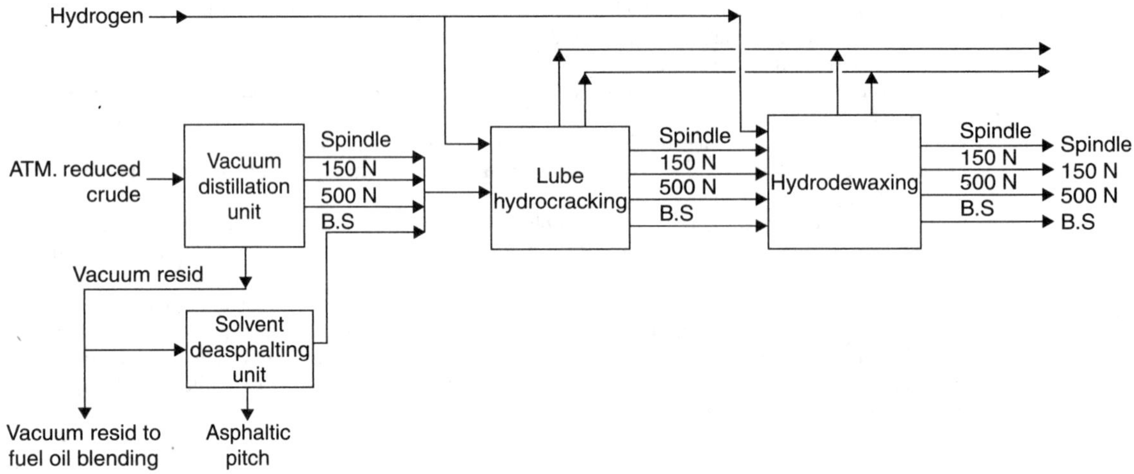

FIGURE 10-10 Lube processing scheme.

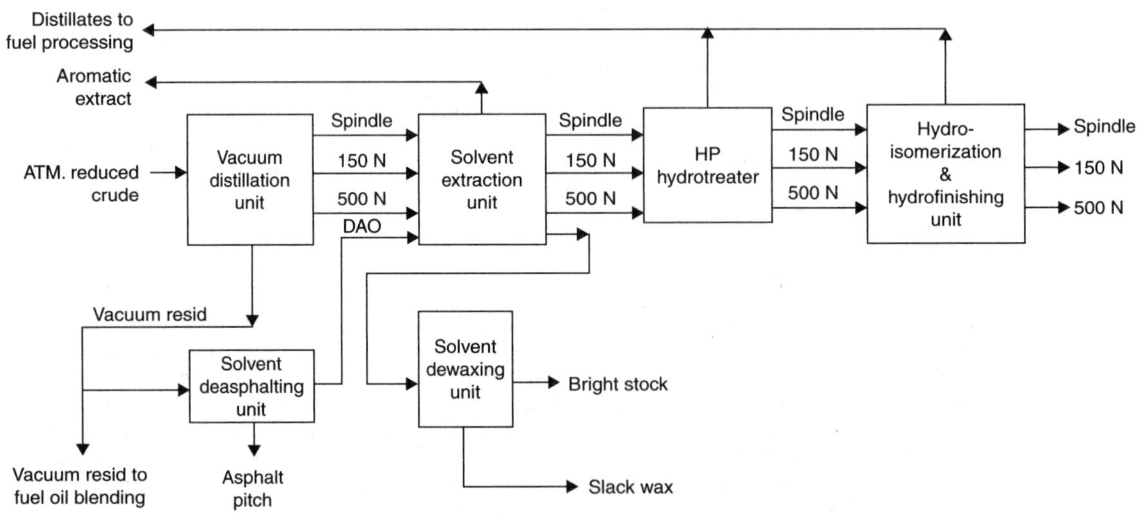

FIGURE 10-11 Lube base stock processing scheme.

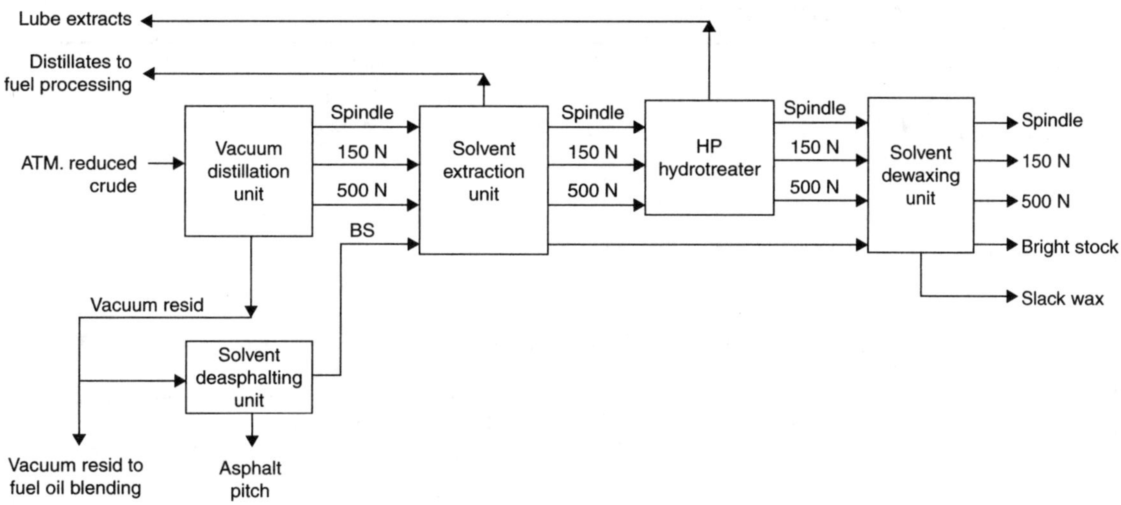

FIGURE 10-12 Lube base stock processing scheme.

followed by deep hydrotreating and hydroisomerization unit to yield group II and group III base stocks. Figure 10-12 shows solvent extraction, deep hydrotreating, and conventional solvent dewaxing to produce high VI lube base stocks.

AMERICAN PETROLEUM INSTITUTE CLASSIFICATION OF BASE OILS

Until recently, conventional lube-making schemes (solvent extraction, dewaxing, and hydrofinishing) dominated the industry. However, original engine manufacturers (OEMs) are now requiring lube base stocks for their high-performance new engines with the following properties:

- Lower viscosity for increased fuel economy
- Lower volatility for reduced oil consumption
- Improved oxidation and thermal stability for longer drain intervals
- Higher VI for improved lubricant performance at low and high temperatures

Changes in the specifications of finished motor oils are requiring an increased use of higher quality lubricants that cannot be produced by conventional lube-making schemes comprising solvent extraction, solvent dewaxing, and hydrofinishing. Lube base stocks for automobile lubricants meeting the requirements just listed can be produced by alternative hydrogen-treating processing options such as lube hydrocracking, deep hydrotreating, hydroisomerization of wax, and so on. In response to these changes, the American Petroleum Institute (API) established a base oil classification system in 1990 to help lube oil formulators to minimize retesting cost when blending licensed engine oils with base oils from different manufacturing sources. The system uses physical and chemical parameters to divide all base stocks into five groups as listed in Table 10-20.

As one moves from group I lubes to group II and group III lubes, the paraffin content of the base stock and the VI of the base stock increases. Only group I lubes can be economically produced by conventional lube technology consisting of solvent dewaxing and solvent dewaxing processes.

TABLE 10-20 API Classification of Lube Base Stocks

Group	Saturates, Wt %	Sulfur, Wt %	Viscosity index
I	<90	<0.03	>80–<120
II	$\geqslant 90$	<0.03	$\geqslant 80$–$\leqslant 120$
III	$\geqslant 90$	$\leqslant 0.03$	$\geqslant 120$
IV	All poly alpha olefins		
V	All base stocks not included in groups I to IV		

Group II and group III lubes base stocks with the very low sulfur, low volatility, and high VI required by modern automobile engines can only be produced by alternative hydroprocessing routes. It is not implied that conventional lube-making scheme is likely to be replaced by alternative processes. High-viscosity lubes can only be produced by the conventional methods. Also lubes for industrial usages, where high VI is not critical, could still be produced by a conventional lube-making method.

REFERENCES

1. U.S. Patent 5149421, "Catalytic Dewaxing Process for Lubes Using a Combination of Silico-Alumino-Phosphate Molecular Sieve Catalyst and an Aluminosilicate Zeolite Catalyst," issued September 22, 1992.
2. U.S. Patent 4440871 "Crystalline silico alumino phosphates" issued April 3, 1984 "Crystalline silico alumino phosphates" issued April 3, 1984.

CHAPTER 11
LUBRICATING OIL BLENDING

Lubricating oils are designed to perform several jobs in engines and other industrial machines. These are lubrication of moving parts, cooling, cleaning and corrosion control. Lube base stocks produced by refineries and petrochemical manufacturers (synthetic lube base stocks) are used to produce a large number of products such as the following:

- Gasoline and diesel engine oils
- Agricultural engine oils
- Marine engine oils
- Aviation and turbine oils
- Hydraulic and transmission oils
- Gear oils
- Automotive and industrial greases
- Metal working oils
- Electrical insulating oils
- White mineral oils
- Process oils

Automotive engine oils, transmission oils, and gear lubricants constitute roughly two-thirds of the total lube oils produced. Most of these products are prepared by blending an appropriate percentage of additives with lube base stocks. For every application, the base stock and additive package has to be carefully selected to meet the compatibility and requirements of the intended application.

CLASSIFICATION OF LUBRICATING OILS

Lubricating oils and greases can be classified in many ways; by their end use, by viscosity grades, by their additives package, or by their producers' brand names. The most popular classification of lubes is according to their usage:

- Engine oils (petrol and diesel engines, aircraft, marine engines)
- Turbine oils
- Gear oils
- Compressor (refrigeration, air) oils
- Quench oils used in metalworking
- Cutting oils (in metal cutting)
- Insulating oils used in transformer and circuit breakers
- Wire rope lubricants
- Chain lubricants
- Hydraulic oils

The type of base oil (naphthenic, paraffinic, or synthetic) and additives may vary to provide the qualities required for a given application.

CLASSIFICATION BY VISCOSITY

Classification according to viscosity has been done by these professional societies and organizations:

- SAE (Society of Automotive Engineers, USA)
- API (American Petroleum Institute)
- AGMA (American Gear Manufacturers Association)
- NLGI (National Lubricating Grease Institute)

INTERNATIONAL STANDARDS

For industrial lubricants, many different classification systems have been used in the past in different parts of the world. In 1972, a worldwide viscosity classification system for industrial lubricants came into place. Table 11-1 shows the current International Standards Organizations (ISO) 3448 viscosity classification system. The classification is based on series of viscosity grades, each being approximately 50 percent more viscous than the preceding grade. Viscosity variation within the grade is plus or minus 10 percent of the nominal viscosity of the grade. For lubricating oils other than automotive oils, the ISO VG classification system is generally used. This classification is in terms of viscosity only; other characteristics are not considered. Each ISO viscosity grade number corresponds to the midpoint of the viscosity range expressed in centistokes (cSt) at 40°C. For example, a lubricant with an ISO grade viscosity of 100 has a viscosity in the range of 90 to 110, with 100 the midpoint of the range.

CLASSIFICATION BY ADDITIVE TYPES

Lube oils may be classified by additive type as follows:

- Inhibited or RO (rust and oxidation inhibitor containing additives)
- Anti-wear (AW) containing lube oils
- Lubes with extreme pressure (EP) additives
- Compounded oils: containing 3 to 10 percent fatty or synthetic fatty oils
- Residual compounds: These are heavy-grade straight mineral oils with asphaltic component for open gear application where tackiness is required to increase adhesion

AUTOMOTIVE ENGINE OILS

Viscosity is considered the most important single property of lubricating oils. Automotive crankcase and gear oils used in gasoline and diesel engine lubrication have been classified by the SAE (Society of Automotive Engineers, USA) in terms of viscosity. The viscosity classification takes into account the temperature at which the oils are to be used. In 1911, the SAE developed the SAE J300 standard that specifies engine oil viscosity classification. The SAE viscosity rating was based on average time, in seconds, for a specific quantity of motor oil at 212°F to flow through a test apparatus or viscometer with an orifice of a specified diameter. An SAE viscosity number, according to SAE J300, was the average time and not an exact measure. For example, any oil that took 5 to 15 s to flow would be

TABLE 11-1 Nonautomotive Lubricating Oils ISO Viscosity Grades

ISO grade	Midpoint viscosity at 40°C cSt	Viscosity at 40°C Min. cSt	Viscosity at 40°C Max. cSt	Viscosity at 37.8°C Min. SUS	Viscosity at 37.8°C Max. SUS	Approximate viscosity at 98.9°C, SUS 95 VI Min.	Approximate viscosity at 98.9°C, SUS 95 VI Max.	Approximate viscosity at 98.9°C, SUS 65 VI Min.	Approximate viscosity at 98.9°C, SUS 65 VI Max.	Approximate viscosity at 98.9°C, SUS 35 VI Min.	Approximate viscosity at 98.9°C, SUS 35 VI Max.
2	2.2	1.98	2.42	32.8	34.4						
3	3.2	2.88	3.52	36	38.2						
5	4.6	4.14	5.06	40.4	43.5						
7	6.8	6.12	7.48	47.2	52						
10	10	9	11	57.6	65.4	34.6	35.7	34.2	35.3	33.8	34.9
15	15	13.5	16.5	75.8	89.1	37	38.3	36.4	37.8	36	37.3
22	22	19.8	24.2	105	126	39.7	41.4	39.1	40.6	38.5	40
32	32	28.8	35.2	149	182	42.9	45	42	43.8	41.4	42.9
46	46	41.4	50.6	214	262	47.1	49.9	45.4	47.8	44.2	46.2
68	68	61.2	74.8	317	389	53	56.9	50.3	53.4	48.6	51.1
100	100	90	110	469	575	61.4	66.9	56.8	61	54	57.2
150	150	135	165	708	869	74	81.9	66.6	72.7	62.1	67.2
220	220	198	242	1046	1283	90.3	101	79.3	87.6	72.6	79.5
320	320	288	352	1531	1878	112	126	95.7	106	86.3	95.3
460	460	414	506	2216	2717	139	158	116	130	104	115
680	680	612	748	3298	4046	178	202	145	162	127	142
1000	1000	900	1100	4885	5994	227	257	181	204	156	175
1500	1500	1350	1650	7385	9063	293	331	229	256	204	219

ISO classification of viscosity is issued by International Organization for Standardization, as ISO 3448, BS 2431.
The system classifies industrial lubricating oils into ranges by their nominal kinematic viscosity at 40°C (104°F).
Each range is identified by an ISO VG, viscosity grade, number that corresponds to midpoint viscosity of its range in centistokes.

TABLE 11-2 SAE J 300 Engine Oil Viscosity Specifications*

SAE viscosity grade	Cold cranking max. viscosity		Cold pumping max. viscosity		Viscosity, cSt at 100°C		Hot/High shear viscosity at 150°C cP
	cP†	at, °C	cP	at, °C	Min.	Max.	
0 W	6200	−35	60,000	−40	3.8		
5 W	6600	−30	60,000	−35	3.8		
10 W	7000	−25	60,000	−30	4.1		
15 W	7000	−20	60,000	−25	5.6		
20 W	9500	−15	60,000	−20	5.6		
25 W	13000	−10	60,000	−15	9.3		
20					5.6	9.3	2.6
30					9.3	12.5	2.9
40					12.5	16.3	3.7
50					16.3	21.9	3.7
60					21.9	26.1	3.7

*Revised December 1999.
†CentiPoise.

rated as SAE 10. Similarly, oil that took 25 to 34 s would be SAE 30, and so on. At that time, only SAE 10, 20, 30, 40, and 50 grades existed. There were no SAE 0 or SAE 60 or winter grades as in the present classification. Over the years SAE J300 standards were amended many times with advances in automobile design. SAE grade 60 was added as the need for thicker oil in aviation and heavy-duty engines became apparent. SAE W grades were added as it became apparent that engines could not be started in cold winter conditions with some SAE 30 oils. Table 11-2 shows the SAE classification for crankcase oils.

Multigrade Engine Oils

A multigrade engine oil such as SAE 10W/30 has the viscosity characteristics of both SAE 10W and SAE 30 oil. Similarly, the multigrade grade oil SAE 80W/90 has the viscosity characteristics of SAE 80W and SAE 90 gear oils. Today, most automobiles and trucks use what is termed "multiviscosity oils." These are in fact multigrade, not multiviscosity, oils. An oil cannot be multiviscosity, but it can be multigrade. For example, an SAE 10W40 oil meets the viscosity requirements of SAE 40 oil and also that of subzero grade SAE 10W. This implies that oil will flow like SAE 10W oil at a low ambient temperature such as during cold starting. Referring to Table 11-2, an SAE 10W oil will have a viscosity not greater than 7000 cSt at −25°C (for cold cranking) and not more than 60,000 cSt at −30°C to facilitate the flow of oil to the engine during cold weather. Multigrade lubricating oils are manufactured by blending a low-viscosity base stock oil with a viscosity index improver additive. These are polymer additives that produce a thickening effect at high temperatures but are dormant at low temperatures. In multigrade oils, polymers are added to light base stock that prevent the oil from thinning out at a fast rate as the oil warms up. At low temperatures, the polymer molecules are coiled up and allow the oil to flow without freezing. As the oil warms up, the polymer molecules begin to unwind into a long chain that prevents the oil from thinning at a rapid rate with the temperature increase. Multigrade oils are one of the greatest advancements in lube formulation, but they should be chosen with caution. It is appropriate to choose a multigrade oil with the narrowest span of viscosity appropriate for the highest and lowest ambient temperature in summer and winter driving conditions. Lubricating oil with a minimum amount of polymers is better for the engine. In actual use in engines, polymers can shear and burn, forming deposits that can cause sticking and other problems in engines. For example, if we compare the two lube oils 10W-40 and 20W-50, both oils have a 30-point spread, but because 20W-50 starts with a heavier base oil, it requires less of a viscosity index improver (polymer) to meet the specifications and thus is preferable. The wide viscosity range oils with their higher polymer content are more prone to viscosity and thermal breakdown. It is the oil that lubricates, and additives do not play any part in it. The oils with the minimum of additives are the best. The optimum viscosity grade for a given application depends on the ambient conditions and the type of engine used.

Neat Engine Oils

In certain applications, multigrade engine oils with a high VI additive content is not desirable. These include high r/min, high-load engines such as lawn mowers, high horsepower racing cars, and diesel engines. Lubricants such as SAE 30 or SAE 40 are recommended because these engines operate under a very high r/min, load, and heat conditions and, in the case of small engines, they are air cooled. VI additives are a weak link in the lube. VI additives under extreme heat, load, and shear condition tend to depolymerize, resulting in failure to reach their intended high-temperature viscosity. As the shearing continues, loss of oil film, increased wear rates, high oil consumption, or engine failure may result. Use of synthetic (PAO, esters) base oil lubes with very high VI (120+) is increasing in this segment.

Optimum Viscosity Grade

The ideal viscosity for motor oil in piston engines operating at normal engine operating conditions is equivalent to SAE 30 (9 to –12 cSt at 100°C). If a thinner oil is used, say SAE 20, there will be less resistance to motion due to reduced viscosity and therefore better fuel economy. However, this gain in fuel economy does not occur without cost due to the following factors:

- Lower viscosity and higher volatility results in greater oil consumption.
- Decrease in engine service life due to increased engine wear under the same operating conditions. It is estimated that the gain in term of better mileage by using SAE-20 instead of SAE-30 may be less than 1 percent, whereas the reduction in engine life may be on the order of 30 percent.

If a heavier oil (SAE 40 or SAE 50) is used under normal operating conditions, there will be more resistance to motion due to increased viscosity, resulting in poorer fuel economy. The loss in fuel economy is somewhat compensated for by the following:

- Decrease in oil consumption due to higher viscosity
- Decrease in oil consumption due to lower volatility
- Increase in engine service life due to reduced engine wear

Most heavy-duty engine manufacturers recommend SAE-40, 15W-40, or 5W-40 oils. If the ambient or the operating temperature is increased from ideal or normal (70°F), oil viscosity must be increased to assure same level of protection from engine wear, thus SAE 40 for 100°F and SAE 50 oils for 120°F ambient temperatures. Heavier grades are also desirable for towing and heavy loads. If the engine speed is increased during long-distance high-speed driving in low ambient temperatures, lower viscosity oils such as SAE-20 could be used in place of SAE-30 (in manual transmission cars where vehicle speed is proportional to engine r/min). If the load is decreased, oil viscosity can be decreased without any penalty. Long-term durability of the engine is related to high temperature/high shear specifications in SAE J300 standards. For SAE-20 it is 2.6 cP minimum at 150°C. For SAE-30, it is 2.9 cP (minimum). For SAE-40 oil, there are two specifications: 2.9 cP, which is the same as that for SAE-30, and 3.7 cP, which is the same as those for SAE-50 and SAE-60. The first specification is for light-duty engines (service life, 100,000 to 150,000 miles), and the second specification is for heavy-duty engines (service life, up to 1,000,000 miles). These oils are labeled as "HD" oils. Heavy-duty oils must meet the second SAE-40 specification of 3.7 cP, minimum, at 150°C. (Refer to Table 11-3 for a list of the optimum motor oil grades.)

TABLE 11-3 Optimum Motor Oil Grades

SAE viscosity grade	Temperature conditions for use, °F	Recommendations for use
5W–30	<0	Provide excellent fuel economy and low-temperature performance.
10W–30	>0	Most frequently recommended viscosity grade for most automobile engines, including high-performance multivalve engines and turbocharged engines.
10W–40	<0	For greater protection against engine wear and oil breakdown from oxidation.
20W–50	>20	Maximum protection for high-performance, high r/min racing engines.
30	>40	For high temperatures and heavy loads, such as driving in the desert or towing a trailer at high speeds for long periods.

EFFECT OF VISCOSITY ON FUEL ECONOMY

For a vehicle to economize on fuel, frictional losses in engine must be cut to minimum. Fuel economy improves when more energy goes into propelling the vehicle and less energy is used in overcoming frictional losses in the engine. This may include oil film resistance between moving parts, and the friction reduction must be maintained through changes in speed, shear stress, and temperature. In passenger vehicles, significant friction losses occur in the valve train, whereas in commercial vehicles using heavy diesel engines, friction in the valve train is not significant but most frictional losses occur in the piston assembly and bearing where rubbing surfaces are separated by an oil film. The viscosity grade of the engine lubricants has a profound effect on friction and fuel consumption, particularly in heavy-duty diesel engines. Higher viscosity oils provide adequate wear protection but create more resistance to motion and high frictional losses. Lower viscosity oils reduce frictional losses, but engine protection may be compromised, more so in high soot conditions unless the lube is formulated with the correct additive package. High temperature/high shear (HT/HS) viscosity is also important in achieving fuel economy and wear protection. In general, the lower the HT/HS rating, the better the fuel efficiency. However, too low an HT/HS rating can compromise wear protection. The level of fuel economy improvement when using lower viscosity lube depends on the engine design, operating condition, drive cycle, and fuel quality.

AUTOMOTIVE OIL ADDITIVES

Very little mineral lube oil without additives is sold in the market as a lubricant. Most commercial lubricants contain additives to enhance their performance, but the largest market for additives is in the transportation field, including engine and drive trains for cars, buses, trucks, locomotives, and ships. Motor oils, unlike many other industrial lubes, operate over a wide range of temperature under which they must provide engine lubrication, minimize engine wear, resist high-temperature lube degradation, and keep contaminants in suspension over a longer time interval. Generally, properties of the lube base stocks are not adequate to meet all these requirements; therefore many additives are added to the base stock to constitute a motor lube suitable for gasoline or diesel engines. On average, a typical motor oil may contain 18 percent additives and 82 percent base oil. Present-day petroleum-based motor oil may contain the following additives:

- Viscosity index improvers
- Detergents
- Dispersants
- Anti-wear/Extreme pressure additives
- Friction modifiers
- Antioxidants/Corrosion inhibitors
- Rust and corrosion control additives
- Pour point depressants
- Antifoam agents
- Other additives (seat swell agents/dyes/biocides, etc.)

VISCOSITY INDEX IMPROVERS

The viscosity of lubricating oil changes with temperature. The rate of change depends on the composition of the oil. A naphthenic base oil viscosity changes more rapidly with a temperature increase compared with paraffinic base stocks. The viscosity of certain synthetic lubricants changes more slowly compared with paraffinic base oils. For a given oil, the viscosity index (VI) is a measure of change of viscosity with temperature. Thus a high VI implies a smaller change in viscosity with temperature, and a low VI implies large change in viscosity with temperature. The VI of a lubricating oil is of immense importance in applications where an appreciable change of temperature of the lubricating oil could adversely affect the startup or operating characteristics of automotive engines and other equipment. Lube manufacture aims at the highest possible VI so the change in viscosity of lube oil during actual use is minimal. The VI of an oil is estimated by the ASTM D 2270 test method. This method estimates the VI of a given oil based on its viscosities at 40 and 100°C. For an oil to be effective at low temperature and yet be thick enough to lubricate at high temperatures, oil must have a fairly flat viscositytemperature relationship. The change in viscosity of lube oil with temperature is defined by the VI of the oil. During the manufacture of petroleum lube base stocks, low VI components (aromatics) in the feed are either removed by solvent extraction or hydrogenated to naphthenes or paraffins to raise the VI of the lube. Raising the VI of lube beyond a limit by processing alone during base stock manufacture becomes uneconomical, and a further increase in VI is achieved by certain additives called VI improvers during motor oil formulation. VI improver additives make it possible to furnish multigrade oils commonly used in most engine oils today. These multigrade oils have a low-temperature viscosity equivalent to the first or a low viscosity number, but on heating they do not drop in viscosity as rapidly as straight mineral oils and have a viscosity of a higher grade at the elevated temperature. The improvement in the slope of the viscositytemperature curve is achieved by the addition of organic polymers. There are two main types of VI improvers: isobutylene polymers (Fig. 11-1) and methacrylate copolymers (Fig. 11-2). Molecular weight is chosen (typically weight average of 180,000 to 250,000). Typically the content of the VI improver in a finished motor lubricant may be between 0.5 and 10 percent by weight (wt %). High molecular

Isobutylene polymers

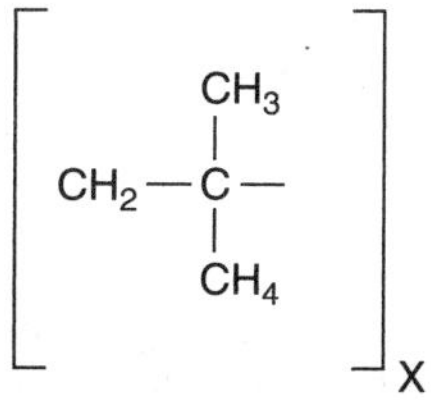

FIGURE 11-1 Lube additive.

Methacrylate copolymer

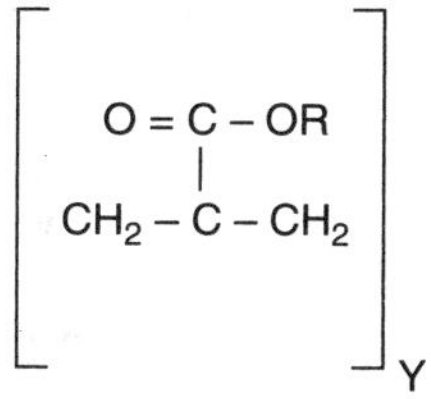

FIGURE 11-2 Lube VI improver.

weight polymers give a higher VI improvement at a given concentration. With increasing molecular weight, polymers are increasingly subject to breakdown under shearing conditions encountered in high-speed rotating engine parts. The use of viscosity improvers makes it possible to meet both the low-temperature viscosity requirements of W grades as well as the high-temperature requirements of non-W grades. For example, the viscosity of 15 W 40 grade with a VI of 135 at a different temperature would be as follows:

Temperature °C	Viscosity cP
−15	3000
40	105
100	14

Many of the organic polymers used as VI improvers are shear-sensitive compounds. This means that under high-shear conditions, there is change in the average molecular weight of the polymer and a reduction in its viscosity. The molecular weight change is believed to be due to the breakage of the chemical bond in the polymer chain. Of course this defeats the very purpose of adding these additives to the oil. Shear stability is a measure of the amount of viscosity an oil may lose during actual usage. Oil experiences very high stresses in certain areas of the engine such as in the oil pump, cam shaft area, piston rings, and in many areas where two mating metallic surfaces squeeze the oil film. As oil passes through these mating surfaces, a fraction of the polymer compound contained in VI additives, under severe mechanical stress, is permanently sheared or depolymerized. This results in a loss of viscosity of the oil and thus the effectiveness of the VI improver. The shear stability of an engine oil is measured by using the ASTM D 5275 test method. First, the viscosity of a formulated engine oil is measured. Next, the oil is exposed to severe shearing conditions by repeatedly pumping it through a specially sized diesel fuel injection nozzle at high pressure. After this operation, the viscosity of the oil is measured again. The percentage viscosity loss is measured by comparing the second viscosity measurement with the original viscosity. Although there are no specifications indicating required level of shear stability for engine oils, oils with lower viscosity loss would be preferable.

Shear Stability of VI Improvers

As stated earlier, many of the organic polymers used as VI improvers are shear-sensitive compounds. This means there is a change in the average molecular weight of the polymer and a reduction in viscosity every time the oil is subjected to high shear. Molecular weight change is brought about by breakage in the chemical bonds in the polymer chain, which defeat the very purpose of adding these additives to oil. It is essential that these compounds be sold in as stable condition as possible. In manufacturing these compounds, the mixture coming out of polymerization reactor may contain a wide variation of molecular weights and with a corresponding variation in resistance to shear. To overcome this problem, the polymer coming from the reactor is passed through a homogenizer. In the homogenizer, the polymer is subjected to an intense energy and shear environment, much more than what it is likely to undergo in the lube circuit of an engine. This ensures that a change in the viscosity of oil due to high shear in the engine is minimal. The product coming out of the homogenizer is remarkably stable with respect to viscosity change. The stability of the polymer additives in actual shear conditions encountered in engines is measured by the shear stability index (SSI),[1] which is defined in Fig. 11-3.

This index can be used to predict the viscosity of an oil after it has been used in an engine. Measurement of the shear stability index is done by the following procedure: The polymer additive is dissolved in oil and next heated from a minimum of 100 to 300°F, usually to 210°F to make it pumpable. The polymer solution is pumped to the homogenizer, which has special wearing parts. The solution is processed in one or more passes at a high homogenizing pressure. The pressure and number of passes used depends on the particular polymer and the desired viscosity reduction. The finished product is added to the oil to obtain the desired viscosity. The principal test is a viscosity

$$SSI = \frac{\mu_i - \mu_f}{\mu_i - \mu_0}$$

Where:

SSI = Shear stability index

μ_i = Initial viscosity in cSt at 100°C

μ_f = Final viscosity after subjection to high sheer, cSt at 100°C

μ_0 = Viscosity of base oil blend with all additives except VI improvers, at 100°C

FIGURE 11-3 Shear stability index.

check immediately after homogenization and again after 24 h of hold time to determine any viscosity regain if this occurs. The viscosity is usually determined at 210°F.

DETERGENT INHIBITORS

Lubricating oils for modern automotive engines are designed not only to provide adequate lubrication under varying temperatures and operating conditions, but also to keep the engine clean and provide protection against chemical corrosion from acidic combustion products. These important properties are added to it by means of alkaline reacting additives referred to as detergents/dispersants. During the combustion of fuel, sulfur is oxidized to SO_2 and SO_3. Part of these sulfur oxides combine with water during the combustion process and form sulfurous and sulfuric acid, which are very corrosive to engine components. To neutralize these acids, bases are needed. Thus oil-soluble bases must be present in the lubricating oils of internal combustion engines. Bases react with acids, sludge, and varnish precursors so as to neutralize them and keep them in solution. The compounds used for this service are generally metallorganic compounds such as sodium, magnesium or calcium sulfonates, salicyclates, phenates, thiophenates, and phosphonates. Oil-soluble bases such as calcium sulfonate and calcium phenates are also excellent deposit control additives.

Sulfonates

Sulfonates are the products of neutralization of sulfonic acid with a metallic base. Petroleum or natural sulfonates are a by-product of white oil manufacture. During white oil manufacture, lube base stock is treated with oleum. The sludge formed is separated by settling, oil is neutralized by an alkali, and then the sulfonates formed in the oil are extracted with a solvent like isopropyl alcohol. This process is repeated a number of times until the aromatics are completely removed. The structure of the organic petroleum sulfonate produced depends on the crude source and can be the aliphatic, naphthenic, or aromatic hydrocarbon group. Due to the high demand for natural petroleum sulfonates, sulfonates have become the main product and white oil the by-product. However, environmental concerns discourage the use of sulfuric acid, and white oil manufacturers are increasingly switching over to the hydrogenation route for white oil manufacture, resulting in a dwindling supply of natural sulfonates. Synthetic sulfonates are increasingly being manufactured by the reaction of alkylated aromatics of the proper molecular weight with sulfur dioxide. Metallic cations of sulfonate detergents are calcium, magnesium, and sodium. Oil-soluble sulfonates containing metals in excess of stoichiometric amounts are called basic sulfonates, and they are helpful in neutralizing acid bodies over a longer time interval apart from acting as dispersants of contaminants.

Preferred alkaline earth sulfonates are alkaline earth metal salts, preferably magnesium or calcium salts of an alkyl aromatic sulfonic acid with a molecular weight of 400 to 700 (Fig. 11-4). A finished lube may contain 2 to 5 percent by mass of these detergents.

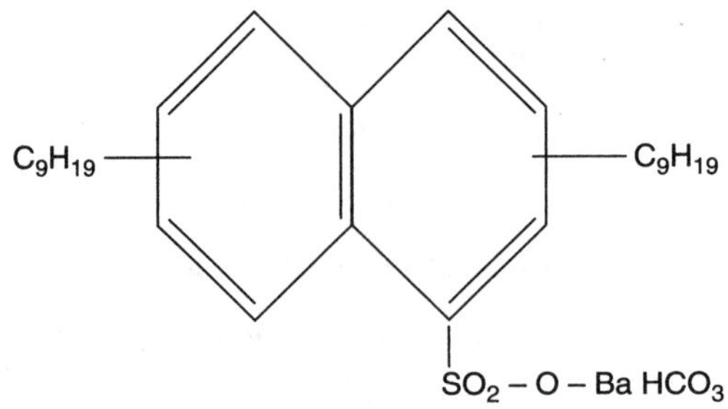

Barium dionylaphthalene sulfonate

FIGURE 11-4 Structure of basic barium sulfonate, a metallic detergent.

Salicyclates

Salicylates are generally prepared from alkyl phenols by Kolbe's reaction. Here phenol is first converted into sodium phenoxide, by neutralization with sodium hydroxide. Next sodium phenoxide is heated with carbon dioxide under pressure. In this process a carboxyl group joins the ring next to the phenoxide group. Finally, this product is heated with dilute acid (e.g., hydrochloric acid) and 2-hydroxybenzoic acid is formed (Fig. 11-5). The salicylates of potassium, calcium, and magnesium are highly basic detergents that are used in diesel engine oil formulations.

$$C_6H_5OH + NaOH \longrightarrow C_6H_5O^- Na^+ + H_2O$$

$$C_6H_5O^- Na^+ + CO_2 \longrightarrow Na^{+-}O\ C_6H_4\ COOH$$

$$Na^{+-}O\ C_6H_4\ COOH + HCl \longrightarrow HO\ C_6H_4\ COOH + NaCl$$

FIGURE 11-5 Manufacture of salicyclates.

Phenates and Phenol Sulfide Salts

Metal phenates include salts of alkyl phenols, alkyl phenol sulfides, and alkyl phenol aldehydes. Oil solubility is provided by reacting or alkylating phenol with C_7+ olefins. Sulfur is incorporated into the phenate by reacting the alkyl phenol with sulfur chloride or elemental sulfur. The introduction of sulfur lowers the corrosivity of the product toward bearing metals and improves their antioxidant characteristics.

DISPERSANTS

The purpose of dispersants is to keep in suspension harmful products such as dirt, fuel, water, lube degradation products such as sludge, varnish, and oxidation products. Contaminants are bonded by polar attraction to dispersant molecules, preventing them from agglomerating. A major development in recent years has been the use of ashless dispersants. These materials may be categorized in two broad types; (1) high molecular weight polymeric dispersants used to formulate multigrade oils, and (2) lower molecular weight additives for use where viscosity modification is not necessary.

These additives are much more effective than the metallic type in controlling sludge and varnish deposits. The compounds useful for this service are again characterized by a polar group attached to a relatively long hydrocarbon chain. The polar group generally contains one or more elements: nitrogen, oxygen, or phosphorus. The solubilizing chains are generally higher in molecular weight than those in detergents. These ashless dispersants may serve the duel function of dispersants and viscosity modifiers. The preferred ashless dispersants are any type of polybutenyl succinimides. Specific examples of such dispersants are mono-type imides and bis-type imides (Fig. 11-6). Polybutenyl succinimides may be obtained by reacting polybutenyl succinate with maleic anhydride. These dispersants are added to lube to give a concentration of 0.1 to 0.2 percent in terms of nitrogen concentration based on total mass of the composition. Typical applications include diesel and gasoline engine oils, transmission fluids, power steering fluids, and some gear oils.

Polybutenyl succinimide type

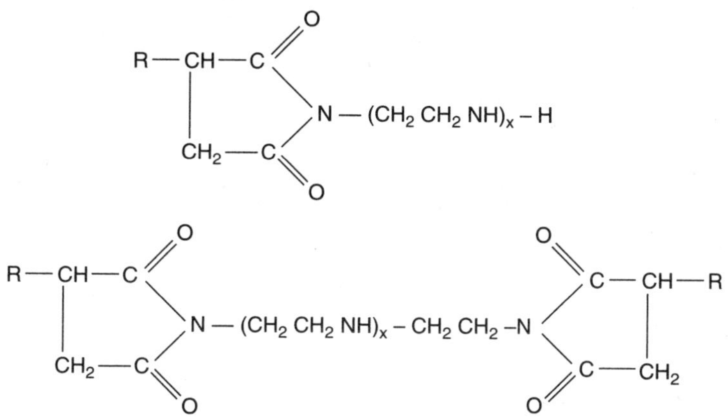

Where:

R is polybutenyl group with mol weight between 1000 and 3000 and x is and integer between 2 and 5.

FIGURE 11-6 Structure of polybutenyl succinimide type ashless lube oil dispersant.

ANTI-WEAR/EXTREME PRESSURE ADDITIVES

Both anti-wear and extreme pressure (EP) additives reduce friction and wear and prevent scoring and seizure. The additives form a protective layer on metal parts by decomposition and adsorption. Anti-wear additives function in a moderate environment of temperature and pressure, whereas EP additives are effective under more severe operating conditions. Wear is the loss of metal with a subsequent change in clearance between surfaces moving relative to each other. If it continues, an attack of particulate matter and corrosive acids can occur at these points. Metal-to-metal contact is prevented by adding film-forming compounds that protect the surfaces either by physical adsorption or chemical reaction. Zinc dithiophosphates, organic phosphates, organic sulfur and chlorine compounds, sulfurized fats, sulfides and disulfide are used for this purpose, and these are very effective in reducing wear. Molybdenum disulfide and graphite additives are special forms of anti-wear additives known as anti-seize agents. They form a protective layer on metal parts by the deposition of molybdenum disulfide or graphite. Anti-seize agents work independently of temperature or pressure. The content in excess of the upper limit can accelerate the poisoning of the catalyst in an exhaust cleanup system, adversely affecting the cleanup of exhaust gases.

FRICTION MODIFIERS

Friction modifiers are lubricant additives blended in lube to reduce friction. Friction modifiers also reduce wear, scoring, and noise. Organic fatty acids and amides, lard oil, and high molecular weight organic phosphorus acid esters are used as friction modifiers. Molybdenum thiocarbamate (Fig. 11-7) is a popular friction modifier. It is added to lube formulation in a concentration of 0.02 to 0.06 percent by mass in terms of Mo concentration in many formulated lubes.

Friction modifier lube additive

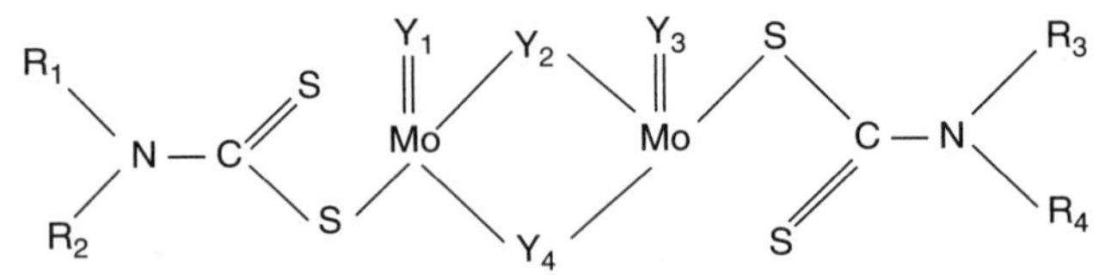

Where:

R_1, R_2, R_3, R_4 are same or different alkyl or aryl groups having 2 to 18 carbon atoms

Y_1, Y_2, Y_3, Y_4 are each independent sulfur or oxygen atoms

FIGURE 11-7 Structure of molybdenum dithiocarbamate.

OXIDATION INHIBITORS

The function of oxidation inhibitors is to retard the deterioration of lubricants associated with an oxygen attack. These additives act by either destroying free radicals (chain breaking) or reacting with peroxides involved in the oxidation mechanism. Among the most widely used antioxidants are phenolic type such as 2,6 ditertiary butyl phenol [$(C_4H_9)_2$ $C_6H_2(OH)]_2$ CH_2, 2,6-di tert-butyl para -cresol (BHT) etc.] (Fig. 11-8) and zinc dithiophosphates. The former are considered to be the

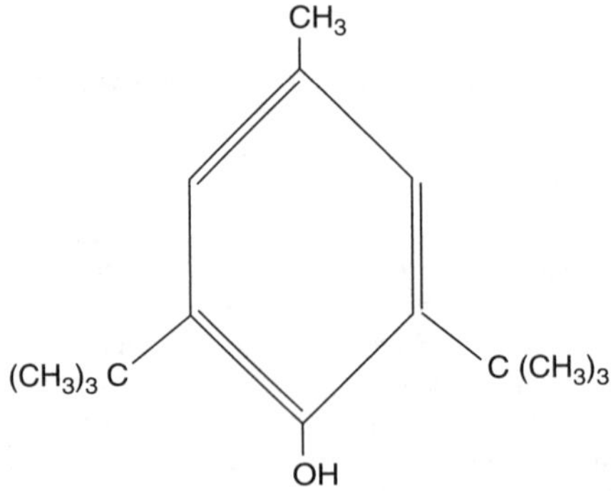

BHT (2, 6 DITERTIARY BUTYL 1-4-METHYL PHENOL)

FIGURE 11-8 Structure of antioxidant BHT.

chain-breaking type; the latter are of the peroxide-destroyer type. Oxidation inhibitors significantly reduce corrosion of bearing metal and lubricant oxidation by combustion products in the engine. Bearing corrosion is also reduced by detergents that neutralize the corrosive acids in combustion products. Certain anti-wear additives in lube such as zinc dithiophosphate, phosphosulfurized olefins, not only inhibit oxidation but also form a protective film on the bearing surface, making it impervious to acid attack.

RUST AND CORROSION INHIBITORS

Rust and corrosion are the result of attack on metal surfaces by oxygen and acidic products. The presence of water and impurities accelerates this attack. These additive work by neutralizing acids and forming protective films. Typical applications include engine oils, gear oils, metal working fluids, and greases. These inhibitors must work in lubricants and on surfaces above the liquid level. Vapor phase corrosion inhibitors (VPIs) provide a simple means of protecting internal components against rust in gear boxes, engines, and other closed surfaces. VPIs fill the closed spaces with vapor that forms a monomolecular protective layer on all metal surfaces. Contrary to what the name suggests, VPIs actually have low vapor pressure and are solids or liquids at room temperature. The most common form of VPI is a salt of an amine and a weak acid. An example is carbonate of dicyclohexyl amine. This VPI, in actual usage, dissociates to an amine and an acid that recombine on metal surfaces.

POUR POINT DEPRESSANTS

Pour point depressant additives prevent the congealing of oil at low temperatures. The phenomenon is associated with the crystallization of paraffin wax, which is present in lube base oils. To provide a low pour point of the finished lube, wax is removed from the feed during manufacture of lube base stock, by a solvent dewaxing or catalytic dewaxing process. Complete dewaxing would reduce the yield of base stock to uneconomic levels, however; therefore certain additives are used that lower the pour point of the lube oil by retarding the growth of wax crystals and thus keep the oil pumpable. In cases where VI improvers such as polymethacrylates are added during lube formulation, supplemental pour point depressants are not required.

ANTIFOAMANT ADDITIVES

Almost all lubricants tend to foam due to the agitation and aeration that occurs during lube usage. Air entrainment due to agitation increases foaming. The presence of detergent and dispersant additives also encourages foaming. Foaming promotes oxidation and reduces the flow of oil to machines. Antifoam agents are added to reduce foaming. These additives act by reducing surface tension, which speeds up the collapse of foam. Antifoamants are added typically in a concentration of 0.0005 to 0.001 percent by weight of the total lube. The chemicals used are silicone polymers (polymethylsiloxanes; viscosity 95,000 to 105,000 cSt at 77°F).

OTHER ADDITIVES

In addition to the major types of additives already discussed, minor amounts of some other additives are added to automotive and industrial lubricant to serve a specific purpose. For example, dyes are added to mark lubricant types and seat swell agents are added to counteract the adverse effect of additives on seals. Seal swell agents react with elastomer to cause seal swell. Organic phosphates and aromatic hydrocarbons are used for this purpose. Metal deactivators reduce the catalytic effect of metals on oxidation rates. These act by forming an inactive film on metal surfaces by complexing with metal ions. Organic complexes containing nitrogen, sulfur, amines, sulfides, and phosphites are used. Biocides are sometimes added to bulk storage tanks to retard or prevent bacterial growth. Usually additive manufacturers present an additive package that is added to base oil to formulate a lubricating oil for a specific use. Additive package suppliers usually have their package tested and approved with a base stock. Additive package may include detergents, dispersants, oxidation inhibitors, antifoamants.

Additive manufacturers recommend the dosage of the additive package to be added and the type of base stock to be used for lube blending. Additive package and base oil can present compatibility problems if the base oil is not of the recommended type. Keeping additives in suspension in base oil can present a challenge. Aromatics in base stocks that provide solubility to additives can be quite low. Additive package manufacturers also provide the chemical analysis (metals, sulfur, nitrogen, TBN, estimated sulfated ash) and physical properties of additive package (specific gravity, viscosities at 40 and 100°C), and maximum handling and storage temperatures. Some additional additives such as VI improvers, dyes, not present in the package but required to meet requirements, may also be included. Additive manufacturers usually have different additive packages for, say, medium-speed diesel engines, railroad diesel engines, marine diesel engines, gasoline engines, gear oils, automatic transmission fluids, and so on, tested with different types of base stocks and end uses that make lube formulation easier.

ADDITIVE DEPLETION

The principal conditions that cause depletion of these additives in lubricating oils are described in the following paragraphs.

Oxidation

Oxidation occurs when the hydrocarbon constituents of lube oil combine chemically with oxygen. Oil in the engine will combine chemically with available oxygen under certain conditions to form a wide variety of oxidation products. Many of these direct oxidation products combine with other materials such as wear metals, solid contaminants, and moisture to form corrosion products. Oil oxidation is accelerated by heat and pressure. Various studies have shown that lube oil oxidation rate is doubled for every 15 to 20°F increase in temperature. Also, engine load that dictates the level of oxygen intake can cause accelerated acid formation, corrosion, oil thickening, deposit formation and accelerated wear.

Nitration

The combustion chamber of the engine provides one of the few environments where there is sufficient heat and pressure for reaction of atmospheric nitrogen in engine intake air with oxygen to form nitrogen oxides. These nitrogen oxide products enter the lube oil through normal blow-by, which reacts with the moisture present in the lube to form a corrosive and powerful oxidant, nitric acid, which rapidly accelerates the oxidation rate of the oil. "Blow By" is escape of compressions and combustions gases past piston and piston rings into the crank case.

Acid Formation

Acids are formed in the lube by several sources. In all forms of fuels used in internal combustion engines, varying amounts of sulfur are present. Sulfuric acid is formed when sulfur molecules react with oxygen in the combustion chamber to form sulfur oxides. These sulfur oxide molecules are blown past the rings and enter the oil. Here the sulfur oxide molecules mix with moisture to form highly corrosive sulfuric acid.

Soot Contamination

One of the major contaminants facing new lube oils today is soot contamination, which is particularly severe in diesel engines. Most of the carbon soot particles generated within the engine are 10 microns or smaller. Engine full-flow filters are designed to remove only 15-micron particles and larger. Thus these filters do little to remove soot contamination. A diesel engine oil used in trucks may last 12,500 miles, burning 1800 gal of fuel at 7 miles per gallon. During this 12,500-mile journey, more than half a pound of soot will enter the oil. Soot will enter the lubricating oil at the rate of 0.0048 ounce for every gallon of diesel burned. In recent years, engine manufacturers have developed new engine designs to meet the new environmental emission standards requiring lower emission of nitrogen oxides and particulate matter. In these new (exhaust gas recirculation, or EGR, type) diesel engines, exhaust gas from the engine is cooled and recirculated through the engine to reduce oxygen concentration within the cylinder, thereby lowering the flame temperature and the formation of nitrogen oxides (NOx). These newer engines emit fewer contaminants, including carbon soot due to exhaust gas recirculation, and therefore higher soot levels are being found in engine oils. Thus the incorporation of EGR technology has resulted in a shorter drain interval for lube oils in diesel engines.

ENGINE OIL FORMULATION

The crankcase oil used in engines is a formulated product consisting of a base oil in which various additives are dissolved. In an engine oil, the base oil components may be 75 to 85 percent of the total formulated lube, and the remaining 15 to 25 percent may be different types of additives. Viscosity modifiers and detergent inhibitors are the most prominent additives. Other additives used in lube formulation are dispersants, AW/EP agents, oxidation inhibitors, antifoamants, rust inhibitors, and demulsifiers. Base oils of a required viscosity are obtained by blending different base stocks such as neutrals and bright stocks. Table 11-4 shows the typical concentration of base oil and various

TABLE 11-4 Automotive Lube Oil Composition

	Range % wt	Typical
Base oil	75–85	
Additives	15–25	
Total	100	
Detergents	1.5–5	Alkaline earth (ca/mg) metal sulfonate, MW 400–700
Dispersants (polybutenyl succinimide)	0.05–0.2*	Polybutenyl succinimide
VI improver	0.5–10	Polymethaacrylate, MW 180,000–25,000
Friction modifier	0.04–0.15†	Molybdenum thiocarbamate
Wear inhibitor	0.04–0.09‡	Dialkyldithio zinc phosphate
Anticorrosion		
Oxidation inhibitors	0.3–2.0	Phenolic oxidation inhibitors such as 2,6 ditertiary butyl phenol
Antofoamants	0.0005	
Pour point depressant		
Metal deactivators	0.005	

*In terms of nitrogen concentration in finished lube.
†Molybdenum concentration in total finished lube.
‡In terms of zinc concentration in finished lube.

TABLE 11-5 Commercial Lubricating Oil Characteristics

SAE viscosity grade		15W–40	10W–30
API service classification	Diesel	CH-4, CG-4, CF-4, CF-2, CF	CG-4, CF-4, CF
	Gasoline	SJ	SH
API gravity		27.9	28.9
Specific gravity		0.8877	0.8822
Flash point	°C	218	210
Pour point	°C	–33	–33
Viscosity at °C			
–15	cP	3000	
–20	cP		2840
–25	cP	25,000	
–30	cP		19,500
40	cSt	120.7	73
100	cSt	15.5	11.0
High-temperature/high-shear viscosity	cP, 150°C	4.4	3.4
Viscosity index		134	141
Zinc	Wt %	0.142	0.142
Nitrogen	Wt %	0.124	0.113
Sulfate ash	Wt %	1.4	1.4
TBN	mg KOH/g	12.2	11.8

additives in automotive engine oils for gasoline and diesel engines. Properties of lube oils 15W-40 and 10W-30 are shown in Table 11-5. Properties of a few other commercial engine oils are shown in Table 11-6.

The additives are added to the base oil to enhance its performance when used in various types of engines. The most important properties of a lubricating oil are its viscosity and alkali reserve (base

TABLE 11-6 Typical Properties of Commercial Motor Oils

	1	2	3	4	5
SAE grade 20W–50					
Viscosity index	122	119	155	121	130
Flash point, °F	440	419	430	432	450
Pour point, °F	–15	–13	–25	–11	–15
Sulfated ash, Wt %	0.85	0.7	0.9	0.74	1
Zinc, Wt %	0.12	0.11	0.15	0.12	0.15
SAE grade 15W–40					
Viscosity index	134	136	135	146	140
Flash point, °F	415	421	399	410	420
Pour point, °F	–15	–27	–11	–25	–10
Sulfated ash, Wt %	1.3	1	0.9	1	0.99
Zinc, Wt %	0.14		0.14	0.13	0.13
SAE grade 15W–30					
Viscosity index	140	150	133	155	130
Flash point, °F	415	401	400	405	410
Pour point, °F	–33	–26	–31	–35	–26
Sulfated ash, Wt %	0.85	0.96	0.85	1	1.2
Zinc, Wt %	0.12	0.11	0.13	0.15	0.2

TABLE 11-7 Lubricating Oil Base Number (BN)

Natural gas operation of spark-ignited (SG) and dual-fuel (DF) engines	
BN 4–7	For use in gas engines
LFO operation of dual-fuel (DF) engines	
BN 12–15	For use in dual fuel engines using light fuel oil
HFO operation of DF engines	
BN 30–55	For use in DF using heavy fuel oils
Diesel engines	
LFO operation	
BN 12–15	For low-sulfur diesel fuel
BN 20–30	For high-sulfur diesel fuels
HFO operation	
BN 30	Share is continuously decreasing. Can be used with low SLOC or when engine is equipped with SCR catalyst
BN 50–55	When engine is equipped with wet lube oil sump and fuel sulfur content is above 2% or more

number, or BN). In addition, many other properties of the oil affect the performance of the engine. Fuel quality significantly affects the kind of lubricating oil to be used. For example, for a low-sulfur fuel such as natural gas, little sulfur dioxide is formed during combustion. In this case a lubricating oil with a low BN, 4 to 7, may be OK. In case the fuel contains high sulfur, 0.5 to 5 percent, a higher alkali reserve is required to prevent corrosion damage. For fuels containing 2 percent or more sulfur, a high BN (50 to 55) lubricating oil gives longer lube oil change intervals compared to lower detergency levels (BN) lubes. Initial cost of the lube may be higher, but overall economics dictates their use. BN values for some diesel and gasoline engines are shown in Table 11-7. After an engine has run certain number of miles; typically 5000 to 10,000, the engine oil is drained and new oil is filled in its place. The reason is that although the base oil remains broadly unchanged, the additives within the oil get depleted during actual engine running because of contaminants, acids, corrosion, and wear products formed in the oil. Lubricating oil ages and undergoes degradation during operation and certain additives are partly consumed. For these reasons, condemning limits have been set for various physical and chemical properties of lubricating oils. Table 11-8 presents the typical condemning

TABLE 11-8 Condemning Limits for Used Lubricating Oils

Property	Units	Condemning limits, gas engines	Condemning limits, medium-speed diesel engines	Test method
Viscosity	cSt, 100°C	–20%/+25%	–20%/+25%	ASTM D 445
Viscosity	cSt, 40°C	–25%/+50%	–25%/+45%	ASTM D 445
Water	Vol % mass, Max.	0.3	0.3	ASTM D 95/D 1744
Base number	mg KOH/g, Max. depletion	50%	20%/50%*	ASTM D 2896
Total acid number	mg KOH/g, Max. increase	2.50%		ASTM D 664
n-Pentane insoluble	Wt %, Max.	1.00%	2.00%	ASTM D 893b
Flash point	PMCC, °C Min.		170	ASTM D 93

*Max. 50% depletion for LFO operation and max. 20% depletion for HFO operation.

limits for used lube oils in gas and diesel engines. The high water content can be reduced by efficient centrifuging, which can extend the life of lubricating oil.

Lube formulators are faced with conflicting demands from customers, original equipment manufacturers (OEMs), emission standards set by government agencies, lube manufacturers' profitability, and the marketability of formulated lube. Customers demand increased oil drain intervals (50,000+ miles), improved fuel economy, longer engine life, and meeting these requirements increase the cost of production.

OEMs desire enhanced engine protection from viscosity increase and soot, low volatility, greater oxidation stability, corrosion protection, and enhanced low temperature performance.

Environmental agencies all over the world are targeting emissions from internal combustion engines, particularly nitrogen oxides (NOx standards per g/BHP/h) and particulate matter from engines (g/BHP/h), are constantly decreasing compared with the engines of the 1990s.

EFFECT OF BASE STOCK QUALITY

Optimal engine oil performance requires proper balance of lube base stock characteristics and additive technology. Paraffinic base stocks are preferentially used to formulate most of the world's automotive and industrial lubricants, including engine oils, transmission fluid, and gear oils due to their better oxidation stability, higher viscosity index, and lower volatility relative to their comparable viscosity grades of naphthenic base oils. Naphthenic base oils have a lower pour point and better solvency characteristics compared to paraffinic base oils, which make them particularly useful in formulating low-temperature hydraulic oils, refrigeration oils, rubber process oils, metal working oils, cylinder lubricants for large engines, and greases. The possibility of longer life automobile engine oils has emerged over the past few decades with availability of polyalphaolefin (PAO) synthetic lubes. These oils came with 25,000 miles+ suggested oil change intervals. Nevertheless, the high cost of PAO base stocks limited their market share. Hydrocracked and hydrotreated lube base stocks constituting group II and group III lubes have very high saturates and very low aromatics. Absence of aromatics make them more resistant to oxidation and thus minimizes sludge and varnish deposits. The service life of oils blended with group II and III base stocks, with VI of 120 or more is estimated to be double or more compared with oils of group I base stocks.

However, despite the advantages of group II and group III base oils, their use in motor oil blending is not without problems, a few disadvantages are:

- Automobile engine oils are blended with a large amount of detergents, oxidation inhibitors, antiwear, and many other additives. In the absence of aromatics in group II and group III base stocks, additive solubility in the base stocks is very poor and the additives tend to separate from base oil. To avoid such a possibility, blending either with group I solvent refined oil or with synthetic ester fluid may be required, thus sacrificing potential longer life. Altered solvency action with paraffinic hydrocracked base oils may adversely affect some gaskets, seals' paint, and coupling components in existing engines.
- Due to cracking, viscosity of lube base stocks produced by hydrocracking is limited. Hydrocracking of aromatic and cycloparaffin molecules results in a severe reduction in viscosity. Hydrocracked base stocks have a maximum viscosity of 320 to 460 cSt at 100°F. Higher viscosity grades are required for certain industrial gearing, large reciprocating compressors, and many other applications. Group III base stocks consisting mainly of paraffins have still lower viscosity, which restricts their usage to 0W and 5W SAE grades. For blending higher viscosity grades, more viscous bright stock from traditional lube making, synthetic polyalphaolefins, or long chain polymeric material must be used.

AMERICAN PETROLEUM INSTITUTE SERVICE CLASSIFICATION

The American Petroleum Institute (API) service classification of motor oils is based on the performance of an oil in certain types of engines. Whereas the SAE classification refers only to oil viscosities at specified temperatures, API specifications refer to oxidation protection, high temperature engine deposits, acid formation, foaming, sludge control, engine emissions, and other properties critical to engine performance, pollution control, and oil consumption. API maintains a system of service classification ("S" rating) for gasoline engines and commercial classification ("C" ratings) for diesel engines. Changes are made to the specifications in consultation with major automotive companies, original equipment manufacturers, oil refining companies, and engineering companies directly involved in automobile manufacture. Periodic changes are made to the specifications to keep them updated with current gasoline and diesel engine technology, refining processes, and emission norms. It may be noted that when a new specification is released, it makes the previous specification obsolete.

Specification revision are driven by the following objectives: General oil improvement includes lower oil volatility and reduced additive depletion. Lower volatility reduces oil consumption rates and piston deposits and help keep viscosity from rising over service life. Reduced additive depletion ensures adequate protection over the course of oils service life.

Rapidly changing emission standards have led to stringent lubricant requirements. Specifications are changing every two or three years. The OEM demand lubes with greater protection from soot-aggravated wear and viscosity increase, reduced consumption, lower volatility, greater oxidative stability, material compatibility, and corrosion protection. The API administers the licensing and certification of engine oils in the United States.

API service classification is listed on each container of lubricating oil sold in the market or at service stations in the United States, and it is easy to match it with the recommended oil as per its service manual. The API system has two general classifications:

- First letter "S' indicates that oil is appropriate for "spark-ignited" or gasoline engines (typical passenger cars and light trucks).
- First letter "C" indicates that oil is appropriate for "compression-ignited" or diesel engines.
- The second letter in each category indicates the performance level of the category. For "S" categories, the performance level increases as the categories go through the alphabet. However, the same is not true for the "C" categories because the types and the intended application range of diesel engines is very great. It is important to refer to the service manual of the engine for appropriate performance recommendations.

GEAR OILS

Gear oils are heavy high-viscosity lubricating oils that are used in enclosed units containing gears. Gear oils perform a number of functions. These oils reduce friction and wear, act as heat transfer agents, and protect against corrosion and rust. They contain additives to minimize oil oxidation, inhibit foaming, and separate water readily. Gear lubricants SAE viscosity grades should not be confused with engine oil SAE viscosity grades. A gear lubricant and an engine oil having the same viscosity will have widely different SAE viscosity grade designations as defined by two viscosity classifications.

Viscosity is the most important property of gear oils. The viscosity of oil should be high enough to protect the gear teeth against direct metal-to-metal contact. High VI is also important. The oil should be able to provide a consistent lubricating film over the operating range of the gear box. Enclosed high-speed gears are formulated with rust and oxidation inhibitors and anti-wear agents. Worm gears are blended with synthetic fatty oils to provide lubricity for sliding motion under heavy pressure. Open gear sets require high-viscosity base oils with extreme pressure and anti-wear additives.

SAE GEAR OIL CLASSIFICATION

Table 11-9 shows U.S. standard SAE J306 (July 1998) for automotive gear lubricants, based on viscosity. Note that viscosities at 100°C range from 4.1 to 41 cSt; crankcase oil viscosities range from 3.8 to 21.9 cSt. Another important classification is the American Gear Manufacturers Association (AGMA) (Table 11-10), which has set up a numbering system to define gear oil viscosities. Wide-span multigrade oils are becoming more and more popular due to their fuel economy and applicability over a wider ambient temperature range. Multigrades like 75W-90, 80W-90, and 85W-140 are gaining in importance in areas with wide swings in ambient temperatures. One problem with many traditional gear lubes is that lubes do not stay in grade after being in use for some time, due to a loss in viscosity.

TABLE 11-9 Viscosity Classification for Axle and Manual Transmission Lubricants

Viscosity classification: SAE J306

SAE viscosity number	Maximum temperature for viscosity of 150,000 cP °F	Maximum temperature for viscosity of 150,000 cP °C	Viscosity at 210 °F minimum		Viscosity at 210 °F maximum	
			cSt	SUS	cSt	SUS
70 W	−67	−55	4.1			
75	−40	−40	4.2	40		
80 W	−15	−26	7.0	49		
85 W	10	−12	11.0	63		
80			7.0		<11	
85			11.0		<13.5	
90			14.0	74	<25	120
140			25.0	120	<43	200
250			43.0	200		

TABLE 11-10 AGMA Gear Oil Standards (AGMA-9005-E02)

AGMA no. lubricant	Equivalent ISO grade	Viscosity range cSt, 40°C	Viscosity range SUS, 40°F
1	46	41.4–50.6	193–235
2	68	61.2–74.8	284–347
3	100	90–110	417–510
4	150	135–165	626–765
5	220	198–242	918–1122
6	320	288–352	1335–1632
7	460	414–506	1919–2346
8	680	612–748	2837–3467
8A	1000	900–1100	4121–5098

The shear stability requirement test in SAE standards assures that the lubricant stays within the acceptable range of each viscosity grade. Formulation of gear oil is done with more shear-stable VI improvers or different or synthetic base stocks to stay in grade. Synthetic gear oils are used in many industries. Synthetic gear oils are known to prevent startup engine wear better and are pumped to critical surfaces much faster than mineral oils. Table 11-11 lists the properties of some commercial synthetic gear oils.

TABLE 11-11 Properties of Synthetic Gear Oils Commercial Grades

	1	2	3	4
Base oil	PAG	PAG	PAG	PAG
Density, 60°F, kg/M^3	1074	1072	1051	1070
Kinematic viscosity, 40°C, cSt	222	460	680	1000
Kinematic viscosity, 100°C, cSt	34.4	73.2	114	160
Viscosity index	203	239	276	286
Pour point, °F	–27	–33	–4	–4
Flash point, COC, °F	568	586	509	509

AUTOMOTIVE LUBRICANTS TEST METHODS

Thin Film Oxygen Uptake

The thin film oxygen uptake (ASTM D 4742) evaluates the oxidation stability of lubricating oils. A mixture of test oil and fuel found in gasoline engine operation (oxidized/nitrated fuel, soluble metals, and distilled water) are placed in a test vessel, which is pressurized with oxygen and placed in a heated bath. Antioxidant breakdown is evident when the oxygen pressure in the vessel rapidly decreases. At this time the induction time (breakpoint) of oil is recorded.

A good quality oil may have an induction time in the order of 500 min. Superior oxidation stability of motor oil allow it to effectively resist the formation of engine deposit and sludge, keeping the engine running clean and extending the oil life. It also resists thickening, maintaining its superior wear protection and lubrication properties and maximizing fuel efficiency.

High-Temperature/High-Shear Test (ASTM D 4683)

The high-temperature/high-shear test measures the lubricant viscosity under severe high-temperature and shear conditions that resemble highly loaded journal bearing in fired internal combustion engines. In order to prevent bearing wear, it is important for a lubricant to maintain its protective viscosity under severe operating conditions. The minimum high-temperature/high-shear viscosity at 150°C for SAE-30 oil is 2.9 cP. The superior viscosity retention of a good quality oil in the face of severe temperatures and shear conditions allows it to provide continuous unsurpassed protection for engine bearings, extending equipment life and preventing wear.

Noack Volatility Test (ASTM D 5800)

The Noack volatility test determines the evaporation loss of lubricant in high-temperature service. The more the motor oil vaporizes, the thicker and heavier the remaining oil becomes, contributing to poor circulation, reduced fuel economy and increased oil consumption, wear, and emissions. Generally a maximum of 15 percent evaporation loss is allowable in motor oils. Low volatility of a well-formulated oil allows it to maintain its superior protective and performance qualities throughout extended drain intervals even under severe operating temperatures. In addition, oil consumption and emission are minimized and fuel economy is maximized.

Pour Point (ASTM D 97)

The pour point determines the lowest temperature at which lubricant will flow when cooled under prescribed conditions; the lower the lubricant's pour point, the better protection it provides in low-temperature service.

The low pour point of an oil allows it to maintain its fluidity in low ambient temperatures, reducing drag on moving vehicle parts, providing critical engine components with quick, essential lubrication, and easing startup in cold temperatures. Wear is greatly reduced and equipment life is extended.

Total Base Number (ASTM D 2896)

New and used lubricating oils generally contain basic constituents such as sulfonates, phenates, salicyclates, and so on, of alkaline earth metals (e.g., calcium, magnesium) to impart alkalinity to lube formulations. BN is a measure of the amount of basic substance in oil. It is equal to the quantity of perchloric acid equivalent to the number of milligrams of potassium hydroxide required to titrate 1 g of sample. In the test, the sample is dissolved in a mixture of chlorobenzene and glacial acetic acid and titrated with perchloric acid ($HClO_4$) in glacial acidic acid with a potentiometric titrimeter.

TBN is the measurement of a lubricant reserve alkalinity that aids in the control of acids formed during the combustion process. The higher the motor oil's TBN, the more effective it is in suspending wear-causing contaminants and reducing the corrosive effect of acids over an extended period of time. The high TBN of an oil provides superior protection and performance over extended drain intervals.

High TBNs are essential to control deposits in four-stroke cycle diesel engines and to neutralize the effect of high-sulfur fuel. Engine burning fuel with more than 3.5 % sulfur requires an engine oil with high TBN of 30–50 whereas engine burning low sulfur fuels, with less than 0.5 % sulfur require a low TBN (6–10) oil.

Sulfated Ash Content

Sulfated ash is the noncombustible residue of a lubricating oil according to ASTM D 582 and D 874. Sulfated ash is thus the solid material left when the oil burns. A high sulfated ash content tends to form more sludge and deposits in engine, which may cause valve failure under certain operating conditions. Because most detergents are metallic compounds, the percentage of sulfated ash has been considered to have a relationship to detergency. As TBN increases, sulfated ash also increases. Thus a motor lubricant with a TBN of 10 or more is most likely to have sulfated ash of more than 1 percent. Lubricating oil with a low sulfated ash content is more appropriate. In the ASTM D 874 test method, the sample is ignited and burned until only ash remains. After cooling, the residue is treated with sulfuric acid and heated at 775°C until oxidation of the carbon is complete. The ash is then cooled, retreated with sulfuric acid, and heated at 775°C to a constant weight. Sulfated ash is used to indicate the concentration of known metal-containing additives in new oils. When phosphorus is absent, Ba, Ca, Mg, Na, and K are converted to their sulfates while tin and zinc are converted to their oxides, but when phosphorus is present with metals, it remains partially or wholly in sulfated ash as metal phosphates. Sulfur and chlorine do not interfere in the test.

Zinc Content

Of great importance in engine lubrication are zinc compounds such as dialkyl-dithio zinc phosphates, used in lubricating oils as an extreme pressure/anti-wear and antioxidant additive. Zinc is only used when there is actual metal-to-metal contact in the engine. Under normal engine operations, lube oil film prevents any metal-to-metal contact; however, under certain abnormal operating conditions (high engine loads, high temperatures, etc.) if metal-to-metal contact occurs, zinc compounds react with metal or get adsorbed on the surface to form a film that prevents scuffing and wear. Typically a level of 0.11 percent zinc is sufficient to protect an automobile engine for extended oil drain interval under normal usage. For air-cooled motorcycles or turbocharged cars, higher zinc is sometimes allowed to give protection for a longer time. Generally, more zinc may lead to more deposit formation and plug fouling.

COLD CRANK SIMULATOR (ASTM D 5293)

The cold crank simulator test determines the apparent viscosity of lubricants at low temperatures and under high shear rates. Viscosity of lubricants under these conditions is directly related to engine cranking and startability. The lower the lubricant's cold cranking viscosity, the easier for an engine to turn over cold ambient conditions. For example, 10 W motor oils are tested at –25°C and must have a viscosity below 7000 cP to pass SAE J300 specifications.

The low cranking viscosity of an oil reduces drag on moving engine parts and allows the engine to achieve a critical cranking speed in extremely low temperatures. Engines turn over quickly, and dependability in cold winter temperatures increases.

Cold Temperature Starting

To start effectively, the engine must reach a critical cranking speed. As the temperature drops, achieving critical cranking speed becomes more of a challenge. Low temperatures cause motor oils to thicken. If they thicken too much and impose excessive drag on the moving parts, critical cranking speed will not be achieved and the engine will fail. Motor oils used in cold climates must maintain a sufficient low cranking viscosity to allow engine turnover at the lowest temperatures. If the motor oil is able to meet this challenge of allowing the engine to turn over, it immediately faces another significant challenge of providing quick, critical lubrication to the engine-bearing and other moving parts. Two types of engine pumping failure can result from cold thickened motor oil: air binding failure and flow limited failure. Air binding failure occurs when motor oil surrounding the pump inlet screen gets sucked into the pump but is not replaced by new oil from the sump. The oil pump inlet screen then gets starved for fluid and the oil pressure becomes erratic as air is entrained and proper oil flow cannot be maintained. Flow limited failure occurs when oil becomes so thick that it cannot be pumped through the narrow passages that deliver oil to the engine moving parts. All motor oils thicken in cold temperatures but how much they thicken is significant to the level of protection an engine receives.

Pour point tests (ASTM D 97) pinpoint the temperature at which a motor oil thickens to the point it cannot flow. When the oil stops flowing, it is useless. For an engine to receive even minimal protection, it is important that it has a pour point lower than typical winter minimal temperatures. Petroleum-based motor oils face significant challenges at low temperatures because paraffinic wax begins to crystallize as the temperature drops. These wax crystals agglomerate into larger structures and the motor oil gels and fail to flow, starving the engine of vital lubrication. To hinder the growth of wax crystals, additives known as pour point depressants are added that lower the pour point. Petroleum-based lube base oils may have a pour point of 5°F, but pour point additives can lower it by another 25°F.

FOUR-BALL WEAR TEST (ASTM D 4172)

The four-ball wear test determines the wear protection properties of a lubricant. Three metal balls are clamped together and covered with test lubricants while a rotating fourth ball is pressed against them in sliding contact. This contact rapidly produces a wear scar. The smaller the average wear scar, the better the wear protection provided by the lubricant. A good quality motor oil must provide protection against engine wear. Thus equipment life is extended[2] and repairs, downtime, and expenses are reduced.

REFERENCES

1. Rohm and Hass Company, Philadelphia, PA, "Acryloid 700 series."
2. M. Khonsari, "New Lubes Last Longer," *Machinery Lubrication Magazine,* May 2004.

CHAPTER 12
SYNTHETIC LUBRICANTS

Synthetic lubricants were developed more than 50 years ago and became widely used in jet engines. Lower than –120°F operating temperatures, 60,000 r/min shaft speed, and 500°F+ exhaust temperatures proved too much for conventional lubes. Synthetic lubes were created specifically for these harsh operating conditions, and at present every jet engine in the world uses synthetic lubricants. A synthetic lubricant base stock is a product made from a chemical reaction of two or more simple chemical compounds. These base stocks are manufactured to meet specific physical and chemical characteristics not found in petroleum lubricant base stocks. This base stock is then used to formulate lubricants by the addition of performance additives. Literally hundreds of types of specialized synthetic base stocks are in existence. Some of the most common synthetic lubricants are as follows:

- Polymerized alpha olefins, olefin polymers, olefin oligomers
- Dibasic acid esters
- Polyols esters
- Polyalkaylene glycol ethers
- Phosphate esters
- Alkylated benzenes, a synthetic hydrocarbon
- Silicons

With the exception of polyglycol fluids, all these base stocks have viscosities in the range of lighter HVI neutral mineral oils. Their viscosity indexes and flash points are higher, however, and their pour points are considerably lower. This makes them a valuable blending component when compounding oils for extreme service for both high and low temperatures. The following performance features are claimed for synthetic lubricants over mineral oil lubes:

- Engine cleanliness
- Improved fuel economy
- Lower oil consumption
- Good low temperature (cold starting) fluidity
- Outstanding performance in extended oil drain intervals
- Superior high-temperature oxidation resistance
- Excellent wear protection

The main disadvantage of synthetics is that they are inherently more expensive than mineral oil and available in limited supply. This limits their use only to specialty oils and greases that command a premium price. Esters suffer the further disadvantage of greater seal swelling tendencies than hydrocarbons, so caution must be exercised in using them in applications where they may contact elastomers designed for use with mineral oils.

POLYALPHAOLEFINS

Polyalphaolefins (PAOs) are the most widely synthetic lubricant[1] in the United States and Europe. They are made by combining two or more decene molecules into an oligomer or short chain length polymer (Fig. 12-1). The resulting oligomers, consisting of dimers, trimers, and tetramers, are hydrogenated and then formulated with additives to yield synthetic lube. PAOs are all hydrocarbon structures and they contain no sulfur, phosphorus, or metals. Because they are wax free, they have low pour points, usually below –40°F. Viscosity ranges from 2 to 100 cSt at 100°F, and the viscosity index (VI) for all but the lowest grade exceeds 140 (Table 12-1). PAOs have good thermal stability, but they require suitable antioxidant additives to resist oxidation. The fluids have a limited ability to dissolve some additives and tend to shrink seals. Both these problems can be overcome by adding a small amount of an ester. Polyalphaolefin oligomers or PAOs have many favorable properties that contribute to their rapidly growing use for a variety of industrial applications. They have low volatility, good natural lubricity, thermal stability, hydrolytic stability, and chemical inertness. The high-temperature stability in the absence of air has led to the use of PAO as a sealing fluid and lubricant for the mechanical seals of pumps handling polystyrene process liquid at 450°F. A nitrogen inert atmosphere is maintained over the sealing fluid. Because PAOs have a low pour point, they are also used in refrigeration compressors (ammonia, fluorocarbons, etc.). PAOs and other synthetic lubricants are less resistant to oxidation than mineral oils. Apparently some of the sulfur present in mineral oil serves as a natural antioxidant. Antioxidants are particularly needed for PAOs to withstand high temperature use and to provide adequate service life without thickening and forming harmful acids. Without antioxidants, rapid oxidation can deteriorate the lubricant quickly. The advantage of PAOs is improved low-temperature properties with less volatility compared with mineral oil base stocks.

$$\underset{\text{1 - DECENE}}{\text{HCH} = \text{CH}_2} \xrightarrow{\text{Catalyst}} \underset{\text{WHERE X = 2, 3, 4, 5, etc.}}{(\text{C}_{10}\text{H}_{20})_\text{X}\ \text{ISOMERS MIXTURE}}$$

FIGURE 12-1 Polyalphaolefin synthesis.

TABLE 12-1 Polyalphaolefin Properties

Hydrogenated alpha decene		
Specific gravity		0.828
Kinematic viscosity		
	cSt, 104°F	28
	cSt, 212°F	5.2
Viscosity index		128.9
Boiling point, °F		601
Flash point, °F		450
Pour point, °F		–65
Vapor pressure, mm Hg, 68°F		0.1

DIESTERS

Diester lubricants were developed during World War II and proved to be excellent for low-temperature lubrication. Further development of the diester was closely linked to aviation gas turbine development. In the early 1960s, neopolyol esters were used in this application because of their low volatility, high flash point, and good thermal stability. Dibasic acid esters are esters of long chain alcohol and acids

TABLE 12-2 Properties of Diesters and Polyester Lubricants

	Diester	Polyester
Viscosity, 210°F, cSt	9.4	11.3
Viscosity, 100°F	106.8	76.7
Flash point, °F	510	545
Autoignition temperature,°F	770	865
Evaporation loss, %, 22 h at 300°F	0.9	0.4
Oxidation corrosion stability, 42 h at 425°F 5 L/h air		
Acid number increase, mg/KOH/g	5.64	1.71
% viscosity increase, 100°F	24	13.5
Four-ball wear test, 1 h, 600 r/min 129°F, 40 kg Load		
Wear scars, mm	0.64	0.37

and are synthesized by reacting an acid and an alcohol, forming water as by-product that must be removed. Under certain conditions, when water is present as a contaminant, a reverse reaction of water with an ester lubricant can occur, again forming parent alcohols and acids. If strong acids (e.g., adipicic or phthalic) are formed as a result of hydrolysis, corrosion can become a problem. Proper additive selection is crucial to avoid this problem. Table 12-2 lists the typical properties of diesters. Certain metals such as iron and zinc can act as a catalyst for the hydrolysis reaction. Diesters have good lubricating properties, good thermal and shear stability, a high VI, and exceptional solvency and detergency. The pour point ranges between –45 and –54°F. They are clean running in that they tend to dissolve varnish and sludge rather than leave any deposit. In fact, diesters can remove deposits formed by other lubricants. For these reasons, diesters have been used until recently for aircraft engines and compressors. Diesters are also used as a base oil or a part of a base oil for automotive engine oils and in some low-temperature greases. Other applications include misting oils, bearing lubricants, and high-temperature hydraulic oils. Diesters are often used in small quantities with PAOs to provide natural seal swell and adequate solvency for additives. Most additives are readily soluble in these esters. This property provides flexibility in formulating lubricants with 100 percent organic esters or with organic esters plus other synthetic fluids. Because of their fire resistant properties and high oxidation stability, made possible with proper additives, diester lubricants are being used for reciprocating air compressor's crank case lubricants.[2] The advantages over the mineral oil lubricants are lower lubricant consumption, less power usage, cleaner valves, and lower fire and explosion risks. The hydrolytic stability of diesters is not as good as that of petroleum oils. Hydrolytic stability refers to how lubricants react in the presence of water. Hydrolytic degradation can lead to acidic products, which in turn promote corrosion. Also, hydrolysis can significantly change the chemical properties of base fluid, making it unsuitable for intended use. Diesters are incompatible with some sealing materials and can cause more seal swelling than that with mineral oil. Diesters have a low molecular weight and a low viscosity, and they are highly polar. These properties make them quite aggressive to elastomer seals. This can be reduced by choosing better elastomers for seals or blending diesters with PAO base stocks that are nonpolars. Proper additive selection is critical to prevent hydrolysis and provide oxidative stability. In addition, chemically resistant seals are recommended.

POLYOL ESTERS

Polyol esters, like diesters are formed by the reaction of an acid and alcohol (Fig. 12-2). "Polyol" refers to a molecule with two alcohol functions in its structure; examples include trimethylolpropane (TMP), neopentylglycol (NGP), and pentaerthritol (PE). The pour point ranges from –22 to –116°F, and the VIs from 120 to 160. The fluids have excellent thermal stability and resist hydrolysis somewhat better than diesters. With proper additives, polyol esters are more oxidatively stable than diesters and PAOs.

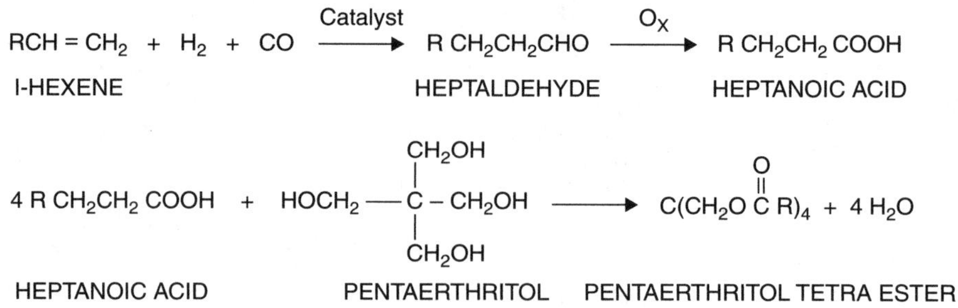

FIGURE 12-2 Polyo ester synthesis.

Seal swell behavior is similar to that of diesters. Polyol esters have excellent high-temperature properties,[3] superior to those of diesters. Also, polyol esters have better hydrolytic stability and better seal swell characteristics. Table 12-2 compares the properties of diesters and polyol ester lubricants. Polyol esters have the same advantage and disadvantages as those of diesters; however, they have much more thermal stability than a diester of the same viscosity. Large quantities of polyol esters are manufactured for use as aviation turbine lubricants operating at high temperatures. Their outstanding high-temperature performance and very low pour point are eminently suited for the wide temperature ranges encountered between cold outside air and jet aircraft turbines operating at high temperatures. Other uses include hydraulic fluids and heat exchange fluids. Modern military and commercial jet aircraft almost universally use lubricants formulated with polyol esters as base fluids. Polyol ester lubricants are preferred for air compressor cylinder lubricants if the discharge air temperature is more than 300°F. For rotary screw air compressors operating above 180°F, polyol esters are a better choice than polyalphaolefins. At 220 to 230°F, the service life of polyol ester lubricant is almost the double that of a polyalphaolefin lubricant.[4] Other polyol ester lubricant applications include chain lubricant for hot air ovens operating at 400°F, lubricant for porous metal bushing in furnace blower fans, and gear and bearing lubricant for 350°F positive rotary air blowers. Polyol esters give a much lower coefficient of friction than that of poly alfa olefin or mineral oil lubricant. Lubricating greases with polyol esters as base fluids are particularly suited to high-temperature applications.

POLYALKYLENE GLYCOLS

Polyalkylene glycols (PAGs) are polymers of alkylene oxides such as ethylene oxide or propylene oxide. Lubricant performance[5] and physical properties of a particular PAG depends on the monomer used to manufacture it, molecular weight, and nature of terminal groups; thus a wide range of properties is possible (Table 12-3). Figure 12-3 shows the molecular structure of PAG.

A more appropriate name for PAGs would be polyalkylene oxide or polyethers. In general, PAGs have good high-temperature stability and high VIs, and they can be used over a wide temperature range. They exhibit low deposit formation and tend to solubilize their decomposition products.

TABLE 12-3 Properties of Polyalkylene Glycols

	1	2	3	4	5
ISO viscosity grade	220	320	460	680	1000
Specific gravity	1.048	1.069	1.05	1.051	1.07
Flash point, °F	527	527	500	509	509
Pour point, °F	–31	–22	–22	–4	–4
Viscosity, cSt, 104°F	220	320	460	680	1000
Viscosity, cSt, 212°F	39	54.6	77.1	114	160
Viscosity index	238	246.5	258.2	275.7	285.5

$$CH_3-(CH_2)_3-(OCH_2\,CH_2)_X-(OCH_2\,CH\,(CH_3)_Y-OH$$

FIGURE 12-3 Polyalkene glycol monoether.

Like other synthetics, PAG require additives to resist oxidation. The non-sludge-forming characteristics of polyalkylene glycols has led to their use for high-temperature applications. For example, polyalkylene glycols can be mixed with graphite for lubrication of 2000°F refractory kiln bearing. The PAG product of degradation flash off or burn clean, leaving the solid lubricant without any tar deposit. Water-soluble PAG are used as water-based fire-resistant hydraulic fluids. One novel application of water-soluble PAG lubricants is the valve actuators of large gas transmission lines located outdoors in subzero climates. In the past, with conventional lubricants, water would condense and freeze up the actuator in wintertime. With polyglycol lubricants, the moisture dissolves in the polyglycol and the resulting solution has a subzero pour point, thus eliminating the freezing problem. The use of polyalkylene glycols lubricants for hydrocarbon gas compressor application is expanding rapidly, particularly for oil-flooded rotary screw compressors. Polypropylene glycol is becoming a popular lubricant in propane refrigeration compressors because of its high VI (200) and low propane solubility in this lubricant. For well gas and rotary screw compressor applications involving both condensable hydrocarbons and water vapors, a polyethylene/polypropylene glycol copolymer lubricant can be used. This lubricant withstands hydrocarbon dilution, and at a lubricant operation temperature above 160°F, it withstands water dilutions. PAGs are not compatible with petroleum oils. PAGs made from ethylene oxide are water soluble and oil insoluble, whereas those made from propylene oxide are water insoluble and oil soluble. Some PAGs have a specific gravity greater than that of water, and with such lubricants, any separated water will float on lubricant. The interaction of PAGs with water can be "customized" with a special combination of ethylene oxide, propylene oxide, and other chemicals. Water-soluble PAGs may separate from an aqueous solution at high temperatures. In metalworking fluids, this can result in PAGs coming out of dilute aqueous solution to provide lubricity at the hot surfaces of the tool and work piece. In quenching fluids, the coating of hot metal surface by PAGs controls the cooling rate. In antifoamants, PAG insolubility at high temperatures increases defoaming properties. The ability to customize the properties of PAGs has led to their use in specialty applications including metalworking fluids and quenchents. In addition, their good high-temperature stability has led to their use in air compressor lubricants, and their insolubility in natural gas makes them useful in gas compressors. PAGs are finding some uses as a gear oil lubricant. Because of their water solubility and biodegradability, PAGs are considered more environmental friendly.

PHOSPHATE ESTERS

Phosphate esters are synthesized from phosphorus oxychloride and alcohols or phenols. They are used both as base oils and as anti-wear additives.[6] Thermal stability is good, and the pour point ranges from –13 to 23°F. However, the VI is extremely low, ranging from 0 to –30. As a result, the viscosity becomes too low at high temperatures to function as a lubricant. Phosphate esters base fluids are particularly used where high fire resistance is desired. The anti-wear function of these fluids is believed to be due to the reaction of phosphate esters with iron surfaces forming a low melting alloy, which under pressure form a smooth polish on the surfaces, reducing friction. The most common application of phosphate esters is as fire-resistant hydraulic fluids.

Phosphate esters are used as fluids and lubricants primarily because of their good fire-resistant properties (autoignition temperature more than 1000°F, flash point higher than 470°F). Uses include fire-resistant lubricants for heavy-duty industrial gas turbines and auxiliary equipment and hydraulic fluids in steel mills, foundries, and other plants where fluids could leak onto hot surfaces. Phosphate esters have a high-bulk modulus, offering an extremely fast response in electrohydraulic servo systems. These systems are commonly used where precise control is required, such as turbine speed control.

Phosphate esters have been used as air compressor lubricants, primarily because of their high autoignition temperature. However, explosions have been reported caused by the excessive amount of lubricants in the compressed air system. As with most esters, hydrolysis is a problem in the presence of water. Contact with moist hot air can gradually cause phosphate esters to break down; the degradation products (strong acids) can cause damage to paints, rubber seals, gaskets, and so on. Also, phosphate esters are resistant to biodegradation and thus are considered undesirable contaminants in waterways.

NATURAL ESTERS

Natural esters are derived from rape seeds and caster plants. These oils may be hydrogenated to increase their thermal stability and oxidative resistance. They find application as lubricants in the agriculture, wood, and food industries where low operating temperatures are encountered. Natural ester lubricants have a short life, but their biodegradability makes them environmentally safe lubricants.

POLYPHENYL ETHERS

Polyphenyl ethers[7] (PPEs) consist of a short chain of benzene rings linked with oxygen atoms. The molecular structure of five-ring polyphenyl ether is shown in Fig. 12-4 and its properties are presented in Table 12-4. These ethers are characterized by their high thermal stability. A six-member PPE is thermally stable up to 836°F, a five-member ring to 900°F. This extremely high thermal stability and their nontoxic characteristics make them a unique class of chemicals. These compounds were used in late 1960s to lubricate turbines of high-performance spy planes. Conventional hydrocarbon lubricants break down at 400°F, and the operating temperature of these aviation turbines was 600°F. Apart from their thermal stability, PPEs have many more unique properties. They have a high optical index of refraction, optical clarity, and radiation resistance. Also, PPE are super lubricants that resist heat and radiation, tend to stay where they were applied, and have low vapor pressure. In the 1970s, earth-orbiting satellites proliferated, and it became necessary to lubricate their moving parts. Lubrication in space encounters three serious constraints; temperatures vary widely, materials are exposed to the evaporative effect of a hard vacuum, and the mechanisms being lubricated are

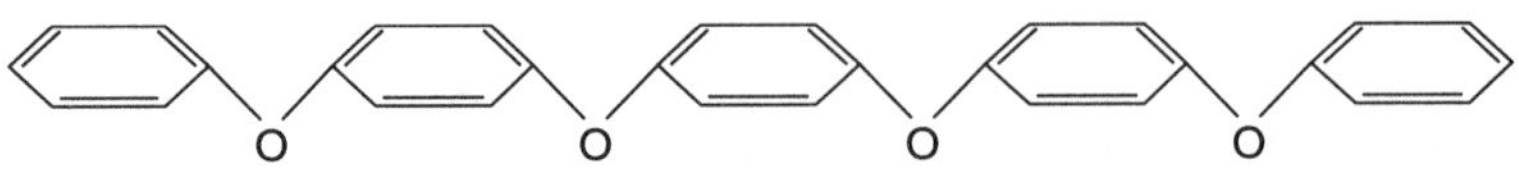

FIGURE 12-4 Molecular structure of 5-ring polyphenyl ether (PPE).

TABLE 12-4 Polyphenyl Ether Properties

Appearance	Crystal clear
Molecular formula	$C_{30}H_{24}O_4$
Molecular weight	449
Specific gravity	1.195–1.201
Viscosity, 77°F, cP	100
Refractive index, 589 nm	1.67
Boiling point, °F, 760 mm Hg	889
Pour point, °F	40
Flash point, °F	550
Autoignition temperature, °F	1135

subjected to ionizing radiation. In space, ionizing radiation comes in the form of high-energy electrons that raise the energy level of a complex molecule to a point where the molecules dissociate into smaller molecules. PPEs have high-resonance energy, implying that these molecules can absorb a great deal of radiation without dissociating. PPEs have been in use in nuclear reactors for the lubrication of electronic connectors, switches, and other equipment functioning in a high-radiation environment. Table 12-4 lists the properties of PPEs. Polyphenyl ethers have radiation resistance to 1010 ergs/g °C, thermal stability up to 850°F, and oxidative stability as high as 600 to 700°F with proper inhibitors. Polyphenyl ethers have very low vapor pressure at ambient temperatures, in the 10 torr (1 torr = 1 mm Hg) range, making them ideal for high vacuum uses. They do have a high pour point (40°F). They are expensive, which limits their use to specialized situations of high radiation, high temperature, and high vacuum.

FLUORINATED LUBRICANTS

The three classes of fluorinated lubricants currently in use in industrial applications are as follows:

- Chlorofluorocarbons[8]
- Per fluoro alkyl polyethers[9]
- Fluorosilicons[10,11]

Chlorofluorocarbons and per fluoro alkyl polyethers are most inert, even to liquid oxygen and chlorine. Applications include liquid oxygen pumps, compressors and vacuum pumps handling oxygen, fluorine and other very reactive chemical vapors, and equipment handling solid products with strong oxidizing characteristics. Chlorofluorocarbons have temperature limitations, with rapid evaporation at 300°F and above. Also, they have poor viscosity temperature characteristics and relatively high pour points compared with per fluoro alkyl poly ethers and fluorosilicons.

Fluorosilicon lubricants have good chemical inertness to a wide variety of harsh corrosive chemicals. One of the earlier applications was the lubrication of gaseous hydrochloric acid compressors and in methyl chloride service. Fluoro silicon lubricants are inert to chlorine gas up to 90 lb/in^2 and 392°F. Fluorosilicon greases made with polytetrafluoroethylene thickeners have improved chemical inertness as compared with conventional hydrocarbon greases. Fluorinated lubricants are very expensive, which limit their use.

SILICATE ESTERS

Silicate esters[12,13] have high viscosity indices, high thermal stability, and good low-temperature properties, which make them good for use in certain greases, torsion dampers, and brake fluids. Silicon brake fluids have excellent temperature stability, but they do not have as good water tolerance as polyglycol brake fluids. Water may get into the silicon brake fluid over a period of time through hydraulic lines, fittings, and breather caps and cause brake failure. These esters have poor hydrolytic stability and break down to form a gel in contact with moisture in the air. Silicon brake fluids and polyglycol brake fluids are not compatible with each other. This characteristic limits their use to closed systems. Silicate esters find uses as low-temperature refrigeration compressor lubricants.

Dibasic Acid Esters

Dibasic acid esters are synthesized by reacting an acid and an alcohol. Diesters have more a varied structure than PAOs. They contain no sulfur, phosphorus, metals, or wax. The pour point ranges between –50 and –65°C. The advantages of diesters include good thermal stability. They are clean running in that they tend to dissolve varnish and sludge rather than leave any deposit. In fact, diesters can remove deposits formed by other lubricants. Proper additive selection is critical to prevent hydrolysis and provide oxidative stability. In addition, chemically resistant seals are recommended.

Alkylated Aromatics

Alkylated aromatics are formed by the reaction of olefins or alkyl halides with an aromatic material such as benzene. The fluids have good low-temperature properties and good additive solubility. The VI is about 50 for fluids with linear molecules and zero or lower for fluids with branched side chains. Thermal stability is similar to that of PAOs, and additives are required to provide oxidation stability.

REFERENCES

1. B. L Cupples, "Polyolefins Fluids and Lubricants," presented at the American Institute of Chemical Engineers 86th national meeting, April 1989.
2. D. R. Fairbanks, "Diesters Compressors Lubricants in Petroleum and Chemical Plant Services," presented at the 30th American society of lubrication engineers (ASLE) annual meeting, May 1975.
3. R. S. Barnes and M. Z. Fainman, "Synthetic Ester Lubricants, Lubrication Engineering," August 1957.
4. G. D. Short, "Development of Synthetic Lubricants for Extended Life in Rotary Screw Compressors," presented at the 38th ASLE annual meeting, April 1983.
5. E. R. Muller, "Polyalkylene Glycol Ethers Accepted for Industrial Applications," presented at the American Institute of Chemical Engineers Material Engineering and Science Division Biennial Conference, February 1970.
6. T. G. Smith, "Phosphate Esters," presented at the American Institute of Chemical Engineers, Material Engineering and Science Division Biennial Conference, February 1970.
7. R. C. Gunderson and A. W. Harr, *Synthetic Lubricants* (New York: Reinhold Publishing Company, 1962).
8. W. Cassanos, "Halocarbon Lubricants Uses in the Chemical Industry," presented at the American Institute of Chemical Engineers 86th national meeting, April 1979.
9. N. D. Lawson, "Perfluoroalkyl Polyethers—A Unique New Class of Synthetic Lubricants," presented at the American Institute of Chemical Engineers Material Engineering and Science Division Biennial Conference, February 1970.
10. R. E. Smith, E. D. Groenhof, and W. O. Winer, "The Behaviour of Fluorosilicones as Lubricants," presented at the American Institute of Chemical Engineers Material Engineering and Science Division Biennial Conference, February 1970.
11. J. W. Miller, "Fluorosilicone Cuts Maintenance Cost," *Hydrocarbon Processing,* October 1967.
12. R. E. Smith, "Silicon Lubricants in the Chemical Processing Industry," presented at the 30th ASLE annual meeting, May 1975.
13. E. D. Brown, "Methyl Alkyl Silicon Lubricants: Unique in Structure and Applications," presented at the American Institute of Chemical Engineers Material Engineering and Science Division Biennial Conference, February 1970.

CHAPTER 13
TURBINE OILS

Turbine oil is a high-quality circulating oil used in steam turbines and many other industrial applications. It is essentially a bearing lubricant, and as such it must be able to lubricate the bearing/gears and protect these machine parts against wear. Also turbine oil acts as a heat transfer medium and keeps bearing temperatures below 150 to 180°F. The function of the turbine and the type of service dictates the viscosity of turbine oil. Turbine may be a direct drive, geared turbine, or hydraulic turbine. Industrial and marine power generation turbines are generally horizontal turbines. Hydraulic (water) turbines are generally vertical units. Gas turbines, similar to aircraft turbines, are often used in naval or other high-speed vessels. In general, steam turbine oils generally meet the requirements of industrial gas turbine systems. The major difference between steam and gas turbine workings is that in gas turbines, oil works in a dry atmosphere, whereas in steam turbines, oil works in the presence of steam. Higher temperatures are encountered in gas turbines. Typical oil temperature at full load may exceed 390°F. Control of oil temperature and flow to critical points within the system can retard oil oxidation, which is monitored by an increase in oil viscosity and an increase in its acid number.

The viscosity of turbine oils ranges from 20 to 450 cSt at 104°F (Table 13-1). Viscosity is largely determined by any associated transmission gearing, as in the case of steam turbines. Turbine oils protect critical system components from rust and corrosion. The oil must have good water separation characteristics to minimize the formation of emulsions. Turbine oils are formulated from high-quality paraffinic base oils and fortified with additives that provide oxidation resistance and minimize the formation of sludge and varnish deposits. A foam inhibitor additive is generally included in the formulation to prevent excessive foam buildup, which interferes with lubrication.

BASE OILS

Turbine oils are formulated from highly refined paraffinic base oils with a high viscosity index (VI). Laboratory tests show there is an approximately threefold increase in the life of oil if group II (hydroprocessed lube) oils are used in place of group I base (solvent refined) oils in turbine oil formulation. Turbine oils are being increasingly compounded with more severely hydroprocessed group III base oils with a VI greater than 120. The use of group II and group III hydroprocessed base oils offer the following advantages:

- Lesser antioxidant requirement
- Longer life
- Possible design and operation of machine at higher temperature
- Less carbon and varnish deposits
- Improved low-temperature fluidity
- More efficient water and foam separation
- Better biodegradability

The life expectancy of oil is extended with paraffinic structure base oils produced by hydrocracking or severe hydrotreating processes (group II and group III base oils). The relationship

TABLE 13-1 Typical Properties of Commercial Turbine Oils

		ISO viscosity grades						
	Units/test method	32	46	68	100	220	320	460
Specific gravity	60/60	0.865	0.866	0.868	0.873	0.889	0.89	0.892
ASTM color	ASTM D 1500	L0.5	0.5	0.5	1.0	3.5	3.5	4.5
Flash point	°F	430	410	469	531	572	572	590
Pour point	°F	10	–27	–27	–20	5	5	16
Viscosity	cSt, 104°F	32	47	64.2	103	203	306	456
Viscosity	cSt, 212°F	5.45	6.9	8.7	11.7	18.3	23.4	32
Viscosity index		105	116	108	105	96	96	96
Acid number	ASTM D 664		<0.1	<0.1	<0.1	<0.1	<0.1	<0.1
Acid number	ASTM D 974, KOH/g	0.15						
Copper corrosion	ASTM D 130	1a	1a	1a	1a	1a	1a	1a
Rust test	ASTM D 665 A, B	Pass	Pass	Pass	Pass	Pass	Pass	Pass
Demulsibility	ASTM D 1401, min to pass		10	10	15	15	20	20
Foam test	ASTM D 892		Pass	Pass	Pass	Pass	Pass	Pass
Oxidation stability	TOST, ASTM D 943, h	10,000+	7500	7500	7500			
Oxidation stability	RPVOT, ASTM D 2272, min	1600+	850	850	850			

between the chemical composition of turbine oils and their oxidation characteristics has been investigated by many authors. In one such study,[1] 12 turbine oil samples from different crudes, blended with a common R&O (rust and oxidation) inhibitor additives package (0.8 percent by weight [wt %]) were studied. Turbine oil lives were estimated by the rotating bomb oxidation test (RBOT) as per the ASTM D 2272 test procedure. In these turbine oils, the R&O package was predominantly antioxidant with rust inhibitor, metal deactivator, and antifoam additives. It was found that oils with a maximum amount of saturates and the lowest amount of tri- and tetra-aromatics and sulfur compounds performed the best. Oils with a low percentage of saturates, high aromatics, and high sulfur performed the worst. As a guideline, turbine oils must have more than 85 percent saturates, aromatics below 15 percent, and sulfur less than 0.4 percent by weight.

Despite their longer life, group II and group III base oils have encountered some problems. One drawback is the lower solubility of additives in oil. The lack of additive solubility can be a problem with group II and group III base stocks in R&O turbine oils. In high-speed rotary compressors, centrifuging action is known to have separated the additives from solution in oil because of their incomplete solubility. To avoid this separation, base oil is modified by the addition of another blending component: synthetic ester fluid or group I solvent refined lube oil. Altered solvency action in paraffinic hydrocracked oils may also affect seals and gaskets. Upper viscosity is limited in hydrocracked base stocks. Unconventional base stocks such as medicinal-grade white mineral oils have been used in turbine oil formulation for an extended life expectancy. The possibility of longer life lubricant oils has emerged with the availability of synthetics oils such as polyalphaolefins (PAOs), despite their higher cost compared with petroleum based oils.

FORMULATION

A turbine oil is expected to remain in the system for long periods: 10 to 20 years. During this long period, it should be capable of retaining its original properties. This can be achieved by providing a high reserve of antioxidant additives in the turbine oil. The finished product must have good oxidation stability, demulsibility, antifoaming, and air release characteristics. Most turbine oils are formulated with API group II base oils[2] and with less than 2 percent additives.

The additives used in the formulation of turbine oils are antioxidants, rust inhibitors, corrosion inhibitors, defoamers, demulsifiers, and anti-wear/extreme pressure (AW/EP) additives. The method of blending additives to base oil is critical. Most additives have solubility limits in base oils and are more viscous than the turbine oil base stock. For complete and thorough mixing into formulation, inline blending equipment is used. Additives constitute generally 1 to 2 percent of the formulated oil. Antioxidant additives are the dominant constituent of the total additive package. AW and EP additives are generally used in geared turbines or specialized applications.

LIFE OF TURBINE OIL

The important factors affecting the service life of turbine oils are operating conditions, temperature of the system, ingress of air and moisture, and oil compatibility with system metallurgy. Oil degradation become more pronounced in turbines of high output and a higher load-to-oil ratio. The base oil composition exerts an appreciable influence on its oxidation characteristics and thus the life of the oil. Over a period of time, oil tends to break down by reacting with dissolved atmospheric oxygen. The oxidation starts a chain reaction that first forms hydroperoxides and then other oxidation products; all with increased acidity, viscosity, and darker colors, and it leaves surface deposits and varnish. By eliminating initial hydroperoxides and by interrupting the chain sequence, oxidation-inhibiting additives slow down the oxidation rate by more than 100-fold. Useful life continues through an induction period as the oxidation inhibitor supply is slowly depleted.

Reclamation and Refortification of Turbine Oils

Over a period of time, turbine oil in use gets contaminated with particulate matter and moisture. Both these contaminants accelerate the aging of oil. Reclamation refers to the cleaning and reconditioning of turbine oil, thus rendering it suitable for continued use in the same application. During the reclamation process, water is removed by settling, vacuum dehydration, or centrifugation. Particulate matter is removed by settling or by filtration. These methods are effective in removing contaminants but do not reduce additive concentrations in turbine oils. However, surface active additives such as antifoamers and demulsifiers tend to accumulate on filter media and can get selectively removed[3] from the formulation. Properties such as foaming and water separation characteristics need to be constantly monitored. Refortification refers to adding a predetermined amount of additives to a clean dry used lubricant to replenish some of the depleted additives. In most cases, reclamation and refortification are done together. These processes do not remove spent additives, and soluble oxidation products remain dissolved in turbine oils. New turbines are normally flushed with new turbine oil. After flushing, the turbine oil is reclaimed. The particulate matter and water contaminants are removed and the oil is retested to ensure that it meets original equipment manufacturer (OEM) requirements. In general, the practice of using flush oil as a system lubricant is discouraged.

Refortification or the addition of additives requires the knowledge of formulation. Turbine oils are balanced formulations, and the addition of different types of additives to the used turbine oil may result in an imbalance in the formulation and thus adversely affect performance. Turbine oils formulated with API group I base oils (solvent refined lube oil) have different solvent characteristics from newer API group II base oils (hydroprocessed lube oils). The addition of a wrong type of additive to a used turbine oil may result in additive insolubility and additive dropout and/or formation of reaction products, deposits, varnish, and filter plugging.

TEST METHODS

Acid Number (ASTM D 664)

The acid number is the quantity of acid expressed in milligrams of potassium hydroxide per gram of oil sample required to titrate a sample. In the test method, the sample is dissolved in a mixture of toluene and isopropyl alcohol containing a small amount of water and titrated potentiometrically with alcoholic potassium hydroxide. The new or used oil may contain acidic component resulting from the degradation of oil in service or from acidic additives. The acid number is used as a guide in assessing the condition of the oil and its remaining life.

Copper Corrosion (ASTM D 130)

The copper strip corrosion test is designed to assess the relative degree of corrosiveness of a petroleum product. In the test method, a polished copper strip is immersed in a given quantity of petroleum product maintained at a specified temperature, typically 100°C for a specified time, typically 2 to 3 h. At the end of the specified time, the copper strip is removed from the sample, washed, and compared with ASTM copper strip corrosion standards. The corrosiveness of the sample is interpreted by matching the appearance of the test strip with one of the standards.

Rust Test (ASTM D 665)

This test determines the rust-preventing characteristics of inhibited mineral oils such as steam turbine oils to aid in preventing the rusting of ferrous parts should water become mixed with oil. This method is also used for testing other oils such as hydraulic oils and circulating oils. A cylindrical steel specimen is completely immersed in 300 mL of oil mixed with 30 mL of distilled water at 140°F. The specimen is kept immersed for 24 h at this temperature. At the end of the period, the specimen is examined for signs of rusting and the degree of rusting. If there is no rust at the end of test period, the oil specimen passes the test. In case any streak of rust is observed, the sample is deemed to have failed the test.

Demulsibilty (ASTM D 1401)

This test method measures the ability of petroleum oils or synthetic fluids to separate from water and provides a guide for determining the water separation characteristics of oils subject to water contamination and turbulence. A 40-mL sample of oil is stirred with 40 mL of distilled water for 5 min at 54°C in a graduated cylinder. The time required for the separation of the emulsion thus formed is recorded. If complete separation does not occur after standing for an hour, the volume of oil, water, and emulsion are reported.

Foam Test (ASTM D 892)

This test method covers the foaming characteristics of lubricating oils at specified temperatures. The tendency of the oil to foam can be a serious problem in systems such as high-speed gearing, high-volume pumping, inadequate lubrication, and cavitation. This method is valuable in the selection of oils suitable for such operating systems. In this test method, the oil sample is maintained at 75°F and blown with air at a constant rate for 5 min and then allowed to settle for 10 min. The volume of foam is measured at the end of both these periods. The test is repeated on a second sample at 200°F and then after collapsing the foam, at 75°F. The volume of foam at the end of the 5-min blowing period is reported as the foaming tendency. The volume of foam at the end of the 10-min settling period is reported as the foam stability.

Turbine Oil Stability Test (ASTM D 943)

The turbine oil stability test (TOST) method is used to evaluate the stability of inhibited steam turbine oils in the presence of oxygen, water, copper, and iron metals at an elevated temperature. In this test method, the oil sample is reacted with oxygen in the presence of water and an iron-copper catalyst at 95°C. The test continues until the measured total acid number of oil reaches 2.0 mg KOH/g or above. The total number of test hours required for oil to reach the acid number of 2.0 mg KOH/g is the oxidation life of the oil.

Oxidation Stability of Steam Turbine Oils by Rotating Pressure Vessel Oxidation Test (ASTM D 2272)

The rotating pressure vessel oxidation text (RPVOT) is test measures the oxidation stability of new and used turbine oils in the presence of water and a copper catalyst coil at 150°C. In this method, the oil, water, and a copper catalyst are placed in a covered glass container and placed in a bomb equipped with a pressure gauge. The bomb is charged with oxygen at a pressure of 90 lb/in^2, placed in a constant temperature bath at 150°C, and rotated axially at 100 r/min at an angle 30° from the horizontal. The number of minutes required to reach a specified pressure drop is the oxidation stability of the oil.

REFERENCES

1. R. T. Mookken, D. Saxena, A. K. Jaiswal, B. Basu, S. Sataphaty, S. P. Srivatsva, and A. K. Bhatnagar, "Significance of High Quality Base Stocks for Formulating Turbine Oils," presented at the International Symposium on Production and Application of Lube Base Stocks, New Delhi, India, November 1994.
2. M. Khonsari, "New Lubes Last longer," *Machinery Lubrication Magazine,* May 2004.
3. W. H. Stein and R. W. Bowden, "Turbine Oil Reclamation and Fortification," *Machinery Lubrication Magazine,* July 2004.

CHAPTER 14
USED OIL RE-REFINING

Lubricating base oils are generally one of the more costly products for a refinery to produce. The lube processing units require a high capital investment. They are very complex units and require high energy inputs. Also, specialized equipment is required to handle and store base stocks produced. These higher operating costs are reflected in the higher costs of the base oils. Typically the base oil prices average twice that of crude oil. After the addition of additives (15 to 20 percent by weight [wt %] of base oil), the price of the finished lubricant may increase as much as 10 to 20 times to that of base oil. Apart from petroleum-based oils, a growing market is developing for synthetic lubricants. These are manufactured by polymerizing olefins from petrochemical sources. Although more expensive than petroleum-based lubricants, they have superior lubricant properties and a longer life in engines. Synthetics are increasingly being used in high-performance engines. The share of synthetic lubes in automotive lubes market is still low.

Millions of tons of petroleum based lubricant products are sold every year, but within a few months of purchase, most of it is either lost or discarded. Only a portion of lubricants used in industry is recoverable. Estimates[1] of recoverable lubes vary for different end uses (Table 14-1).

Where the lubricant is used in two-stroke engines or as a process oil, the product is either used up or so diluted that it cannot be economically recovered. Because of the high cost of lubricating oil, there is every incentive to re-refine the used lubricant. In automotive engine oils and agricultural engine oils, approximately 65 percent of used oil is collectable. Lube oil in engines does not undergo any chemical change; it gets contaminated with corrosion products, dirt, soot, fuel, and so on, and additives get depleted. Table 14-2 shows the typical composition of used crankcase lubricating oil. However, the composition of used oils can vary widely depending on the crude, processing sequence, viscosity classification, and original additive composition.

Basically there are two alternatives for the disposition of used lubricating oils:

- Burning as fuel
- Re-refining to produce lubricating base oils

Many problems are likely to be encountered in both options because of the presence of high amounts of metals and organometallic compounds in used oils, originating from lube oil additives. On combustion these yield a high ash content (up to 1.5 wt %), resulting in severe furnace fouling with inorganic ash. This, in turn, increases furnace operating costs and shortens tubes and refractory life. Also, burning of these oils can cause unacceptable air pollution. Used lube oils may contain lead, phosphorous, and chlorine, which are emitted to the atmosphere on combustion. Lead is particularly hazardous because submicron-size lead emissions can injure humans via respiration. For these reasons, legislation in many countries has now prohibited the combustion of used lube oils in power plants. The inappropriate disposition of used lubricating oils on land and in water has already emerged as a major cause of environmental damage in many industrialized countries.

BURNING AS FUEL

As heavy hydrocarbons, used lubricating oils have a high calorific value, a low pour point, and a low sulfur content. However, these oils have high water and light hydrocarbon content. Preflashing removes water and light hydrocarbons. The combustion of used oil can be done in specially designed burners. Combustion in conventional burners can be done after suitable dilution in virgin fuel oil. Usually the proportion of used lube oil is kept below 5 percent.

TABLE 14-1 Estimated End Uses of Lubricants and Recoverable Used Lubricating Oil

	Fraction of total	Percentage collectable
Gasoline and diesel engines	0.3157	65
Agricultural engines	0.0190	65
Other engines	0.0092	0
Marine engines	0.0477	25
Aviation and turbine oils	0.0028	50
Total engine oils	0.3944	
Hydraulic and transmission	0.1219	80
Other gear oils	0.0681	80
Total gear/transmission oils	0.1900	
Automotive greases	0.0061	10
Industrial greases	0.0088	10
Other greases	0.0001	0
Total greases	0.0149	
Metal working neat/soluble	0.0364	20
Others	0.0086	20
Total metal working oil	0.0450	
Turbine and electrical oils	0.0343	95
General machine lubricants	0.0193	50
Nonlubricating industrial oils	0.0149	10
Other industrial oils	0.0138	20
Total other oils	0.0823	
Process oils	0.1369	0
White oils	0.0274	0
Total processing oils	0.1644	
Total	1.0000	50

TABLE 14-2 Used Crankcase Lubricating Oil*

Specific gravity, 60/60	0.888
Viscosity, cSt	
100°F	120.7
210°F	15.5
Viscosity index	134
Pour point, °F	–28
Flash point, COC, °F	420
Ash content, Wt %	1.3
Total acid number (TAN), mg KOH/g	3.4
Conradson carbon, Wt %	1.8
Color	Black
Metals	
Fe, ppmw	150
Al	27
P	800
Ba	1000
Ca	1120
Pb	115
Zn	820
Contaminants	
Water, Wt %	4.0
Diluents (diesel), Wt %	2.0

*Most used motor oils are SAE 20, SAE-30, and SAE-40 and their variants. Used lube oil composition can show wide variation depending on its viscosity grade and additive content.

RE-REFINING

Used lubricating oil can be considered a suitable lube oil feedstock that already has the correct viscosity, viscosity index, and pour point for manufacture of lube base stock. Thus minimum processing is required. In fact, the only processing required is the removal of contaminants (water, light ends, ash, asphalt), which in turn also improves the color and the neutral number of the oil. The cost of manufacturing lube base stock from used lube oil is much less than that of making lube from crude oil. However, the difficulty of used oil re-refining has increased in recent years for the following reasons:

- Higher amount of additives are being used in lube formulations due to the ever-increasing performance requirements of automotive engines.
- A significant increase has taken place in the average interval between two replacements of oil in an automobile crankcase. The corresponding distance has increased from 7,000 to 10,000 km or more. As a consequence, the average used oil today has a higher asphaltic material compared with used oil composition a decade ago.

RE-REFINING PROCESSES

One of the main difficulty in re-refining used lubricating oils is the collection of used oil, which means collecting small quantities of used oils from a large numbers of sites. For this reason, small capacity low-cost re-refining plants using an acid-clay treatment have been in operation and are economically viable. To reduce sulfuric acid consumption and acid sludge volumes, some additional procession steps to the conventional process have been added. Preflashing consists of heating oil to about 600°F and flashing it under vacuum to remove water and diluents. This step may be followed with propane extraction. Dehydrated oil is treated with propane that dissolves only the oil, rejecting any asphaltic residue and metals to asphaltic resid fraction. Propane treatment reduces metals and asphalt in used oil by 80 to 90 percent. After these steps, used oil may be processed in an acid-clay plant or in a hydrofinishing unit to yield regenerated lube base stock. Preflashing and propane extraction reduce acid consumption and sludge production in the acid-clay process, and in the case of hydrofinishing, process severity and catalyst consumption are significantly reduced.

BATCH ACID-CLAY PROCESS

The batch acid-clay process is suitable for small-scale units, and many such plants are in operation in several countries. In this process, used lube oil is heated to a high temperature and flashed to remove water and diluents such as gasoline and diesel. This step also causes the thermal degradation of many additives (detergents and dispersants) contained in the oil. The dehydrated oil is next treated with sulfuric acid. Sludge is settled and removed after which sour oil is treated with active clay. Clay is filtered off, and clear oil is collected as regenerated base stock for reformulation with additives. Figure 14-1 presents the details of the acid-clay process.

Raw Material Storage and Water Removal

Referring to Fig. 14-1, Used oil received either through tanker deliveries or in barrels is stored in underground tank V-101. A metered amount of used lube oil is transferred from tank V-101 by pump P-102 to batch vacuum dryer V-103. Here oil is heated to 575°F, with thermic fluid and under vacuum with continuous stirring for about 2 to 4 hours. Water vapor and diluents (gasoline, diesel, etc.) are distilled over and condensed in water-cooled heat exchanger E-101 and collected in V-104. The dried oil is transferred to V-105, equipped with cooling coils, where the temperature is brought down to 100°F.

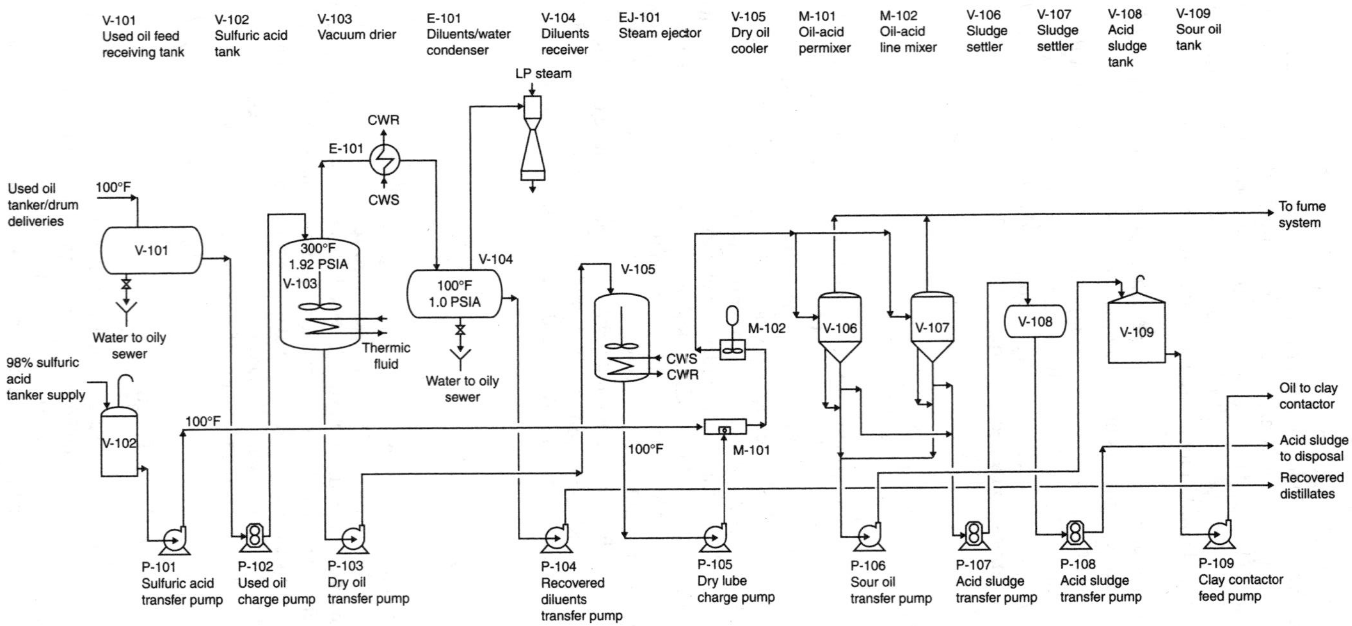

FIGURE 14-1 Used oil regeneration plant (acid treatment).

Acid Treatment and Sludge Settling

Dried oil from V-105 is contacted with 98 percent sulfuric acid in acid-oil premixer M-101 and next in line agitator M-102 where oil and acid are thoroughly mixed. Sulfuric acid is pumped from storage tank V-102 through pump P-101, and oil is pumped from V-105 via P-105 to a number of sludge-settling tanks, V-106 and V-107, after acid mixing. The volume of sulfuric acid used is about 7 percent of the oil volume. The sludge settling time is about 40 to 48 h. The sludge settled is collected drawn from tank bottoms and transferred by a gear pump to sludge tank V-108. Sludge may be neutralized with lime in the form of an aqueous solution to a neutral pH before disposing it to a landfill. Clear sludge-free oil is transferred by pump P-106 to intermediate storage vessel V-109.

Clay Treatment

Clay treatment is also done batchwise (refer to the process flow diagram in Fig. 14-2). A fixed volume of acid treated oil is transferred from V-109 via pump P-109 to clay treatment vessel V-110, equipped with an agitator and steam heating coils. Oil is heated to 176°F, and next the activated clay is added to the tank. Clay added is generally 6 to 8 wt % of oil treated. The oil-clay mixture is kept in contact for about an hour after which the oil is filtered in plate and frame-type filter

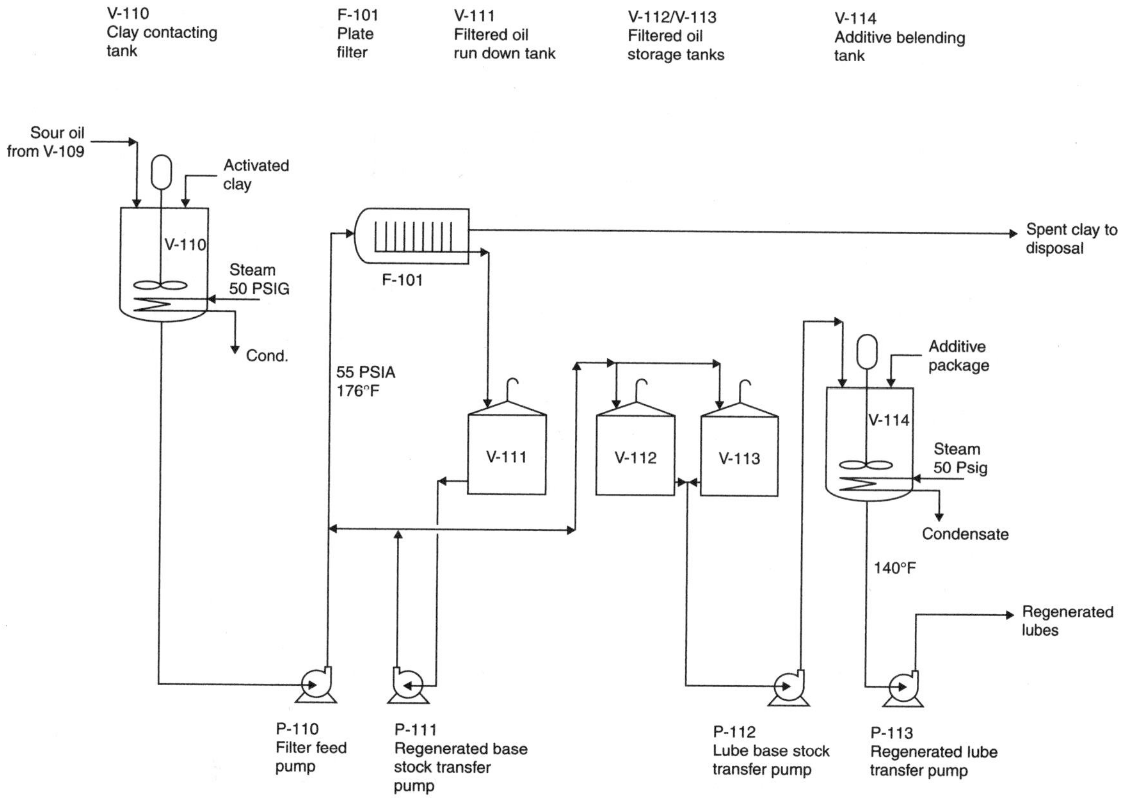

FIGURE 14-2 Used oil regeneration plant (clay contacting and refinishing).

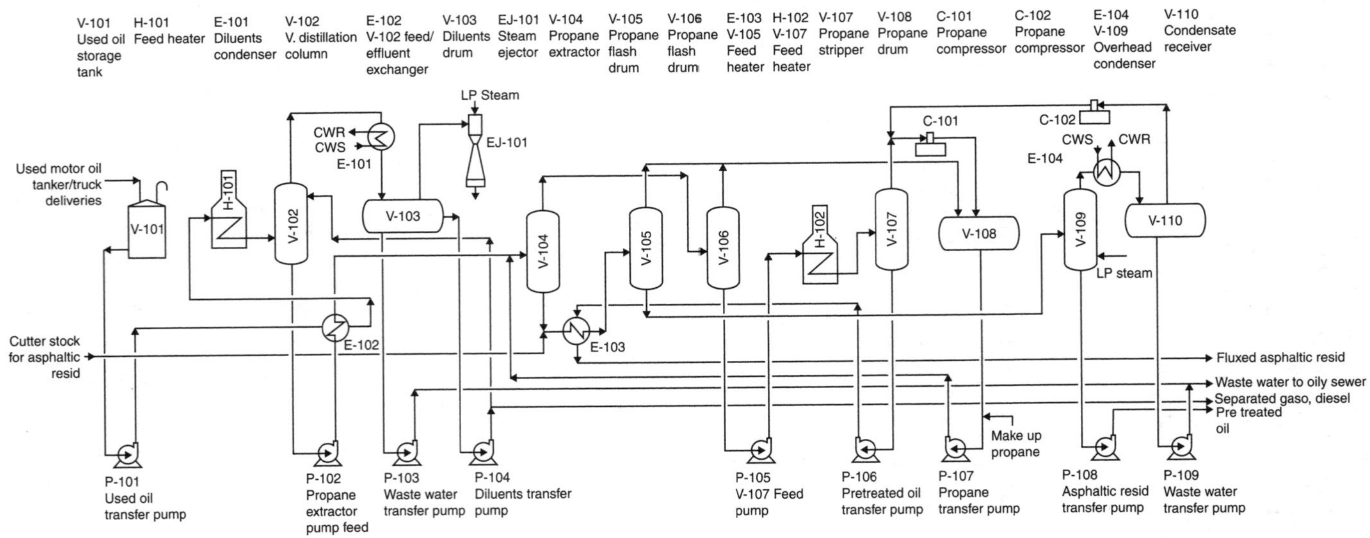

FIGURE 14-3 Used oil feed pretreatments.

press F-101. Feed to the filter is kept at a pressure of 55 lb/in^2 by means of pump P-110. Normally more than one filter is used to maintain continuous operation. Hot clear filtrate from the filter flows by gravity to down tank V-111. If the lube oil filtrate is not clear, it can be recycled back to the filter press. After the filtration, oil is pumped to V-112 and V-113, the base stock storage tanks, by pump P-111.

Addition of Additives

Samples of base stock in tanks V-112 and V-113 are tested, and the amount of various additives needed are worked out. Additives (typically 0.5 to 12 wt %) for individual batches are done in steam-heated tank V-114 where the temperature is maintained around 120°F.

PRETREATMENT

Used lubricating oils contain a high percentage of metals originating from lube oil additives. These contaminants can be removed by certain pretreating[2] steps that make the subsequent processing of oil in acid-clay units or in hydrofinishing units much easier.

The pretreating steps consist of preflashing and propane extraction. After removal of diluents and water, the oil is treated with propane to remove mainly asphaltic components in an extraction tower similar to the one used in lube-making propane deasphalting (Fig. 14-3). With propane extraction, approximately 80 percent or more of the metals originating from the lube additives are removed. After pretreatment, used oil can be processed in acid-clay type units where acid consumption and sludge production are significantly reduced.

HYDROFINISHING PROCESS

In the hydrofinishing process (Fig. 14-4), the feed is preflashed to remove diluents and next extracted with propane. After propane extraction, the feed is sent to a hydrofinishing unit. Operating conditions are a function of feed metals and other impurities. Propane treated feed has a low metal content, and the operating conditions of a hydrofinishing unit are much less severe compared with those for untreated feed. Without pretreatment, feed has a very high content of metals, and the catalyst consumption in the hydrotreating unit is very high, with a high operating cost. In the hydrofinishing unit, feed is treated with hydrogen over a conventional hydrotreating-type catalyst with a high-pore volume that adsorbs metals on its surface and thus demetallizes the feed, but in the process the catalyst in the bed is deactivated. The hydrofinishing reactor effluent may still contain some contaminants

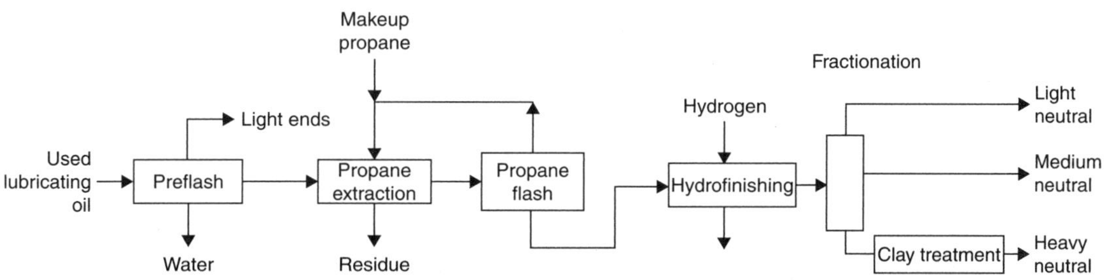

FIGURE 14-4 Used oil regeneration with pretreatment and hydrofinishing.

that get concentrated in the heaviest lube fraction. After the hydrofinishing step, the effluent is distilled under vacuum to fractionate the oil into light neutral, intermediate neutral, and heavy neutral stock. Heavy lube is next clay treated with about 3 percent active clay to remove metals and other heavy contaminants. The quality of the product is comparable with that of virgin lube base stock. The hydrofinishing route for used oil processing suffers from a number of drawbacks, however (e.g., variable feed quality, high operating cost, and high minimum plant throughput for an economically sized unit).

REFERENCES

1. "UK Waste Oils Market 2001," report by Oakdene Hollins Ltd. for Department of Environment Transport and Regions.
2. "Oil Refining and Petrochemicals Production," IFP technology symposium on oil refining and petrochemical production 1980.

CHAPTER 15
LUBRICATING GREASES

The need for lubricating greases has existed since ancient times. As far back as 1400 B.C., the Egyptians used animal fat such as tallow to reduce axle friction in chariots. Until the 19th century, greases were made by mixing animal or vegetable fat with lime. This solid grease was known as "briquette." Towards middle of 1900's, greases made from mineral oils were available. Compared with earlier lubricants, these were found to be extremely efficacious and were used in miners' carts and other slow-moving industrial machines used at that time. With the development of steam engines, motor-driven vehicles, and industrial and agricultural machines in 1920s, greases with bases of metallic soaps such as sodium, aluminium barium soaps were developed.

Lithium soap greases were first introduced in the 1950s. Today, lithium soap greases are the most widely used greases. The primary function of a grease, like any other oil lubricant, is to reduce friction between moving metal surfaces such as those found in gears and bearings. Many applications are encountered where it is more practical and economical to lubricate the moving surface with grease rather than with lubricating oil. Because grease does not flow readily, it is used where extended lubrication is required and where oil would not be retained. Lubricating grease is preferable under extreme operating conditions, such as high temperature, extreme pressures, low speeds, shock loading and bearings that operate intermittently or reversing. When machine parts are badly worn, lubricating grease is the only means of providing lubrication. Lubricating grease will ensure some lubrication when a bearing is neglected for some time. With lubricating grease, less frequent application is required. Lubricating grease acts as a seal against the entrance of dust and dirt into bearing and fittings. Less expensive seals are required for grease-lubricated bearings. Dripping and spattering are practically eliminated when a machine is grease lubricated. In certain cases, the use of lubricating grease in place of lubricating oil simplifies the design. Grease lubrication does have some negative features. Lubricating greases are not as good a coolant as oils are. Oils flush out contaminants more readily than greases. Also, changing oil is less labor intensive than changing grease in a machine.

GREASE COMPOSITION

Grease consists of three main constituents:

- Base oil
- Thickening agent
- Performance additives

Grease consists of a thickening agent dispersed in a lubricating oil. Performance-enhancing additives are added to impart such additional properties as antiaging, rust prevention, anti-wear, and extreme pressure resistance. The lubricating oil may be petroleum based or synthetic oil. The lubricating oil is commonly a mineral oil from paraffinic, naphthenic, or aromatic base crude oils. Synthetic oils are usually used for grease formulation in extreme temperature conditions. USP or medicinal-grade mineral white oils are used for machinery, where lubricants may come in direct or indirect contact with food. The thickening agent or gellants include alkali metal soaps, clays, polymers, carbon black, colloidal silica, and aluminium complexes. The most common type of grease is soap-based grease. Soap comes from animal or vegetable fats or fatty acids, wool grease, rosin, or petroleum acids. Other components of grease include unreacted fat, fatty acids, alkali, unsaponifiable matter (including glycerol and fatty alcohols), rosin or wood grease, and water. Additives used in grease are antioxidants, rust and corrosion inhibitors, color stabilizers, metal passivators, water repellants, and viscosity index improvers.

BASE OIL

Lubricating oil is the largest component in a grease formulation; typically 80 to 95 percent of total grease mass. The lubricating oil used must be of high quality and of the proper viscosity and pour point for the intended application. The base oil selected for grease formulation must have the same qualities as if the equipment was to be lubricated with oil alone. The lubricating oil used in grease formulation may be a natural mineral oil or a synthetic oil such as diester oil, silicon oil, fluorocarbon oil. For petroleum oils in general, naphthenic oils are a better choice than paraffinic oils because of their better solubility characteristics for metallic soaps and additives. The property of any grease is determined by the properties of the base oil. Greases with a low-viscosity base oil are best suited for low-temperature and high-speed applications, whereas greases made from a high-viscosity base oil are best suited for heavy loads. A low-viscosity oil is used for low-temperature, low-load, and high-speed applications. A high-viscosity oil must be used for high-temperature, low-speed, and heavy-duty applications. For use in the food and beverage processing machinery, food-grade greases are formulated with medicinal grade or USP-type white mineral oils. Synthetic oils are more expensive than corresponding petroleum base lubricating oils, but they have superior low- and high-temperature characteristics. Synthetic oils such as polyalphaolefin (PAO)-based fluids are used where low-temperature operating capabilities are important. Polyalkylene glycol base fluids are used in greases where temperatures exceed 400°F, up to 600°F. Bearings, chains, and reducer gears are typical applications for these. With growing environmental concerns, vegetable oils are being used in applications requiring biodegradable lubricating oils.

GREASE THICKENERS

The thickener may play as important a role as oil in lubrication. Thickener constitutes 5 to 20 percent of total grease mass. The thickener in grease may be metallic soap, clay, or organic compounds.

Metallic Soaps

Soap thickeners are formed by reacting (saponifying) a metallic hydroxide or alkali with a fat, fatty acid, or ester. The most common alkalis used are the hydroxides of earth metals such as lithium, sodium, calcium, barium, and aluminium. Fatty acids can be derived from animal fat such as beef, tallow, lard, fish oil, or butter or from vegetable oils such as caster, soybean, olive, palm, or peanut oils. The fatty acid usually contains a 16 to 18 carbon atom chain. Table 15-1 shows the properties of fatty acid salts used as thickeners.

Generally, nonedible, low-cost oils are preferred. Soap is formed when a long carbon chain fatty acid reacts with the alkali (metallic hydroxide). The metal is incorporated into the carbon chain, and the resultant molecule develops a polarity. The quality of the soap formed also depends on the fatty acid from which it is formed.

TABLE 15-1 Properties of Fatty Acid Salts Used as Grease Thickeners

Compound	Carbon number	Molecular weight	Melting point °F	Water solubility
Lithium stearate	18	291	428	Insoluble
Lithium 12 hydroxystearate	18	307		
Lithium docosanoate	22	347		
Dilithium azelate	9	202		
Lithium salt of hydrogenated Castor oil	>16	>260		
Lithium sals of fatty acids, C16–22	16–22	263–347		
Calcium 12 hydrostearate	18	640		
Calcium stearate	18	608	354	0.004 g/100 cc, 59°F
Calcium salts of tallow	14–18	>490		
Sodium stearate	18	308		Slowly soluble
Sodium oleate	18	305	453	10 g/100 cc, 54°F
Sodium palmitate	16	278	518	Insoluble
Potassium oleate	18	322	464	25 g/100 cc, 54°F
Magnesium stearate	18	591		0.003 g/100 cc, 59°F
Zinc stearate	18	632	266	Insoluble

Complex Metallic Soaps

To improve the high-temperature capabilities of greases, instead of simple metallic soaps, complex metallic soaps are used in grease blending. The normal metallic soap manufacturing process is modified to include an additional ingredient: a short-chain organic or inorganic acid. The short-chain acids are typically acetic, oxalic, or lactic, and the inorganic acids employed are carbonates or chlorides. A complex soap is formed when apart from the normal saponification reaction of metallic alkali with a fatty acid, metallic alkali also reacts with a short-chain acid to form a metallic salt or complexing agent. The dropping points of greases blended with complex soaps are significantly higher than that of corresponding simple soap greases. The type of soap depends on the grease property desired. Calcium (lime) soap greases are highly resistant to water but unstable at high temperatures. Sodium soap greases are stable at high temperatures but may get washed out in moist conditions. Lithium soap greases resist both heat and moisture. A mixed-base soap is a combination of soaps, offering some of the advantages of each type. Complexing agents usually increase the dropping point of greases. A common form of soap-based grease uses lithium 12 hydroxystearate as a thickener. To thicken the grease properly, soap must be in the form of fibers of suitable size dispersed in lubricating oil. The crystalline fibers are usually in the size range of 1 to 100 microns, with a diameter 0.1 to 0.01 of their length. For good shear stability, the fiber should have a large ratio of length to diameter, and for good oil retention, the fiber should be as small as possible. Therefore grease needs a mixture of two types of fiber. Also, there must be a balance between the solvency of the fluid and the solubility of the soap to get suitable thickening. The melting point of grease made with various soaps differs appreciably. The grease dropping point indicates the temperature limitation of grease. Table 15-2 shows the effect of different thickeners on the dropping point of grease. The dropping temperature of soap-based grease can be increased by using soap complexes. The complex consists of a soap salt thickener. For example, the aluminium complex might consist of aluminum with a fatty acid, a nonfatty acid, and an alkyl. Then each molecule of thickener consists of aluminum complexed with stearate, benzoate, and hydroxide. The dropping point of complex greases is at least 100°F higher than the dropping point of corresponding soap greases. The range of application of greases is also extended by multipurpose greases that consist of a mixture of soap bases of different metals and soaps. Calcium sulfonate thickened greases are characterized by their excellent resistance to rust and

TABLE 15-2 Types and Characteristics of Sodium, Calcium, and Aluminum Lubricating Greases

	Sodium grease mineral oil	Calcium grease (cup grease) mineral oil	Aluminum grease mineral oil
Base oil thickener	Sodium soap	Calcium soap	Aluminum soap
Dropping point, °F	300–355	175–195	45–195
Application temperature range, °F	–4–265	–4–158	15–175
Water resistance	Not good	Good	Good
Pressure resistance	Good	Not good	Good
Applications	*	†	‡

*Some of the grease is emsulfied when mixed with water; excellent high-temperature resistance.
†Excellent in water resistance; inferior in heat resistance low speeds and heavy load uses.
‡Suitable for vibrating bearings.

corrosion, good mechanical stability, and high-heat and load-carrying capability. These may be formulated with high-viscosity synthetic base oils for high-temperature applications such as paper machine felt roll dryers and fan bearings in kiln vents. The most common thickener used for food-grade greases is aluminium complex. This thickener withstands elevated temperatures and is highly water resistant. This is a very vital characteristic for food and beverage processing equipment, where equipment is frequently washed for hygienic reasons.

Inorganic Clays

In many grease formulations, nonsoap thickeners such as clays, silica gel, polytetrafluoroethylene (PTFE), graphites, and various other synthetic organic materials are used. The clay, such as bentonite or attapulgite, is reacted with a quaternary amine to change the clay from hydrophilic (water-loving) to hydrophobic (water-rejecting) and oleophilic (oil-attracting). Effective thickening is achieved by combining the clay with a polar activator or dispersant such as acetone, methanol, or ethanol with small amounts of water and by delaminating and reducing the platelets to a small size. This process increases the total surface area of the dispersed clay, which immobilizes a very high percentage of oil based on the weight of clay, thereby thickening the grease. The solid additives thicken grease by the nature of their fine dispersion of electrochemically charged particles throughout the fluid and by their particle-to-particle interactions. Solid additive greases extend the operating temperature range over that of soap greases. The solid-type greases do not have a melting point, and their upper temperature limit is that of the oil being used. The base oil in such greases becomes more prone to evaporation, oxidation, and thermal degradation, and frequent relubrication may be necessary.

Organic Thickeners

Polyurea is the most important organic thickener. It is a low molecular weight organic polymer formed by reacting amines with isocyanates, which results in an oil-soluble thickener. Polyurea grease has excellent oxidation resistance and thus has a long life. It effectively lubricates over a wide temperature range (–4 to 350°F). It works well with many elastomers and has good water resistance. It is used in all types of ball bearings, and its durability makes it suitable for sealed-for-life bearing applications.

ADDITIVES

Additives may constitute 0 to 15 percent of the total grease mass. Additives used in lubricating greases are similar to those used in lubricating oils with a few exceptions. Various additives are added to greases to enhance the various properties desired in the finished grease. These are antioxidants, anti-wear, extreme pressure (EP) additives, antirust, and anticorrosion additives, tackifiers, and dyes. Antioxidants are added to enhance the life of a grease. For comparatively high operating temperatures or in applications where grease cannot be replenished for long periods, EP agents are added to guard against scoring and galling under heavy loads. EP additives are used in greases formulated for bearing subjected to heavy loads. Anti-wear additives prevent abrasion and metal-to-metal contact; anticorrosion additives are added to protect metal against water, sulfide, and other corrosive element attack. Tackifiers are added to most greases to increase the tackiness or sticking of the grease to metal parts and enhance its water washout resistance. Most grease formulators add dye to their greases to distinguish between different end-use products.

Tackifiers

Industrial and automotive fluids sometime require an additive that improves the adhesion of fluid to the surface that is lubricated with lubricant or grease. Tackifiers are additives that confer a tack or stringiness to a substance and are typically used to provide adherence in fluid lubricants and stringiness in greases. Tackifiers along with thickeners provide drip resistance to grease and serve to inhibit stray mist in pneumatic system lubricants. Polyisobutylene, natural rubber, and latex dissolved in mineral oil or esters are known to be excellent additives that provide such properties. Tackifiers find application in many areas, such as automotive chassis grease, industrial EP greases, automotive undercoatings, bar and chain saw oil, pneumatic drill oil, aluminium forming oil, slideway lubricants, oil bases rust preventives, gear oils, spindle oils, machine oils, cutting oils (anti-mist grade), and textile machinery oil. Spindle oils normally require tackifiers of a very light color and with high tackifying properties as well as with low viscosity. Viscosity is not the determining feature for tackifying properties.

Viscosity and color are the other important properties, which in turn may be a function of the molecular weight and chemical composition of the tackifier additive. Tackifiers are shear-sensitive high molecular weight polymers, and dissolving them in low-viscosity diluent oils can pose a problem. They can drop out of solution if the polarity of the solvent is changed or when other additives are added. The best blending sequence is to dissolve the tackifier completely in oil with good mixing before adding the other additives. Also reducing the viscosity of tackifier by heating before blending facilitates the blending process.

Tackifiers may be added to grease in the concentration range of 0.1 to 1.0 percent by weight. In certain applications where biodegradability of grease is a consideration, the tackifier may be dissolved in vegetable oil as a solvent carrier.

GREASE MANUFACTURE

Grease can be made either in a batch or in a continuous process. The batch process is the most common manufacturing method. The first step involved in soap-based grease making is the saponification of fatty acid with an alkali to produce a metallic soap. This soap acts as a thickener for the grease. This step however is not required for making clay-based greases.

Metallic Soap Manufacture

The saponification reaction is carried out in a batch saponification reactor or contactor. The contactor is a closed mixing vessel equipped with heating coils, jacket, and a powerful mixer (Fig. 15-1). An impeller inside the contactor pulls the liquid contents of the contactor through a circulation tube. The discharge from the impeller is directed up through the annular space formed by the contactor wall and annular tube. The contactor thus provides intense mixing and rapid circulation of the reacting

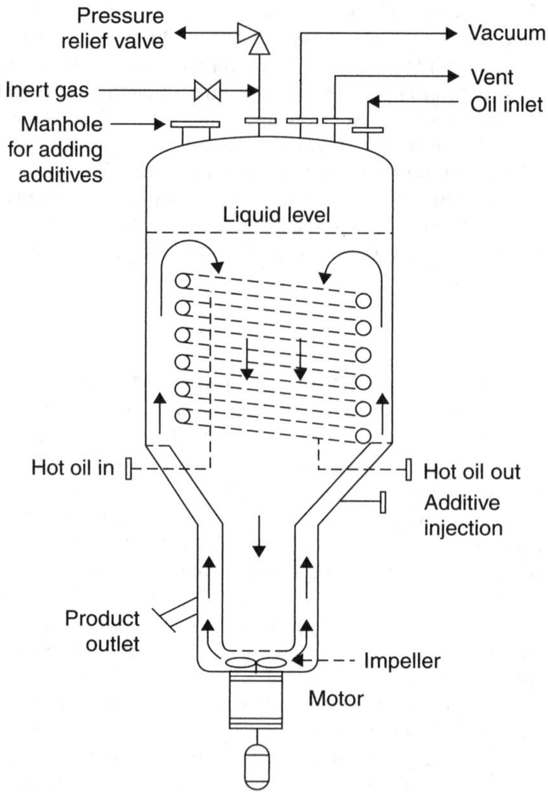

FIGURE 15.1 Contactor for grease manufacture.

mixture. The contactor design has one objective of maximizing both the heat transfer and mixing. Because of non-newtonian properties and the high viscosities of most greases, it is difficult to achieve good mixing and high heat transfer rates in conventional grease mixer equipment equipped with countercurrent paddles and scraper blades. A metered amount of base oil is charged to the contactor by a pump. The contactor is equipped with a hinged opening at the top, used for charging active chemical ingredients used in grease compounding. Most fatty acids and alkalis are received in solid form. The specified amount of each of these raw materials is delivered into the reactor, after which the charging door is closed and the contactor is started. After the circulation has been established, heating of the contactor is started. Heating is done either by medium-pressure steam or by a hot oil heating system. High-heat transfer coefficients are achieved by high circulation rates across heating surfaces. Hot oil temperature is maintained at 500°F. Hot oil is circulated through the external heating jacket or internal heating coil. Very rapid heating of the contents of the contactor occurs, and the temperature rises to 425°F in less than an hour. At this temperature, the saponification reaction is complete. Water is formed during the saponification reaction, and the pressure in the contactor rises to 75 to 90 lb/in^2. At the end of the heating cycle,contactor is vented that removes the water formed during the saponification reaction. Vacuum is next applied to the contactor to further dehydrate the soap. At this stage, the metallic soap has been formed. The next step in grease manufacture is the addition of more base oil to the contactor. The addition of this dilution oil results in the cooling of the oil/soap mass in the contactor. When the oil/soap solution reaches a certain temperature, the soap gels, and a viscous concentrated soap is formed. Additional dilution oil is added until the desired soap concentration is reached. Additives can be pumped into the bottom of the contractor while it is still under pressure. The grease at this stage is called semifinished grease.

Grease Mixing

The semifinished grease from the contactor is pumped to one or more finishing kettles. These are large mixing vessels, usually the open or nonpressure type equipped with large slow-speed mixers. The objective is to provide a gentle blending action and cooling. In the classic grease mixing kettle design, intimate mixing within the kettle is achieved by two countercurrent rotating assemblies. The central assembly is provided with multiple-blade impellers arranged to pump in one direction, and the outer assembly pumps it in the opposite direction. The outer rotative assembly is designed to carry, in addition to impeller blades, multiple scraper blades to scrape the interior of the vessel thoroughly throughout the length of the straight shell and the curved part of the lower head. The mixer and sweep arm speeds are typically in the range of 30 and 15 r/min, respectively. The blades of these agitators are twisted to give not only radial mixing but also mix the grease in the axial direction, along the vertical axis of the kettle. The grease kettle is provided with a jacket that covers the bottom head and straight side of the vessel. Cooling can be affected by cold oil or cooling water. The outside sweep arm of the agitator is fitted with scraper blades that make positive contact with the wall of the kettle and greatly increase the heat transfer coefficient. This scraping action is particularly important so rapid cooling rates and also high production rates can be achieved.

When the final amount of finishing oil has been added and the grease is cooled to an acceptable temperature level (typically 130 to 150°F), additives are added to the grease and blended. Most additives are temperature-sensitive compounds and therefore cannot be added to grease while it is hot. The cooling and finishing of grease in the finishing kettles takes from 2 to 4 h. The estimate of the quantity of final finishing oil to be added is made only after the oil content and also the physical properties of grease, particularly penetration, has been compared with the desired physical properties of the grease.

Homogenization

After mixing in the kettles, the grease is sent to a homogenizer. The homogenization step is very important because it will produce uniform gel structure which will not change when the grease is used. The homogenization of grease breaks down solid particles or fibers that will disperse the resultant small particles in the liquid. It also breaks down lumps, eliminates graininess, and produces a smooth product. The homogenization of certain types of greases stiffens the grease, producing a lower penetration value product. Homogenization can improve texture and brighten a grease's appearance. In many cases, homogenization is carried out at a temperature greater than 200°F. After homogenization, the grease is further cooled, deaerated, and packed.

LUBRICATING GREASE QUALITY

Consistency (ASTM D 217)

This test method covers four procedures for measuring the consistency of lubricating greases by penetration of a standard cone. These test methods include procedures for the measurement of unworked, worked, prolonged worked, and block penetrations. Penetrations up to 475 may be measured. The U.S. National Lubricating Grease Institute (NLGI) classifies greases according to their consistency as measured by worked penetration. The classification is shown in Table 15-3. The unworked penetration does not generally represent the consistency of grease as does the worked penetration. Penetration of block greases can be obtained on samples that are sufficiently hard to hold

TABLE 15-3 National Lubricating Grease Institute (NLGI) Classification of Greases

Grade (consistency)	Penetration (after 60 strokes at 25°C, 0.1 mm)	Appearance
000	445–475	Fluid
00	400–430	Fluid
0	355–385	Very soft
1	310–340	Soft
2	265–295	Moderately soft
3	220–250	Semifluid
4	175–205	Semihard
5	130–160	Hard
6	85–115	Very hard

their shape. Penetration is determined at 77°F by releasing the cone assembly from the penetrometer and allowing the cone to drop freely for 5 s.

The worked penetration results may be used to establish the consistency of lubricating grease within the NLGI consistency grades. The change in penetration due to working is believed to be a measure of its shear stability under conditions of the test.

In this test, a cup of standard dimension is filled with the lubricant, which is brought to a temperature of 77°F. Its surface is smoothed. The instrument used, called a penetrometer, is designed to measure in tenths of a millimeter the depth to which the standard cone penetrates the grease. The cone weighs about 102.5 g. Provision is made on the scale of the penetrometer to measure the depth in tenths of a millimeter that the cone penetrates into the sample.

For worked penetration determination, the sample is transferred from a container to an apparatus called a "grease worker," which is provided with a metal disk held by a plunger that can move through a stuffing box in the cover for the cup. The disk has a series of closely spaced holes 0.25 inches in diameter. The disk is forced through the lubricating grease for 60 double strokes, and lubricant is then smoothed off and a penetration value obtained. The result is reported as worked penetration. Sufficient sample, at least a pound, to overfill the cup of standard grease worker is required. The test is reported as a worked penetration.

Note that the lower the worked penetration value and the higher the consistency number, the stiffer the lubricant. Stiff products such as NLGI grades 4, 5, and 6 act as better seals and have less of a tendency to drip than softer lubricant grease grades. However, harder grades may channel and leave bearing elements dry, and hence they are not good general-type lubricants. In an average bearing of a modern machine, a No. 1 or 2 NLGI grade of lubricant grade provides the most satisfactory lubrication. When lubricating greases are applied by handguns or pressure systems, No. 2 grade or a softer product can generally be forced through the lines and fittings even in cold weather.

Dropping Point (ASTM D 566)/ASTM D 2265)

The dropping point of a lubricating grease is the temperature at which the lubricant becomes fluid enough to permit a drop to separate from a small sample. The sample is held in a small cup with a bottom hole, and the temperature is increased at a controlled rate during the test. The dropping point is the temperature at which the grease passes from a semisolid state to a liquid state under the conditions of the test. This change of state is typical of greases containing soap thickeners. Grease containing nonsoap thickeners may separate oil without a change in state. Although there is no direct relationship between the dropping point and service performance, a lubricating grease cannot be expected to provide satisfactory performance at temperatures higher than its dropping point. The test is useful in identifying the grease type and maintaining a benchmark for quality control.

Evaporation Loss (ASTM D 972)

In this test, a sample of lubricant in an evaporation cell is placed in a bath maintained at the desired temperature (210 or 300°F) while heated air is passed over the sample for 22 h. The evaporation loss is calculated as percentage loss in weight of the sample. This method can be used to determine the loss of volatile materials from grease or oil at any temperature in the range of 200 to 300°F.

Low-Temperature Characteristics (ASTM D-1092)

This method determines the apparent viscosity of lubricating greases. The sample is forced through a capillary by means of a piston actuated by a hydraulic system. From the predetermined flow rate and force developed in the system, the apparent viscosity is calculated by means of Poiseuille's equation. A series of determinations are made with different capillaries and pump speeds. The results are expressed as a log-log plot of apparent viscosities versus shear rates. This information can be useful in predicting pressure drop in distribution systems under steady-state flow conditions at constant temperatures. NLGI has an alternative method in which the flow properties of the grease are compared with that from a grease-dispensing pump at low temperature. Satisfactory delivery from pump is taken as a minimum of 2 oz per minute. It is shown that delivery is a function of grease viscosity at the operating temperature.

Shear Resistance

The consistency of lubricating grease as it is received by a user does not necessarily reflect its consistency in service. In bearings, the lubricant grease is subject to shear or working for prolonged periods. In general, such a treatment will result in softening of the product. The ability of lubricating greases to resist softening after working is referred to as shear resistance. The shear resistance of a grease sample is calculated by measuring penetration of the grease before and after 10,000 double strokes. In the case of most multipurpose greases, such as No. 2 NLGI, with a worked penetration of 265 to 295, the softening will be in the range of 10 to 25 points. When the seals are inadequate, undue softening of lubricant will cause leakage.

Water Resistance (ASTM D-1264)

Lubricating greases are frequently subjected to wet conditions, and therefore the resistance to washout by water is important. This test method covers the evaluation of the resistance of a lubricating grease to washout by water from a bearing when tested at 100 and 175°F. In this test, the grease is packed in a ball bearing, which is then inserted in a housing. The assembly is rotated at approximately 600 r/min. Water at a specified controlled temperature impinges on the bearing housing at the rate of 5 mL/s. The amount of grease washed out in 1 h is a measure of resistance of grease to water washouts. This test method estimates the resistance of grease to water washouts from ball bearings under the conditions of the test.

Rust Prevention (ASTM D-1743)

This method covers the determination of the corrosion preventive properties of lubricating greases using grease-lubricated tapered roller bearings stored under wet conditions. In this test method, new cleaned and lubricated bearings are run under a light thrust load for 60 s to distribute the lubricant in a pattern found in service. The bearing is exposed to water and then stored for 48 h at 125°F and 100 percent relative humidity. After cleaning, the bearing cups are examined for evidence of corrosion. Because all lubricating greases do not provide full protection against rust, rust inhibitor additives are included in the lubricant grease formulation. Sodium-based thickeners in the grease lubricant have better rust preventive properties than that of other thickeners.

Service Life: High-Temperature Life (ASTM D 3527) and Oxidation Stability (ASTM D 942)

Most lubricant greases are used where frequent application occur; the older lubricant is forced out and new lubricant grease is added in its place. This is the situation with automobiles, farm machinery, and in most industrial uses. However, in many applications such as electric motors, bearings are packed at an interval of two years or more. Some antifriction bearings are even packed for life. Bearing lubricant in such applications must be inhibited with antioxidant additives. A test method, ASTM D-942-50, is available for determining the oxidation stability of lubricating greases by the oxygen bomb method. In this method, a sample of lubricating grease is subjected to an atmosphere of oxygen at a temperature of 210°F, permitting absorption of oxygen to be determined at a 100-h interval. A lubricant that will cause pressure drop of less than 5 lb/in^2 under these conditions will have a service life of over 2 years. Generally, all lubricating greases, especially for ball and roller bearing lubrication, contain oxidation inhibitors.

Oil Separation (ASTM D-1742)

This test method covers the determination of the tendency of lubricating grease to separate oil during storage in either normally filled or partially filled containers from which some grease has been removed leaving a crater. This is most notable in the case of softer grades. However, a little oil separation is desirable because this will permit oil to seep into the cracks and cover metal surfaces when lubricant is used. In this test method a sample of grease is supported on a 75-micron (200-mesh) screen. It is then subjected to 0.25 lb/in^2 air pressure for 24 h at 77°F. Any oil seepage that occurs is collected in a beaker and weighed. This test method is not suitable for grease grades softer than NLGI No. 1 consistency because of the tendency of the grease to seep through the screen.

EP Characteristics: Four-Ball EP Test (ASTM D 2596) and Timken OK Load Test (ASTM D 2509)

Lubricating greases do not possess EP characteristics, but additives can be included if service demands this quality. Various methods are available for measuring these characteristics. The Timken EP tester consists essentially of a steel test cup rotating against a steel test block loaded from below. The test cup is attached to a horizontal spindle mounted on two roller bearings and driven by a 2-hp motor.

The rotating speed is 405.88 ft/min, which is equivalent to a spindle speed of 800 r/min. Fluid samples are preheated to 100°F before starting the test. A lever arm on which weights are suspended causes measured force against the ring as lubricant is fed to the rotating member. After a period of 10 minutes, if no scoring of block is evident but only a smooth scar is present, the lubricant is said to have passed the test. The load is given either in terms of load on the lever arm or in pounds per square inch calculated from the width of the scar and the projected load of weight times the lever length. The latter figure may be in the range of 20,000 to 30,000 lb/in^2 for a 33-lb load or 27,000 to 38,000 lb/in^2 for a 43-lb load and an average scar width.

Two determinations are made. One is for the minimum load that will rupture the lubricant film being tested between the rotating cup and stationary block and cause scoring or seizure. The second determination is for estimating the maximum load (OK value) at which the rotating load will not rupture the lubricating film and cause scoring or seizure between the rotating cup and the stationary block. This method is used widely for specification purposes and to distinguish between lubricants having a low, medium, or high level of extreme pressure characteristics.

Four-Ball Wear Test (ASTM D 2266, ASTM D 2783)

This test method covers the determination of the load-carrying properties of a lubricant fluid. The following two determinations are made:

1. Load wear index
2. Weld point by means of a four-ball EP tester

The load wear index is an index of the ability of a lubricant to minimize wear at applied loads. Under the conditions of the test, specific loading in kilogram force or newtons, having intervals of approximately 0.1 logarithmic units, are applied to three stationary balls for 10 runs prior to welding. The load wear index is the average of the sum of corrected loads determined for 10 applied loads, immediately preceding the weld pair. The weld point is the lowest applied load in kilograms at which the rotating ball welds to three stationary balls, indicating the EP level of lubricant (in newtons) has been exceeded. The four-ball EP tester can also be used for determining the load-carrying capacity of lubricating greases. The tester is operated with one steel ball under load rotating against three steel balls held stationary in the form of a cradle. Test lubricant covers the lower three balls. The rotating speed is 1760 ± 40 r/min. The machine and test lubricants are brought to 65 to 95°F, and a series of tests of 10-s duration are made at increasing load until welding occurs. Ten tests are made below the welding point. After a 10-s run, the scar diameter on the bottom balls are measured, and from these figures and load, the EP value is determined.

Deleterious Particles (ASTM D-1404)

This test method describes a procedure for the detection and estimation of deleterious particles in lubricating greases. A deleterious particle by this test method is one that will scratch a polished plastic surface. In this method, the lubricating grease being tested is placed between two clean, highly polished acrylic plastic plates held rigidly and parallel to each other in metal holders. The assembly is pressed together and the lubricant is squeezed into a thin layer; any particle in the lubricant larger than the distance of separation of the plates and harder than plastic will become embedded in the opposite plastic surfaces. Provision is made to rotate one of the plates 30° while pressure is maintained. This will cause any embedded particles to form an arc-shaped scratch on the plate. The number of scratches is some indication of the relative number of hard particles in the lubricant. Note that particles abrasive to plastic are not necessarily abrasive to metals.

Tackiness

Tack improver additive effectiveness is measured by its threadiness (ductility). Test methods for measuring threadiness are not covered by accepted standards.

In the test method, a 200-cc solution of tackifier (0.25, 0.5, and 1.0 percent by weight) in mineral oil is prepared. A Brookfield No. 1 needle is weighed. Next it is submerged in the additive solution maintained at 68°C. The solution is stirred for 2 min. The spindle is next rinsed with solvent or oil without additive and allowed to drain for 2 more minutes. Next the spindle is inserted into a stirrer and rotated at 400 r/min. The spindle is next removed from the stirrer and weighed. The weight increase of the spindle is recorded. Before testing any tacky additive, a blank reading is taken with straight oil, without additive.

Tackifier polymers, like other polymers, can drop out of solution if the polarity of the diluent is changed. When blending a polymeric additive and polar additive, it is best to completely dissolve one additive in diluent oil before adding another additive. Tackifiers and viscous additives can be difficult to dissolve in low-viscosity solvents. Poor mixing can cause incompatibility when other additives are added. If agitation is insufficient for good blending, heating the tackifier additive can avoid the incompatibility problem. Tackifier polymers are all high molecular weight compounds. Subjecting the additive to high shear can eventually break down any tackifier. Shear that occurs during pumping frequently leads to the loss of tackiness. Also, many tackifiers break down when heated to a temperature above 212°F. The water spray test (ASTM D 4049) indirectly measures the tackiness of the grease.

Compatibility

Grease substitution may be necessary under certain conditions when the performance of the originally recommended grease in an application falls short of requirements. In such cases, the old grease is completely purged and cleaned out of the appliance before the new grease is introduced. The mixing of lubricant greases of a different soap and nonsoap bases is not desirable and may result in impaired performance of the lubricant. If two different greases are in contact, the oil may migrate from one grease to the other and basically alter the structure of both the greases, and the combination may be softer than either of the original lubricants. If the thickeners of two greases are incompatible, the mixture of two greases will fall short of the properties of the two greases. A higher temperature operation can even increase incompatibility. In the absence of actual performance data, any two greases with different thickeners should be considered incompatible.

AUTOMOTIVE GREASES

Greases are used in cars and trucks at many points in the chassis and in wheel bearings. Chassis lubricant greases are used at a number of points in the vehicle's chassis requiring lubrication, such as kingpins, pedal shafts, spring shackles, control arms. Wheel-bearing greases, as the name suggests, are mainly used in automobile wheel bearings. Quality requirements for the two end uses are different. The main requirements of chassis greases are lowest ambient temperature pumpability and the ability to withstand heavy loads and shock loads without rupture of lubricant film over metal surfaces. Another important characteristic of chassis grease is water washout resistance. For this reason, sodium soap greases are not suitable. For a given ambient temperature range, chassis lubricant greases are formulated with the heaviest lubricating oil, metallic soaps, fillers, and additives such as graphite, molybdenum compounds, and tackifiers. The quality requirements of wheel-bearing greases include long service life, stability to working, and resistance to temperatures of 250°F or above. Brakes of heavy vehicles, trucks, and trailers, especially on hilly roads, can develop enough heat to cause wheel bearings to encounter temperatures of 250°F or more.

Automotive Grease Service Classification

The NLGI provides standard rating and certification for greases. Chassis lubricants are identified with the letter designation "L" and wheel-bearing lubricants with the letter designation "G." The performance categories within the two categories result in two-letter designations for chassis grease (LA, LB) and three for wheel-bearing greases (GA, GB, and GC). The highest present performance category in each group (LB and GC) is suitable for service lubrication. NLGI GC-LB is the highest performance category for multipurpose greases. Chassis grease usually contains an EP additive. Wheel-bearing greases are of two types: the low-temperature grease and the high-temperature grease. For most modern cars, only the high-temperature grease is suitable. Table 15-4 shows the performance requirements of various grades.

TABLE 15-4 NLGI Automotive Greases Service Classification

Designation	Use	
LA	Chassis Noncritical applications Lubrication interval < 2000 mi	With antioxidants, anti-wear and Antirust, anticorrosion additives
LB	Chassis Relubrication interval > 2000 mi Mild to heavy duty, vibration, Water resistance	With antioxidants, antiwear, EP and Antirust, anticorrosion additives; Service temperatures: 100–300°F
GA	Wheel bearing Frequent relubrication Mild duty	Service temperatures: –5–160°F
GB	Wheel-bearing Cars and trucks	With antioxidant and anti-wear Antirust and corrosion additives; Service temperature: –40–250°F
GC	Wheel-bearing Mild to severe duty, Heavy trailers, mountain driving	With: antioxidant and anti-wear, EP Antirust and corrosion additives Service temperature: –40–320°F

AIRCRAFT GREASES

Greases are used in aircraft in the lubrication of moving parts of the fuselage, hot zones of the engine, wheel bearings, generators, magnetos, starters, landing gear assemblies, main and tail rotor bearings of helicopters, and in gearboxes. Greases used in aircraft are subject to a wide temperature range (–65 to 350°F). Other qualities required are similar to normal high-performance greases, such as high water resistance (typical loss, 0.5 percent), anti-wear and heavy load-carrying ability, long life, NLGI grades 1 or 2 (penetration, 260 to 300), are suitable. For this reason greases formulated with synthetic fluids and clays (dropping point higher than 500°F) are more popular than normal petroleum-based greases. Table 15-5 shows properties of some commercial aircraft greases.

TABLE 15-5 Characteristics of Commercial Aircraft Greases

	1	2	3	4	5
Base oil	Synthetic diester	Syntheic polyol/ mineral oil	Mineral oil	Silicon oil	PAO
Thickener operation	Lithium soap/clay	Clay	Lithium	Calcium complex	Clay
Temperature range, °F	–100–250	–65–350	–40–212	–75–300	40–300

HEAVY MACHINERY GREASES

Wire Rope Lubricants

Wire ropes form an important part of many heavy machines and structures. Wire ropes are used in drag lines, cranes, elevators, shovels, drilling rigs, suspension bridges, and cable-stayed towers. A wire rope is comprised of continuous wire strands wound around a central core. There are many types of wire ropes designed for different end uses. Wire ropes are generally made of steel wires made into strands and then wound into each other. The core can be made of steel wires or ropes or even plastics. Typical wire ropes/cables are identified by several parameters, including size, grade of steel used, whether or not it is preformed, and by its "lay," the number of strands and number of wires in each strand. For example, a wire made of 6 strands with each strand made of 19 wires will be designated as 6 × 19. Different strand sizes and arrangements allow for varying degrees of rope flexibility and resistance to crushing and abrasion. A small diameter wire is better suited to being bent sharply over small sheaves/pulleys. Large outer wires are preferred when cable will be rubbed or dragged through an abrasive medium. The lay of the wire is the direction that the wire strands in a cable twist. The four common lays are right lay, left lay, regular lay, and lang lay. In a right lay, the rope strands twist to the right as it winds away from the observer. A left lay twists to the left. A regular lay rope has the wires in the strand twisted in the opposite direction from the strands in the cable. In a lang lay rope, the twists of the strands and the wires in the strands are both twisted in the same way. Lang lay ropes are known to have better fatigue resistance due to the flatter exposure of the wires. Wire ropes are made mostly from high carbon steel for strength, versatility, resilience, and cost considerations. Wire ropes can be uncoated or galvanized. Wire ropes can also have fiber cores made from sisal fiber or from plastics. However, the fiber core ropes have less strength than steel core ropes.

Wire rope lubrication has two important functions:

- To reduce friction as individual wires move over each other
- To provide corrosion protection and lubrication in the core and inside wires and on the exterior surfaces

There are two types of wire rope lubricants; penetrating and coating. Penetrating lubricants contain petroleum solvents that carry the lubricant to the core of the wire ropes and then evaporate, leaving behind a heavy lubricating film to protect and lubricate each strand. Coating lubricants penetrate slightly, sealing the outside of the cable from moisture and reducing wear and fretting corrosion from contact with external elements.

Both types of wire lubricants are used on every cable. Most wire cables fail from the inside. It is important to ensure that the cable core receives adequate lubricant. A penetrating lubricant is used first to saturate the core, followed by a coating lubricant to seal and protect the outer surface.

Lubricants used for wire rope lubrication can be petrolatum, asphalts, mineral lubricating oils, vegetable oils, or greases. Petrolatums compounded with proper additives provide excellent corrosion and rust resistance. In addition, because petrolatums are translucent, they permit visible inspection of the cable or rope. Petrolatum lubricants can drip off at higher temperatures but maintain their consistency under cold temperature conditions. Asphaltic compounds dry to a very dark hardened surface that makes inspection very difficult. They adhere well for extended long storage but crack and become brittle in a cold climate. Asphaltic compounds are the coating type. Various types of greases are used for wire rope lubrication. These are the coating type that penetrate partially but usually do not saturate the wire rope core. Common grease thickeners include sodium, lithium, lithium complex, and aluminium complex. Greases used for this application generally have a soft, semifluid consistency. They coat and achieve partial penetration if applied with pressure applicators. Petroleum lubricating oils and vegetable oils formulated with proper additives are the best penetrating-type lubricants for wear and corrosion resistance to wire ropes. The fluid properties of these lubricants help wash off external contaminants from the wire ropes. Wire ropes are generally lubricated during the manufacturing process. If the rope has fiber core, the fiber will be lubricated with a mineral lubricating oil or petrolatum with proper additives to saturate the fiber core. The fiber core acts as a lubricant reservoir for prolonged lubrication of the rope in service. In case of steel wire rope, the lubricant, oil, or grease type is pumped in a stream just ahead of the dye that twists the wires into a strand. This allows complete coverage of all wires. After the cable is put in service, relubrication is required due to loss of original lubricant from the loading, bending, and stretching of the cable. The fiber core cables dry out over time due to heat from evaporation and often absorb moisture. Field relubrication is necessary to minimize corrosion and to protect and preserve the rope core and wires, thus extending the service life of the wire rope. If a cable is dirty or has accumulated layers of hardened lubricant or other contaminants, it must be cleaned with a wire brush, petroleum solvent, compressed air, or steam cleaner before relubrication. Wire rope must next be dried and immediately relubricated to prevent rusting. Field lubricant can be applied by spray, brush, dip, drip, or pressure boot. Lubricants are best applied at a drum or sheave where rope strands have a tendency to separate slightly due to bending. This facilitates maximum penetration to the core. If a pressure boot application is used, the lubricant is applied to the rope under slight tension in a straight condition. Cable life cycle and performance are influenced by several factors, including type of operation, care, and environment. Cables can be damaged by worn sheaves, improper winding, high stress loading, shock loading, jerking heavy loads, rapid acceleration or deceleration, and improper storage. Abrasive wear occurs on the inside and outside of wire ropes. Individual strands inside the rope move and rub against each other during normal operation causing internal wear. Abrasive wear reduces rope diameter and can result in core failure and internal wire breakage. The outside of the cable accumulates dirt and contaminants from sheaves and drums. Wire lubricants reduce both abrasive wear inside and keep out outside contaminants and thus reduce corrosion. The life of the wire rope can be considerably extended by proper lubrication of the core and external surface during use. Properties of some commercial wire rope lubricants are shown in Table 15-6.

TABLE 15-6 Properties of Wire Rope Lubricants

NLGI grade		2
Penetration	Unworked ASTM D 217	275 ± 25
Density	g/cc, 77°F	0.969
Dropping point	°F	400
Base oil properties; type		Petroleum
Viscosity	cSt, 104°F	
Flash point	°F	601
Usable temperature range	°F	15–115
Four-ball EP test	ASTM D 2596	
Load wear index	kg	34
Water washout	ASTM D 1264	<1
Salt spray	U.S. Federal Standard 791	Pass

Open Gear Greases

Open gears form an important part of heavy machinery such as coal mining, clay product machines, cement plants, agricultural machines, sugar mills, and so on. Machines function in near ambient temperatures and are exposed to dust, dirt, and atmospheric conditions; rain, snow, sleet, and water from washing. Finished grease must adhere to metal surfaces between gear teeth to prolong the life of gears, reduce noise, resist moisture, and provide anti-wear and EP protection. For many years open gear greases were formulated with heavy lubricating oils (approximately 450 cSt at 100°F), asphalts, diluent solvents, tackifying, and other additives (antirust, anti-wear and EP agents) Asphaltic open gear greases have poorer adhesion to metal surfaces and tend to flake off in very cold weather. Present-day open gear greases have replaced asphalt with metallic soaps, bentonite clay, silica, graphite, and so on. Synthetic fluids are also being used. Many of these lubricant greases are available as an aerosol for ease of application.

Fifth Wheel Greases

A fifth wheel is a circular or wheel-shaped mechanism fixed on the rear of the chassis of a truck-trailer that engages the trailer's kingpin with a spring lock device. It supports the weight of the front end of the trailer. This mechanism, apart from providing a coupling to the truck-trailer provides support and stability to the vehicle during turns. Without lubrication, the fifth wheel and trailer bolster plate will weld to each other due to heavy loads (friction welding), resulting in lack of control. Fifth wheel grease is exposed to the elements and thus susceptible to "water washouts" where the grease may be totally washed out in rainstorms. In cold regions the grease may be exposed to road salt, corrosive anti-icing chemicals used in snow-clearing operations, dirt, and other solid contaminants caught in the grease. Fifth wheels typically experience extreme fretting, which is the removal of finely divided

TABLE 15-7 Properties of commercial Fifth Wheel Grease

		1	2	3
NLGI grade		1	1	2
Penetration	ASTM worked, 77°F	325		
	unworked		315–325	265–270
Density	g/cc, 77°F	0.969		
Dropping point	°F	210	Nonmelting	
Thickener		Calcium soap	silica	
Filler, graphite	Wt %	10–12		
Base oil properties				
Type		Petroleum	Synthetic	
Viscosity	cSt, 104°F	430	600–650	
Usable temperature range	°F	15–115		
Water washout	ASTM D 1264	<1	<5	
Water sprayoff test	ASTM D 4049			0.2
Rust test	ASTM D 1243	Pass	Pass	
	ASTM D 1743			Pass
Copper corrosion	ASTM D 4048			1b
Pumpability	g/min, 60°F			100

metal particles from rubbing surfaces, oscillation, sliding, and vibration. Fifth wheel greases are generally formulated with heavy base oil, typically heavy naphthenic bright stock, approximately 10 to 15 percent solids (calcium soap, graphite, asphalt, limestone, silica, etc.), a high percentage of tackifiers, and EP additives (molybdenum disulfide [MoS_2]) to handle severe loads associated with fifth wheel applications. It is formulated to repel water and resist washouts. Tackifier performance is reflected in the water spray-off test (ASTM D 4049) where it must achieve less than a 1 percent spray-off rating. Table 15-7 shows the typical properties of commercial fifth wheel greases. Synthetic lubricant greases for fifth wheel applications use PAO fluids with silica thickeners and various performance-enhancing additives and are delivered as an aerosol spray product, making application quick, neat, and economical. Synthetic greases can be used over a wide temperature range. Fifth wheel greases can also be used for open gears, wire rope lubricants, chains, and other applications where a heavy lube is required. Conventional formulations for these applications are based on heavy petroleum lube, asphalt, tackifiers, and additives. A diluent solvent is used to ease application. In cold-temperature applications, greases formulated with asphalt, typically flakes off, leaving surfaces unprotected against wear and causing relubrication intervals to increase. In warm weather, asphaltic material softens and loses film strength, exposing metal surfaces to wear. Synthetic fifth wheel greases do not suffer from such limitations and can be used over a wide ambient temperature range. Open gears, wire ropes, and fifth wheels require a similar quality of grease lubricants.

MARINE GREASES

Marine greases are used in many applications on marine vessels, offshore drilling platforms and on other equipment in seaports (e.g., cranes and winches). Grease has to perform its lubricating function in a highly corrosive atmosphere saturated with salt water in heavy storms and high winds. Although the temperatures encountered are relatively low, heavy loads and exceptional water resistance is required. Greases provide sealing and retention of the lubricants on lubricated parts. In open systems, greases are also used to coat surfaces for rust/corrosion prevention. Marine greases are formulated with lithium or calcium sulfonate thickeners in a concentration range of 15 to 30 percent. The base lubricants used are mineral, paraffinic oils or synthetics. A combination of additives is employed. These include antioxidants, rust and corrosion inhibitors (which protect the lubricated metal surfaces against rust from salt water), anti-wear additives that strengthen the lubricating film, and tackifiers that enhance the adhesion of grease to lubricated metal parts. Solid friction modifiers

TABLE 15-8 Properties of Marine Greases

NLGI grade		2
Thickener type		Calcium sulfonate complex
Penetration	ASTM worked, 77°F	287–290
	Unworked	
Density	g/cc, 77°F	
Dropping point	ASTM D 2265°F	>500
Filler, graphite	Wt %	
Base oil properties type		Mineral oil
Viscosity	cSt, 104°F	205
	210°F	20
Usable temperature range	°F	
Water washout, %	ASTM D 1264	2.7
Water sprayoff test	ASTM D 4049	
Rust test	ASTM D 1243	
	ASTM D 1743	Pass
Copper corrosion	ASTM D 4048	1A
Timken EP test, lb	ASTM D 2509	60
Four-ball EP test	ASTM D 2596	
Weld load, kg		500
Load wear index		65
Four-ball wear test	ASTM D 2266	
Scar diameter, mm		0.4

such as graphite and MoS_2 are added so that grease can tolerate very heavy and shock loads. Blending components are chosen so that finished grease density is more than that of water. This results in very good resistance to water washouts. Table 15-8 shows the typical characteristics of some commercial marine greases.

HIGH-TEMPERATURE GREASES

In many industrial operations, moving equipment operates under high temperatures, heavy loads, and in corrosive and erosive environment. Furnace car wheel bearings, steel mills, coke oven batteries, ceramics/pottery kilns, and cement mills are many such examples. Grease in such applications may experience temperatures in the range of 500 to 750°F. For continuous service near 400°F, it may be necessary to relubricate frequently to prevent the deterioration of petroleum base fluid. Additional grease must be applied before the grease in the bearing stiffens. Most high-temperature greases are blended with high-viscosity petroleum base oils of group II or III with high oxidation stability. PAO and synthetic esters are other important base fluids. A limited amount of silicon oils is also used in high-temperature grease formulation. Petroleum oils begin to decompose at 530 to 600°F. A high-temperature grease thickener may be a metallic complex soap such as a lithium, aluminium, or calcium sulfonate complex. An organic thickener often used is polyurea. It has a temperature range limit similar to metal soap thickened greases but additionally has antioxidant and anti-wear properties that come from the thickener itself. Inorganic thickeners employed are clays or fumed silica (silicon dioxide). Metal soaps or complex soap thickeners have thermal degradation limits of 250 to 430°F. The calcium sulfonate complex system is an exception. It has inherent antioxidant, antirust, and EP/anti-wear properties. Also, the calcium sulfonate complex thickener has a high drop point. Additives used in these greases are MoS_2 and graphite. Long after the base oil has evaporated, these additives provide a laminated film that protects sliding and rolling elements. These greases are used in kiln bearings, steel mills, pulp and paper mills, and in other demanding services where temperature, water washout and contamination, extreme loads, or a combination of these factors are present.

Food-Grade Greases

Food-grade greases are used in the lubrication of food, beverages, pharmaceutical, and drug manufacturing plants such as meat and poultry processing, soft drinks and bottling, bakeries, vegetable and food processing, cosmetics, drinking and potable water treatment, and so on. In these plants, there is always a possibility that lubricant or grease may come in contact with the food or another edible product being processed. Lubricants used in these plants must meet the requirements of the U.S. Food and Drug Administration (FDA), such as US FDA 21CFR 178.3570, 1783620(b), and 573.680. These greases are blended with the highest quality, highly refined and purified, nontoxic, pure paraffin base white oils or with medicinal-grade white mineral oils. Aluminium complex is the most common thickener and produces a very shear stable grease. Aluminum complex thickened greases can withstand elevated temperatures. These greases are highly water resistant, which is very vital for food and beverage processing equipment that requires very frequent washing. Food-grade additives such as tackifying agent, antioxidant, and EP additive are generally added to the formulation to increase anti-wear, antirust, and corrosion capabilities and enhance oxidation stability. Blended food-grade grease must be colorless, odorless, and tasteless to qualify for use in food and beverage plants. Formulation also includes an antimicrobial agent.

Synthetic food-grade lubricants are primarily PAO-based fluids. They provide superior oxidation resistance compared with petroleum-based oils. These also provide significantly better low-temperature operating capabilities. In combination with food-grade additives, PAO-based fluids are outstanding lubricants for air compressors, oil recirculating systems, hydraulic systems, and gear reducers. Polyalkylene glycol based food-grade synthetic fluids are becoming popular for applications where temperatures exceed 400°F, up to 600°F. Bearings, chains, and reducer gears subjected to these temperatures are candidates for polyalkylene glycol fluid lubrication. Table 15-9 shows the properties of food-grade greases/lubricants.

TABLE 15-9 Properties of Commercial Food-Grade Grease

		1	2
NLGI grade		2	2
Thickener type		Aluminium complex	Aluminium complex
Four-ball wear	ASTM D 2266		
(40 kg, 1200 r/min, 1 h, 167°F)			
Scar diameter	mm	0.55	0.5
Four-ball EP	ASTM D 2596		
Weld point	kg	400	315
Load wear index	kg	55	41
Timpken EP	ASTM D 2509		
OK load	lb	45	40
Penetration	ASTM worked, 77°F	265–295	265–295
Dropping point	°F	510	540
Base oil properties			
Type		Petroleum	PAO
Viscosity	cSt, 104°F	561	
	cSt, 212°F	66	
Viscosity index		108	
Usable temperature range	°F		
Water washout	ASTM D 1264	7.7	1.67
Water sprayoff test	ASTM D 4049	9.2	
Rust test	ASTM D 1243	Pass	
	ASTM D 1743		
Copper corrosion	ASTM D 4048	1A	

REFERENCES

1. E. Armstrong, U.S. Patent 2704, 363, "Preparation of lithium calcium grease compositions" issued 1955.
2. J. E. Turner and C. Barnes, "Lubrication Basics for Wire Ropes," *Machinery Lubrication Magazine* (July 2002).
3. American Iron and Steel Institute, "Wire Rope User's Manual," 1985, pp. 5–17.
4. J. T. Ronan, W. A. Graham, and C. F. Carter, "New Equipment Shortens Grease Processing Cycle," *NLGI Spokesman* 31 (1968).
5. A. T. Polishuk, "Physical and Chemical Properties of Complex Soap Greases," *Lubrication Engineering* 19 (1963).
6. J. L. Drcher, T. H. Koundakjian, and C. F. Carter, *NLGI Spokesman* 29 (1968).
7. L. Claude, J. Jhonson, and J. A. Altshuler, *NILGI Spokesman* 11 (1948).
8. C. J. Bonner, *Manufacture and Application of Lubricating Greases* (New York: Reinhold, 1954).
9. "Additive Agents for Lubricants," *Refining Engineer* (1956).
10. E. N. Klemgard, *Lubricating Greases, Their Manufacturer and Use* (New York: Reinhold, 1957).

CHAPTER 16
WAXES

The word "wax" generally refers to a variety of organic substances that are solid at ambient temperature but become free-flowing liquids at a slightly higher temperature. The chemical composition of waxes is complex, but normal alkanes are always present in high proportions, and molecular weight profiles tend to be wide. The main commercial source of wax is crude oil, but not all crude oils when processed produce wax. Wax is also produced from lignite, plants, and animals. Even insects produce a material sold commercially as wax.

NONPETROLEUM WAXES

Beeswax

Beeswax has been traded for over 2000 years. Any reference to "wax" before the 19th century typically meant beeswax. Yellow beeswax is secreted by bees to build honeycombs. After honey recovery, the empty comb is melted in boiling water to recover the wax. Yellow wax can be bleached with oxidizing agents to white beeswax, a product favored in the cosmetic industry. Smaller amounts of wax are used in the pharmaceutical industry and in candle making. The composition of beeswax varies widely with geography and the diet of the bees forming the combs, but typical components are C25 to C31 hydrocarbons, esters of C30 to C32 alcohol with C16 acids, and free C25 to C31 carboxylic acids. Beeswax typically has a viscosity of 1470 mm^2/s at 98.9°C, an acid number of 20, and a saponification number of 84. Table 16-1 shows the properties of beeswax.

Lanolin

Lanolin is obtained from the wool of sheep. It is a complex mixture of esters, diesters, hydroxyesters, and polyesters of high molecular weight. It has the physical consistency of soft grease with a melting point of 100 to 108°F. Lanolin is used in cosmetics and pharmaceuticals as a cream to smooth skin, in rust preventives, and in lubricant manufacture.

Carnauba Wax

Carnauba wax is recovered from a variety of palm trees that grow almost exclusively in northeastern Brazil. Carnauba wax forms on the fronds of the tree and is recovered by cutting and drying the fronds rather than mechanically removing the wax. Impurities are removed from the wax by melting and filtering or centrifuging.

Carnauba wax is distinguished by its hardness and high melting point, combined with its ability to disperse pigments such as carbon black (Table 16-2). These properties make carnauba wax useful in printing inks. It is also used to gel organic solvents and oils as component of solvents and paste formulations. Carnauba polishes to a high gloss and is used to polish items such as leather, candles, and metal surfaces. The major components of carnauba wax are aliphatic and aromatic esters of long chain alcohols and acids with smaller amounts of free fatty acids, alcohols, and resins.

TABLE 16-1 Properties of Beeswax

Property	Units	Value
Acid value	mg KOH/g	17–24
Ester value	mg KOH/g	72–79
Melting point	°F	143.6–149.0
Penetration	dmm, 25°C	20
Ratio number	Ester value/acid value	3.3–4.2
Saponification		
Cloud point	°F	149.0 max.

TABLE 16-2 Properties of Carnauba Wax

Property	Units	Value
Acid value	mg KOH/g	2–8
Heavy metals	Wt %, Max.	0.004
Melting point	°F	177.8–186.8
Penetration	dmm, 25°C	2
	dmm, 43.3°C	3
Saponification value	mg KOH/g	80–95
Viscosity	98.9°C, mm^2/s	3960

Carnauba wax is very hard with a penetration of 2 dmm at 25°C and only 3 dmm at 43.3°C. Carnauba has one of the highest melting points among natural waxes at 183°F, with a viscosity of 3960 mm^2/s at 98.9°C, an acid number of 8, and a saponification number of 80. The hardness and high melting point when combined with its ability to disperse pigments such as carbon black allow carnauba wax an increasing use in thermal printing inks. Carnauba is also widely used to gel organic solvents, making the wax a valuable component of solvent and oil paste formulations. Carnauba polishes to a high gloss and is thus widely used as a polishing agent for items such as leather, candles, and pills. Other uses include cosmetics and investment casting applications.

Candelilla Wax

Candelilla wax is harvested from shrubs grown in the Mexican states of Coahuila and Chihuahua and in Texas. The entire mature plant is uprooted and immersed in boiling water, acidified with sulfuric acid. The wax floats to the surface and is separated. Principal markets for candelilla wax include cosmetics, food, and pharmaceuticals.

The major components of candelilla wax are hydrocarbons, esters of long chain alcohols and acids, long chain alcohols, sterols, natural resins, and long chain acids. Typically, candelilla wax has a melting point of 158°F, a penetration of 3 dmm at 25°C, an acid number of 14, and a saponification number of 55 (Table 16-3). The principal uses of this wax include cosmetics, foods, and pharmaceuticals.

TABLE 16-3 Properties of Candelilla Wax

Property	Units	Value
Acid number	mg KOH/g	14
Melting point	°F	158
Penetration	dmm, 25°C	3
Saponification number	mg KOH/g	55

Other Plant Waxes

Japan Wax. Other plant waxes include Japan wax, produced on the berries of a small tree native to Japan and China. Japan wax is composed of triglycerides, primarily tripalmitin. Japan wax typically has a melting point of 127°F, an acid number of 18, and a saponification number of 217 (Table 16-4). Principal markets are the formulation of candles, polishes, lubricants, and as an additive to thermoplastic resins. The product has some food-related applications.

TABLE 16-4 Properties of Japan Wax

Property	Units	Value
Acid number	mg KOH/g	18
Melting point	°F	127.4
Penetration	dmm, 25°C	
Saponification number	mg KOH/g	217

Ouricury Wax. Ouricury wax is obtained from the fronds of another palm tree growing in Brazil. It is difficult to harvest because it does not flake off the fronds like carnauba wax; rather, it must be scraped free. Ouricury is sometimes used as a replacement for carnauba wax in applications that do not require a light-colored wax. Table 16-5 shows the properties of ouricury wax.

TABLE 16-5 Properties of Ourcury Wax*

Property	Units	Value
Acetone solubles, 59°F	Wt %, Max.	9.0
Acid number		8–18
Benzene solubles, 77°F	Wt %, Max.	18
Flash point	°F, Min.	277–530
Melting point	°F, Min.	82.5–180.5
Paraffin hydrocarbons	Wt %, Max.	3.0
Volatile matter	Wt %, Max.	1.0

*Crude, original virgin product.

Rice Bran Wax. Rice bran wax is extracted from crude rice bran oil. It can be de-gummed, the fatty acid content reduced by solvent extraction, and bleached. It is primarily composed of esters of lignoceric acid, approximately 43 percent by weight, behenic acid (16 percent by weight), and C22 to C36 alcohols (28 percent). Properties of rice bran wax are shown in Table 16-6.

TABLE 16-6 Properties of Rice Bran Wax

Property	Units	Value
Free fatty acids	Wt %, Max.	10
Iodine number	Max.	10
Melting point	°F	167–176
Saponification number	mg KOH/g	75–120

Jojoba Wax. Jojoba oil is obtained from the seeds of the jojoba plant, grown in semiarid regions of Costa Rica, Israel, Mexico, and the United States. The oil is made of approximately 97 percent by weight of esters with the balance consisting of free alcohols, free acids, and steroids. Jojoba wax is used primarily in the formulation of cosmetics. Hydrogenated jojoba oil is a wax used in candles and in other low-volume speciality applications.

Castor Wax. Castor wax is catalytically hydrogenated castor bean oil. The wax has a melting point of 86°C, an acid number of 2, a saponification number of 179, and an iodine number of 4. Caster wax is primarily used in cosmetics. Derivatives of castor wax are used as surfactants and plastic additives.

Bayberry Wax. Bayberry wax is removed from the surface of the berry of the bayberry shrub by boiling the berries in water and skimming the wax from the surface of water. The wax is green and made up primarily of lauric, myristic, and palmitic acid esters. The wax has a melting point of 113°F, an acid number of 15, a saponification number of 220, and an iodine number of 6. The wax has an aromatic odor and is used primarily in the manufacture of candles and other products where a distinct odor is desirable.

Mineral Waxes

Montan Wax. Montan wax is derived from the solvent extraction of lignite. Earliest production on a commercial scale was in Germany during the second half of the 19th century, and Germany continues to lead the world in the production of montan wax. Some montan wax is produced in the United States from the Ione lignite bed in California. The largest traditional use of montan wax is the formulation of carbon paper inks. The decrease in the use of carbon paper resulted in montan wax being further refined for use in the formulation of polishes and as plastic lubricants. Montan wax is hard and brittle and has a high melting point; its properties are similar to those of natural plant waxes such as carnauba, which it can replace. The composition of montan wax differs with the region but includes a varying amount of wax, resin, and asphalt. Black montan wax may be further processed to remove resins and asphalt, after which it is known as refined montan wax. White montan wax has been reacted with alcohols to form esters. The wax components of montan wax are a mixture of long chain C24 to C30 esters (62 to 68 percent by weight), long chain acids (22 to26 percent by weight), and long chain alcohols, ketones, and hydrocarbons (7 to 15 wt %).

Synthetic Waxes

Polyethylene Waxes. Polyethylene waxes are low molecular weight polyethylene (molecular weight less than 10,000) having waxlike properties. It is manufactured by either low-pressure or high-pressure polymerization of ethylene using a Zeigler-type catalyst. All such waxes have the same basic structure, but various production processes yield products with distinctly different properties. Major uses include hot melt adhesives for applications requiring high-temperature performance, additives to improve processing of plastics, and slip and rub additives for inks, paints, and cosmetics.

Fischer-Tropsch Wax. Fischer-Tropsch (FT) wax is synthetically produced by reaction of carbon monoxide and hydrogen. The hydrocarbon product of FT reaction is distilled to separate mix into fuel products and waxes with a melting point ranging from 45 to 106°C. Currently, FT waxes are commercially produced in large volumes in South Africa and Malaysia. Uses are similar to polyethylene waxes, including hot melt adhesives and additives for inks and coatings.

PARAFFIN WAXES

The three general categories of petroleum waxes obtained from lube oil refining are paraffin, microcrystalline waxes, and petrolatums. Paraffin waxes are derived from light lubricating oil distillate processing. Paraffin waxes contain predominantly straight chain hydrocarbons with an average chain length of 20 to 30 carbon atoms. Paraffin waxes are characterized by a clearly defined crystal structure and have the tendency to be hard and brittle. The melting point of paraffin wax falls between 120 and 160°F.

Wax properties are determined by molecular weight, chemical composition, and oil content. Paraffin wax consists mostly of straight chain hydrocarbons, with 80 to 90 percent normal paraffin content. The balance consists of branched paraffin (isoparaffin) and cycloparaffins. Paraffin waxes have the following general properties:

- Nonreactive
- Nontoxic
- Good water barrier
- Clean burning fuel
- Colorless

The properties of wax that can be measured and controlled are as follows:

- Melting point
- Congealing point
- Hardness
- Oil content
- Viscosity
- Color
- Odor and taste

However, these physical properties alone do not completely define the suitability of wax for a particular application. The important functional properties of wax are translucency or opaqueness, solid appearance (dry, waxy, mottled, or shiny), and flexibility. It is the combination of physical and functional properties that ultimately determines whether a particular wax is suitable for a given application.

Paraffin waxes are composed mainly of straight chain hydrocarbon molecules originating from crude petroleum. Through the refining process, the composition and properties of wax from various manufacturers can have varying characteristics.

The three main grades of paraffin wax are fully refined, semirefined, and scale wax. These differ mainly by the degree to which the entrapped oil has been removed during refining and the color of the wax.

Fully refined paraffin waxes are hard, brittle, white, and odorless materials with less than 0.5 percent oil. They have good gloss properties and a melting point between 115 and 155°F.

Semirefined paraffin waxes contain between 0.5 and 1.0 percent by weight oil. This additional oil detracts from the gloss, and the wax is softer, light colored, and also has a slight odor and taste.

Scale or slack waxes are unrefined waxes obtained during the dewaxing step of lube distillates. It has an oil content between 1 and 3 percent depending on dewaxing process conditions. It is a softer material with a color varying between white and yellow.

PROPERTIES

The properties of petroleum wax depend on its hydrocarbon composition and degree of refining. Some important properties of petroleum waxes are described next (Table 16-7).

TABLE 16-7 Typical Properties of Paraffin Waxes

Property		Paraffin wax
Average mol wt		350–420
Carbon atoms/molecule		20–36
Flash point	COC, °F	399.2
Melting point	°F	115–155
Needle penetration	77°F	11–15
Oil content	Wt %	0.5
Refractive index	210°F	1.43–1.433
Viscosity	cSt, @ 210°F	4.2–7.4

Melting Point

The melting point is the temperature at which wax liquefies when heated or reverts back to a solid state when cooled. Paraffin waxes have a sharp melting point.

Oil Content

The oil content of paraffin wax is reduced during the refining process. All petroleum waxes contain oil. Oil in paraffin wax adversely affects the gloss, hardness, strength, and color. A small amount of oil can be beneficial in certain situations such as for improving burning qualities of wax in candles, but proper balance must be made to improve the best overall properties for a given application.

Gloss

Gloss is determined either visually or by light reflectance. Gloss is obtained by chilling molten wax and freezing the wax in a supercooled (semiglass-like) state having very small crystals. Gloss is important in maintaining the good appearance of the wax product.

Hardness

The hardness of wax varies with the composition, oil content, and temperature. Hardness of wax is measured by an ASTM needle penetration test, with a higher needle penetration indicating a softer wax.

Taste, Odor, and Color

Taste, odor, and color depend on the degree of refining. Food-grade petroleum waxes are white with little or no odor or taste. The color of highly refined paraffin wax is reported as the Saybolt color. Wax with a +30 Saybolt color is water white; lower numbers progressively tend to be off-white and light pale yellowish white colors.

Paraffin wax is macro-crystalline and brittle and composed of 40 to 90 percent normal alkanes with the remainder C18 to C36 isoalkanes and cycloalkanes. Paraffin wax has little affinity for oil; fully refined paraffin wax has less than 1.0 percent by weight oil; crude scale wax, 1 to 2 percent by weight; and slack wax, more than 2 percent oil content.

Within these classes, the melting point of wax determines its grade. The typical range for the melting point of paraffin waxes is 115 to 160°F.

Separation of paraffin wax from distillate occurs during the lube oil dewaxing step. It is then decolorized usually by hydrogenation, but percolation through bauxite is also used.

In case of residual fraction, after solvent deasphalting, heavy deasphalted oil recovered is actually a crude petrolatum, a dark-colored unctuous material containing oil and microcrystalline wax.

The percentage of each may vary but are usually 40 percent wax and 60 percent oil by weight. This wax is recovered during solvent dewaxing. However, microcrystalline wax has a great affinity for oil. The oil content of wax is 1 to 4 percent, depending on the grade of wax. Unlike paraffin wax, oil is held tightly in the crystal lattice of microcrystalline wax and does not migrate to the surface. Microcrystalline waxes obtained from petrolatums are generally known as plastic grades, with penetration greater than 11 dmm at 25°C.

TEST METHODS

Melting Point

Selection of a proper melting point method depends on the characteristics of the wax. Drop melting point (ASTM D 127) is suitable for amorphous waxes (e.g., microcrystalline), but it is not reliable for higher viscosity synthetic waxes, for which ring and ball softening point (ASTM D 36) should be used. ASTM D 87 may evaluate waxes whose time-temperature cooling curve exhibits a plateau (e.g., paraffin wax). Open or closed capillary tubes are used to measure the melting point of many natural waxes.

Congealing Point

The congealing point (ASTM D 938) is the temperature at which melted wax cease to flow, and it is more consistent than melting point for some waxes.

Hardness

The standard test for the hardness of waxes is the penetration test (ASTM D 1312). This test measures the depth in tenths of a millimeter that a needle of a certain configuration under a given weight penetrates the surface of a wax at a given temperature. A series of penetrations measured at different temperatures, rather than at a single temperature, is preferred.

Color

On solidification of a wax and depending on factors such as the rate of cooling, the amount of occluded air, and surface finish, the color of solidified samples of the same wax may be different. For this reason, the color of most waxes is judged only while molten. There are a few exceptions, however. For example, standards for carnauba wax are based on the color of solid wax. The accurate measurement of color in light-coloured (amber to off-white to white) is difficult but very important. The two most widely used color standards for numerical measurements are ASTM D 1500, which is used to measure dark brown to off-white color, and ASTM D 156, which is used to measure off-white to pure white color.

Oil Content

The production of petroleum wax involves the removal of oil. The oil content of wax is thus an indication of the quality of the wax. Oil content here implies oil and low molecular weight fraction. The oil content is determined by ASTM D 721. The test actually involves the determination of the percentage of wax soluble in methyl ethyl ketone solvent at −31.7°C.

Viscosity

Viscosity is an important test for mineral and synthetic waxes. ASTM D 88 is used to measure the time in seconds required for a specified quantity of wax at a specified temperature to flow by gravity through an orifice of specified dimensions. This viscosity is expressed in Saybolt Universal Seconds (SUS).

Acid Number

Acid number (ASTM D 1386) is the milligrams of potassium hydroxide necessary to neutralize 1 g of wax. It indicates the free carboxylic acid present. The test is widely used for vegetable, insect wax, and synthetic waxes containing carboxylic acid groups.

Saponification Number

The saponification number (ASTM D 1387) is the milligrams of potassium hydroxide that react with 1 g of wax at an elevated temperature and indicates the amount of free carboxylic acid plus any ester materials that may be saponified. Both acid number and saponification number are generally provided to give an indication of the free carboxylic acid and ester content of vegetable and insect waxes and synthetic waxes containing carboxylic acids and/or esters.

Microcrystalline Waxes

Microcrystalline waxes are malleable solids, ranging in color from white to dark yellow and with a melting point above 140°F. Compared with paraffin waxes, microcrystalline waxes have a higher proportion of hydrocarbons with a branched and ring chain. Also, their molecular weight is higher than that of paraffin waxes.

Microcrystalline waxes are produced during lubricating oil manufacture. Heavy distillates from vacuum distillation after solvent extraction are processed in a dewaxing unit where high melting micro waxes separate. These are refined by hydrotreating, clay, or bauxite filtration to yield a white odorless product that complies with food regulations.

Crude oil also contains some high molecular weight fractions, which are soluble in oil at high temperature and pressure found in underground oil reservoirs, but not very soluble at ambient conditions once the oil is produced. These high molecular weight fractions precipitate into the walls and floor of storage tanks and are known as crude oil tank bottoms. Crude oil tank bottoms are essentially crude oil with a very high wax content. Microcrystalline wax obtained from crude oil tank bottoms are generally known to be harder grades, with penetration less than 11 dmm at 25°C.

Microcrystalline waxes consist of odorless, tasteless nonpolar hydrocarbons with a relatively high melting point. The color of waxes varies from white to yellow. Hardness and melting point also vary from grade to grade. Properties of microcrystalline wax are shown in Table 16-8. The FDA standards are FDA CFR 172.886 and 178.3710.

TABLE 16-8 Typical Properties of Microcrystalline Waxes

Property		Microcrystalline wax
Average mol wt		600–800
Carbon atoms/molecule		30–75
Flash point	COC, °F	500
Melting point	°F	140–199
Needle penetration	77°F	3–10
Refractive index	210°F	1.435–1.445
Viscosity	CST, 210°F	10.2–25
Polycylic aromatics*	UV extinction	
	280–289 nm	0.15 Max.
	290–299 nm	0.12 Max.
	300–359 nm	0.08 Max.
	360–400 nm	0.02 Max.

*Solution of PCAHs (polycyclic aromatic hydrocarbons) from 25 g wax in 25-mL isooctane shall show UV extinction not more than above limits to comply with USP requirements for wax quality for direct use in food, food processing, and in pharmacy.

Microcrystalline waxes act as bases for chewing gum and cold flexible coating for cheese wheels. They are natural bases for lipsticks, cold creams, and ointments where they harden, lubricate, carry pigments and medications, and protect against moisture.

Food-grade waxes play an important role in food packaging. Wax-impregnated paper, film, foil, and corrugated board protect food from moisture. Wax helps hot melt adhesives flow readily and form flexible bonds.

Many wax products are used daily, from crayons, candles, and caulk to sealants and fine polishes. Microcrystalline wax acts as a carrier for compounds that prevent rust in a wide range of industries.

Microcrystalline waxes have better adhesive and elastic qualities and are used extensively in hot melt adhesives as well as in packaging, printing inks, and many other industries. Microcrystalline wax is refined from petrolatum, which comes from bright stock oil. It typically contains 41 to 50 carbon atoms.

Microcrystalline waxes differ from paraffin waxes in that they have a poorly defined crystalline structure, dark color, and a generally higher viscosity and melting point. Microcrystalline waxes tend to vary much more widely than paraffin waxes with regard to physical characteristics. Microcrystalline waxes can range from being soft and tacky to being hard and brittle, depending on the composition.

PETROLEUM WAX MANUFACTURE

Petroleum waxes are produced as a by-product during conventional lube base stock manufacture. Raw lube cuts from vacuum distillation have a high pour point that must be lowered to produce lube base stock of acceptable quality. This is achieved in a solvent dewaxing unit. Here feed is mixed with a dewaxing solvent and cooled to a low temperature by means of ammonia chillers when wax separates out. For a detailed process flow diagram and operating conditions of solvent dewaxing unit, see Chap. 10.

Wax Hydrotreating

Waxes produced from petroleum refining processes, such as the solvent dewaxing or the solvent deoiling process, contain polynuclear aromatics and may have an odor, taste, or color not acceptable in food or pharmaceutical applications. Waxes may require further treatment to improve their properties. Traditionally, acid/clay and percolation treatments of wax have been used to produce food and medicinal grades of waxes (color, ultraviolet absorption, and sulfur). For large plants, this process is replaced by a wax hydrotreating process. Wax hydrotreating has a lower operating cost, higher product yield, and eliminates environmental concerns.

In the wax hydrotreating process, wax feed is treated with hydrogen over a hydrotreating catalyst at an elevated temperature and pressure. The catalytic hydrotreating process typically uses two reactors. Hydrogen and feed are heated upstream of the first reaction zone (containing a special presulfided Ni-Mo-alumina catalyst) and are separated downstream of the reactor into main product and by-product (hydrogen sulfide and light hydrocarbons). Use of a stripping column permits adjustment of the product specification for technical grades or feed to the second hydrogenation stage. In the second reactor, feed is passed over a highly active noble metal hydrogenation catalyst to achieve a very low level of aromatics, especially the polynuclear compounds. Yield after the first stage ranges from 85 to 99 percent depending on feedstocks. Yield from the second hydrogenation stage is approximately 98 to 100 percent.

The operating conditions for a wax hydrotreating unit are shown in Table 16-9. The process flow diagram of a wax hydrotreating unit is shown in Fig.16-1.

TABLE 16-9 Wax Hydrofinishing

Feed: SAE 30 stock, kirkuk crude		
API: 37.3		
Specific gravity: 0.8383		
Operating conditions		
Reactors		
Reactor 1, inlet pressure	lb/in^2	1337
Reactor 1, outlet pressure	lb/in^2	1318
Reactor 2, inlet pressure	lb/in^2	1318
Reactor 2, outlet pressure	lb/in^2	1303
Reactor 1, feed inlet temperature	°F	644
Reactor 1, effluent outlet temperature	°F	644
Reactor 2, feed inlet temperature	°F	644
Reactor 2, effluent outlet temperature	°F	644
Feed outlet temperature	°F	666
Space velocity, LHSV	Hr^{-1}	1.0
Flash drums		
HP flash drum pressure	lb/in^2	1290
HP flash drum temperature	°F	446
LP flash drum pressure	lb/in^2	110
LP flash drum temperature	°F	446
Recycle gas rate	NSCF/BBL	3625
Makeup gas (95% purity)	NSCF/BBL	196
Stripper column		
Feed temperature	°F	446
Column top temperature	°F	446
Number of plates		13
Colunm top pressure	mm Hg A	70
Column bottom pressure	mm Hg A	122
Recycle compressor		
Recycle gas inlet pressure	lb/in^2	1270
Recycle gas inlet temperature	°F	136
Recycle gas outlet pressure	lb/in^2	1432
Recycle gas outlet temperature	°F	156

Wax Applications

Waxes are used as a feedstock for many products and for a wide range of applications: candle making, packaging, coatings, foods, cosmetics, adhesives, inks, casting, crayons, polishes, and chewing gums. Here are a few uses of wax.

Very High Viscosity Index (VHVI) lube base stock manufacture. Petroleum waxes obtained from Solvent Dewaxing Unit can be a feed stock for a wax catalytic isomerization[1] unit which catalytically isomerizes wax to isoparaffins. The isoparaffins have a high VI, low pour point and an excellent resistance to oxidation. The process converts or isomerizes n-paraffins to isoparaffins over a shape selective noble metal catalyst in catalytic isomerisation reactor in the presence of hydrogen. With a slack wax feed (5 to 25 percent weight oil content), the yield of lube base stock is 50 to 70 percent by weight. Lube base stock produced has a very high viscosity index, less than 130.

Chlorinated Paraffins. Chlorinated paraffins are manufactured by chlorination of paraffin wax. A large application of chlorinated paraffin is as a plasticizer in PVC plastics, in paint, sealants and adhesives. Higher chlorine content grades of chlorinated paraffin wax are used as flame retardants in rubber and polymer industry and in some metal working fluids.

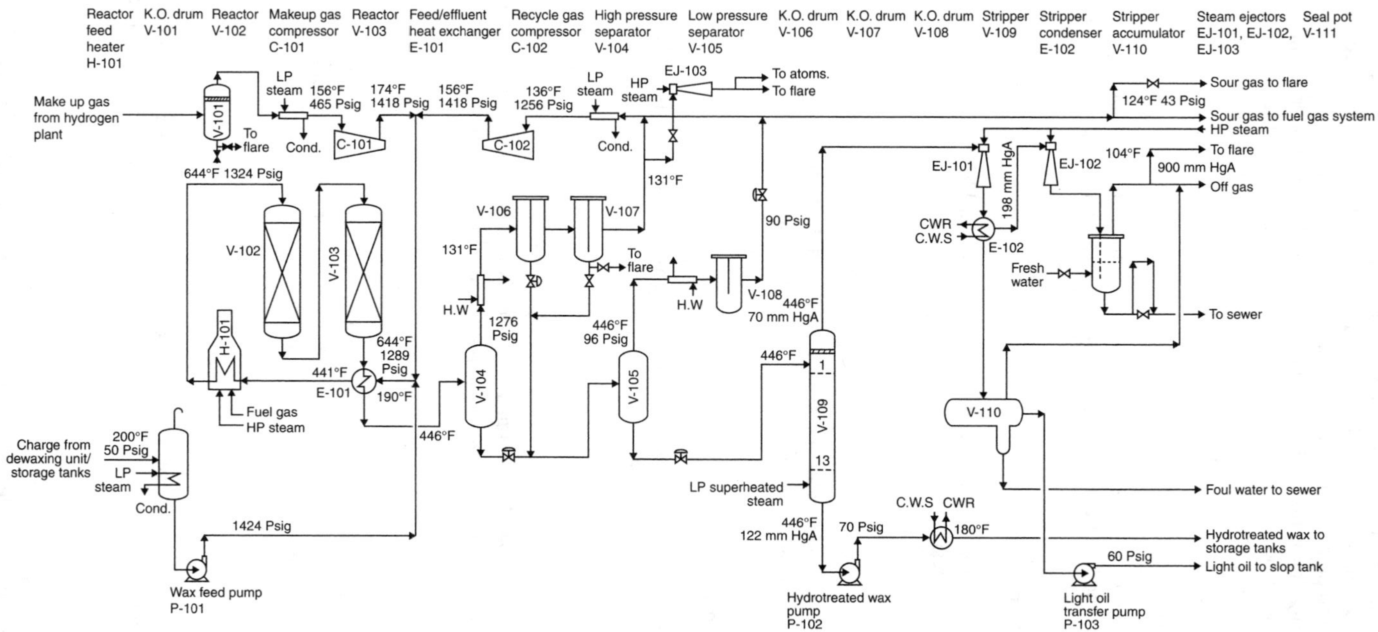

FIGURE 16-1 Wax hydrofinishing unit.

Petrolatum Manufacture. Petrolatums have a wide range of applications. Technical grade petrolatums are used in rubber processing, polishing compounds, corrosion preventive formulation, printing inks, etc. Medicinal grades are used in cosmetics (creams, lotions, hair products) and pharmaceuticals (ophthalmic ointments, topical ointment). Petrolatums are used in food industry as coating for fruits and vegetables, food packaging material and in animal feed. Petrolatums are blended from white mineral oils and petroleum waxes. Medicinal grades of petrolatums require high quality medicinal grade white oils and medicinal grade petroleum waxes.

Adhesive Manufacture

Petroleum waxes can be formulated into numerous hot melt adhesive systems to modify viscosity, flow, and "set time" properties. Waxes can also influence an adhesive's bond strength and flexibility.

Asphalt Modification

Petroleum waxes are incorporated into asphalt blends and emulsions to modify flow and enhance moisture resistance.

Bags

Petroleum-based waxes and wax blends can be applied to paper bags to reduce the unwanted migration of numerous liquids and enhance moisture resistance.

Barrel (Wooden Lining)

Paraffin wax is used to coat the interior of wooden barrels to prevent wood from being attacked by its contents (vine, vinegar, etc.).

Barrier Coating

Petroleum-based waxes and wax blends can be applied to numerous surfaces to reduce unwanted penetration or interaction of numerous liquids (water, greases, etc.) and gases (oxygen, etc.).

Board Sizing

Petroleum-based waxes are applied in a molten or emulsion form to provide moisture repellency to wood composite boards (oriented strand board, particle board, hardboard, gypsum, and medium-density fiberboard).

Bottle Cap Liner

Petroleum waxes can be used to coat and secure cap liners into plastic caps. The wax serves to form a water tight seal between the bottle and outer cap to reduce leaking if the bottle is overturned.

Bottle Sealing

Petroleum wax blends are used for sealing corked wine bottles, oil candles, and vinegar bottles. This dipped wax coating creates a professional finished look and a clean tamper-free seal.

Braid Packing

Paraffin and crude scale waxes are used in yarn-braiding applications as a lubricant and can provide mildew protection on natural yarn such as flax.

Butcher Paper

Petroleum waxes are a safe and effective moisture barrier coatings for paper used to wrap cuts of meat. The treated substrate prevents dehydration, prolonging shelf life.

Cable Filling

Crude scale wax and petrolatums are employed as a moisture barrier, fillers separating electrical wires from sheathing. The wax serves to occupy air spaces, reducing the potential for water seepage into the cable.

Candle Making

Petroleum waxes and wax blends provide fuel to the candle's wick, permitting burning, and they impart dimensional stability and shape to the candle. The oil content of wax must be less than 0.5 percent by weight to make good quality candles. Candle making is one of the oldest use of paraffin wax and still one of the most important end use of wax. This industry is one of the largest consumer of paraffin wax. Though not required for illumination, most candle production are for decorative purposes.

Candy Wrappers

Petroleum waxes are a safe and effective moisture barrier coating for paper used to wrap candies. The treated substrate protects the content from external sources of damage (moisture, gas, etc.) and protects the consumer by reducing dehydration and outward migration of oil/gases. Specially designed waxes offer excellent dead fold capabilities.

Canning and Preserving

Paraffin and crude scale wax can be used effectively to seal the top openings of canning jars to protect preserves from dehydration and mold formation. Low melting paraffin waxes are used for this application.

Carton Coating

Petroleum waxes and wax blends enhance moisture and grease resistance (protecting the contents), aesthetics (control gloss), coefficient of friction (ease plant handling), and sealing capabilities (heat seal) of cartons.

Chalk and Pencil Fabrication

Graphite/lead pencils are made of a crystallized form of carbon (soot) that is mixed with wax and pressed into pencil molds. A reduction in wax levels produces a harder pencil. Similar processes are used for the manufacture of chalk.

Cheese Coating

Petroleum wax blends provide a suitable moisture barrier on cheese to prevent rapid drying.

Chewing Gum Base

Paraffin wax is used to bind ingredients and serve as a rheology modifier to impart "chew" characteristics. Occasionally microcrystalline waxes are blended to modify these properties as well.

Corrosion Inhibitors

Petroleum waxes possess excellent film-forming and water-repellent properties, making them valuable components in rust formulations.

Cosmetics

Petroleum waxes serve as a safe base material, solvent, or carrier to provide moisture barrier, modify viscosity, or solidify formulations. Medicinal grade microcrystalline waxes are used in lipsticks, cold creams and ointments formulation.

Crayons

Paraffin waxes serve as a pigment binder. They assist in the transfer of the pigment from the crayons to the desired substrate. Generally, mild melt paraffin is useful for this application.

Dental Wax

Paraffin waxes and wax blends are formulated to high-quality premium modeling waxes for use in dentistry. The waxes are provided by speciality businesses that utilize paraffin and microcrystalline wax with other materials to make a finished dental wax.

Drink Cups

Petroleum waxes provide the structural rigidity and reduce liquid penetration through the paper substrate.

Emulsions

Paraffin and microcrystalline waxes are used in emulsion formulation, primarily to provide moisture resistance to a substrate. Wax emulsions are widely used in the manufacture of textiles, paper, and engineered wood products.

Explosives and Fireworks

Paraffin wax is used in dynamite wrappers as a moisture barrier. Microcrystalline waxes are used as a desensitizer in the manufacture of detonating fuses.

Fabric Finishing

Paraffin waxes and wax blends can be applied to fabric to add desired characteristics such as drape and wrinkle or moisture resistance.

Fabric Waterproofing

Petroleum waxes and wax blends due to their hydrophobic nature can be applied to numerous substrates (paper, wood, canvas, etc.) to impart waterproofing characteristics.

Fertilizer Coating

Petroleum waxes and wax blends are used to control release rates of active ingredients in sulfur-coated urea fertilizers.

Fire Logs

Petroleum waxes (primarily slack waxes) and wax blends are used to bind compressed wood particles and maintain shape. Waxes serve to ease ignition of product and support burn properties.

Floor Polishes

Petroleum-based polishes enhance the gloss and buffing characteristics of polishes, particularly wooden floors.

Food Wrappers

Petroleum waxes and wax blends enhance moisture and grease resistance (protecting the contents), aesthetics (control gloss), coefficient of friction (plant handling), and sealing capabilities (heat seal) of food wrappers.

Fruit Coating

Waxes are applied as a safe and effective surface coating to various fruit and vegetables to prevent dehydration and prolong shelf life.

Hot Melt Adhesives

Petroleum waxes and wax blends enhance moisture and grease resistance (protecting the contents), aesthetics (control gloss), coefficient of friction (plant handling), and sealing capabilities (heat seal) of various substrates.

Ink Manufacture

Petroleum waxes (paraffin, microcrystalline, and petrolatums) are incorporated in ink formulations to improve scuff (mar) and slip-resistant properties in ink. Waxes also serve as a carrier of pigments and other additives.

Investment Casting

Paraffin and microcrystalline waxes serve as binders in investment casting "waxes" and assist in removal of the blend after the mold is produced. Investment castings waxes offer exceptional dimensional stability required in the precision casting of metals.

Leather Treating

Crude scale and paraffin waxes are employed to soften and impart water-repelling characteristics to leather goods.

Lubricants

Paraffin and microcrystalline waxes are formulated into some industrial lubricants due to their low viscosities and film-forming properties.

Matches

Paper and wooden matches are tipped with crude scale wax to transfer the flame to the matchstick and to support a smooth, consistent burn.

Meat Wrappers

Waxed cloth or paper reduce dehydration of meat and prevents freezer burn.

Metal Coating

Paraffin and crude scale waxes can be used on the surface of metals during production and shipping to provide lubrication and serve as a protective barrier (acid etching or moisture).

Mold Release

Paraffin and crude scale waxes can be incorporated into a resin matrix or applied externally to the mold surface to facilitate the release of polymeric or cellulose materials.

Paper Coating

Paraffin wax and paraffin wax blends can be designed to enhance moisture and grease resistance (protecting the contents), aesthetics (control gloss), coefficient of friction (plant handling), and sealing capabilities (heat seal) of paper.

Paper Making

Paraffin wax and paraffin wax emulsion are used for internal and surface sizing in paper manufacture. Waxes impart a number of surface properties related to hydrophobic properties, release, scuff resistance, friction control, and antilocking.

Physiotherapy

Low-melt paraffin and crude scale wax are used in physiotherapy to provide heat to an injured area. Waxes permit a greater release of heat for a longer period than water. Also, the wax completely surrounds the part being treated and the patient does have to remain in a fixed position.

Plasticizers for Polyethylene

Petroleum waxes can be used to depress the glass transition temperatures, reduce stiffness and brittleness, and improve the processability of polyolefins.

Protective Coating

Petroleum waxes and wax blends can protect numerous substrates (paper, wood, rubber, metal, etc.) from physical scuff and chemical (e.g., moisture, gas, acid) damage.

Release Agents

Petroleum waxes and wax blends, as a result of tremendous wetting capabilities and barrier properties, can be formulated to provide release from molds and processing equipment.

Rope Wax

Crude scale wax provides flexibility to finished ropes and twines.

Rubber Processing

Wax continuously migrates to the surface of rubber components/tires providing a protective film against ozone degradation.

Sculpture Wax

Microcrystalline waxes and wax blends due to their malleability and thermal properties are used to create relief sculptures from which molds are prepared.

Sealants (General)

Paraffin and crude scale wax can be used to seal seams effectively on numerous substrates from unwanted liquid or gas migration. Waxes can also be employed in sealant formulations to modify rheology and viscosity.

Shoe Polishes

Petroleum-based waxes enhance the gloss and buff characteristics of a polish paste.

Stop-Off Wax for Etching

Paraffin or crude scale wax are used to prevent "acid etching" of metals by forming a protective covering that can easily be removed.

Tackifiers

Petroleum waxes can be formulated into tackifier resins to modify viscosity, flow, and other properties.

Tissue Embedding

Paraffin wax has been used as an embedding medium in the study of tissue for over 100 years. It is a good embedding medium for routine histology because it can thoroughly permeate the tissue in liquid form (when warm) and it solidifies (when cooled) with little damage to tissue.

Vegetable Coating

Waxes are applied as a safe and effective surface coating to various fruit and vegetables to prevent dehydration and prolong shelf life.

Vine Grafting

Vine grafting is a process of bringing two different grapevines together. After a graft is made, some covering must be used to keep it from drying out. A special elastomeric material (grafting tape) can be wound around the graft, or a more economical approach is to apply a wax blend.

Waxed Paper

Petroleum waxes offer a safe and effective moisture barrier treatment for paper. The treated substrate protects the contents from external sources of damage (moisture, gas, etc.) and protects the consumer by reducing dehydration and outward migration of product components (oil, grease, etc.).

Waterproofing

Petroleum waxes can be applied to electrical wires to provide lubrication during processing or to protect against moisture penetration.

REFERENCES

1. M. Wilson, T. Muller and G. Kraft, "Commercialization of Isodewaxing—A new technology for dewaxing to manufacture high quality lube base stocks," FL-94-112, November 1994.

CHAPTER 17
METALWORKING FLUIDS

Metalworking fluids (MWFs), more popularly known as cutting oils, have been in use since automatic metalworking machines were introduced in factories almost a century ago. The primary function of cutting fluid is to allow higher cutting speeds and to prolong the life of cutting tools. The fluid is used to cool the tool and work piece, to reduce friction at the sliding contacts, and to prevent welding on the contact edges of the tool and work piece. Cutting fluids also help prevent rust and flush away chips. These fluids are generally applied as a pumped flow or via an oil mist using compressed air. The majority of MWFs used, however, are liquids. Metal cutting oils are blended from highly refined petroleum lubricating oil base stocks, animal fats, vegetable oils, deionized water, and a significant proportion of property-enhancing additives. MWFs are expensive to produce. After cooling the tool and work piece, the fluid is collected, filtered, or decanted to remove metal swarfs and filings. Some make-up water, lube oil, or additives may be added and then recycled back to the machine. Over a period of time, contaminants and bacteria build up in the circulating cutting fluid, making it malodorous or discolored. At this stage the contaminated cutting fluid is purged from the circulating circuit and disposed of. If this discarded cutting fluid contains hazardous materials or chemicals, disposal in a municipal sewer is not allowed. For this reason, blending and maintenance of metal cutting fluids is done for a long service life to avoid the use of any hazardous additive that can pose problems in the disposal of used metal fluid.

TYPES OF MWFs

MWFs are usually a blend of high-quality lubricating oils of various viscosities together with additives to provide specific characteristics depending on the type of material being machined. Cutting oils can be subdivided into four categories; straight cutting oils, soluble cutting oils, semisynthetic cutting oils, and synthetic cutting fluids. Approximately a third of all cutting operations use straight cutting oils, two-thirds use water-soluble coolants, and the share of soluble coolants is increasing. Of the soluble cutting oils, about two-thirds of all coolant applications involve mineral oil based coolants, and of the remaining third, the share of biodegradable vegetable oil based coolant is a few percent; the rest are synthetic coolants.

Straight Cutting Oils

Straight cutting oils are the oldest class of cutting oils. These are called straight because they do not contain any water. These fluids are basically mineral lubricating oils and often contain polar lubricants such as animal fats, vegetable oils, and esters. Animal or vegetable fat is added to improve the wetting and lubricating properties of the cutting fluid. These oils provide a high degree of lubrication to help reduce frictional heat as the cutting tool penetrates the metal. These oils also help reduce the energy consumption of machines. However, neat oils are not very effective in dissipating the heat generated by metal cutting operations, and their use is limited to processing softer metals, low-speed operations, or where water must be kept out for corrosion protection and a high-quality surface finish is a consideration. Straight cutting fluids are generally a mixture of mineral oils and animal, vegetable, or marine oils to improve wetting or lubricating properties. Petroleum/mineral lubricating oils used are generally low-viscosity base stocks with viscosity in the range of 15 to 50 cSt at 100°F

TABLE 17-1 Properties of Straight Cutting Oils

		1	2	3	4
Viscosity	100°F, cSt	13	26	33	46
	100°F, SUS	70	133	155	214
Flash point	°F	335	400	410	410
Sulfur	Wt %	2	0.1	0.3	1.9
Chlorine	Wt %	1.1	2	1	2.6
Fatty esters	Wt %	5	0.5	3	5

(77 to 232 Saybolt Universal Second [SUS]). High viscosity index (VI) hydrocracked lubes with a VI higher than 105 are preferable because of their better resistance to oxidation. Low-viscosity oils accelerate heat removal from the tool and work piece because of better heat transfer coefficients, and they facilitate the rapid settling of chips and swarfs from circulating oil. The animal fats used are lard and sulfurized lard. Vegetable oils such as soybean and rapeseed oils are becoming popular because of their biodegradability. Sulfur, chlorine, and phosphorus additives are used to improve the extreme pressure (EP) qualities of cutting oils. The additive package can vary according to the specific end use of the formulated oil. Table 17-1 shows the typical properties of some commercial neat cutting oils. Straight cutting oils provide the best lubrication and the poorest cooling characteristics among all cutting fluids.

Active and Inactive Oils

Straight cutting oils can be divided into two main classes according to their corrosion characteristics; inactive straight cutting oils and active straight cutting oils. Inactive straight cutting oils are those oils that will not darken a copper strip immersed in it for 3 h at 212°F. If the cutting oil darkens the copper strip, it is termed *active oil*. Staining or corrosion of the copper strip is due to sulfur EP additives present in cutting fluids. Inactive oil contains sulfur that is firmly attached to the oil. Very little sulfur is released during the machining process to react with the work piece. These are suitable for nonferrous metals such as aluminium, brass, and magnesium. Active oils contain sulfur that is released during machining operations to react with the work piece. These are recommended for use with tough low-carbon or chrome alloy steels.

Soluble Cutting Oils

Soluble oils are mineral lubricating oils to which an emulsifier is added. The percentage of oil is in the range of 60 to 90 percent. The emulsifier enables it to form an emulsion with water. These products are sold as a concentrate to which water is added at a ratio of 1 part concentrate to 20 or more parts water. The size of the emulsified particles is large enough to refract light, which creates a milky appearance. These fluids provide good lubrication and excellent heat transfer performance, and they are the least expensive among the cutting oils. Soluble cutting fluids are used where rapid heat removal is a major consideration. These are used for both ferrous and nonferrous metals. Depending on the job requirements, the ratio of oil and water can be varied to provide lean or rich emulsions. Lean emulsions with more water provide better cooling, whereas rich emulsions with more mineral oil provide better lubrication, better tool life, and a smoother surface finish. However, for certain application such as machining magnesium, water-based cutting fluids cannot be used because of the fire risks. Table 17-2 shows the properties of some soluble cutting oils.

Synthetic Fluids

A major motivation to develop synthetic and semisynthetic cutting fluids in the 1940s was the carcinogenic properties of mineral oil/petroleum-based cutting fluids that caused a high incidence of disease among workers in metal workshops. These new fluids contained less mineral oil or no mineral

TABLE 17-2 Properties of Soluble Cutting Oils

		1	2
Viscosity	100°F, cSt	36	95
	100°F, SUS	170	440
Sulfur	Wt %	0.2	1.1
Chlorine	Wt %		11.8
Fatty esters	Wt %	4.1	1
Water-to-oil mix	Ratio	10:1/30:1	

oil, but they contained many additives, including chlorinated paraffins, to provide the required physical properties. Synthetic fluids contain no petroleum or mineral oil. They can be classified into three types; simple, complex, and emulsifiable. When diluted with water (1 to 10 percent synthetic fluid concentration), the simple and complex form a transparent solution, whereas the emulsifiable type forms an opaque solution. Simple synthetic fluids consist of organic and inorganic compounds dissolved in water. They offer good corrosion control and cleaning action, along with efficient heat removal. Simple synthetics are mainly used as coolants in grinding operations. The complex fluids, in addition to the ingredients of the simple synthetics, also contain water-soluble synthetic chemical lubricants, making this product capable of moderate to heavy-duty operations. Simple and complex synthetics because they are transparent solutions enable the operator to see the work. Some of these are fire resistant for machining specific metals. Other advantages offered by synthetic MWFs are less mist production, better corrosion and microbial control, and a good cooling capability. These synthetics are also stable in hard water. Disadvantages of synthetics are the lack of lubricity in the absence of oil. Some synthetics have a tendency to foam under moderate agitation. Compared with petroleum-based MWFs, they are more expensive. Hydroxy-alkyl amines are used in water-based MWFs, especially emulsions. Members of this class of compounds, such as triethanol amine, are excellent emulsifying agents. These amines have antimicrobial properties. Propanol amine is a particularly effective biocide. These amines impart an alkaline pH to these metal cutting fluids. Chlorinated paraffins are used in MWFs to enhance their viscosity and lubricity. Chlorinated paraffins for this use are made by chlorinating C23+ paraffins. Low molecular weight chlorinated paraffins are not used because these are known carcinogens. Other additives in synthetic cutting oils are phosphates and borates for water softening, soaps and wetting agents for lubrication, phosphorus, chlorine, and sulfur compounds as EP additives, glycols to act as blending agents, and germicides to act as bacteria control additives. Table 17-3 shows the properties of some of the commercial synthetic cutting fluids.

TABLE 17-3 Synthetic Cutting Fluids

Specific gravity	1.08
Boiling point, °F	212
Composition	
Monoethanol amine, Vol %	5–10
Triethanol amine, Vol %	10–20
Solubility in water	Complete
Dilution ratio, v/v	10:1 to 30:1

Semisynthetic Fluids

Semisynthetic cutting fluids are essentially a blend of synthetic and soluble cutting oils and have characteristics common to both types. These have a lower oil content (20 to 30 percent). Normally light naphthenic, high VI oils are used. More emulsifiers such as mono, di- or triethanol amines (20 to 30 percent) are used that make oil droplets much smaller. They are almost transparent. Lubricity improvers such as chlorinated paraffins (5 to 10 percent) are common in these formulations, apart from minor amounts of EP and other additives. Table 17-4 shows the properties of some commercial semisynthetic fluids.

TABLE 17-4 Semisynthetic Cutting Oils

Viscosity, 100°F, cSt	65
Specific gravity	1.01
Boiling point, °F	215
Mineral oil content,* Wt %	25
Chlorinated paraffins, Wt %	5–10
pH @ 5% dilution	9.9
Dilution ratio, v/v	10:1 to 30:1

*Light naphthenic, hydrotreated LBE base stocks.

Biodegradable MWFs

The need for biodegradable and nontoxic MWFs has been realized for quite a long time because the use of mineral oil in MWFs represents a potential health risk to workers exposed to these fluids in the form of vapor or liquid contact. The fine mist of mineral oils generated in metal processing shops is known to cause respiratory problems. These oils are not readily biodegradable and are toxic to aquatic organisms.

Biodegradable MWFs (cutting and grinding oils) are blended with rapeseed, soya, and many other types of vegetable oils. These oils have high lubricity and are readily biodegradable. These are particularly suitable where recycling is impossible and contact with the environment is likely. Biodegradable soluble cutting and metal grinding coolant fluids are prepared by mixing vegetable oil (rapeseed oil, etc.) and emulsifying agent or surfactants (ethylene oxide/propylene oxide copolymers, rapeseed oil alkoxylates, along with minor amounts of glycerol esters and polyethylene glycols) approximately in the ratio of 2 to 1. A bacteriocide, approximately 0.1 percent, may be added to this concentrate. For metalworking functions, this may be diluted with deionized water to 3 to 5 percent concentrations. For metal grinding coolant, a more dilute solution, approximately 2 percent, is used.

FUNCTIONS OF MWFs

Cooling

Heat produced during machining has a direct effect on tool life. Reducing cutting tool temperatures is important to tool life. Cutting tools may experience a temperature as high as 950°F, and reducing the temperature even by 50°F may increase tool life by as much as five times. Water is the most effective agent for removing heat generated during machining. However, water if used alone can cause rusting and corrosion. Soluble oils, oil emulsifiers, and anticorrosion and other additives are added to water to convert it into a cutting fluid. A higher water content also implies a greater corrosion rate and a higher fouling rate due to bacterial growth.

Lubrication

The lubricating function of a cutting fluid is as important as its cooling function. The effective life of a cutting tool is greatly increased if the friction and heat generated at the tool and work piece interface during the cutting process is decreased. In a typical machining operation, two thirds of the total heat generated is due to the resistance of work piece atoms to shear, and one third of the total heat generated is due to the friction of chips sliding over the tool face. Lubrication actually changes the shear angle and produces a thinner chip. Normally, the higher the oil content of the metal cutting fluid, the better the lubricity of the fluid. However, a higher oil content implies less water content and a lesser cooling capacity. The function of fatty components is to prevent welding of freshly cut metal surfaces to the cutting tool. This is due to the reaction of sulfur or chlorine in the cutting fluid with the metal to form a protective sulfide or chloride film. Fatty oils form metallic soap under conditions of high temperature and pressure.

Rust Control

Cutting fluids used on machine tools should inhibit rust from forming; otherwise machine parts in contact with cutting fluid will be corroded. Water is the best and most economical coolant, but it causes iron and iron alloys parts to rust unless rust inhibitors are added to it. All cutting fluids formulated with water must also contain rust inhibitors.

Rancidity Control

Soluble cutting oils are a complex mixture of mineral oils, animal fats, vegetable oils, and other organic molecules. This mixture when mixed with water forms a stable emulsion. These emulsions are used as coolants and lubricants in the cutting and grinding of metals. But these fluids unfortunately also support the growth of many types of bacteria that over a period of time cause spoilage of the coolant.[1] The spoilage is due to complex biochemical reactions carried on by a heterogeneous microbial population under various conditions. These reactions result in the breaking of emulsion with or without resultant lower pH, production of hydrogen sulfide with or without discoloration of emulsion, and production of foul odors other than that of hydrogen sulfide (H_2S). Components in soluble oils vary in their susceptibility to microbial oxidation. Fatty oils, vegetable oils, and their derivatives are readily oxidized by bacteria; mineral oils are more resistant.

BLEND COMPONENTS OF CUTTING OILS

Mineral Oils

Lubricating oils used are preferably naphthenic, highly refined, high VI base stocks with a viscosity of 30 to 100 cSt at 100°F. The reason for using high VI oils is to have oils with high oxidation stability and the minimum possible polynuclear aromatics content. Polynuclear compounds with five or six fused rings are among the most potent carcinogens in unrefined mineral oils. High VI oils are normally manufactured by solvent extraction or deep hydrotreating, which can almost completely remove these compounds and thus reduce risks to personnel exposed to the metal cutting fluid vapors.

Fatty Oils

Animal fats such as lard oil are generally used as lubricity/metal wetting agents in straight cutting oils. Lubricating and cooling functions are thus enhanced compared with oil without this additive. Blends of mineral oils and fatty oils give a better finish than that from straight cutting oils when machining mild steel, copper, aluminium, and hard brasses. The percentage of fatty esters varies from 3 to 5 percent or more in neat cutting oils.

Emulsifiers

Emulsifiers are the basic building blocks for formulating soluble cutting oils and semisynthetic water-based MWFs. They provide both emulsification and corrosion inhibition properties. A wide variety of chemical compounds can be used, such as sodium sulfonates, synthetic sodium sulfonate, sodium salts of mixed succinic and sulfonic acids, oleyl alcohol ethoxylate, diisopropanolamine tall oil fatty acid amide. Sodium sulfonates are the most popular emulsifier and corrosion inhibitor in use. In the case of natural sulfonate derivatives, a concentration of 10 to 15 percent is generally sufficient to emulsify the oil, depending on the type of sulfonate used. Other popular emulsifiers are hydroxy-alkyl amines such as triethanol amine. The amines also have antimicrobial properties. The degree of hardness of water can have a significant effect on its emulsion-forming properties. Also, hard water is resistant to saponification.

EP Additives

Effective EP characteristics are incorporated into the fluid in the form of elemental or combined sulfur and combined chlorine additives. Sulfur-containing additives perform an anti-wear function in straight cutting oils by forming a chemical film between the cutting tool and work piece, thereby keeping the tool from coming in direct contact with the metal being cut. Straight cutting oils are classified as active or inactive oils. Straight mineral oil containing sulfurized fatty oils is an inactive oil. Active oils contain alkyl polysulfides, sulfurized polyisobutenes, and many other additives. The objective of the active sulfur is to prevent welding of the tool to the work piece at the high temperatures generated during metal cutting and promote longer tool life.

The inactive sulfur additives are used for cutting low-carbon steel and copper and will not stain softer metals. Active oils are used for cutting harder varieties of steel such as chrome alloys and thread cutting and grinding operations.

Chlorine

Chlorine additive works (typically 0.5 to 3 wt %) in the same way as a sulfur additive and tends to complement the function of a sulfur additive in cutting oils. Therefore the two additives are generally used together. Chlorine tends to be liberated at elevated temperatures and is not particularly useful for very high temperature applications. The typical compounds used are chlorinated paraffins.

Corrosion Inhibitors

Corrosion inhibitors are essential in all water-based MWFs. Sodium nitrite is one of the oldest and cheapest corrosion inhibitors. Sodium nitrite in itself is safe to use in cutting fluids, but its use in any system using secondary or tertiary amino compounds is hazardous. These compounds have a propensity to react and form n-nitrosamines, a known class of carcinogens. Sodium sulfonate, calcium sulfonate, fatty acid soaps, amines, and boric acid are other popular anticorrosion agents. The high salt content in water can also make soluble cutting oils more corrosive and more prone to bacterial infestation.

Anti-Mist Agents

Mists are aerosols comprised of liquid particles smaller than 20 microns. During the machining process, considerable heat is generated at the tool and work piece interface or the cutting zone, causing vaporization of a part of the cutting fluid. The nonaqueous components such as oil vapor and biocides on condensation may form a mist or aerosol. Mist may also be generated due to the splashing of cutting fluid on a rotating machine element. The mist from cutting fluids is a serious health risk to workers exposed to it. One of the ways to reduce mist formation is to add anti-mist additives to MWFs. Additives used for this purpose are polymethacrylate polymer, polyisobutylene, poly n-butane, and poly (1,2, butane oxide). The concentration of these additives in MWFs is approximately 0.2 wt % of the formulation.

Biocides

Biocides are added in a small concentration (0.1 percent) to all metal cutting fluids containing water in the final formulation to retard or prevent bacterial growth, rancidity leading to foul odors, and other undesirable changes in fluid characteristics. Use of alkanol-amines as a biocide has already been mentioned, although its principal use is as an emulsifying agent in metal cutting fluid. A number of other agents, so-called formaldehyde releasers, are made by the reaction of formaldehyde with ammonia or amines. A popular biocide of this class is hexahydro-1,3,5-tris-(2 hydroxyethyl)-triazine. Other biocides include o-phenyl-phenol, morpholine derivatives of nitroparaffins known as biobans, alkyl thiocarbamates, pine oil, and sodium benzoate.

Miscellaneous Additives

Small amounts of many other additives are added to enhance their sales:

- Dyes/coloring agents.
- Fragrances.
- Defoamers used are silicones and polyorganic siloxanes to reduce surface tension and release air trapped in circulating oil. A concentration of approximately 50 to 100 ppm is typically used in the case of silicon-based defoamers.
- Dispersants.
- Stabilizers.
- Tackiness additives

CUTTING FLUID FORMULATION

Cutting fluids are formulated with the following objectives in mind:

- Nontoxic, with minimum exposure risks to machine operators
- Low cost
- Resistance to rancidity for a long period of time/long operating life
- Biodegradable and easy disposal of used metalworking fluid
- Good cooling capacity
- Good lubrication characteristics
- Rust resistance
- Low viscosity, to permit chips and dirt to settle quickly
- Transparent, to allow the operator to see the job clearly during machining
- Nonflammable/nonsmoking
- Prevent clogging of machine cooling channels.

CUTTING FLUID MAINTENANCE AND DISPOSAL

After a period of time, the cutting fluid within a machine coolant system degrades due to bacterial growth[2] and contamination from tramp oil and metal cutting/swarf generated from various machining operations. When it becomes uneconomical to maintain the fluid by regular make-up, it is dumped.

However, prior to dumping into a sewer system, it must be treated to bring the fluid composition to safe disposal levels. The key to extending coolant operating life is proper monitoring and maintenance. Monitoring includes determination of concentration using a refractometer or chemical titration and water quality. Maintenance involves the adjustment of fluid concentration, tramp oil removal, chips, and fines removal through filters, settlers, sump, and cleaning of this equipment, regular biocide addition, and pH adjustments.

Water Quality

The quality of water used to dilute cutting oil concentrates is vital to the overall performance of the cutting fluid. Cutting fluid corrosion characteristics, salt deposit tendency, emulsion stability, foaming tendencies, tool life, and fluid useful life are significantly affected by water quality. During usage, the evaporation of water from cutting fluid increases the concentration of the working fluid. Make-up water is added to restore working fluid concentration, but this also brings in mineral salts dissolved in make-up water. The rate of salt buildup in cutting fluid will depend on the salt content of the make-up water. Hard water has a high salt content. As the chloride and sulfate concentration increases, the corrosivity of water and thus the cutting fluid also increases. Sulfates also promote the growth of sulfate-reducing bacteria that cause fluids to become rancid. These problems can be minimized by the use of deionized water from an reverse osmosis (RO) or distillation unit. The quality of water has a significant effect on the performance of cutting fluid. Deionized water is preferred for soluble oil dilution. In case hard water is used (Table 17-5), some soluble fluids may form a precipitate that can build up on machine parts and filters. Water with a high mineral content can cause rust, stains, and corrosion of machines and the work piece. Because water can support bacterial growth, a rancidity problem, foul odors, and a short sump life can occur if proper biocides and anticorrosion additives are not present.

TABLE 17-5 Water Quality for Soluble Metalworking Fluids

Water type	Dissolved salts, ppmw
Soft	0–50
Moderately soft	50–100
Slightly hard	100–150
Moderately hard	150–200
Hard	200–300
Very hard	Over 300

Coolant Concentration

The concentration of coolant must be monitored. Concentration is the measure of amount of active ingredients present in the coolant. A dilute concentration below the proper level can result in a shorter tool life, an increased risk of machine and machined part rust, and increased bacterial growth in the fluid. A higher than optimum coolant concentration can cause increased foaming and higher operating costs by way of increased coolant consumption.

Coolant pH

The pH is a measure of the hydrogen ion concentration or the acidity of the coolant. The higher the pH, the more alkaline the solution. Coolant pH is maintained within a narrow range of alkalinity: 8.5 to 9.5. If the pH of the coolant falls below 8.5 (less alkaline), it loses efficiency and become more likely to cause corrosion and promote biological activity. A sudden drop in pH value is indicative of increased biological activity or a sudden change in coolant concentration due to contamination. If pH suddenly falls off, at constant coolant concentration, biological activity could be the more likely cause.

Tramp Oil Removal

Coolants are used in machines to remove heat generated during the machining process. Coolant is collected in a sump and recirculated to the machines. During use, coolant collects lubricating oils from the machine lubricating system and also metal chips and fines that settle down in the sump and get removed. This oil, called *tramp oil*, floats on the coolant surface and must be removed and not allowed to recirculate with the coolant. Many devices are available, such as belt and disk skimmers. Oil is removed from the sump periodically, monthly or weekly depending on the system design. Biocides are periodically added to the circulating fluid to control bacterial activity. Also, the sump is cleared of chips and fines at regular intervals. Failure to do so can cause an explosive growth of bacteria in the sludge at the bottom of the sump because biocides do not reach the fluids mixed in the fines at the bottom of the sump, making it toxic and unfit for disposal of spent coolant in a municipal sewer system.

REFERENCES

1. L. Liberthson, "Bacterial Deterioration of Cutting Oils," *Lubrication Engineering* 1 (1945): 103–105.
2. L. R. Sabina and H. Pivnic, "Oxidation of Soluble Oil Emulsions and Emulsifiers by *Pseudomonas oleovorans* and *Pseudomonas formicans*," *Applied Microbiology* 4 (1956): 171–175.

CHAPTER 18
METAL FINISHING QUENCHANTS

Since the beginning of Bronze Age, nearly five thousand years ago, metallurgists have used heat treatment to control the final properties of metal objects. Heat treatment operation is done after machining or other metalworking operations. Before heat treatment, the metal is relatively soft and easy to machine. After heat treatment, the metal becomes hard. For this reason heat treatment is done after machining or other metalworking operations. Heat treatment is the process of heating the metal to a certain temperature and then quenching to obtain the desired crystalline structure. Alloys of iron or steels of different types are the most common metals subjected to heat treatment. Different crystalline states of steel yield different physical properties. There are many crystalline form of steel. Ferrite is the iron portion of steel. Cementite is the carbon portion of steel. Austenite is a mixture of cementite (carbon) and ferrite (iron) formed at approximately 1400°F. The temperature at which austenite is formed is called the austenizing temperature. Martensite is the hard crystalline form of steel. Pearlite is a soft form of steel, good for impact absorption and shock tolerance. It is so named because it looks like mother of pearl under a microscope. Bainite is steel in its spongiest form. The objective of heat treating carbon steel is to transform the soft austenitic structures to the much harder martensitic form. A fast cooling rate produce a harder steel but also increases brittleness and may cause cracking and distortion. Size distortion arises due to volume changes resulting from phase transformations. Shape changes may arise due to other reasons such as steel composition and quenching operating conditions. But if the metal is allowed to cool too slowly, undesirable intermediate structures known as pearlite and bainite are formed. Heat treating offers manufacturers control over hardness, toughness, and corrosion resistance.

HEAT TREATING PROCESSES

The term *heat treatment* as applied to metals refers to any thermal treatment the metals are subjected to, with the objective of modifying their physical properties. It therefore covers many processes such as annealing, tempering, hardening, or quenching. Annealing and quenching make use of oils to a certain degree.

QUENCHING/HARDENING

Hardening of steels involves heating the metal part to a certain minimum temperature above what is called its critical range (characteristics for each steel) and then rapidly cooling (quenching) it in some liquid. Hardenability is defined as the ability for a ferrous alloy to form martensite when quenched from its austenizing temperature. The hardness of a particular alloy is a function of its cooling rate. The time temperature transformation (TTT) curve of an alloy is used to determine the amount of transformation product present. Quenching begins at a temperature above the austenization temperature. Cooling must be rapid so that the cooling curve does not intersect with the stable phase regime on a TTT diagram. The primary and critical role that quenching plays is to control the rate of cooling of metal. If it is cooled too slowly, it can become brittle. If it is cooled too rapidly, there is likelihood of distortion and cracking. Another important function of a quenchant is to maximize the uniformity of the cooling process over the surface of the part during quenching, which may lead to increased distortion and even cracking. Quenching helps to influence hardness and reduce the level of residual stress and distortions created during the manufacturing process.

TYPES OF QUENCHANTS

A wide variety of liquid quenchants are available for metal heat treating operations. These are water, brines, aqueous caustic soda solutions, mineral oils, and aqueous polymer solutions.

Water

Water can be used to quench some forms of steel but does not produce good results with alloy steels. Water absorbs large volumes of atmospheric gases, and when a hot piece of metal is quenched, these gases form bubbles on the surface of the metal. These bubbles tend to collect in holes or recesses and can cause soft spots that can later lead to cracking. The temperature of the quench water should not exceed 65°F. The quenching power of water falls rapidly as its temperature is raised. At 93 to 104°F, water gives a much slower quench than water at 68°F.

Brine

Brine solution is obtained by dissolving common salt, 7 to 10 percent by weight (wt %), in water. This mixture reduces the absorption of atmospheric gases in water, which in turn reduces the amount of bubbles. As a result, brine wets the metal surface and cools it more rapidly than water. The quenchant temperature is maintained between 65 and 100°F. Low alloys and carbon steels can be quenched in brine. Because of the corrosive action of salt, nonferrous metals cannot be quenched in brine.

Caustic Soda

A 10 percent by weight solution of caustic soda in water has a higher cooling rate than water. A caustic soda solution is used only for those types of steels that require extremely rapid cooling. Caustic soda solution is not used for nonferrous metal because of its corrosive properties.

Mineral Oils

Mineral oils are excellent quenchants and are valued for their ability to offer rapid cooling over a wide range of temperatures. One of the most favorable characteristics of quenchant oils is their ability to thin when heated, with high-heat transfer rates. They offer more uniform heat transfer, resulting in much less distortion and cracking in metals. The flash point of quenchant oils is 270 to 550°F. For operation safety, the quenchant bath is operated at 120 to 160°F below the flash point of the quenchant oil. .

An important difference between water and oil quenching media is that the quenching power of water falls rapidly as its temperature is raised, whereas the quenching power of oil remains fairly constant. Thus at 93 to 105°F, the quenching power of water is much slower than at 68°F, whereas the quenching power of oil quenchant over the same range and even up to 140 or 175°F remains fairly constant. This is because an oil temperature rise produces a sharp drop in oil viscosity that assists cooling and offsets the effect of the temperature rise.

THREE STAGES OF HEAT REMOVAL

During quenching, the rate of cooling of a metal work piece is determined by the surface cooling rate. This cooling rate can be subdivided into these three distinct stages:

- Film boiling or vapor blanket cooling (stage 1)
- Nucleate boiling (stage 2)
- Convective cooling (stage 3)

Film Boiling or Vapor Blanket Stage (Stage 1)

In this stage, the metal piece being quenched is at a high temperature compared with the quenching medium. The quenchant liquid in immediate contact with the hot metal is immediately vaporized, producing a thin vapor film that surrounds the metal piece. This insulating vapor film prevents the liquid quenchant from making direct contact with the metal piece. The heat transfer and cooling rate is very slow because cooling is only by convection and radiation. The quenching system is designed to minimize the time at this stage.

Vapor Transport Stage (Stage 2)

Cooling in the vapor blanket (stage 1) continues until the rate of heat flow from the metal pieces to the liquid quenchant is less than that needed to evaporate the quenchant and maintain a vapor blanket around the metal piece. When this occurs, nucleate boiling (stage 2) commences, in which the vapor blanket of stage 1 is no longer there. Quenchant liquid comes in direct contact with the hot metal and is vaporized. Very rapid heat removal and cooling of the metal occurs. This is the fastest cooling stage. To achieve high hardness and tensile strength properties, this stage time interval is maximized.

Liquid Stage (Stage 3)

When the temperature of the metal falls below the boiling point of the liquid quenchant, boiling stops and further cooling occurs by convection. This cooling stage is called stage 3. The entire surface of the metal part is in direct contact with the quenchant liquid. The cooling rate is much slower than that in stage 2 and takes place by conduction and convection. This in turn is strongly influenced by viscosity and the temperature of the liquid quenchant: The higher the viscosity, the lower the cooling rate. To minimize metal cracking, distortion, and residual stress, the stage 3 cooling rate should be kept to a minimum. The liquid stage is of great importance to many alloys because the cooling rate influences the structural transformation and resulting cracking and/or distortion in these alloys.

In film boiling, the surface temperatures are sufficiently high to produce a stable vapor film around the part. The cooling rate during this stage is very low due to the slow-heat transfer through the vapor barrier. The transition temperature from vapor blanket to nucleate boiling is the "Leidenfrost temperature." As the surface temperature decreases, the vapor film collapses and nucleate boiling begins. During nucleate boiling, the liquid contacts the hot surface, producing vapor bubbles that in turn produce strong convective currents, resulting in high-heat transfer rates. When the surface temperature drops below the boiling point of the liquid, the surface is permanently wetted by the liquid, resulting in convective cooling. Convective cooling rates are low and are primarily a function of the viscosity of the fluid.

If the fluid convection is strong, the vapor film may be ruptured by convection currents, resulting in reducing or eliminating film boiling. If there is poor quenchant bath temperature control, the film boiling process may be nonuniform and lead to uneven hardness, an increase in residual stress, possibly unacceptable distortion or cracking. The quenchant bath is generally maintained within 2 to 4°F. Quenchant temperatures are maintained generally below 85°F because the film boiling regime starts to become more stable above this temperature and leads to a nonuniform quench and consequent problems with distortion. It is important to note that the three stages occur simultaneously[1] on the metal surfaces throughout the quenching process.

Because these three cooling mechanisms exhibit different cooling rates, the potential for formation of thermal gradients and resultant distortions exists. This is minimized by the proper choice of quenchant composition and the various process variables.

Aqueous polymer quenchants moderate heat transfer by formation of film around hot metal, which provides uniform heat transfer relative to vapor film in the case of water and oils. Upon initial immersion, the polymer film encapsulates the vapor formed around the hot metal. At approximately the Leidenfrost temperature, the vapor film explosively ruptures, resulting in a pseudo nucleate boiling process. The thickness of this insulating film is a function of the polymer concentration and also the quench bath temperature.[2] Typically, quench severity decreases with increasing polymer concentration, increasing the quenchant temperature and decreasing flow rates.

ACCELERATED QUENCHING

Generally, plain carbon steels are quenched in water, but they are sometimes quenched in oil when the degree of hardness so obtained is adequate for the application in view. Except for very small pieces, quenching in oil entails a considerable sacrifice in hardness. Quenching rates necessary to harden plain carbon and low alloys steels are so high that in specimens larger than 1.5 cm in diameter, only a shallow surface layer reaches full hardness; below this, hardness rapidly falls. With accelerated quenching oils, this hardness falling off is less steep than with normal quenching oil. However, the use of accelerated quenching oils is only of benefit for small work pieces, up to about 5 cm in diameter. With large work pieces, the cooling rate is governed by the rate of flow of heat through the steel and not by the rate of heat removal from the surface. The increased speed of quenching is obtained by the incorporation of special additives in the quenching medium. These additives function by suppressing the vapor blanket, thus reducing the duration of stage 1 in the quenching process and increasing the rate of heat removal in stage 2. The effect in stage 2 is achieved due to the greater affinity of additive-treated oil with metal surfaces.

MARQUENCHING

Marquenching involves quenching steel in hot oil, holding it in the quench bath until the temperature throughout the work piece becomes uniform, and then cooling it in air. The main advantage of this process over normal quenching is that it reduces distortion and cracking. The oil bath used in marquenching is maintained at a temperature of 302 to 392°F. To use oil at these temperature, the oil must have high flash point, low volatility, and high oxidation stability.

MINERAL QUENCHING OILS

The properties of mineral oils that affect mineral quenchant performance are described in the following points:

Viscosity

The viscosity of mineral oils is important. A viscous oil produces a more stable vapor blanket and thus slows down the cooling rate. A low viscosity oil (approximately 20 cSt at 104°F) is preferable because of faster cooling rates are attainable and less drag out occurs. However, very low viscosity oils also have low flash points, which cannot be used as quench oils.

Boiling Point

The boiling point of quench oil is a critically important characterization factor. The boiling point of quench oil affects the timing of transitions from film boiling to nucleate boiling and to convective cooling.

Flash Point

The flash point of oil is of little significance in quenching oils. The bath is maintained at a relatively low temperature (176°F maximum), which is far below the flash point of oils used for quenching. When the red-hot work piece is rapidly immersed in oil, it is surrounded by a jacket of oil vapor containing no air, so combustion cannot occur. Moreover, the temperature of the work piece falls rapidly to a safe value.

Oxidation Stability

Oxidation of quenching oils occurs because of the chemical reaction of oil with atmospheric oxygen. The rate of oxidation increases greatly as the temperature of the oil rises. The presence of metallic contaminants acts as a catalyst for oxidation reactions. The products of oxidation are complex and may include acidic and insoluble components. These oxidation products can affect quench rates and may cause corrosion, metal staining, and sludge formation.

Thermal Stability

An important consideration in oil quenching is the chemical stability of oil. When red-hot steel is plunged into oil, some oxidation of oil is inevitable. In course of time, the resulting changes cause thickening of oil and formation of sludge. For this reason, hydrotreated paraffinic oils (API class II or III lube base stocks) are preferable.

Quenching Speed

The behavior of quenching oils may be modified by the incorporation of additives and is also affected by contaminants. However, it is its physical properties that determine the cooling rates in successive stages of cooling. These properties are specific heat, thermal conductivity, heat of vaporization, and viscosity. Mineral oils vary in types. The selection of oils for quenching is basically based on their viscosity and volatility. An increase in viscous resistance to flow tends to hinder convective turbulence in the boiling or vapor transport stage. To a certain degree, this reduces the risk of cracking of oil, but the effect of the vapor transport stage is to reduce the quenching power of the oil. Oil quenchants can be classified into three groups:

- Conventional
- Fast
- Martempering or hot

Conventional oils are mineral oils that may contain antioxidants. They are characterized by a long vapor phase with relatively slow cooling, very fast cooling in their boiling range, and again very slow cooling during the convection stage. The use of these oils is limited to steels with medium to high hardenability.

Fast quenching oils are mineral oils that contain proprietary additives, antioxidants, and wetting agents. These are characterized by high initial quenching speeds that can approach that of water, a fast cooling rate during stage 2, and a cooling rate similar to that of conventional oils during the convection period. Martempering oils are paraffinic mineral oils that offer excellent thermal and oxidation stability. They are primarily used at high temperatures (200 to 450°F) for the martempering of ferrous metals. Many additives are added to modify cooling and quenching rates to suit various requirements.

POLYMER SOLUTIONS

Polymer quenchants are the most versatile and stable among the aqueous quenchants. The first widely used polymer quenchant was polyvinyl alcohol, which is no longer used because of the health concerns its use posed to workers. The most widely used quenchants use polyalkylene glycols (PAGs), polyvinyl pyrolidines (PVPs), and polysodium acrylates (PSAs). PAGs are the most popular polymer quenchants in use. Compared with water, they offer more stable vapor blanket cooling to facilitate uniform cooling. PAGs help reduce quench cracking and distortion. They are safer to use because they do not burn or smoke when heated. They are nontoxic, and spills can be cleaned up with water without creating any disposal or environmental risks.

Polymer solutions are typically a series of nonflammable aqueous solutions (10 to 30 percent by weight) of organic polymers such as polyalkylene glycols (with different molecular weights and viscosities). A corrosion inhibitor such as sodium nitrite is included to resist the corrosive effects of water. The organic polymers are completely soluble in water, producing a clear homogeneous solution at room temperature. However, at elevated temperatures, the polymers separate from water as an insoluble phase. On cooling the polymer redissolves to form a homogeneous solution. When a hot metal is quenched in a dilute aqueous solution of polymer quenchant, a film of liquid polymer is deposited on the surface of the hot metal. This film thus modifies the cooling rate or quench speed. Quench speed can be varied by choosing different polymer concentrations, agitation rates, and quenchant temperatures. Thus the same quenchant may be used for various quenchant applications.

QUENCH SYSTEM DESIGN

The most important process variables for quenching are as follows:

- Equipment design
- Type of quenchant
- Concentration of quenchant
- Bath temperature
- Agitation rate

The factors that affect cooling curve shape are agitation, bath temperature, and quenchant viscosity. In case of polymer solutions, the thickness of the polymer film surrounding the hot surface during cooling affects the heat transfer rates; the thicker the film, the slower the cooling rate. Uniformity of flow rate around the cooling surface is of great importance to minimize undesirable gradients during quenching. Increasing agitation increases the cooling rate (higher H).

Bath Temperature

Control of the bath temperature or temperature rise is important because heat transfer coefficient and cooling rates during quenching are a function of the temperature differential between the metal temperature and the quenchant temperature. Temperature rise tends to reduce this, resulting in lower cooling rates.

Agitation

Optimal quench uniformity is essential to minimize the potential for cracking, distortion, residual stresses, and spotty hardness. This means that heat transfer during film boiling (vapor blanket) and the nucleate boiling stage for vaporizable liquids (water, oils, polymer quenchants, etc.) must be as uniform as possible. Agitation is one of the most critical areas of system design. Increasing agitation increases heat transfer and thus the cooling rate. Agitation of quench oil is necessary to destabilize film boiling and nucleate boiling processes to achieve uniform heat transfer rates. Although decreasing the oil temperature provides some improvement in hardening, it is considerably less effective than increasing the agitation rate. Studies on the relative efficiency of immersion and spray quenching with aqueous polymer and mineral oil quenchants have shown that the depth of hardness increases[3] with increasing agitation. In commercial quenching systems, the agitation of both oil and water quenchants is provided by a mixer, which significantly improves quenching uniformity and produces parts with less distortion and cracking. Cracking and distortions are known to result from nonuniform cooling rates and are not due to any property of the quenchant used. Thus the design of the agitation system is one of the largest contributors to the uniformity of the quench process.

Quench Severity

Quench severity depends on the cooling rate, which in turn depends on agitation rate, size of the tank, and fluid viscosity. Agitation thus affect the hardness and depth of hardening during the quenching process. Agitation mechanically ruptures the unstable vapor film during film boiling when vaporizable quenchants such as oil, water, and aqueous polymer solutions are used.

Quench value is measured in terms of H value, which is the quench speed in relation to water.[4] Water has an H value of 1. Oil may have an H value of 0.3; brine may have an H value of 2.0.

Cooling Rate Characterization

One of the oldest tests to quantify the quench severity of an oil is the GM Quenchometer test (ASTM D 3520). In this test, a 7/8 in (22-mm) nickel ball is heated to 1625°F (885°C) and then dropped into a wire basket containing 200 mL of quench oil at 70 to 80°F. A timer is activated as the glowing nickel ball passes a photoelectric sensor. A horseshoe magnet is located outside the beaker as close as possible to the nickel ball. As the ball cools, it passes through the curie point of nickel (354°C or 670°F), the temperature where nickel becomes magnetic. This in turn activates the sensor, which stops the timer. The time required for the nickel ball to cool from 1625 to 670°F is thus recorded. Table 18-1 provides some typical GM Quenchometer times for some quench oils.

Castrol Index

The Castrol Index uses data from the standard ISO 9950 cooling curve to evaluate a quenchant. The Castrol Index is used for evaluating mineral oil quenchants (see Fig. 18-1).

TABLE 18-1 Quenching Speeds of Oil Quenchants

Quenchant	Time, s
Fast oil	8–10
Medium oil	11–14
Slow oil	15–20
Martempering oils	18–25

$$CI = K \times \frac{V_{max}}{(T_{MAX} - T_{BATH})}$$

Where:

CI = Castrol index

K = Equipment constant

V_{max} = Maximum cooling rate

T_{MAX} = Temperature corresponding to maximum cooling rate

T_{BATH} = Bath temperature

FIGURE 18-1 The castrol index.

OTHER HEAT TREATING PROCESSES

Tempering

Tempering consists of reheating the hardened steel to some temperature below the critical range, maintaining the temperature for a period, and then cooling it again in air. Sometimes the cooling is done more rapidly, for example, in water. The objective of tempering is to relieve the stresses set up by the hardening/quenching processes and increase the toughness of steel by reducing the hardness slightly. Steel are tempered at a temperature ranging from 300 to 1200°F. Mineral oils form a convenient medium in which to conduct operations at the lower end of this range. It is necessary to use oils of a very high viscosity range. At working temperature, oils become very fluid and heat transfer rates are not impaired. The only other property of importance in a tempering oil is resistance to thickening and sludging (high oxidation stability). These effects lead to the uneven tempering of the parts immersed in the bath, and an oil of high stability is required.

Annealing

Annealing is the process of heating metals to soften them. Bright annealing is annealing in a special atmosphere that will not oxidize the metal surface. When metals are fabricated by cold work (cold rolling, wire drawing, etc.), they need to be annealed, either to permit further cold work or to give them certain physical properties for use. It is important especially in case of bright annealing that metal be free from contamination by substances that might burn up and stain its surfaces during the annealing process. This applies particularly to lubricants used in cold working. Such lubricants should be formulated so as to give the least possible surface deposits.

REFERENCES

1. M. H. Tensi, "Fundamentals of Quenching by Submergence," in *Proceedings of the ASM International Conference*, 1993: 243–254.
2. J. Grum, "Handbook of Residual Stresses and Deformation of Steel," in G. Totten, T. Inoue, and M. Howes, eds., *Proceedings of the ASM International Conference*, 2002: 220–247.
3. G. E. Totten, C. E. Bates, and N. A. Clinton, "Quench System Design," in *Handbook of Quenchants and Quenching Technology.* (Materials Park, OH: ASM International, 1993), pp. 339–411.

CHAPTER 19
HYDRAULIC FLUIDS

The primary purpose of any hydraulic fluid is to transmit power mechanically through a hydraulic power system. The mechanism by which this is accomplished involves a pump, fluid–motor assembly, piping, fittings, and controls to transmit power at whatever speed variations are desired. Such a system can transmit power subject to reversal, irrespective of direction of rotation of the driving electric motor. Also this power can be multidirectional in its application. Hydraulic power transmission frequently is favored over mechanical transmission owning to the elimination of chains, belts, gears, and friction clutches that normally require more detailed lubrication. It is important to note, however, that hydraulic power transmission is not devoid of lubrication requirements. The very close tolerances built into the pump and fluid-motor elements necessitate adequate lubricity in the fluid to prevent undue wear. Petroleum-based oils possess a natural lubricating ability that may not be found in some non-petroleum-based hydraulic fluids. Many hydraulic system components such as control valves operate with tight clearances where seals are not provided. In these applications, hydraulic fluid must provide the seal between the low-pressure and high-pressure side of the valve port. The amount of leakage will depend on the tolerance between adjacent surfaces and fluid viscosity. The circulating hydraulic fluid must be capable of removing heat generated in the system.

Among the factors for hydraulic fluid selection are lubrication and viscosity. Lubrication refers to the fluid's ability to reduce friction and heat between hydraulic equipment metal components that slide against each other. The hydraulic fluid must form a thin film between these components to minimize wear damage. Film strength is determined by viscosity. To figure out the best viscosity for hydraulic fluid, the hydraulic system's maximum and minimum operating temperatures must be examined. At the highest temperature, the fluid must be sufficiently thick for lubrication and low internal leakage. At the minimum temperature, it must be thin enough for easy flow. Hydraulic fluids must provide, apart from lubrication, the qualities necessary to protect all hydraulic system components against friction, wear, rust, oxidation, corrosion, and demulsibility. These protective qualities are usually incorporated through the use of additives.

Hydraulic fluids must also protect against component wear. Water getting into hydraulic fluid is the primary culprit for corrosion. Water can get into hydraulic fluid due to contact with humid air. Corrosion is sometimes caused due to the chemical reaction of hydraulic fluid with system components. To prevent such corrosion, hydraulic fluid must be compatible with materials in the components. Additives such as defoamers and demulsifiers are used in corrosion prevention. They not only help emulsify free water but also resist air retention in the fluid. Air in hydraulic fluids can cause harmful cavitation.

Anti-wear additives are added to fluids where the hydraulic fluid circuit contains gear or vane pumps. In these pumps, metal-to-metal contact is a design feature. The fluid must not only be a lubricant but must also have anti-wear properties. Anti-wear additives form a protective coating over metal parts.

Other considerations include elastomer compatibility. Hydraulic equipment such as seals, hoses, and accumulator bladders are made of elastomer materials that must be compatible with the fluid through the entire temperature range and also through the entire age of the fluid.

Safety is another concern because hydraulic systems often operate close to open flames or hot surfaces in applications such as furnaces, forges, and steel mills. If the leakage occurs in these applications and the hydraulic fluid is flammable, like all petroleum oil fluids, a fire can result. For these applications, hydraulic fluids developed for increased safety must be considered.

PHYSICAL PROPERTIES

The physical characteristics of hydraulic oils are similar to those of lubricating oils. Important properties of hydraulic fluids are discussed next.

Viscosity

Viscosity is the most important characteristics of hydraulic fluid and has a significant impact on the hydraulic system. If the viscosity is too high, then friction, pressure drop, power consumption, and heat generation increases. Also the sluggish operation of valves and servos may result. If the viscosity is too low, increased internal leakage may result under high operating temperatures. The oil film may be too thin to prevent excessive wear or possible seizure of moving parts. Pump efficiency may decrease, and sluggish operation may be experienced.

Compressibility

Compressibility is a measure of volume reduction due to pressure. Compressibility is sometimes expressed by "bulk modulus." Petroleum fluids are relatively incompressible, but volume reduction can be approximately 0.5 percent for pressure ranging from 6900 to 27,600 kPa (1000 to 4000 lb/in^2). Compressibility increases with pressure and temperature and has a significant effect on high-pressure fluid systems. Compressibility can result in power loss because the volume reduction due to compressibility cannot be recovered.

Stability

The stability of a hydraulic fluid is the most important property affecting service life. The properties of a hydraulic fluid can be expected to change with time. Factors that influence the changes include mechanical stress and cavitation, which break down the viscosity improvers and cause viscosity reduction, and oxidation and hydrolysis, which cause chemical changes leading to formation of volatile components, insoluble materials, and corrosive products. The types of additives used in hydraulic fluids are selected carefully to reduce potential damage due to chemical breakdown at high temperatures. Compatibility of additives is important. An additive may enhance one property while adversely affecting another property.

Factors Affecting Hydraulic Fluid Quality

The quality of a hydraulic fluid is an indication of length of time that the fluid's essential properties continue to perform as expected. The primary factors affecting quality are oxidation stability, rust prevention, foam resistance, water separation, and anti-wear properties. Many of these properties are achieved through the use of chemical additives.

Oxidation

Oxidation is the chemical reaction of oxygen present in air with oil. Oxidation affects all hydraulic fluids except water. During oxidation, fluid can make sludge, varnish, and acids. The chemical reaction products result in increased corrosion rates and increase in fluid viscosity. These reaction products, insoluble gum, sludges, and varnishes, plug lines and valves, increase wear, reduce clearances, and cause sluggish operation. Oxidation reactions are promoted by an increase in temperature, pressure, contaminants, water and metal surfaces, and agitation. The most effective preventive measure is to keep the temperature below 135°F. Another alternative is to add antioxidant additives to fluids, which allow fluids to function for a long time at high temperatures. Like any other reaction, the rate of oxidation doubles approximately every 18°F. Oxidation reaction once started in a local area where temperature is high is accelerated by the metal surfaces that act as a catalyst. Increased pressure also increases the oxidation rate. As the pressure increases, the fluid viscosity also increases, causing an increase in friction and heat generation. As the operating temperature increases, the rate of oxidation increases. Also, due to the increased pressure, the amount of entrained air and dissolved oxygen in the fluid increases, this in turn increases the oxidation rate.

Contaminants

Contaminants that accelerate the rate of oxidation may be dirt, moisture, paints, or sealant materials. A 1 percent sludge concentration in a hydraulic fluid is sufficient to reduce the useful life of the hydraulic fluid by 50 percent. Particulate contamination and water can have serious adverse effects on the physical and chemical properties of fluids. Water is much less effective as a lubricant and can result in decreased viscosity, load-carrying ability, and dynamic film thickness. Presence of free water in the system can lead to degradation of the performance and malfunction during subzero ambient temperatures. Several methods are available for removing particulate contamination and water from the hydraulic fluid circuit. Heavily contaminated fluids are subjected to an external purification system prior to use. In-line contamination control is achieved by the incorporation of a particulate filter, a water-absorbing filter, and also by incorporation of a vent dryer to minimize the ingression of water from environment into the system. Use of these devices[1] can cause a marked increase in the service life of hydraulic fluids, sometimes by a factor of 4 to 6.

Water and Metals

Certain metals such as copper are known to be a catalyst for oxidation reactions, especially in the presence of water. Due to the production of acids during the initial stages of oxidation, the viscosity and neutralization number increases. The neutralization number provides a measure of the amount of acid contained in the fluid. The most commonly accepted oxidation test for hydraulic fluids is ASTM D 943. This test measures the neutralization number of oil as it is heated in the presence of pure oxygen, a metal catalyst, and water.

Air Entrainment and Foaming

Air enters a hydraulic system through the reservoir or through the air leaks within the hydraulic systems. Air entering through the reservoir contributes to surface foaming on the oil. Air entrainment is a dispersion of very small air bubbles in a hydraulic fluid. Oil under low pressure absorbs approximately 10 percent air by volume. Under high pressure, the percentage is even greater. When the fluid is depressurized, the air produces foam as air is released from fluid. Foam and high air entrainment in a hydraulic fluid cause the erratic function of servos and contributes to pump cavitation. Cavitation can cause metal fracture, corrosive fatigue, and stress corrosion. Oil oxidation is another problem caused by air entrainment. As fluid is pressurized, the entrained air is compressed and increases in temperature. The increase in air temperature in turn promotes oxidation. The amount of foaming in a fluid depends on the viscosity of the fluid and the nature of the chemical compounds present in it. Foam depressants are commonly added to hydraulic fluids to aid foam breakup and the release of dissolved air.

Demulsibility or Water Separation

Water that enters a hydraulic system can emulsify and promote the collection of dirt, dust, and grit, which can adversely affect the operation of valves, servos, and pumps, increase wear and corrosion, promote fluid oxidation, deplete additives, and plug filters. Highly refined mineral oils permit water to separate readily. However, certain additives such as anti-rust actually promote emulsion formation to prevent separated water from settling and breaking through the anti-rust film.

Anti-Wear Properties

Conventional hydraulic fluids are satisfactory for low pressure and low speed applications. However, hydraulic fluids for high pressure (more than 1000 lb/in^2) and high speed (more than 1200 r/min) applications that use vane or gear pumps must contain anti-wear additives. These applications do not permit formation of full fluid film lubrication to protect contacting surfaces. Anti-wear additives provide a protective film at the contact surfaces to minimize wear. Use of hydraulic fluids without anti-wear additives will cause premature wear of pumps and cause inadequate pressure or pump failure.

Quality assurance of anti-wear properties is determined through standard laboratory testing. Laboratory tests to evaluate anti-wear properties of a hydraulic fluid are performed in accordance with ASTM D 2883. This test is generally conducted at a variety of high speeds with high-pressure pumps. The pump is operated for a specified speed and at the end of each period, the pumps are disassembled and specified components are weighed. The weight of each component is compared with its initial weight. The difference reflects the amount of wear experienced by pumps for the operating period. The components are also inspected for visual signs of wear and stress.

BIODEGRADABILITY

Environmental issues such as biodegradability and toxicity of the hydraulic fluids are important considerations in certain applications. Vegetable oils such as rapeseed oil form the most common base for these fluids. These fluids are readily biodegradable, which normally implies that 60 percent decomposes within 28 days of exposure to air and they are nontoxic. Vegetable oil based fluids have lubrication and anti-wear properties that match those of petroleum oils. However, these fluids are susceptible to oxidation and perform poorly in cold weather during which they can congeal and present problems in cold starting.

Use of Additives

Many of the qualities and properties just discussed are achieved by product manufacturers by careful blending of additives with base oils. Because of incompatibility problems and complex interaction between various additives, selection of an additive package is critical to the overall performance of the formulated fluid. Hydraulic fluids are carefully formulated for the specific area of application. The additive package consists of dispersants, detergents, antioxidants, anticorrosion additives, anti-wear, extreme pressure (EP) agents, and viscosity improvers.

BASE OILS FOR HYDRAULIC FLUIDS

Petroleum

Petroleum base oils (Table 19-1) are the most commonly used stock for hydraulic applications. Stock properties are so chosen that the base oils present minimal risk of fire or leakage, withstand wide

TABLE 19-1 Petroleum Base Hydraulic Oils

	ISO viscosity grade						
	32	46	68	100	150	220	460
Average viscosity, 104°F, cSt	32	46	68	100	150	220	460
Viscosity index	90	90	90	90	90	90	90
Flash point, COC,°F	374	374	410	410	446	446	500
Pour point, °F	26	26	32	32	32	32	32
Neutralization number, mg KOH/g max.	1.5	1.5	1.5	1.5	1.5	1.5	1.5

temperature fluctuations, and present no environmental risk. Composition of a hydraulic fluid used in aircrafts is shown in Table 19-2.

TABLE 19-2 Aircraft Hydraulic Fluid Composition

	Wt %
Highly treated light gas oil	60–80
Highly treated heavy gas oil fraction	15–30
Polyalkyl methacrylate	4–8
Oxidation inhibitor	0.1–0.5
Dye	Trace

Fire-Resistant Hydraulic Fluids

The International Organization for Standardization (ISO) currently recognizes four major groups of fire-resistant hydraulic fluids:

- High water-containing fluids (HFAs)
- Invert emulsions (HFBs)
- Water glycols (HFCs)
- Water-free fluids, including synthetic fluids (HFDs)

HFA Fluids

HFA fluids are mostly made of water; typically consisting of a 5 percent emulsion of oil in water. These fluids consist of very small oil droplets dispersed in a continuous water phase. These fluids have low viscosities, excellent fire resistance, and good cooling capability due to the large proportion of water. Additives must be added to improve their inherently poor lubricity and protect against rust. HFA fluids have the strength and weakness of plain water. Although they are extremely fire resistant, they need equipment that has been specially designed for use with water. HFA-type hydraulic fluids cannot replace petroleum oils in typical hydraulic equipment. HFA fluids are commonly used in steel rolling mills and coal mines.

HFB Fluids

HFB fluids consist of an emulsion of water in oil with 60 percent oil content. These fluids consists of very small water droplets dispersed in a continuous oil phase. The oil phase provides good lubricity while water provides the desired level of fire resistance and enhances the cooling capability.

Emulsifiers are added to improve stability. Additives are added to minimize rust and to improve lubricity as necessary. These fluids are compatible with most seals and metals common to hydraulic fluid applications. The operating temperature of water in oil fluids must be kept low to prevent evaporation and oxidation. The proportion of oil and water must be monitored to ensure that proper viscosity is maintained, especially when adding water or concentrated solutions to fluids to make up for evaporation. To prevent phase separation, fluid should be protected from repeated cycles of freezing and thawing. HFB-type fluids are not popular because of their inherent instability and maintenance needs.

HFC Fluids

HFC or water glycols are the most popular of the fire-resistant hydraulic fluids. These fluids consists of 35 to 45 percent water. Water glycol fluids contain an antifreeze such as ethylene, diethylene, or propylene glycol that are nontoxic, biodegradable, and a thickener to provide the required viscosity. These fluids also contain anti-wear, antifoam, rust, and corrosion inhibitors additives. These fluids can run most equipment that has been designed for oil, but pump speeds, temperatures, and pressure may have to be modified. Their water content, which provides fire resistance, must be monitored and maintained by adding make-up water to compensate for evaporation losses. In applications where fire hazards or environmental pollution is a concern, water-based fluids offer a distinct advantage. Fluids consist of water-glycol or water-in-oil fluids, with emulsifiers, stabilizers, and additives. Due to their low lubricity, piston pumps used with these fluids are limited to a maximum of 3000 lb/in^2 pressure. Also, vane pumps should not be used with water-based fluids unless they are specifically designed for use with such fluids. Operating temperature for these fluids is below 120°F to prevent evaporation and deterioration of fluids. To prevent separation of fluid phases and other adverse effects on additives, the minimum temperature should not drop below 32°F. Viscosity, pH, and water hardness monitoring is important in water glycol systems. If water is lost due to evaporation, fluid viscosity, friction, and the operating temperature of fluid increases, resulting in sluggish operation of the hydraulic system and increased power consumption. If fluid viscosity is permitted to drop due to excessive water content, internal leakage at actuators increase and cause sluggish operation. Also, thin fluid can cause turbulent flow that will increase the potential for erosion of the system components. Make-up water added to the system must be distilled or deionized. Calcium and magnesium present in potable water will react with lubricant additives and cause them to floc or come out of solution and compromise the fluid's performance. To prolong fluid or component life, water added to the system should have a maximum total dissolved solids (TDS) of 5 ppm.

HFD Fluids (Synthetic)

HFD category hydraulic fluids do not contain any water or petroleum oil; hence these are considered synthetic or synthetic organic compounds. These high-performance fluids are fire resistant, compatible with system materials, and are also completely biodegradable. The most popular are polyol esters and water-free polyalkylene glycols (PAGs). They are fire resistant, environmentally friendly, water free, oxidation resistant, and have good low-temperature performance. They are more expensive than any other type of hydraulic fluid, and their use is limited to situations where fire resistance and biodegradability is important.

Three types of synthetic fire-resistant fluids are popular:

- Phosphate esters
- Halogenated hydrocarbons
- Synthetic base

These fluids do not contain water or volatile material, and they provide satisfactory operation at high temperatures without the loss of essential elements. These fluids are also suitable for high-pressure applications. Synthetic hydraulic fluids have low viscosity indexes, so their usage is restricted to relatively constant operating temperatures. Synthetic fluids have high specific gravities so pump inlet conditions must be carefully selected to prevent cavitation. Phosphate esters have a high flash point (more than 400°F) and auto ignition temperature above 900°F, making them less likely to ignite and sustain burning. Halogenated hydrocarbons are inert, odorless, nonflammable, noncorrosive, and have a low toxicity. Seal compatibility is very important when using synthetic fluids. The most commonly used seal is nitrile (Bauna rubber). Neoprenes are not compatable with these fluids.

BRAKE FLUIDS

Brake fluid is the hydraulic fluid used in automobile braking systems. Brake fluid transmits power from the brake pedal through a master cylinder to brake drums that stop the rotating wheels. Important requirements for brake fluid are that the fluid must be incompressible and not suffer any changes in physical properties such as compressibility over the operating temperature range. The fluid is designed not to boil even when exposed to wide operating temperature ranges. During operation, temperature of the brake fluid can rise to more than 212°F. If the brake fluid boils or there is a leak in the system, the incompressibility of the fluid will be lost and the brake pedal travel will increase. Also during actual usage, brake fluid absorbs moisture from the atmosphere, which lowers the boiling point of the brake fluid drastically. Moisture in the brake system can cause corrosion in the brake components.

The base fluids used for brake fluid formulations are petroleum oils, ethylene glycol, or silicon fluids. Most present-day brake fluids have ethylene glycol, which is hygroscopic, as its base oil with corrosion inhibitors. Ethylene glycol has a high boiling point and has good lubricating properties for elastomer parts in the brake system. The boiling point is reduced when moisture is present. Another popular brake fluid is silicon oils. They have a high boiling point, approximately 700°F and are not hygroscopic. When moisture is absorbed by the system, beads of moisture move through the brake line, collecting in the calipers, where temperature may be more than 212°F. The result may be vapor lock and brake failure. Silicon is also highly compressible due to aeration and foaming under normal braking conditions.

Specifications

Most brake fluids are highly hygroscopic, implying that these fluids absorb water. Even a small amount of water dissolved in brake fluid can reduce its boiling point by hundreds of degrees. Brake fluid boiling in the system can cause a spongy feel at the pedal and increased stopping distance. The most important specifications of brake fluids are the dry and wet boiling points of the fluid. Dry boiling point is the normal boiling point of fluid; wet boiling point is the boiling point of brake fluid in the presence of water. The U.S. Department of Transportation (DOT) has specified requirements for brake fluids (see Fig. 19-1).

DOT 2 fluid is for drum brakes and is now obsolete. DOT 3 and DOT 4 fluids are glycol based; DOT 5 fluid is silicon based. DOT 3 and DOT 4 are hygroscopic and readily absorb moisture. Moisture causes a sharp drop in the fluid's boiling point. A moisture content of 1 percent in DOT 3 fluid can reduce the boiling point from 401 to 369°F; a moisture content of 2 percent; to 320°F;

	DOT 2	DOT 3	DOT 4	DOT 5
Dry boiling point, °F	374	401	446	500
Wet boiling point, °F		284	311	356

*DOT = Dept of transportation (USA).

FIGURE 19-1 DOT specifications for brake fluids.

and a moisture content of 3 percent, to 293°F. DOT 5 is a silicon-based fluid; it does not absorb water. Water that enters the hydraulic system will settle at low points where it can cause corrosion. DOT 5 is normally used in motorcycles because of the higher temperature required of the brake fluid.

REFERENCES

1. P. Madhvan and N. C. Werner, "Service Life Extension of Hydraulic Fluids Through Contamination Control," *Practicing oil analyst magazine,* Pall Corporation, USA, no. 200503, March 2005.
2. "Synthetic Oil and Lubricant Additives: Advances Since 1979," *Chemical Technology Review* 207 (Park Ridge, NJ: Noyes Data Corporation, 1982).

CHAPTER 20

PETROLEUM PRODUCTS AS PESTICIDES

Pesticides are chemicals that control insects. Control may result from killing the insect or preventing it from its destructive activities. It is estimated that human beings have been living on this earth for over 3 million years, whereas insects have existed for at least 250 million years. Petroleum oils have long been employed as insecticides in agriculture, but the injury they may cause, especially to the foliage of plants, constitutes a serious objection to their use. Literature on the subject has been reviewed by Kelly[1] and by Johnson and Green.[2]

The present-day use of petroleum-derived oils for agricultural crop protection dates back to 1800s when a kerosene and soap emulsion was used with some success against scales and aphids in citrus trees. The search for something more effective led to the use of emulsified crude oil sprays, which were soon found out to be too phytotoxic. Eventually, researchers concentrated on distillates in the range of diesel and lubricating oils. In 1915, de Ong and colleagues[3] showed that tree injury is caused by aromatics and other unsaturated hydrocarbons present in petroleum products. This was a significant discovery because as it was known that aromatics and other unsaturated components could be easily removed from petroleum fractions by treatment with sulfuric acid. In 1932, Smith[4] discovered the relation between pesticidal efficiency, phytotoxicity, and oil heaviness (molecular weight, viscosity, and boiling range). The importance of an emulsifier used with the oil was determined to be due to its influence on spray deposit, which affects its spray control efficiency. Pest control today is a multibillion dollar industry and petroleum products play a major part in it.

SPRAY OILS

Several physical properties have been found to affect insecticidal activity of spray oils. These are discussed in the following paragraphs.

Viscosity

Optimum viscosity for spray oils[5] has been found to be in the range of 10 to 20 cSt (approximately 60 to 100 Saybolt Universal Second [SUS]) at 100°F. The increasing viscosity of spray oil seems to reduce the effectiveness of the oil, probably due to the ineffectiveness of heavier oils to form a uniform film. Higher viscosity also increases phytotoxicity. However, viscosity alone is not sufficient to characterize an oil's suitability as a pest control agent.

Hydrocarbon Type

Paraffinic oils have been found to be more effective[6] than naphthenic or aromatic-type hydrocarbons of the same viscosity. Aromatic and unsaturated hydrocarbons contribute little to pesticidal activity of spray oils; moreover, they are phytotoxic and should be reduced to as low a level as possible (less than 8% v/v).

Unsulfonated Residue

Phytotoxicity is the tendency of a product to harm a plant, which may vary from minor discoloration of foliage to complete killing of the plant. The aromatic component of oil contributes mostly to this action. A property of petroleum oil termed "unsulfonated residue," or UR, has been devised to estimate the phytotoxicity of oil. It is an inverse function of the aromatic content of oil; thus the higher the aromatic content of oil, the lower the UR of the oil. It is a very important specification of horticulture oil, as shown by Volck.[7] Thus the higher the UR, the safer the oil is for use on plants. An oil with high UR (more than 92) is generally considered free from phytotoxic material and safe for use on plants.

Determination of UR

The method used to determine UR[8] (ASTM 483) involves shaking the oil several times with a mixture of concentrated fuming sulfuric acid. The most reactive compounds such as aromatics and unsaturates are absorbed by the acid. The percentage of oil not absorbed is the UR.[9]

Twenty cm^3 of 37 N (100 percent) sulfuric acid is added in small portions to 5 cc of a sample of oil in an ASTM sulfonation bottle, the bottle is immersed in an ice bath and shaken continuously until the temperature no longer rises. It is then put in a water bath at 100°C and shaken at 10-minute intervals for 1 hour. At the end of this time, sulfuric acid having a specific gravity (SG) of 1.84 is added until the clear unsulfonated portion of the sample rises in the graduated neck of the flask. It is then centrifuged, and the portion that has been sulfonated is calculated as volume percent of the original sample.

Molecular Weight

The pesticidal efficiency of petroleum-based spray oils starts to decline below an average molecular weight of 310, regardless of structural composition, and rises as the paraffinicity of oil increases. Suitable spray oils can be selected[10] from petroleum fractions having a molecular weight between 290 and 330. Lighter oils are more suited for use on more sensitive plant species.

Boiling Point and Boiling Range

Most spray oils have a boiling point (ASTM 50 percent distillation) in the range of 410 to 480°F. The boiling range (ASTM 10 to 90 percent) of an individual spray oil is generally narrow, about 50 to 79°F.

In addition to the boiling point, boiling range, hydrocarbon type and viscosity an important factor in insecticidal efficiency is the amount of oil deposited on a tree. This in turn is a function of oil concentration in the spray and the kind and type of emulsifying agent used.

Manufacture and Formulation

Both agricultural mineral oils (AMOs) and horticultural mineral oils (HMOs) are manufactured from paraffinic crude oils. The manufacturing process is almost identical to the light lubricating oil process, involving a vacuum distillation unit, solvent extraction, and dewaxing to meet the pour point specifications of spray oil, and finally acid and clay treating or hydrogenation to remove aromatics and other active compounds. It should be mentioned, however, that as the UR is increased or aromatics are decreased, the solvent properties of the oil decrease, which makes the oil more difficult to emulsify. This could adversely affect the pesticidal efficiency of the oil. Table 20-1 shows the properties of some commercial spray oils.

TABLE 20-1 Typical Properties of Spray Oils

		Commercial products					
		1	2	3	4	5	6
API gravity	°C	35.1	32	32	31	34.8	33
Specific gravity		0.8493	0.8654	0.8654	0.8708	0.8509	0.8602
ASTM 50% distillation	°F	716	685.4	687.2	716	759.2	716
10–90% range	°F	69.84	64.98	91.94	80.1	84.96	54.9
Pour point	°F	6.08	10.04	20.12	20.12	10.04	5
Viscosity	SUS, 100°F	74	68	70	100	105	86
Viscosity	cSt, 100°F	14.1	12.5	13.1	20.5	21.7	17.2
Unsulfonated residue (UR)	Min, Vol %	92	92	92	92	99	93
Classification	ASTM 2887	HMO	HMO	HMO	HMO	AMO	HMO

Emulsification

An emulsion consists essentially of small droplets of one liquid suspended in another. Simple agitation of water with a small quantity of oil will result in formation of globules of oil that eventually rise to the surface because of buoyancy and recoalesce into a thin layer. However, the addition of a small amount of an emulsifier will prevent or delay the coalescence process by forming a thin film around the oil droplets. These droplets also carry a negative charge,[11] which contributes to their tendency to disperse evenly throughout the emulsion.

Potassium soaps have been preferentially used in the past. Fish oils, cotton seed oils, olive oil, and turkey red oil have also been used with varying success. Also, many inorganic emulsifiers are used. These are fine clays such as kaolin, bentonite, Fuller's earth, silica, alumina. Currently, most spray oils contain nonionic surfactants dissolved in oil (0.3 to 0.5 wt %). The mixture of spray oils and water is agitated in a tank and pumped under high pressure, exiting through a nozzle, which emulsifies the oil in water.

Types of Emulsions

Hot Emulsions. In this method, oil, soap, and water are mixed in a tank, heated to boiling, and pumped through a fine nozzle while still hot. The advantages of these emulsions are their stability over a relatively long period of time. Hot emulsions are uneconomical, however, because of the equipment required for heating and energy cost.

Cold Stirred Emulsions. In this method, small quantities of oil are added to soap and vigorously stirred after each addition. The advantage of this method is the low cost of the equipment. The dispersion of oil droplets in soap depends on the nature and viscosity of the soap used and the soap moisture content and temperature.

Cold Pumped Emulsions. These are generally prepared with nonsoap emulsifiers. The method consists of simply adding emulsifier to water followed by oil and then pumping the mixture through a fine nozzle under pressure. The emulsifier commonly used is calcium caseinate. Kaolin clay is also used.

Miscible Oils

In these formulations, emulsifier is dissolved in oil and the emulsion contains a relatively small quantity of water, typically less than 10 wt %. Water is added just before application. The formulation is delivered through a pump and nozzle combination. The advantages of these emulsions are their low water content, ease of handling, and stability. The disadvantage is their higher cost compared with oil emulsions.

Dormant Oils

Dormant oils, or winter oils, are applied to plants during the period before foliage begins to grow in the spring. The term used to refer to heavier, less refined oils that were unsafe to use on plants after plants had matured. New dormant oils are more refined, and they have a high UR that can be used on plant foliage.

Summer Oils

Summer oils, or verdant oils, are applied to the plant after the foliage has emerged. As with dormant oils, the term now refers to the time of application rather than to the properties of the oil.

Supreme Oils

Supreme oils are highly refined oils that distill at slightly higher temperatures and have a wider boiling range than the narrow-range oils. Most supreme oils meet the characteristics of a superior oil.

Superior Oil

Superior oils are highly paraffinic narrow boiling range petroleum mineral oils that can be used on plants all year round without phytotoxicity. Most superior oils are now known as "narrow-range" oils.

Combination Sprays

Often other materials in small concentrations (chemical herbicides and fungicides, etc.) are added to emulsions to prevent fungus and weed growth in farms.

Horticultural and Agricultural Mineral Oils

Following an international conference on spray oils in Australia in 1999, new names were recommended to replace such terms as *summer oils, superior oils, petroleum spray oils, narrow-range,* and *broad-range oils*. The new classification is based on the degree of refinement and recommends the following three categories:

- Mineral oils
- Agricultural mineral oils (AMOs)
- Horticultural mineral oils (HMOs)

Only HMOs and AMOs are used as spray oils for plants. The benefits of spray oils are that they are as effective or more effective than broad-spectrum chemical pesticides for a wide range of pests. Many pests can be controlled simultaneously. Pests are not known to develop resistance to them. Oils used in sprays are biodegradable and break down into harmless products within a few weeks. These are not toxic to humans or other animals and birds.

Specifications of HMOs and AMOs

HMOs and AMOs are straight run petroleum distillates boiling in the range of light lubricating oils. The use of recycled lubes is not permitted. The paraffin content must be greater than 60 wt %. The concentration of unsaturated molecules (aromatics and olefins) must not be greater than 8 percent.

The UR should be 92 percent or greater. The oils with a normal paraffin carbon number of 21 to 24 (ASTM D 2887) would be termed HMOs; oils with a carbon number in the range of 24 to 25 would be termed AMOs.

Herbicides

Herbicides or weed killers are of two general types; selective and nonselective. Selective herbicides destroy the weeds but do not affect the crops. Nonselective herbicides are those that kill all types of plants. Herbicides can cause acute toxicity and rapid burning or chronic toxicity that is the cause of yellowing of leaves.[12]

Weeds have long been the bane of farmers. Losses by weeds exceed those caused by all other pests, insects, rodents, fungi. They compete with crops for water, sunlight, and mineral nutrients.

Petroleum-based herbicides were earlier used extensively for both agricultural and nonagricultural uses. Among the nonagricultural uses are the railroad and telephone and electric utility companies that use herbicides to clear weed growth from the right of the way over vast narrow stretches of land.

The fact that petroleum distillates can kill plants has long been known. Million of gallons of such products are used as general herbicides for railroad right of ways and along roadside fences and along the path of high-tension electric wires. Light aromatic oils cause acute toxicity; heavy oils such as diesel and fuel oils cause chronic toxicity. The first aromatic products used as herbicide were waste materials from refineries such as heavy aromatic extracts and other by-products of processing.

Less volatile materials are desirable because they penetrate into the roots and kill the plant. By remaining in the ground for some time, they prevent regrowth. High viscosity is desirable for greater herbicidal action, but it must be low enough so that the oil can be sprayed. The most important specification for herbicides is its aromatic content. These usually range from 50 to 90 percent for general-purpose weed killers and lower for the selective type. A study of the aromatics shows that benzene is a poor weed killer, but the herbicidal activity increases with the size of the side chain. Higher aromatics such as naphthalenes are also quite active. Naphthenes were found to possess more herbicidal activity than paraffins. Olefins are also thought to contribute to herbicidal action. Often petroleum herbicides are fortified with chemical toxicants to increase their activity. Oil toxicity to plants results from the denaturation of the plasma membrane resulting from the solubility of oil.

Because of environmental concerns to halt the release of hydrocarbons into the atmosphere, petroleum oils are no longer predominantly used as herbicides, but they are still used as adjuvants in herbicide formulations.

Fungicides

For fungicidal applications, oil is applied in the form of a fine mist (50 to 100 microns), approximately a gallon to an acre, every two or three weeks. Both light portable mist sprayers and aircraft have been used to apply the oil.

Both high- and low-viscosity white oils have been used successfully to control the fungus on tobacco plants.[13] Oil is applied as a 50 percent emulsion to the top of the stem at the time the plant is topped. About 2 to 3 mL are used. Oil prevents the growth of suckers that normally grow in the axils of leaves below the portion topped. Petroleum spray oils have been used for the control of powdery mildew in vineyards. Oil sprays with 0.5 to 1 v/v % oil concentration have been used successfully to control fungus.

CHEMICAL INSECTICIDES

Until the beginning of World War II, insecticides in use were limited to petroleum oils, nicotine, pyrethrum, rotenone, sulfur, hydrogen cyanide gas, and a few other inorganic compounds. A new synthetic organic compound called dichloro diphenyl trichloroethane (DDT) was developed for use as insecticide. DDT revolutionized the fight against pests, and billions of pounds were used throughout the world, beginning in 1940, for the fight against malaria, yellow fever, and a host of other insect-vectored diseases. In 1939, Dr. Paul Muller, the Swiss entomologist, was awarded the Nobel Prize in medicine for his lifesaving discovery of DDT. It remained the most popular insecticide for use against a large number of pests until 1973 when some adverse environmental effects of its use were discovered. It was banned by environmental protection bodies of most developed countries because of its poor biodegradability. Residue of the pesticide was found to pollute water bodies and enter the food chain.

Direct use of petroleum products as insecticides is decreasing, but the use of petroleum solvents as the solvent or spray medium for chemical pesticides for agricultural and nonagricultural purposes is increasing. Significant profits are realized by the adoption of pest control measures in crops, orchards, livestock, household furnishing, and woodwork, by preventing losses caused by insects, fungi, and weeds.

The term *pesticide* is now usually applied to all those products that are used for killing insects and other arthropods, weeds, and fungi. Individually, the control products are known as insecticides, herbicides, and fungicides. As stated earlier, the term *herbicide* is also used for products that inhibit plant sprouting and defoliate plants.

Chemical Pesticides

A large number of chemicals[14] are used for the control of harmful insects. Most of these are sold in the market under different trade names. A majority of these chemicals are toxic to mammals and birds and have the potential to pollute bodies of water or enter the food chain. For these reasons the manufacture and use of these chemicals are carefully controlled. Only chemicals that are relatively less toxic, have high biodegradability, and the least potential for environmental damage are allowed. Many of the popular insecticides used in 1950s and 1960s, such as DDT, Lindane, Aldrin, and so on, were withdrawn as soon as their harmful effects to humans and environment were known. The following categories of chemical pesticides and chemicals of plant origin are most popular for various applications.

Organochlorines

Organochlorines are insecticides that contain carbon, hydrogen, and chlorine. Examples of these chlorinated hydrocarbons are given next.

DDT

DDT was used for the control of malaria, yellow fever, and many insect-vectored diseases throughout the world, until 1973 when its use was banned in most developed countries. The chemical name of DDT is 1,1,bis (parachlorophenyl)-2,2,2 trichloroethane. Its chemical formula is $C_{14}H_9C_{15}$ with a molecular weight of 354.49. Figure 20-1 shows the structure of the DDT molecule. The mode of

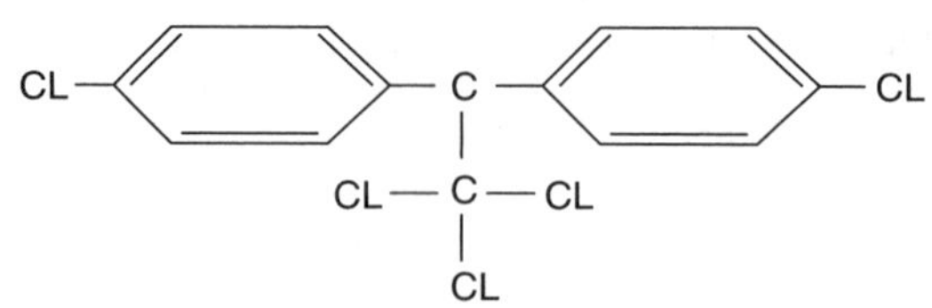

(DDT)
p-p Dchlorodiphenyltrichloroethane

FIGURE 20-1 Structure of DDT, an organochlorine-type insecticide.

action of this insecticide has never been clearly established, but some complex mechanism destroys the delicate balance of sodium and potassium ions within the axons of neurons both in insects and mammals, causing death. DDT has a negative temperature correlation; the lower the temperature, the more toxic it becomes to insects.

Hexachlorocyclohexane

Hexachlorocyclohexane (HCH) is also known as benzene hexachloride, or BHC. Its chemical formula is $C_6H_6CL_6$. Structure of the molecule is shown in Fig. 20-2. There are, in fact, four isomers; beta, gamma, delta, and epsilon. The technical grade containing all these isomers has a

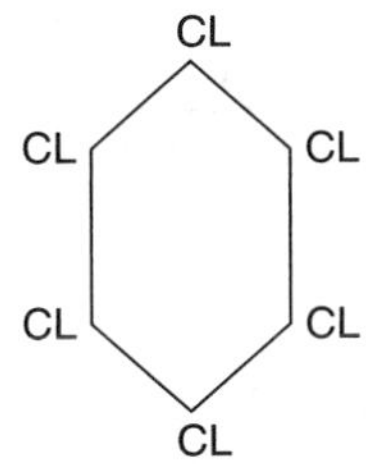

$C_6H_6CL_6$

Gamma 1α, 2α, 3β, 4α, 5α, 6β, hexachlorocyclohexane

FIGURE 20-2 Structure of Lindane, an organochlorine.

strong musty odor and flavor. But only the gamma isomer has insecticidal properties. During its manufacture, only the gamma isomer is isolated and sold as an odorless insecticide called Lindane. The properties of DDT and Lindane are listed in Table 20-2. Lindane has been used on a wide variety of crops to control plant-eating insects, in warehouses, in public health to control insect-borne diseases, and with fungicides for seed treatment. Lindane is used in lotion, creams, and shampoos for the control of lice and mites (scabies) in humans. The use of this insecticide was withdrawn in the United States in 2002 for all food-related uses, although it is still being used in limited quantities for clinical purposes.

TABLE 20-2 Properties of Organochlorine Insecticides

DDT = 1,1 Bis (P-Chlorophenyl)-2,2,2 Trichloroethane	
Formula	$(CLC_6H_4)_2$ $CHCCL_3$
Molecular weight	354.49
Melting point, °F	228.2
Lindane (HCH) = Gamma 1,2,3,4,5,6 Hexachlorocyclohexane	
CAS number	58-89-9
Formula	$C_6H_6CL_6$
Molecular weight	290.85
Melting point, °F	235.4
Water solubility, mg/L, 77°F	7.3
Vapor pressure, 68°F, mPa	5.6

The mode of action of HCH resembles that of DDT but occurs more rapidly. The gamma isomer is a neurotoxicant whose effects are seen within hours as increased activity, tremors, and convulsions leading to prostration. Like DDT, it too has a negative temperature coefficient.

TABLE 20-3 Properties of Cyclodiene Insecticides

Endosulfan*	
Common name	Endosulfan
Synonym	6,7,8,9,10, 10 hexachloro-1,5, 5A,6,9,-9A-hexahydro-6,9-methano-2, 4, 3-benzodioxathiepin-3-oxide
Chemical formula	$C_9H_6CL_6O_3S$
Molecular weight	406.9
Melting point,°F	223
Boiling point,°F	Undefined (decomposes)
Density, 77°F, g/cm^3	1.735
Solubility in water, 77°F, μg/L	60–100

*Technical Endosulfan is sold as brown crystalline flakes, with a faint odor of sulfur dioxide.

Cyclodienes

Cyclodiene insecticides were developed in the 1950s under various trade names such as Chlordane, Dieldrin, Heptachlor, Endrin, Mirex, Endosulfan, and Keponon. Properties and the structure of Endosulfan are listed in Table 20-3 and Fig. 20-3. Most of these cyclodiene insecticides are persistent insecticides and are stable in soil and relatively stable to sunlight. As a result they were used in large quantities as soil insecticides (particularly aldrin, dieldrin, chlordane, and heptachlor) for the control of termites and other soil-borne insects whose larvae feed on the roots of the plants. Cyclodienes are the most effective, long-lasting, and economical termitecides ever developed. However, due to their poor biodegradability, most agricultural uses of cyclodienes were prohibited in the 1980s.

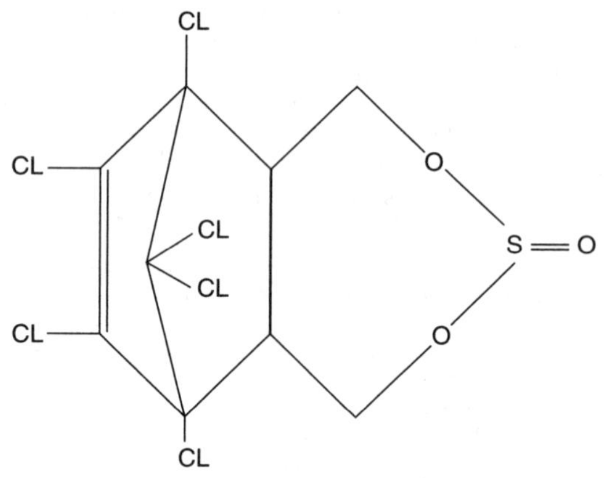

Endosulfan $C_9H_6CL_6O_3S$

FIGURE 20-3 Structure of Endosulfan, a dicyclodiene-type insecticide.

Organophosphates

Organophosphate (OP) is the term that includes all insecticides containing phosphorus. All organic phosphates are derived from one of the phosphorus acids and are as a class the most toxic of all pesticides to vertebrates. Because of the similarity of the organic phosphate chemical structure to the nerve gases Sarin, Soman, and Tabun, the mode of action is similar. OPs have two distinctive features. They are generally more toxic to vertebrates than other classes of insecticides, and they are chemically unstable or nonpersistent. It is due to their biodegradability that they were brought into agriculture and other uses as a substitute for organochlorines. However, due to their high toxicity to mammals, their use was restricted beginning in the 1990s.

All OPs are esters of phosphorous having varying combinations of oxygen, carbon, sulfur, and nitrogen attached. OPs work by inhibiting certain important enzymes of the nervous system. The enzyme becomes attached with the OP, resulting in paralysis. OPs are generally divided into three groups; aliphatic, phenyl, and heterocyclic derivatives.

Malathion, an OP, has been used for the control of sucking and chewing insects on fruits and vegetables and in the control of mosquitoes, flies, household insects, animal parasites, and head and body lice. The properties and structure of Malathion are shown in Table 20-4 and Fig. 20-4.

TABLE 20-4 Properties of Organophosphate Insecticides

Malathion*	
Common name	Malathion
Synonym	1,2,DI (ethoxycarbonyl) ethyl o, o dimethyl phosphorodithioate
Chemical formula	$C_{10}H_{19}O_6PS_2$
Molecular weight	330.36
Melting point, °F	37.22
Boiling point, °F	313.7
Density, 77°F, g/cm^3	1.23
Flash point, °F	350.6
Solubility in water, 68°F, mg/L	145

*Technical malathion is a clear amber liquid, with a garlic-like or mercaptan odor, miscible with most organic solvents; alcohols, esters, ketones, ethers, and aromatics, and with limited solubility with paraffinic hydrocarbons. CAS number: 121-75-5.

Aliphatic OP's have a carbon straight chain structure. Examples of this group are TEP, Malathion, trichlorfon (Dylox), monocrotophos (Azodrin), and acephate (Orthene). The phenyl OPs contain a phenyl ring, with one of the rings of hydrogen displaced by an attachment to a phosphorus moiety and the other hydrogen displaced by Cl, NO_2, CH_3, CN, or S. The phenyl OPs are more stable than aliphatics; thus the residue is longer lasting. Examples of phenyl OPs are parathion (ethyl parathion) in 1947, methyl parathion, sulprofos, and famphur.

Heterocyclic OPs contain different atoms (sulfur, nitrogen, oxygen) in the carbon ring. Examples are diazinon, anziphos-ethyl, chlorpyrifos, and isazophos.

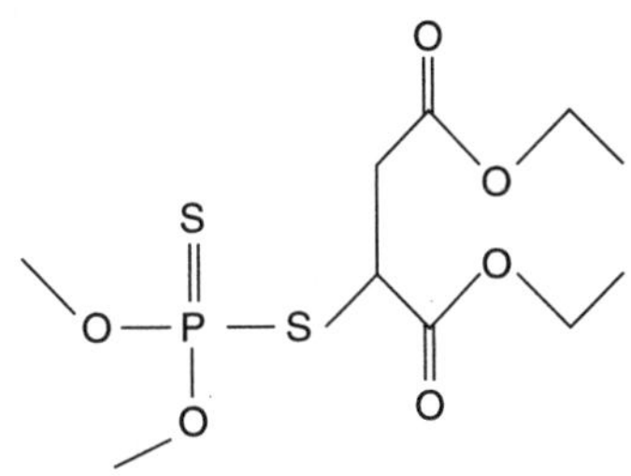

$C_{10}H_{19}O_6PS_2$

FIGURE 20-4 Structure of malathion, an organophosphate.

Carbamates

The carbamate insecticides are derivative of carbamic acid, $HOC(O)NH_2$. They have the general formula shown in Fig. 20-5. Their mode of action inhibits the vital enzyme cholinesterase. The first successful and very popular insecticide carbaryl (Sevin) was introduced in 1956. It was popular because of its very low mammalian oral and dermal toxicity and exceptionally broad spectrum of insect control. Other carbamate insecticides are methomyl (Lannate), carbofuran (Furadan), methiocarb (Mesurol), and many others. Carbamates are relatively unstable compounds that break down in the environment within weeks or months. Carbamates are generally used as surface sprays in the control of household pests, by pest control operators for control of cockroaches and other pests in restaurants, kitchens, and homes, and also as a lawn and garden insecticide. Table 20-5 shows the properties of carbofuran.

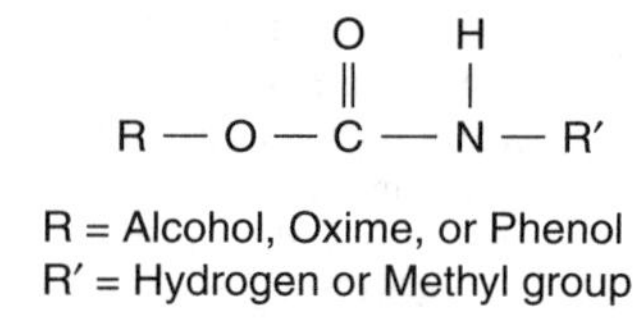

Carbamate general structure

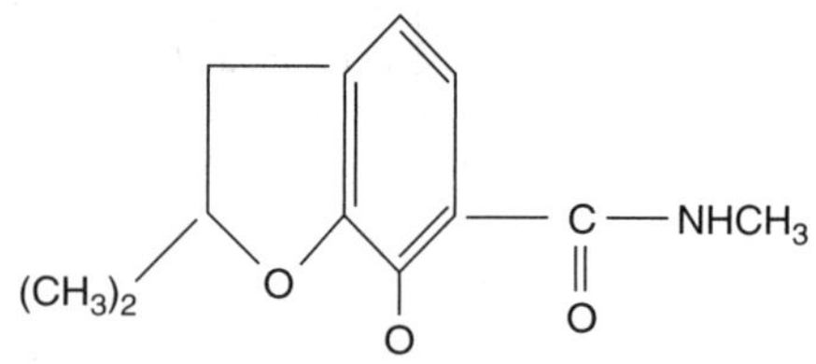

Carbofuran insecticide

FIGURE 20-5 Structure of carbofuron, a carbamate insecticide.

Natural Insecticides

Natural insecticides are toxicants derived from plants. Historically, plant materials have been in use longer than any other material with the exception of sulfur. These include tobacco, pyrethrum, derris, hellebore, quassia, camphor, and turpentine.

Botanical insecticides are of great interest, for they are natural insecticides. Some of these are discussed next.

TABLE 20-5 Properties of Carbamate Insecticides

Carbofuran*	
Common name	Carbofuran
Synonym	2,3 dihydro-2,2 dimethylbenzofuran-7, y1 methyl carbamate
Chemical formula	$C_{12}H_{15}NO_3$
Molecular weight	221.3
Melting point, °F	303.8
Boiling point, °F	
Density, 77°F, g/cm^3	1.118

*Carbofuran is a colorless white crystalline solid. It is noncorrosive and nonflammable. CAS number: 1563-66-2.

Pyrethrins

Pyrethrins are natural insecticides produced from a certain species of chrysanthemum flowers. In general, insecticides derived from plants are low in toxicity to humans. The flowers of the plant are harvested shortly after blooming and are dried and powdered or the oil in the flowers is extracted with a solvent. The resulting pyrethrin extract usually has an active ingredient content of about 30 percent. These active insecticidal components are collectively known as pyrethrins. Pyrethrins are a mixture of mainly four compounds; pyrethrins I and II and cinerins I and II. The typical chemical structure of the pyrethrin molecule ($C_{19}H_{26}O_3$) is shown in Fig. 20-6. Properties of pyrethrins are shown in Table 20-6. Pyrethrins are viscous brown resins or liquids that inactivate rapidly in air. The molecular weight ranges from 316 to 374.

Two pyrethrins are most prominent; pyrethrin I and pyrethrin II. Pyrethrin acts on the insect with phenomenal speed, causing immediate paralysis, thus its popularity in fast knockdown household aerosols. However, unless formulated with one of the synergists, most of the paralyzed insects recover

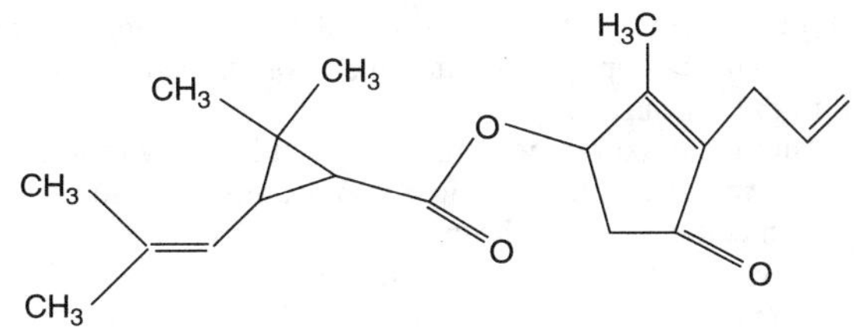

$C_{19}H_{26}O_3$

Pyrethrin (allethrin) structure

FIGURE 20-6 Typical structure of pyrethrin.

TABLE 20-6 Properties of Pytrthrins and Pyrethroid Insecticides

Pyrethrin		
CAS number		8003347
Formula		$C_{19}H_{26}O_3$
Molecular weight		316–374
Appearance		Viscous brown resin, liquid
Water solubility		Insoluble
Solubility in other solvents		Soluble in organic solvents
Boiling point		294.8–302
Pyrethrin 1,°F		377.6–379.4
Pyrethrin 2,°F		Kerosene, alcohol, ether, etc.
Permethrin		
Chemical class	Pyrethroid	
Chemical name	3-(2,2-dichloroethenyl)-2,2-dimethylcyclopropaneecarbobylic acid 3-(2,2-dichloroethenyl)-2,2-dimethylcyclopropaneecarbobylic acid (3-phenoxyphenyl) methyl ester	
Formula		$C_{21}H_{20}CL_2O_3$
Trade name	Ambush, Ectiban, Permethrin, Pounce, Wasp and Hornet Killer II	
CAS number		52645-53-1
Molecular weight		391.29
Color/form	Colorless crystals to pale yellow viscous liquids	
Odor	Odorless	
Water solubility, 86°F, mg/mL		0.2
Partition coefficient K_{ow}		3.0×10^3
Phenothrin		
Chemical class	Pyrethroid	
Chemical name	2-2 dimethyl-3-(2 Methyl-1-Propenyl) cyclopropanecarboxylic acid (3-Phenoxyphenyl) Methyl Ester	
Formula		$C_{23}H_{26}O_3$
Trade name	S-2539, Sumethrin, Sumitrin	
CAS number		26002-80-2
Molecular weight		350.46
Color/form	Colorless crystals to pale yellow to brown liquid	
Odor		
Water solubility, 77°F, mg/mL		1.06

once again to become pests. Natural pyrethrins are quickly detoxified by enzymes in the insects. To delay the enzyme action, an OP, carbamate, or another synergist, for example, piperonyl butoxide, is included in the formulation.

Pyrethrum is an axonic poison, as are synthetic pyrethroids and DDT. Axonic poisons are those that in some way affect the electrical impulse transmission along the axons, the elongated extension of the neuron cell body. Pyrethrum and some pyrethroids have a greater insecticidal effect when the temperature is lowered; they have a negative temperature coefficient, like DDT. Pyrethrum initially stimulates nerve cells to produce repetitive discharges, leading eventually to paralysis. Such effects are caused by their action on the sodium channel, a tiny hole through which sodium ions are permitted to enter axons to cause excitation.

Pyrethrin compounds have unique properties that make them effective against crawling and flying insects. Pyrethrins are used primarily to control human lice, mosquitoes, cockroaches, beetles, flies, in grain storage and in poultry pens, without danger to humans or mammals. Cartons and packages treated with pyrethrins are insect proof.

Pyrethroids

Natural pyrethrum has seldom been used for agricultural purposes because of its cost and instability in sunlight. In recent decades, many synthetic pyrethrin-like materials have become available. They are referred to as pyrethroids (Fig. 20-7). They are more stable in sunlight and generally effective against most agricultural insects when used in smaller concentrations. Pyrethroids are synthesized to be similar to pyrethrins, and yet they are more stable in the environment. Commercial pyrethroids products commonly use petroleum distillates as carriers. Synthetic pyrethroids are among the newest pesticides to enter the market, and they account for a large percentage of the pesticide market. Despite their extensive use, few cases of human poisoning have been reported. Two popular pyrethroids are Permethrin and d-phenothrin.

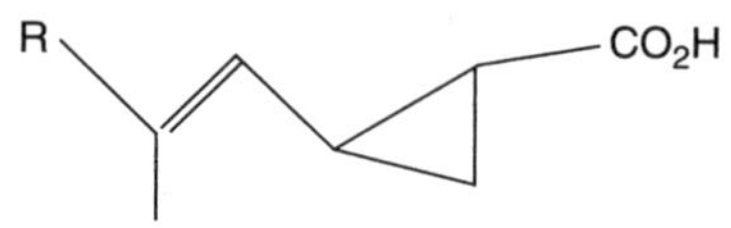

$R = CO_2CH_3$ Pyrethric acid
or
$R = CH_3$ Chrysanthemic acid

FIGURE 20-7 Structure of core of pyrethroids.

Permethrin

Permethrin is a useful insecticide that has proved to be effective in a variety of environmental and clinical settings. It is used extensively as an insect repellent on army clothing. It is very effective in protecting individuals from tick bites (by spraying on battle uniforms and bed netting). Table 20-6 shows the properties of permethrin.

Phenothrin

The compound d-phenothrin is labeled as an indoor use aerosol insecticide intended for purposes such as spraying bed netting, killing insects trapped inside an installation, or spraying inside aircraft to prevent the transport of insects. The application rates are one 10-s spray per 1000 cubic feet in aircraft.

Table 20-7 shows an aerosol composition for aircraft spray. Earlier pyrethroids developed around 1949 contained only one pyrethroid, allethrin. The second generation (1965) included tetramethrin (trade name Neo-Pynamin), followed by resmethrin (Synthrin) in 1967, which was 20 times more effective than pyrethrum. Then bioresmethmethrin (50 times more effective than pyrethrum), than bioallethrin (1969), and finally phenothrin (Sumithrin) in 1973. The third generation includes

TABLE 20-7 Aerosol Sprayss Formulation with d-Phenothrin Spray Preformulated for Buildings, Vans, Tents and Aircrafts

Composition, Wt %	
d-Phenothrin	2.00%
Propellant	98.00%

fenvelerate (Pydrine, Tribute, Bellmark) and permet (Ambush, Astro, Dragnet, Flee, Pounce, Prelude, Telcord, Torpedo), which appeared in 1972 to 1973. These became the first agricultural pyrethroids because of their exceptional insecticidal activity (0.1 lb/acre) and their photostability. They are virtually unaffected by ultraviolet in sunlight, lasting 4 to 7 days as residue on crop foliage. The fourth and current generation is effective in the range of 0.01 to 0.05 lb/area of active material. These are photostable and do not undergo photolysis (splitting) in sunlight. Because they have minimal volatility, they provide extended residual effectiveness up to 10 days under optimum conditions. Pyrethroids' mode of action resembles that of DDT, and they are considered axonic poisons. Pyrethroids affect both the peripheral and central nervous system of the insect. They initially stimulate the nerve cells to produce repetitive discharges and eventually cause paralysis. The stimulating effect of pyrethroids is much more pronounced than that of DDT.

Nicotine

Nicotine is an alkaloid contained in the leaves of several species of plants. Commercial extraction is from the dried leaves of the tobacco plant. It is effective against most types of insect pests but is used particularly against aphids and caterpillars, soft-bodied insects. Nicotine is from a chemical class of heterocyclic compounds containing nitrogen with the chemical formula $C_{10}H_{14}N_2$ and a molecular weight of 162.23 (Fig. 20-8) and prominent physiological properties. Other well-known alkaloids that are not insecticides are caffeine (in coffee and tea), quinine (cinchona bark), morphine (opium poppy), cocaine (coca leaves), ricinine (a poison in castor oil beans), strychnine, connine (spotted hemlock), and LSD (a hallucinogen from ergot fungus attacking grain).

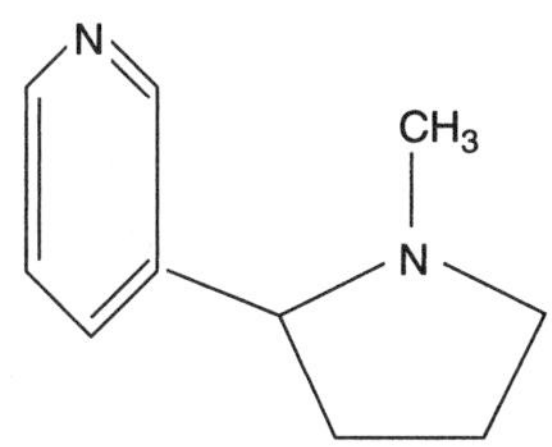

$C_{10}H_{14}N_2$

Nicotine structure

FIGURE 20-8 Structure of nicotine.

Nicotine action was one of the first classic modes of action identified by pharmacologists. Drugs that act similarly are said to have a nicotine response. Nicotine mimics acetylcholine (Ach) at the neuromuscular junction in mammals, resulting in twitching, convulsions, and death, all in that order. In insects, the same action is observed but only in the central nervous system ganglia.

Nictonoids

Nictonides are a newer class of insecticides with a new mode of action. They were previously referred to as nitro-quanidines, neo-nicotinyls, neo nicotinoids, chloronicotines, and chlornicotyls. Just as synthetic pyrethroids are similar to and modeled after natural pyrethrins, similarly nictonides are similar to and modeled after natural nicotine. The first nictonide (Imidacloprid) was introduced in Europe and Japan in 1990 and is currently marketed worldwide under a variety of proprietary names (Admire, Confidor, Gaucho, Merit, and many others). It is one of the most popular pesticides worldwide that is currently in use. Nictonoids are systemic insecticides and used as a soil, seed, or foliar treatment in cotton, rice, peanuts, potatoes, vegetables, and fruits for the control of sucking insects, soil insects, termites, white flies, and many other insects. Nictonoids act on the central nervous system of insects, causing paralysis.

Rotenone

Rotenone has been known to the natives of South America living in the deep forests of Brazil and Peru for centuries. They collected roots of the shrub *Lonchocarpus nicou* and threw the crushed roots into small streams and pools. The chemical in the roots stunned the fish and caused them to float to the surface for easy capture. In Southeast Asia, particularly in Java and Sumatra, a closely related legume, Derris Elliptica, which also contains rotenone, was used as an arrow poison.

Rotenone is a rotenoid plant extract obtained from such species as barbasco, cub, haiari, nekoe, and timbo. These plants are members of the pea (Leguminoseae) family. Rotenone-containing extracts are taken from the roots, seeds, and leaves of various plants. Formulations include crystalline preparations (approximately 95 percent pure), emulsified solutions (50 percent pure), and dusts (0.75 to 5 percent pure). It is a terpene with the chemical formula $C_{23}H_{22}O_6$. Structure of the molecule is shown in Fig. 20-9. Properties of rotenone are shown in Table 20-8.

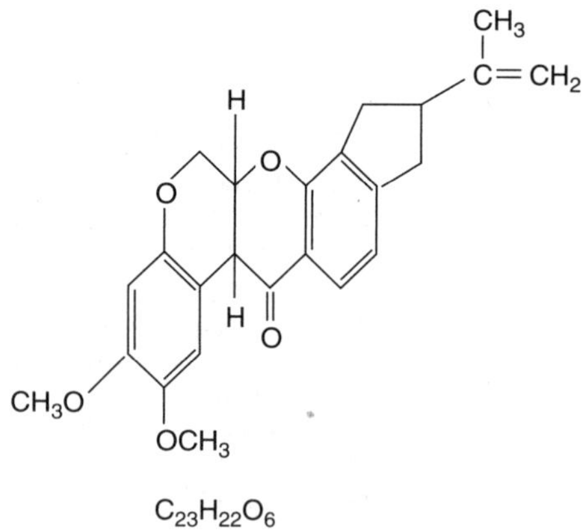

FIGURE 20-9 Structure of rotenone molecule.

TABLE 20-8 Properties of Rotenone

CAS number	83-79-4
Formula	$C_{23}H_{22}O_6$
Molecular weight	394.43
Water solubility	15 mg/L @ 212°F
Solubility	Soluble in acetone, CS2, CCL4, and in alcohols
Melting point, °F	325.4

Rotenone is used in home gardens for insect control, for lice and tick control on pets, and for fish eradication as a part of water body management. It is highly toxic to fish even in low doses. The 96-hour LC_{50} for different fish species ranges from 0.02 to 2 mg/L.(LC_{50} means the lethal concentration in an environment that kills 50 percent of a group of test animals within a prescribed time.) It is a potentially lethal toxin for aphids, cockroaches, houseflies, mosquitoes, and many other insects.

Rotenone is rapidly broken down in soil and in water. The half-life in both these environments is between one and three days. It does not readily leach from the soil and is not expected to be a groundwater pollutant.

Limonen or d-Limonene

Limonen belongs to a group often called floral or scented plant chemicals. Extracted from citrus peel, it is effective against all external pests of pets, such as fleas, lice, mites, and ticks, and it is virtually nontoxic to warm-blooded animals. Several insecticidal substances occur in citrus oil, but the most important is limonene, which constitutes 98 percent of orange oil by weight.

Clove Oil

Eugenol (oil of cloves) and cinnamaldehyde (derived from Ceylon and Chinese cinnamon oils) are used on ornamental plants and many crops to control various insects. Its mode of action is similar to pyrethrum. It affects sensory nerves of the peripheral nervous system but is not a cholinesterase inhibitor.

Neem Oil

Neem oil is extracted from the seeds of the Neem tree and contains active ingredient azadirachtin, a nortriterpenoid belonging to lemonoids. Azadirachtin has shown remarkable insecticidal, fungicidal, and bactericidal properties, including insect growth regulating qualities. Azatin is marketed as an insect growth regulator, and Align and Nemix are marketed as stomach/contact insecticides for greenhouse and ornamental plants. Azadirachtin disrupts molting by inhibiting biosynthesis or the metabolism of ecdysone, the molting hormone.

Synergists or Activators

Synergists in themselves are not insecticides but are materials used as insecticides. The first was introduced in 1940; since then, many materials have appeared. Synergists are a part of all household, livestock, and pet aerosols to enhance the action of fast knockdown insecticides such as pyrethrum, allethrin, and resmethrin against flying insects. Current synergists are piperonyl butoxide (Fig. 20-10 and Table 20-9), which contain methylenedioxy phenyl moiety, a molecule found in sesame oil, sesamin. Synergists inhibit cytochrome P-450 dependent polysubstrate monoxygenases (PSMOs), enzymes produced by microsomes, the subcellular units found in the liver of some mammals and some insects. PSMOs are the enzymes that degrade foreign substances, such as pyrethrums, allethrin, resmethrin, or any other synergized compound. Synergists simply bind the oxidative enzymes and prevent them from degrading the toxicant.

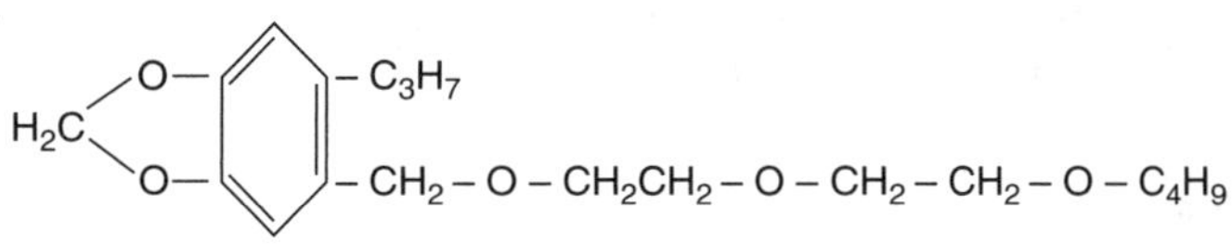

Piperonyl butaoxide

FIGURE 20-10 Structure of piperonyl butaoxide, an organic synergist.

TABLE 20-9 Properties of Piperonyl Butaoxide (Synergist)

Common name	Butacide
Synonym	3,4 methyl enedioxy-6-propylbenzyl (heptyl) diethylene glycol ether
Chemical formula	$C_{19}H_{30}O_5$
Molecular weight	338.4
Boiling point,°F	356
Density, 77°F, g/cm^3	1.04
Flash point, °F	340
Solubility in water, 68°F, mg/L	Not soluble

Fumigants

Fumigants are small volatile organic molecules that become gases above 40°F. They are usually heavier than air and commonly contain one or more of halogens (Cl, Br, or F). Most are highly penetrating, reaching through a large mass of material. They are used to kill insects, insect eggs, nematodes, and certain microorganisms in buildings, warehouses, grain elevators, soils, and greenhouses and in packaged products such as dried fruits, beans, grains, and cereals.

Until recently, methyl bromide was the most heavily used fumigant; the dominant usage was in preplanting soil treatment, in quarantine, to treat perishable products such as flowers and fruits, for nonperishable products such as nuts and timber, and for the dry wood termite fumigation of infected buildings. Its use is being phased out in most countries, under the Montreal Protocol, because of the adverse effects on the earth's ozone layer.

Some of the other common fumigants are ethylene dichloride, hydrogen cyanide, sulfural fluoride, chlorothane, ethylene oxide, and familiar home use moth repellent naphthalene and paradichlorobenzene crystals. Phosphine gas (PH_3) has replaced methyl bromide in a few applications, primarily for insect control in grain and food commodities. Treatment requires the use of aluminum or magnesium phosphide pellets that react with atmospheric moisture to produce phosphine. Phosphine is very damaging to fresh commodities, however, and it is readily adsorbed onto soil; thus it does not perform as a soil fumigant.

Fumigants, as a group, are narcotics. As narcotics, their effect is more physical than physiological. The fumigants are liposoluble or fat soluble. They induce narcosis, sleep, or unconsciousness, which in fact is their action on insects. Liposolubility appears to be an important factor in their action because these narcotics lodge in lipid-containing tissues throughout the insect body, including the nervous system.

Insect Repellants

Historically, repellants have included smoke, plants hung in dwellings, rubbed on the skin as fresh plant or its brew oils, pitches, tars, and various earths applied to the body. In recent times, the repellents used have been dimethyl phthalate, benzyl benzoate, normal butyl acetanilide (military cloth repellent), dimethyl carbate, and diethyl toluamide (DEET). Except diethyl toluamine, all other repellants have been phased out because of environmental concerns. DEET is now used worldwide for biting flies and mosquitoes.

Solvents for Pesticides

The most important attribute of a solvent for a pesticide is its ability to dissolve a sufficient quantity of the pesticide for effective use. It must retain the solute, the pesticide, at temperatures that are likely to be encountered. This varies with the pesticide employed. The solvent must be safe for recommended use. The solvent must not be toxic to humans. If the solvent is to be used in household aerosol or spray, this is very important. But if the product is used as a solvent for a very toxic pesticide, any slight toxicity of solvent would probably be unimportant. The product should not present a fire hazard. In household sprays or aerosols, the flash point should be relatively high, at least 125°F. This is likewise true of airplane applications where the possibility of a fire hazard is increased.

Type of Pesticide Solvents

Three types of pesticide solvents are employed in the manufacture of pesticidal emulsion concentrates:

- Low boiling solvents such as xylenes
- Middle boiling solvents (350 to 550°F boiling range)
- High boiling solvents (450 to 675°F)

The low boiling solvent for pesticidal emulsion has a boiling range of 320 to 360°F. This product has a high flash point of above 100°F and is therefore safer to handle.

Middle boiling aromatics such as kerosene extracts are very satisfactory pesticidal solvents and have a large share of market for these products. Because of rapid evaporation, this product was found to be nonphytotoxic when used as an emulsion on most plants.

High boiling solvents are usually obtained from catalytically cracked gas oils. In the late 1940s and early 1950s, catalytically cracked gas oils were used as a solvent for DDT. However in some cases, it was found that product formulated with these high boiling solvents was more phytotoxic, probably because of a slower evaporation rate.

Characterization of Pesticide Toxicity

Toxicity of pesticides is expressed in terms of two parameters: LD_{50} (lethal dose) and LC_{50} (lethal concentration).

LD_{50}. LD_{50} is the dose that kills half, or 50 percent, of the animals tested. The animals tested are usually rats or mice, rabbits, or guinea pigs. In all tests, the dose is calculated relative to the body weight of the animal. The most common units of chemicals used are milligram per kilogram of animal weight. Table 20-10 gives LD_{50} values for some pesticides. LD_{50} values give only a rough indication of risk to humans. LD_{50} can differ considerably depending on the way the pesticide is administered to the animal (i.e., whether fed, injected, or applied to the animal's skin). Actual LD_{50} values may be different for a given chemical depending on the route of exposure (oral, dermal, or inhalation). In general, the smaller the LD_{50} value, the more toxic the chemical.

TABLE 20-10 LD_{50} Values for Some Common Insecticides and Chemicals

Popular name	Other/commercial names	Class	Dermal LD_{50}	Oral LD_{50} mg/kg
Azadirachtin	Neem	Botanical		5000
Boric acid		Inorganic		10,000
Carbaryl	Sevin	Carbamate	4000	850
Chlordane		Organochlorine	580	283
Chlorpyrifos		Organophosphate	2000	135
Cypermethrin	Vikor	Pyrethroid	2000	247
d-Limonene		Botanical		5000
DDT		Organochlorine	1931	87
DEET	Insect repellent	Repellent		2000
Diazinon		Organophosphate	379	66
Imidacloprid	Grub-X, Merit	Chloronicotinoyl	5000	450
Imidacloprid	Provado, Admire	Chloronicotinoyl	5000	450
Linalool		Botanical	5976	2810
Lindane		Organochlorine	500	76
Malathion		Organophosphate	4100	1842
Nicotine	Black Leaf 40	Botanical	50	55
Permethrin	Ambush, Pounce	Pyrethroid	2000	2200
Permethrin	Intercept HG	Pyrethroid	5000	4810
Piperonyl butaoxide	Butacide	Synergist	7500	7500
Propoxur	Baygon	Carbamate	1000	95
Pyrethrins	Pyrethrum	Botanical	1800	1350
Resmethrin		Pyrethroid	3000	2500
Rotenone liquid		Botanical	1000	60
Sodium chloride	Common salt	Inorganic		3000

LC_{50}. LC_{50} stands for "lethal concentration." LC values refer to the concentration of chemical in air or water that kill 50 percent of test animals in a specified time, usually 4 hours.

Pesticide Applications

Chemical pesticides can be divided into four categories depending on their field of application. These are as follows:

- Household applications
- Nonagricultural uses such as mosquito control
- Industrial; in warehouses, factories, and storage spaces
- Agricultural for use on crops

Household Applications

Household pests that frequent dwellings cause personal annoyance and discomfort. These include ants, bedbugs, cockroaches, houseflies, mosquitos, silverfish, and so on. Color and odor are of considerable importance in household products. The product formulated must be of low toxicity to mammals. The first commercial fly spray consisted of pyrethrum extract dissolved in deodorized kerosene. Although pyrethrum fly sprays produced excellent knockdown (i.e., quick paralysis), ultimate kill was only in the range of 40 to 70 percent depending on the concentration of the active pyrethrins.

The action of the modern liquid insecticide is of two types; contact and residual. A contact toxicant is one that acts on the insect when the insect is directly sprayed with the pesticide. A residual insecticide is one that stays on the surface of the floor or wall after the base oil has evaporated. The insecticide residue remains deadly to any insect crawling on the surface.

Contact insecticides known as surface sprays contain two types of insecticide ingredients: a knockdown agent (pyrethrins, pyrethroids, etc.) and a killing agent (organophosphates, carbamates, etc.). Usually a small amount of killing agent is required to ensure the death of insects that have been knocked down. Synergists are widely used in contact insecticides. The pesticides are dissolved in a petroleum base oil, which constitutes typically 98 to 99 percent of the total weight of the formulation. Compositions of some commercial insecticides for various applications are presented in Tables 20-11 to 20-14.

TABLE 20-11 Commercial Antimosquito Sprays

	w/w %
Pyrethrins	0.05
Malathion	1.00
Perfume	0.05
Kerosene	98.90
Total	100.00

TABLE 20-12 Antimosquito Repellent

	Wt %
Prallethrin	1.60
BHT (antioxidant)	1.00
Perfume	1.00
Deodorized kero	96.40

TABLE 20-13 Permethrin Formulation Aerosol Spray

Use	Apply to battle dress uniforms, bed net, head net, inside tent
Permethrin	0.50%
Propellant	99.50%

TABLE 20-14 Roach and Ant Killer

	Vol %
Petroleum distillate	92.00
Tetramethrin	1.00
Permethrin	1.00
Piperonyl butaoxide	1.00
Carbon dioxide (propellant)	5.00

Nonagricultural Uses

Pesticidal concentrates are extensively used for nonagricultural purposes. Examples of such uses are mosquitoes or fly control and many insects: spruce bud worm, gypsy moth etc. Mosquito or fly control is accomplished by extensive area spraying of a solution of pesticide. This may be in the form of an emulsion or very fine mist. A spray machine is used for this purpose. A 5 percent concentration of a chemical insecticide, organophosphate or carbamate type, using kerosene as a solvent, is used. Airplane spraying is also used, employing 5 percent concentration of the pesticides dissolved in diesel oil, heating oil or kerosene.

Most biologically active chemicals such as insecticides, fungicides, disinfectants, and deodorizers give best results if applied in small droplets, between 10 and 100 microns in size, according to the chemical type. However, droplets in this size range are highly susceptible to drifting long distances by the wind. For this reason most outdoor spraying is done with droplet sizes of 200 to 400 microns. In closed buildings, smaller droplets can be used to maximize the efficiency of the chemical and reduce application cost.

Thermal Fogging

A thermal fogger is a device that uses heat to produce a fog without a degrading active ingredient. Thermal fogger produces a range of droplet sizes, including a large number of very small droplets. This makes the thermal fogger the preferred type of equipment to reach air spaces in areas obstructed by vegetation or buildings. The large number of very small drops also makes the fog highly visible.

Cold Fogging

In cold fogging, fog is generated by using high volumes of air at low pressure. Such a system enables droplets of more precise size to be generated. The absence of a large number of very small droplets limits the penetration of fog into highly obstructed areas. Cold sprayers can dispense the formulation in a more concentrated form because less diluent is required.

Mosquito Larvicide

Oils were the first material used for control of mosquito larvae, dating back to the 18th century. In 1892, Harward[15] recommended the use of kerosene and paraffin for mosquito control. Although oil probably kills mosquito larvae by forming a layer on the surface of water and suffocate the larvae, there are indications that the volatile constituents are responsible for faster action. They are believed to penetrate the tracheal tissue and kill the insect.

A wide variety of oils have been used as mosquito larvicides. These include kerosene, diesel, heating oil, Stoddard solvent, aromatic fractions, lubricating oils, and used lubricating oils. An important property of a larvicide oil is that it readily spreads on the surface of water. Certain emulsifying agents were found to greatly increase the spread of oil on water.

With the advent of DDT, the volumes of mosquito oils declined rapidly. Most larvicidal oils are now fortified with synthetic pesticides. From the 1940s to the 1970s, DDT was the dominant pesticide used for the control of mosquitoes. From 1970s onward, after the use of DDT was prohibited, DDT has been replaced with pyrethroids (such as delta methrin and lambda methrin) for spraying. In many malaria eradication programs, a 5 percent solution of delta methrin is employed. The dosage is typically 20 mg/m^2 of the target area, applied twice a year. In addition, in most affected areas, impregnated mosquito nets are encouraged. Nets are treated by dipping with permethrin, 0.3 g/m^2 of net area, twice a year. Other chemicals for this use are alphacypermethrin and deltamethrin.

Agricultural Uses

There are some agricultural applications where a solvent must be low in aromatics. If the pesticide is to be sprayed on the crops in relatively large doses, an aromatic solvent cannot be used because of its phytotoxicity. It has been found that an isoparaffin fraction obtained from alkylate bottom, which is much less phytotoxic, can be used. Such products have been recommended for the use on crops. Isoparaffins have a poor solubility compared to aromatics, however, and therefore can dissolve only a small percentage of an active insecticide. When a higher viscosity oil is necessary, an oil with a high UR is used. White oils are used for certain applications, such as carriers for the insecticide for the corn ear worm. A typical product would contain 0.1 percent pyrethrins dissolved in 23 cSt viscosity (100°F) USP-grade white oil.

The product should not injure the plant in recommended dosages. One of the chief disadvantages of the use of oil emulsions or oil-based aerosol formulations is the fact that if they are not properly applied, some phytotoxicity or burning of the plant may result. All aromatic petroleum products are phytotoxic if used in excessive concentrations. Two factors affect the phytotoxicity of oil: the percentage of aromatics and the boiling range of the petroleum cut. Lower boiling products tend to be less phytotoxic but of less importance in the case of agricultural formulations. Strong odors may affect the flavor of crops. Low specific gravity is desirable. In case of airplane applications, weight is important.

REFERENCES

1. V. K. Kelly, "Effect of Certain Hydrocarbon Oils on Respiration of Foliage of Dormant Twigs of the Apple," *Illinois Agricultural Experiment Station Bulletin* (1930): pp. 348–371.
2. A. H. Johnson and J. Green, "Effect of Petroleum Oils on Respiration of Bean Leaves," *Plant Physiology* 6 (1931): pp. 149–169.
3. E. R. de Ong, H. Knight, and J. C. Chamberlain, "A Preliminary Study of Petroleum Oils as Insecticides for Citrus Trees," *Hilgardia* 2 (1927): pp. 351–384.
4. R. H. Smith, "The Tank Mixture Method of Using Oil Sprays," *Agricultural Experiment Station Bulletin* 527 (1932).
5. N. A. Davison, J. E. Dibble, M. L. Flint, P. J. Marer, and A. Guy, "Managing Insects and Mites with Spray Oils," Publication 3347, University of California-Oakland (1991).
6. G. W. Pearce and P. J. Chapman, "Insecticidal Efficiency of Petroleum Fractions and Synthetic Isoparaffins," *Advances in Chemistry Series* 7 (1952): pp. 12–24.
7. Volck, U.S. Patent 1707, "Solvent system for pesticide products," 465-1707, issued April 2, 1929, p. 468.
8. W. G. Marshall, *California Department of Agriculture Special Publication* 94 (1929), p. 78.
9. J. R. Green, "Chemical and Physical Properties of Petroleum Spray Oils," *Montana Agricultural Experimental Station* 8 (June 1932).
10. M. A. Agnello, "Spray Oil Beyond 2000," *Cornell University Agricultural Experiment Station* publication.
11. C. H. Richardson and R. C. Burdette, "Relation of the Size of Oil Drops to Toxicity of Petroleum—Oil Emulsions to Aphids," *Journal of Agricultural Research* 34 (1927): pp. 727–738.
12. W. W. Robbins and W. W. Crafts, *Weed Control*, 2nd ed (New York: McGraw-Hill, 1952).
13. Esso Standard Oil Company, "Tobacco Sucker Control-Bayel N-33," 1955.
14. G. W. Ware and D. M. Whitacre, *Pesticide Book* (Willoughby, OH: Meister Media Worldwide, 2004).
15. L. O. Howard, "An Experiment Against Mosquitos," *Insect Life* 5 (1892): pp. 12–14, 109–110, 199.

CHAPTER 21
HYDROCARBON SOLVENTS

Hydrocarbons derived from crude oil have been used as solvents and thinners in the industry since the first barrel of crude was distilled through a still. Over a period of time these petroleum solvents have partially or completely displaced many nonpetroleum solvents such as the turpentine or coal tar used earlier. The principal reason for the change to petroleum solvents was their larger production and lower cost. Nevertheless, some nonpetroleum solvents have superior performance in certain select applications. The use of nonpetroleum solvents continues, although their share in overall solvent usage has declined considerably.

NONPETROLEUM SOLVENTS

Gum Turpentine

Gum turpentine is made from distilling the resinous gum from pine trees. It has a distinctive strong odor and a fast evaporation rate, releasing harmful vapors that are absorbed through healthy skin. Turpentine is an effective solvent for oil, alkyd-based paints, varnishes, and for tar, grease, and tree sap. Turpentine has been the solvent used in oil painting for centuries. It blends beautifully with oil-based paints, varnishes, and enamels and ensures a smooth, even application. Gum turpentine usually contains a small amount of sticky residue, which can be imparted to painting if this kind of turpentine is used in large quantities. It may remain in layers of paint, inhibiting proper drying and in time causing discoloration. For these reasons, natural turpentine is becoming less commonly used in painting and art industries. Rectified or artist-grade turpentine is double distilled to remove the last traces of residue from pine tree gum. This product is ideal for oil paints and media because it does the job and then evaporates from paint film without a trace. Turpentine is also an essential ingredient in the mixture to create an antique oil finish on fine furniture.

Many substitute products have arrived in the market that essentially perform the same functions with less noxious vapors. These are generally petroleum distillates that evaporate quickly. They can cause filming of color in fine art painting and an uneven coverage if used as a thinner for house painting. These are less expensive alternatives, but to achieve excellent paint consistency and avoid rapid drying of other thinners, nothing else can outperform pure turpentine, which is still considered best for color quality, durability, and easy maintenance. Properties of gum turpentine and distilled turpentine are shown in Table 21-1.

PETROLEUM-BASED SOLVENTS

Petroleum-based solvents are widely used in a large number of industries because of their ready availability and lower cost. N-hexane is used as a solvent for the extraction of edible oils from oil seeds. Petroleum distillates are used as a solvent in the dry cleaning of garments and leather. The paint, varnish, and lacquer industry is a major consumer of petroleum distillates. Petroleum distillates are used in agriculture, as a vehicle for pesticides, insecticides, herbicides, defoliants, and wood preservatives. Table 21-2 shows some of the uses of petroleum-based solvents.

TABLE 21-1 Properties of Gum Turpentine and Distilled Turpentine

	Gum turpentine	Distilled turpentine
Specific gravity	0.864	0.864
Distillation range, °F	311–345	311–345
Flash point, °F	91	91
Evaporation rate (n-Butyl acetate = 1)	0.38	0.38
Kauri-butanol value (ASTM 1133)	64	64

TABLE 21-2 Typical Applications of Petroleum Solvents

Adhesives	Alcohol denaturant
Asphalt compounding	Chemical intermediates
Dry cleaning	Leather goods degreasing
Machine cleaning	Paper making
Polishes	Printing inks
Resin solutions; paints and varnish manufacture	Rosin extraction
Rubber industry	Silicon compounds
Textile manufacture	Aluminum rolling oils
Textile printing	Vegetable oil extraction

Properties

The unusually wide range of quality requirements demanded by the industry is met by products manufactured specially for each industry, with different boiling range, density, color, solvent strength, aromatic content, sulfur and evaporation rate. With few exceptions these are straight run products from crude oil distillation with the boiling range adjusted to the required flash point and end use. These specific cuts may be subjected to further treatment such as hydrotreating, solvent extraction, and/or acid/clay treatment to improve color, odor, taste, remove aromatics and sulfur, and make the solvent suitable for various end uses.

Color

Color of industrial naphtha or distillate is a partial indicator of its degree of refining. Typically industrial naphthas have colors ranging from water white to straw color and are measured by ASTM D156 (Saybolt color). Color of an industrial naphtha is an extremely important parameter for some end users. Thus a colored naphtha is totally unacceptable for use in dry cleaning because of its negative effect on clothes if used. For identical reasons many industrial users such as in the paint, varnish, and lacquer industries cannot compromise on naphtha color.

Volatility

Volatility of naphthas is characterized by its flash point and ASTM distillation. Flash point provides a qualitative index of fire and explosion hazard in the shipping and handling of the naphtha. Also it is required for choosing the correct shipping container, labeling, classification, and transportation of the product. An appropriate distillation range of naphtha is of immense importance for a given end use. The distillation range of naphthas is indicative of the relative evaporation rate. The distillation range of naphthas is measured by ASTM D 86. Aromatics such as benzene, toluene, or xylenes are either pure single compounds or very close boiling monomers. These have a very narrow boiling range and are measured by ASTM D 850.

Evaporation Rate

The evaporation rate is another important characteristic of petroleum solvents. It is related to volatility and is of great importance in many consuming industries such as paints, varnishes, dry cleaning, and other cleaning jobs because it provides information on the time required for any specific solvent to dry completely. Many variables are involved in the determination of evaporation rates, such as ambient temperature, relative humidity, atmospheric pressure, vapor pressure of solvent, and heat of vaporization. The test method for measurement of evaporation rates is ASTM D 3539. A thin film evaporation apparatus is used to determine the evaporation rates of volatile organic solvents. The instrument measures weight loss as a function of time under specific environmental conditions. Evaporation rates are generally reported in comparison with two organic liquids; butyl acetate and ethyl ether.

Solvency

Large volumes of industrial naphthas are used primarily as solvents. Solvency of a naphtha is a function of its composition. Naphthas are derived from diverse crudes, and naphtha composition varies greatly depending on its crude source and distillation range. Aliphatic hydrocarbons have a lower solvency compared with naphthenic and aromatic hydrocarbons. Aromatics have the highest solvency. Solvency is generally measured by the aniline Point (ASTM D 611) or Kauri-butanol (ASTM D 1133) value.

Aniline Point (ASTM D-611)

The aniline point is the minimum equilibrium solution temperature for an equal volume of aniline ($C_6H_5NH_2$) and the sample. A specified amount of aniline and the sample or aniline plus heptane and the sample are placed in a test tube, and the mixture is heated at a controlled rate until the two phases become miscible. The mixture is then cooled at a controlled rate, and the temperature at which the two phases separate is recorded as the aniline point or mixed aniline point

The aniline point or mixed aniline point is useful as an aid in the characterization of pure hydrocarbons or mixtures. Aromatic hydrocarbons exhibit the lowest values and paraffins, the highest. Naphthenes and olefins exhibit values that are between aromatics and paraffins. In a homologous series, the aniline point increases with increasing molecular weight. The aniline point is often used to provide an estimate of the aromatic hydrocarbons content of the mixture.

The mixed aniline point is determined for samples having an aniline point below the temperature at which aniline crystallizes from the mixture (20.8°F). In the test, 10 mL of aniline is mixed with 5 mL of the sample and 5 mL of heptane, and the aniline point of the mixture is determined.

Kauri-Butanol Value (ASTM D 1133)

The Kauri-butanol (Kb) value of a solvent is defined as the number of milliliters of the solvent at 20°C that can be added to 20 g of a standard 20 percent stock solution of kauri resin (polar nature) in butyl alcohol without causing cloudiness. Kauri resin is readily soluble in butyl alcohol but not in hydrocarbon solvents. The resin solution will tolerate only a certain amount of dilution. Stronger solvents such as toluene can be added in greater amounts and thus have a higher Kb value than a weaker solvent like hexane. Kauri-butanol value is a measure of solvent power of hydrocarbon solvents. A high kb value indicates relatively strong solvent power.

Aromatics

The higher the aromatic content of an industrial naphtha, the greater its solvency. Therefore the PONA (paraffin, naphthenes, olefins, and aromatic content) of naphthas is of importance in determining the suitability of the naphtha or distillate for a specific application. However, aromatics are generally known human carcinogens and so their concentration is kept to the lowest feasible level. Benzene is particularly toxic and thus undesirable in any industrial naphtha. The distillation range of naphtha is so chosen to exclude benzene, for example in food-grade hexane. Aromatics can also be eliminated by acid and clay treatment or severe hydrotreating of feed naphtha.

Polycyclic Aromatic Hydrocarbons

Polycyclic aromatic hydrocarbons (PCAHs) are known human carcinogens, which are present in higher boiling petroleum fractions. Any petroleum products that are likely to be used directly in pharmaceuticals, in food, or come in indirect contact with food or in animal feed such as hexane for edible oil extraction, medicinal-grade white oils, waxes, and so on, must be free from these compounds. Measurement of PCAHs concentration (more than 0.01 ppm) is done by ultraviolet (UV) absorption. Government regulations in most countries normally lay down test methods and standards for the permissible levels of PCAHs allowed in food, pharmaceuticals, and personal care products.

Solvent Toxicity

There are no completely safe solvents. Contact with liquid solvent or inhalation of vapors they emit into the air can be hazardous. Solvents can irritate and damage the skin, eyes, and respiratory tract, cause a narcotic effect on the nervous system, and can damage internal organs such as the liver and kidneys. Solvent toxicity is measured in terms of the two factors described next.

TLV-TWA. The threshold limit value is an 8-h the time-weighted average (TWA) set by the ACGIH (American Conference of Government Industrial Hygienists). The threshold limit is expressed in parts per million. TLV-C, is similar to an OSHA permissible limit (PEL), or a workplace environmental limit (WEEL), from the American Industrial Hygiene Association, is also used.

OT. OT is odor threshold, in parts per million. OT is generally stated as a range of the amount that normal people can detect. Solvents whose odor cannot be detected until the concentration is above the threshold limit are particularly hazardous. Evaporation rates, TLV, and OT values for some petroleum and organic solvents are shown later in Table 21-16.

MAJOR APPLICATIONS OF PETROLEUM SOLVENTS

n-Hexane

Hexane finds its most important use in vegetable oil extraction processes. Normal hexane is used in the extraction of peanut, soybean oil, rapeseed, sunflower seeds, cottonseed oil, castor oil, corn oil, linseed, coca beans, sugarcane, and in the production of a very large number of vegetable oils, mainly for human consumption and also for nonedible oils. It is also used in the extraction of fats and oils from wool and meat scraps. Table 21-3 lists the properties of various hexane isomers. Hexane is manufactured from a paraffinic crude. A narrow cut, typically 145 to 158°F, is separated by fractionation from light naphtha broad cut to contain most of C6 hexane isomers. This cut may also contain benzene, a known human carcinogen, which must be reduced to 500 ppm or lower.[1] Because the relative volatility of benzene is very close to that of hexane isomers, complete separation of benzene by fractionation is generally not economical without a large number of separation

TABLE 21-3 Properties of Hexane Isomers

Component	Formula	Molecular weight	Boiling point,°F	Density
n-Hexane	C_6H_{14}	86.2	155.7	0.664
2-methyl pentane	C_6H_{14}	86.2	140.5	0.658
3-methyl pentane	C_6H_{14}	86.2	145.9	0.669
2,2-dimethyl butane	C_6H_{14}	86.2	121.5	0.654
2-3-dimethyl butane	C_6H_{14}	86.2	136.4	0.666
Benzene	C_6H_6	78.1	176.2	0.884

plates. Benzene and other polynuclear aromatics are removed from the cut by one of the following processes:

- By sulfuric acid and clay treatment. Hexane cut is treated with 5 percent v/v 95 percent sulfuric acid. Acid sludge formed is separated. Hexane is treated with approximately 2 percent by weight activated clay and next filtered.
- By extraction with a solvent for aromatic separation, such as furfural, n-methyl pyrrolidone.
- By severe hydrotreating to saturate aromatics.

The properties of food-grade hexane[2] are presented in Table 21-4.

TABLE 21-4 Food-Grade Hexane

Color	Saybolt, Min.	+30
Density	68°F, Max.	0.687
Distillation		
Initial boiling point	°F, Min.	145
Dry point	°F, Max.	158
Nonvolatile residue	ppm (w/w)	10
Aromatic	Vol %, Max.	1
Saturates	Vol %, Min.	98.5
n-Hexane content	Wt %, Min.	85
Sulfur content	ppm(w/w), Max.	75
Water-soluble acid	mg/g	0.0003
Lead	g/L	0.0005
Phosphate	ppm (w/w)	20
Chlorides	(as Cl) ppmw, Max.	20
Bromine number	Max.	1
UV absorbance per cm optical path length, Max.		
210 nm		1.00
220 nm		0.20
230 nm		0.10
240 nm		0.04
250 nm		0.02
280–400 nm		0.01

Vegetables Oil Extraction

A metering bin meters the seeds into the process system. The seeds are next conveyed to a roller mill that reduces the seeds to flakes 0.012 to 0.015 in thick. The flakes are then conveyed to the dryer and brought to a temperature of 220°F at approximately 4 percent moisture content. The kernels are conditioned at 240°F and the moisture is reduced to 3 to 4 percent before entering the expeller press.

As the next step the oilseed is cleaned and dried. Foreign materials like stones, glass, and metal are taken out by sieving and magnets and disposed of. Drying is done with hot air or combustion gases using natural gas.

Preparation of the seeds before extraction depends on the kind of seeds and the required quality of the meal. Seeds with a high oil content like rapeseed and sunflower seed are usually mechanically pressed in an expeller after a preheating step in indirectly heated conditioners. The expeller cake is next further treated in the extractor. In some cases the expeller cake containing some oil is sold as such for animal feed.

Oilseeds with a low oil content such as soybean is mechanically and thermally treated and sent directly to the solvent extraction unit. Some oil seeds such as soybean and sunflower seeds may be dehulled after the cleaning step. After dehulling, the meal will have a lower crude fiber content and hence a higher protein content. The hulls can separately be used for feeding purposes as such or in pelletized form.

The preprocessed seeds (soya flakes), the expeller cake, is extracted in an extractor in a multistage countercurrent procedure with the help of a solvent (normal hexane) until the remaining oil is reduced to the lowest possible level. The hexane-containing meal is treated in a desolventizing toaster where it is heated indirectly and with steam. The desolventizing toasting process serves three purposes; to separate the solvent from the meal, to increase the nutritional value of meal by reducing the content of glucosinolates or trypsin inhibitors, and to minimize the risk of biological contamination.

To obtain a stable and transportable animal feed material that is suitable for storage, the meal is subsequently dried and cooled. The moisture content of the meal is reduced from 18 to 20 percent to less than 1.5 percent. The meal is next sent to silos for storage. In order to prevent sticking of the oil meals to the walls of the silos, an anti-caking agent (mineral clays, bleaching clay, etc.) is added. The anti-caking agents are those limited by foodstuff legislation.

The miscella, a mixture of oil and solvent, is separated by distillation into two components; oil and solvent (hexane). The solvent is recycled to the extraction process.

The use of hexane in solvent extraction has some negative features too. Even with the most efficient desolventizer/toaster units, some hexane may remain in the meal, which can run as high as 0.5 percent. The known mild toxicity of this trace contaminant in animal feed precludes the use of solvent extracted meal for a number of end uses. Also, hexane is "too efficient" a solvent; it extracts virtually every oil-soluble fraction, including up to 5 percent of unwanted but nutritionally very valuable materials, referred to as "unsaps." This means that the meal, now contaminated with hexane, no longer contains these oil-soluble critical nutrients (phytosterols, vitamin E family tocols, phosphatidyl choline, and other natural lecithins). Additional chemicals, particularly caustic soda, and severe thermal processing is required, implying higher processing costs and degrading more unsaps to sludge.

Heptane

Heptane is also a paraffinic naphtha, and it is used in place of hexane where a higher flash solvent with lower evaporation rate is required. Properties of n-heptane are shown in Table 21-5.

TABLE 21-5 Properties of n-Heptane

Mol wt	100.2
Boiling point, °F	209.2
Specific gravity	0.6882
Aniline point, °F	157.5
Kauri-butanol value, ASTM 1133	
Evaporation rate (n-Butylacetate = 1)	
Kinematic viscosity, 100°F, cSt	0.521
Research octane	0
Vapor pressure, 100°F, kPa	11.16
Heat capacity, LIQ., 1 atm Btu/lb, °F	0.5276
Heat of vaporization, Btu/lb, at normal B.P, 1 atm	135.99

Light Naphthas

Rubber Solvent. This low boiling aliphatic solvent is used in the manufacture of rubber cements and adhesives, rubber tires, brake linings, intaglio inks, leather degreasing, paints. and lacquers. Typical properties of rubber solvent are listed in Table 21-6. The boiling range is 154 to 275°F. Aromatics are generally restricted to below 5 percent by weight.

TABLE 21-6 Rubber Solvent

Specific gravity, 60°F	0.697
Distillation, ASTM	
IBP,°F	154
Dry point, Max.,°F	275
Aromatics, Wt %, Max.	5
Aniline point, °F	140
Kauri-butanol value	33 (Min.)
	45 (Max.)
Vapor pressure, kPa, 68°F	12

Lacquer Solvents. The principal use of this product is as a diluent in the preparation of lacquer and synthetic coating, where a quick drying material is required. Its evaporation rate is comparable with that of toluene, and it is frequently used in conjunction with toluene.

Varnish Makers' and Painters' (VM & P) Naphtha. It is typically a straight run naphtha cut (235 to 285°F) from crude oil. It has a flash point of 40 to 50°F. Its principal use is in thinning spray paints and for thinning oil-based enamels, alkyd paint, and varnish. It is also employed in the manufacture of rubber cements, adhesives, waxes, and polishes. It is also referred to as "light naphtha" and as "spotting naphtha." Typical properties of this naphtha are shown in Table 21-7.

TABLE 21-7 Varnish Makers' and Painters' (VM & P) Naphtha

Specific gravity, 60°F	0.758
Aromatics, Wt %	<1
Kauri-butanol value	36
Distillation range, °F	242–300
Flash point, °F	50
Evaporation rate, n-Butyl acetate = 1	2

Petroleum Distillates

Mineral Spirits. Mineral spirits, petroleum spirits, and Stoddard solvent are all different types of hydrocarbon solvents or petroleum distillates used in the coating industry and in dry cleaning. Stoddard solvent was originally developed for the dry cleaning trade. Mineral spirits find application in many other industries such as metal cleaning and degreasing.

Petroleum distillates have been the most widely used solvents in dry cleaning. At the beginning of the 20th century, gasoline was the solvent of choice; however because of the fires and explosion associated with the use of gasoline, in 1928 a new solvent called Stoddard solvent with a minimum flash point of 100°F was introduced in the United States. Stoddard solvent is a kerosene cut with a boiling range of 300 to 415 °F. Stoddard solvent remains popular to the present day. However, the flash point of the cut was raised to 140°F (Table 21-8) for enhanced safety. Also, the aromatic content was lowered to a low level (less than 4 percent) to make it less toxic.

TABLE 21-8 Stoddard Solvent

Color	+25 (Saybolt, Min.) 25 (Pt-Co)
Specific gravity	0.787
Aromatics, Wt %	<4
Doctor test	Negative
Kauri-butanol value,	30
Distillation range, °F	370–410
Flash-point, °F	140
Evaporation rate (n-Butyl acetate = 1)	0.08

Petroleum solvents in dry cleaning face severe competition from perchloroethylene (PCE), which was introduced as a dry cleaning solvent in the 1950s. The majority of the commercial dry cleaners prefer PCE because of its nonflammability and superior cleaning ability. However, compared with petroleum solvents, PCE is expensive, and dry cleaning plants must use a solvent recovery system. Solvent recovery is not practiced in petroleum solvent based dry cleaning units, due to the low cost of petroleum solvents. Also the use of chlorinated hydrocarbons in the dry cleaning industry is being viewed with increasing concern about the possible health risks to humans and the damage to the environment due to their lower biodegradability.

Dry Cleaning Process

The dry cleaning industry consists of coin-operated machines (4- to 12-kg capacity), commercial plants (14- to 27-kg capacity), and industrial units (200- to 230-kg capacity). Coin-operated units typically use synthetic solvents. Commercial units can use either synthetic or petroleum solvents. About 50 percent of industrial-scale cleaners use petroleum solvent; another 50 percent use PCE (perchloroethylene). Other popular nonpetroleum dry cleaning solvents in use are 1,1,1 trichloroethane. Use of trichlorofluoroethane (CFC-113), once a popular solvent, is prohibited because of ozone layer depletion concerns.

For synthetic solvent cleaners, the garments are first washed in "charged" solvent, which is used solvent with a small portion of detergent for better cleaning. Next the garment is rinsed with pure solvent and sent to a dryer. The solvent is filtered, and a part of this solvent is returned to the charged solvent tank. The remaining solvent is distilled to remove oils and greases, and the distilled solvent is sent to the pure solvent tank. The solids removed from the filter are heated to remove the remaining solvent. The vapors are sent to a refrigerated condenser and separator to recover the solvent. Vapors from the dryer are also sent to a refrigerated condenser. Petroleum plants are almost identical to those using synthetic solvents except that small plants do not recover solvent vapors from various operations.

Mineral Turpentine Oil

The trade name MTO, or mineral turpentine oil, refers to a colorless petroleum solvent with a boiling range of 285 to 400°F or 260 to 400°F. It is mainly used as a solvent for textile printing, dry cleaning, metal degreasing, and insecticidal formulations.

Jute Batching Oil

Jute is a natural fiber produced mainly in India and Bangladesh and is used for the packaging industry for food grains and other agricultural produce. Jute batching oil (JBO) is used to make jute fiber pliable. It also finds a limited application as wash oil for coke oven gases in the steel industry. It is typically a straight run middle distillate (465 to 700°F) cut from paraffinic crude. Polycyclic hydrocarbon content of JBO used for food packaging and thus indirect contact with food is regulated by a UV absorption test and the pyrene content of JBO (Table 21-9).

TABLE 21-9 Jute Batching Oil

Color, ASTM, Max.	L 7.0
Flash point, °F, Min., cc	212
Distillation,	
Density, 60°F, g/mL	0.85
Kinematic viscosity, cSt, 100°F, Max.	15
Initial boiling point, °F, Min.	464
FBP, °F, Max.	700
Residue, Vol %, Max.	2
UV Absorbance Test	
Pyrene content ppm, Max.	25
Absorbance per cm optical path length in the range of UV wavelength, Max.	
280–299 nm	2.3
300–319 nm	1.2
320–359 nm	0.8
360–400 nm	0.3

Aromatic Solvents

Benzene. Benzene is the lowest boiling material in the aromatic series of naphthas. Benzene is a known human carcinogen, and therefore its use is restricted as a solvent. Its principal use is now as a raw material for the manufacture of a large number of organic chemicals such as phenol, Caprolactam, benzene hexachloride (insecticide), styrene, monochlorobenzene, and various intermediates.

Toluene. When an industrial naphtha with high solvency is required, toluene is the preferred choice. It is the solvent for heavy synthetic resins and a thinner/diluent for the coatings of synthetic resins. It is one of the major components in the preparation of lacquers. Toluene is particularly recommended for situations where fast drying is required. Nitration-grade toluene is used for the manufacture of tri-nitro-toluene (TNT) explosives, saccharin, benzaldehyde, benzoic acid, and many dyes and intermediates.

Xylene. Solvent-grade xylene is in fact a mixture of ortho, para, and meta xylene isomers, and ,like toluene, has high solvency power. It is used in many types of synthetic coatings. It has a slower evaporation rate than that of toluene. Intermediate drying rates can be met by using a blend of xylene and toluene. The properties of benzene, toluene, and xylenes are listed in Table 21-10.

TABLE 21-10 Properties of Benzene, Toluene, and Xylenes

	Molecular weight	Boiling point at 1 ATM, °F	Vapor pressure at 100°F, PS1	Specific gravity, 60°F/60°F	Kinematic viscosity, 100°F	Aniline point, °F	Heat capacity liq, 1 ATM	Heat of vaporization at normal B.P, I ATM, Btu/lb
Benzene	78.108	176.18	3.224	0.8845	0.587	<22	0.4098	169.34
Toluene (methyl benzene)	92.134	231.12	1.032	0.8719	0.5584		0.4017	156.2
Alkyl benzenes								
Ethyl benzene	106.16	277.13	0.371	0.8717	0.6428	< –22	0.4118	145.7
1,2 Dimethyl benzene (O-xylene)	106.16	291.94	0.264	0.8848	0.74	< –4	0.4418	149.1
1,3 Dimethyl benzene (M-xylene)	106.16	282.39	0.326	0.8687	0.591	–22	0.4045	147.4
1,4 dimethyl benzene (P-xylene)	106.16	281.03	0.342	0.8657	0.613	–22	0.4083	146.1

White Petroleum Distillates

Petroleum solvents are highly refined distillates with high flash points. These grades are manufactured from selected crude oils by close fractionation of middle distillates (kerosene/diesel) cut to adjust the boiling range. Aromatics and unsaturates are next removed by treatment with oleum or by severe hydrotreating. The processes used for the manufacture of white petroleum distillates are similar to those used for the manufacture of white oils. The refined petroleum distillates so produced resemble very light white oils. These are colorless, tasteless, and nearly odorless. Chemically and biologically these are inert and stable and do not support pathogenic bacterial growth.

These petroleum distillates typically have a viscosity at 100°F in the range of 1.5 to 3 cSt. The initial boiling point is high, ensuring a high flash point. Boiling range is generally narrow; C_{10} to C_{15} range hydrocarbons or 380 to 450°F. Sulfur and other odorous compounds present in the feed are almost completely removed and aromatics are reduced by processing, to below 1 wt %. Properties such as solvency (aniline point and Kauri-butanol value) and pour point are a function of the feed source.

White petroleum distillates are direct substitutes for high volatile organic compound (VOC) solvents in consumer and institutional products. They lower the VOCs in such aerosol systems as air fresheners, furniture polishes, and insecticides. Vapor pressure of these solvents is generally less than 0.1 mm Hg at 68°F and are not regulated as VOCs in consumer products. Typical properties of white petroleum solvents are listed in Table 21-11.

TABLE 21-11 White Petroleum Distillates

		1	2	3	4
Specific gravity		0.786	0.795	0.800	0.812
Kinematic viscosity	100°F, cSt	1.66	2.18	2.28	4.30
Colour, saybolt	Min.	+30	+30	+30	+30
Flash Point	COC, °F	165	210	215	260
Distillation, ASTM					
IBP	°F	375	430	450	510
FBP		460	500	545	595
Aniline point	°F	170	175	181	192
Pour point	°F	–40	–15	–0.4	25
Aromatics	Wt %	<1	<1	<1	<1

Nonaerosol use of these white distillates include the following uses. Highly refined, light, odorless hydrocarbon solvents are used as a defoamer in processing beet sugar and yeast (more than 1 ppm), in modified hop extract of beer, as a float on fermentation fluids in the manufacture of vinegar and wine, in froth flotation cleaning of vegetables, as a coating on eggshells, in pest control formulations, and as insecticide bases. For use in these applications, the hydrocarbon solvent must comply with the various regulations of food and drugs administrations of various countries. White petroleum solvents are also used in fragrance bases, nail polish removers, waterless hand cleaners, as lamp oils and liquid candles where odorless and clean burning fuel is desired, charcoal lighter fluids, automotive and furniture polishes, floor polishes, industrial cleaning solvents, dust control agents, and the manufacture of paper and board.

Aluminum Rolling Oils

Cold rolling of aluminum produces products such as household foil, packaging foil, beverage cans, and tread plates. Aluminum rolling oils are used as coolants that cool, lubricate, and clean aluminum in sheet and foil rolling in cold rolling operations. Wide coils of foil are usually annealed in an atmospheric furnace before shipping. During annealing, the rolling lubricant on the coil is vaporized or burned, but some oil may remain on the aluminum foil/sheet after rolling that may come in indirect contact with food if the foil or sheet is used for packing food and beverages. For this reason, aluminum rolling oils are expected to comply with food and drug administrations regulations of various countries. These relate mainly to color, odor, and aromatic specifications of the oil.

Aluminum rolling oils are manufactured from paraffinic or naphthenic crude. A narrow diesel cut is chosen, with a boiling range of 430 to 480°F for light duty and 535 to 590°C for heavy-duty operations. Depending on the feedstock quality and product specifications, the distilled narrow cut is next severely hydrotreated to remove sulfur completely and reduce aromatics to below 1 wt %. Important characteristics of aluminum rolling oils are as follows:

- Color and odor: The product must be colorless and odorless
- High initial boiling point (IBP) and narrow boiling range: to reduce evaporation rate and thus rolling oil consumption
- Nonstaining and noncorrosive: Removal of sulfur by the hydrotreating process ensures these characteristics. Ultra-low sulfur imparts excellent nonstaining properties to the product.
- Low aromatics: For manufacture of aluminum foil for food packaging or any other use where foil is to come in contact with food, aromatics in the oil must be reduced to below 1 percent and polynuclear aromatics in oil must pass a UV absorbance test for compliance to national standards (such as FDA 178.3620, part B: indirect food contact).
- Low viscosity: These are generally low-viscosity oils (1.5 to 4.0 cSt at 100°F).

The typical properties of aluminum rolling oils are shown in Table 21-12.

TABLE 21-12 Properties of Aluminum Rolling Oils

	1	2	3
Color, Saybolt	+30	+30	+30
Specific gravity, 60°F	0.761	0.797	0.824
Distillation range, °F	420–490	450–520	500–560
Sulfur, ppm (w/w)	<1	<5	<1
Kinematic viscosity, cSt, 104°F	1.7	2.4	3.3
Aniline point, °F	190	183	180
Flash point, °F	194	220	250
Pour point, °F	21.2	–11	–49
Aromatics, Wt %	0.1	1	0.5

Printing Ink Oils

Printing inks contain four basic components: pigments, solvent, resinous binder, and performance additives. The pigments used in printing inks include both inorganic pigments such as carbon black, titanium dioxide, and organic pigments that are generally dyes rendered insoluble by complexing with a metal ion. Most organic pigments are prepared from azo, anthraquinone, triarylmethane, and phthalocyanines dyes.

There are five main printing processes and the ink for every process has to be specifically designed. Lithography is used in magazines, books, labels, newspapers, and so on. Popularity of this method comes from its low cost and high-quality reproduction with a minimum environmental impact. Lithography and letterpress are collectively known as the paste ink processes. These processes use inks that are essentially nonvolatile at room temperature.

Flexography is used in corrugated packaging, magazines, flexible films, envelopes, and newspapers. Flexography and gravure are known as liquid ink processes that are based on volatile solvents that evaporate readily at room temperatures. Screen printing uses ink that fall in between these two groups.

The choice of vehicle (solvent plus resins) for a printing ink depends on the printing process, how the ink will be dried, and the substrate on which the image is to be printed. In lithography and letterpress, where inks are dried by absorption and oxidation, vehicles are generally a mixture of mineral oils or vegetable oils and resins. Flexographic inks, which are designed to dry quickly by evaporation, can be either water based or based on an organic solvent such as ethanol, ethyl acetate, n-propanol, or isopropanol with a wide variety of resins. Vehicles for gravure inks, which also dry by evaporation, may also contain aromatic or aliphatic hydrocarbons and ketones as solvents. Inks for screen printing use organic solvents that are less volatile than those used for flexography or gravure (higher glycol ethers and aromatic/aliphatic hydrocarbons). Additives in inks include dryers, wax, and plasticizers. UV radiation cured inks commonly based on acrylates are used in all of printing processes to varying degrees.

The manufacture of ink consists of dissolving or dispersing pigment or dyes in organic solvents or oils to produce the vehicle (varnish), mixing and dispersing the pigment or dye into the vehicle, introduction of any additive, and packaging. Table 21-13 shows the properties of printing ink oils.

TABLE 21-13 Properties of Printing Ink Solvents

	Hydrotreated light distillate	Technical white oil
Specific gravity	0.7952	0.8146
Boiling range, °F	440–465	525–570
Avg. Mol wt	199	230
Color, Saybolt	+30	+30
Flash point, °F	205	260
Aromatics, Wt %	0.8	1
Pour point	–20	+20
Cloud point	–10	+27
Heat of vaporization, Btu/lb	136	130
Specific heat	0.485	0.492
Viscosity cSt, 100°F	2.01	3.9
Vapor pressure, mm Hg, 70°F	0.05	0.01
Kauri-butanol number	26.5	22
Aniline point,°F	175	191

Halogenated Hydrocarbon Solvents

Halogenated hydrocarbon solvents are usually not flammable and have no flash point. However, some can react explosively on contact with certain metals, and heating or burning them can create highly toxic decomposition products including phosgene gas. Hazardous amount of these toxic gases can be created by working with chlorinated solvents in an enclosed environment where a pilot light is burning. Clearly, all solvents should be isolated from sources of heat, sparks, flame, and static electricity.

Carbon Tetrachloride

Carbon tetrachloride was among the first chlorinated solvent used in dry cleaning. It was commonly blended with other solvents for use as a dry cleaning solvent. Because of its high toxicity and ozone-depleting properties, use of carbon tetrachloride was phased out under the Montreal Protocol in 1996.

Trichloroethylene

Trichloroethylene was introduced as a dry cleaning solvent in 1958. It caused the bleeding of acetate dyes at temperature exceeding 75°C. Trichloroethylene was never used as a primary dry cleaning solvent. Trichloroethylene is still used as a dry side precleaning and spotting agent.

Perchloroethylene

Perchloroethylene is a member of a family of aliphatic halogenated hydrocarbons. It is a colorless volatile liquid that is essentially nonflammable and has no measurable flash point.

PCE is also known as tetrachloroethylene, or "perc," and it has been used in the industry for over 50 years. It is an effective nonflammable solvent that does not contribute to the formation of smog (ground-level ozone) or to the depletion of the stratospheric ozone. PCE is the primary solvent used in commercial and industrial dry cleaning. Since being introduced in the industry in the late 1930s, it has replaced most other solvents because of its relatively low toxicity and nonflammability. Its other major uses are as a metal cleaning and degreasing solvent and as a solvent in automotive aerosols. It is also a chemical intermediate for the production of several fluorinated compounds.

Properties of perchloroethylene are shown in Table 21-14. PCE is sold under various trade names, such as DowPer, PerSec, and Perklone.

TABLE 21-14 Perchloroethylene Properties

Synonyms	Ethylene tetrachloride, 1,1,2,2 Tetra-chloroethylene
Formula	C_2CL_4
Molecular weight	165.8
Specific gravity	1.619
Boiling point, °F	250
Boiling point, solvent-water, °F	190
Specific heat, 25°C, cal/g	0.209
Viscosity, 25°C, cP	0.84
Heat of vaporization at B.P cal/g	50.1
Vapor density (air = 1)	5.76
Flash point	None
Water solubility, g/100 g, 77°F	
H_2O in solvent	0.0105
Solvent in H_2O	0.015
Refractive index, 25°C	1.503
Surface tension Dynes/cm, 25°C	31.8
Kauri-butanol value	90

It is very stable solvent and not normally corrosive. However, in the presence of heat and moisture and other chlorinated hydrocarbons, acids can form and cause corrosion problems. The presence of other chlorinated compounds such as trichloroethane, which is widely used as a spotting agent, can also contribute to acid formation in PCE.

The popularity of PCE in the dry cleaning industry can be attributed to the following properties:

- Virtual nonflammability, which led to its wide use in decentralized dry cleaning plants.
- Its high solvency (Kb-value of 90), which dissolves out most oil, greases, and fats.
- Low viscosity (0.84 cSt at 60°F) and low surface tension (32 dynes/cm) that allow it to penetrate rapidly in fibers to dissolve soils.
- The high density of perc not only facilitates mechanical action in the wash cycle, but it also helps in the separation of solvent from water in the recovery cycle.
- Perc's high evaporation rate (1.5 compared to butyl alcohol) provides for reasonable drying times at moderate drying temperature.
- Safe to use on all common textiles, fibers, and dyes.

The textile industry uses PCE as a spotting agent for removal of spinning oils and lubricants. It is also used in wool scouring and as a solvent carrier in dyes and water repellents.

TABLE 21-15 Halogenated Hydrocarbon Solvents

Solvent	Formula	Molecular weight	Specific gravity	Vapor pressure, mm Hg, 68°F	Freezing point, °F	Boiling point, 1 ATM, °F	Water solubility, mg/L 77 °F	Kauri-butanol number	Evaporation rate, Bu O Ac = 1	Viscosity, 77°F cP
Carbon tetrachloride	CCL_4	153.82	1.594	90	–9.4	169.8	800			
Trichloroethylene	$C_2H\ CL_3$	131.39	1.5		–121	188.6		130	6.4	0.53
1,1,1 Trichloro-ethane	$C_2H_3CL_3$	133.41	1.3376		–22.72	165.2	0.13			
Perchloroethylene	C_2CL_4	166	1.623			250	0.4			

TABLE 21-16 Solvents and Their Toxicity

Solvent	TLV-TWA, ppm	Odor threshold, ppm	Evaporation rate butyl acetate = 1
Ethyl alcohol	1000	49–716	
Methyl alcohol	200	4–6000	
Propyl alcohol	200	0.03–41	
Isoamyl alcohol	100	0.03–0.07	
n-Butyl alcohol	50	0.1–11	
Kerosene	None		
n Heptane	400	230	
VN & P naphtha	300	1–40	2
Stoddard solvent	100	1–30	0.02
n-Hexane	50	65–250	
Gasoline	300	0.3	
Ethanol amine	3	2–4	
Diethanol amine	0.46	0.27	
Ethyl benzene	100	0.1–0.6	
Xylene	100	20	0.8
Toluene	50	0.2–37	1.9
Styrene	50	0.017–2	
1,1,1 Trichloroethane	350	390	
Methylene chloride	50	160	
Trichloroethylene	50	82	6.4
Perchloroethylene	25	47	
Ethylene dichloride	10	6–185	
Carbon tetrachloride	5	140–584	
Ethyl acetate	400	6.4–50	
Methyl acetate	200	180	
Isoamyl acetate	100	0.22	
Ethylene glycol	39.4 C	0.1–40	
Diethylene glycol	50 (wheel)	UNK	
Butyl cellosolve	25	0.1	
Cellosolve	5	2.7	
Acetone	750	3.6–630	
Methyl isobutyl ketone	50	0.013	
Methyl ethyl ketone	200	2–85	
Turpentine	100	50–200	0.38
Cyclohexane	300	780	

Chemical Intermediate

PCE is used as a basic raw material in the manufacture of hydrofluorocarbon (HFC) 134, a popular alternative to chlorofluorocarbon (HCFC) 123, 124, and 125.

Automotive Aerosols

Perchloroethane has replaced 1,1,1 trichloroethane in aerosol formulations for the automotive aftermarket, particularly for brake cleaning. These formulations provide auto repair shops with highly effective, nonflammable products.

Metal Cleaning/Degreasing

Many industries, including aerospace, appliances, and automotive manufacturers, use PCE for vapor degreasing of metal parts during various production stages. Its high boiling point and resultant longer cleaning cycle are advantageous in removing difficult soils, such as waxes with high melting points. The ability of this chemical to remove water during vapor degreasing is useful to jewelry manufacturers and other metal finishers.

Miscellaneous Uses

PCE is used as an insulating fluid in some electrical transformers as a substitute for polychlorinated biphenyls (PCBs). Relatively small quantities of PCE are used in printing inks, aerosol specialty products, adhesive formulations, paper coatings, and silicones. In addition, PCE is a component of chemical maskant formulations used to protect surfaces from chemical etchants used in aerospace and other industries.

1,1,1 Trichloroethane

1,1,1 trichloroethane (TCA) was introduced as a dry cleaning solvent in the 1980s. It is particularly used in leather cleaning operations. TCA is not a very stable solvent. It is used as a precleaning and spotting agent in dry cleaning. TCA is also used for degreasing motor stator coils and other machine cleaning operations using a vapor degreaser. TCA has been classified as an ozone-depleting chemical, and its use is being discouraged in the industry.

Other Dry Cleaning Solvents

1,1,2 Trichloro-1,2,2 trifluoroethane was introduced by DuPont in 1960s under the trade name Valclene. This chlorofluorocarbon (Freon 113) is used as a dry cleaning solvent. Because the vapor pressure of Valclene is approximately 20 times that of PCE, clothes cleaned in Valclene can be dried at a lower temperature, and it was promoted as a solvent of choice for the dry cleaning of delicate fabrics. Freon 113 is one of the fluorochlorocarbons subject to the Montreal Protocol, and its use has been prohibited in most countries.

Dipropylene glycol tertiary butyl (DPTB) ether has also been used as an alternative to PCE in dry cleaning. The properties of halogenated hydrocarbon solvents are shown in Table 21-15. Solvents and their toxicities are listed in Table 21-16.

REFERENCES

1. Joint FAO/WHO Expert Committee on Food and Additives, MMR Minor. Met. Review 1998, 48B.
2. Hexane specs; Council Directive 88/384/EEC on the extraction solvent used in the production of foodstuffs.

CHAPTER 22
REFRIGERATION GASES

Refrigerant gases are used in domestic refrigerators, air conditioners, industrial and commercial cooling machines for heat extraction from an enclosed space. In vapor compression refrigeration system, refrigerant enter the compressor as vapor and after adiabatic compression is heated to high temperature. The superheated vapor is cooled in a condenser which removes superheat and condenses the refrigerant to a liquid. The liquid refrigerant flows through as expansion valve where its pressure abruptly decreases through flash evaporation of approximately half the liquid. This results in a mixture of liquid and vapor at a low temperature and pressure. This cold liquid-vapor mixture then travels through the evaporator coil and is completely vaporized by cooling the warm air in the space being refrigerated. The refrigerant vapor next returns to compressor and the cycle is completed.

The operating system temperature and pressure must be well below the critical temperature and pressure of the refrigerant used. Freezing point of refrigerant gas used for a system must be well below the minimum operating temperature of refrigeration cycle.

All refrigerant gases eventually leak into atmosphere. Because of the very large usage of these gases worldwide, the impact on the environment is a matter of concern.

FREONS

Refrigerators from 1900s used toxic gases like ammonia, methyl chloride (CH_3Cl), and sulfur dioxide (SO_2) as refrigerants. Fatal accidents were quite common as a result of the leakage of methyl chloride and other gases from refrigerators. In 1928, Thomas Midgley and Charles Franklin, looking for a safer refrigerant for use in refrigerators, figured out that a group of chlorofluorocarbons (CFCs) were ideally suited for the job. These were named "Freons." The CFCs are a group of aliphatic organic compounds containing carbon and fluorine and sometimes chlorine and hydrogen. The word *Freon* is a registered trademark belonging to DuPont company, United States of America. Freons are colorless, odorless, nonflammable, noncorrosive gases or liquids. Because Freons are nontoxic and nonflammable, they very soon became the refrigerant of choice all over the world replacing others. Freon 12, also known as "R-12" (dichlorofluoromethane; chemical formula CCl_2F_2) was the most popular refrigerant before it was banned in many countries.

R-134 or R-134a is the most common replacement for R-12. Unlike R-12, which is a CFC, R-134 is a hydrofluorocarbon. The chemical formula is CH_2FCF_3, and the chemical name is 1,1,1 2 tetrafluoroethane. Properties of some of these gases are shown in Table 22-1.

REFRIGERANTS' NAME AND NUMBERING CONVENTION

CFC

The name chlorofluorocarbon (CFC) is given to a substance formed by replacing all the hydrogen atoms in a hydrocarbon molecule with chlorine and fluorine and sometimes with bromine. Some serious negative properties of some CFCs were discovered in the 1980s.

TABLE 22-1 Properties of Refrigerant Gases

Refrigerant number	Type	Name	Formula	Molecular weight	Boiling point, °F	Density Kg/m^3, 77°F	SG air = 1	CP/CV	Critical temperature, °F	Critical pressure kPa
11	CFC	Freon 11	CCL_3F	137.37	74.48		4.78	1.136		
12	CFC	Freon 12	CCL_2F_2	120.93	–21.62	1.311	4.2	1.137	233.6	4116
13	CFC	Freon 13	$CCLF_3$	104.46	–114.52		3.744	1.172		
22	HCFC	Freon 22	$CHCLF_2$	86.47	–41.08		2.978	1.184		
23	HFC	Genetron	CHF_3	70.014	–119.90		2.383			
113	CFC	Freon 113	$C_2CL_3F_3$	187.39	118.40		6.04	1.08		
114	CFC	Freon 114	$C_2CL_2F_4$	170.926			5.698	1.088		
116	PFC	Freon 116	C_2F_6	138.01			4.681			
134A	HFC	SUVA 134A	CH_2FCF_3	102.03	–15.70	1.210			213.9	4060
290	HC	Propane	C_3H_8	44.096	–43.68		1.5503	1.131		
600	HC	N-Butane	C_4H_{10}	58.124	31.10		2.076	1.096		
600A	HC	Isobutane	C_4H_{10}	58.124	10.90		2.01	1.097		
		Dimethyl ether	$(CH_3)_2O$	46.07	–10.66		1.5602	1.11		
717	NIK	Ammonia	NH_3	17.031	–28.12		0.597	1.269		
744	NIK	Carbon dioxide	CO_2	44.01	–109.30		1.521	1.31		
764	NIK	Sulfur dioxide	SO_2	64.063	14.00		2.264	1.29		
744	NIK	Nitrous oxide	N_2O	44.013	–131.26		1.53	1.415		

Chlorofluorocarbons are a class of manufactured chemicals known by such trade names as Freon, Genetron, and Isotron. CFCs have been used in a wide variety of manufacturing industries and products, such as a solvent in the electronics industry, a foaming or blowing agent, aerosol propellant, fire extinguisher agent, dry cleaning solvent, a key component in making rigid foam insulation for houses and household appliances, foam packaging insulation material (known by the trade name Styrofoam), auto and household air conditioners, refrigerators, and so on.

CFCs are highly volatile compounds contributing to air pollution. CFCs are highly non-biodegradable and do not break down when released into the atmosphere. They rise slowly through the atmosphere, taking 6 to 8 years to rise to the stratosphere. Here CFCs can reside more than 100 years. CFCs are implicated in two major threats to the global environment; the greenhouse effect and the reduction of the earth's ozone layer. CFCs contributes to the greenhouse effect, warming the atmosphere by trapping heat that is radiated back to the earth's atmosphere. CFCs are known to be 10,000 times as effective at trapping the radiated heat as carbon dioxide.

In 1985, scientific evidence about damage to the ozone layer prompted governments to adopt an international legal framework (Montreal Convention) for action on phasing out production and consumption and banning the export and import of all ozone-depleting substances (ODS). Production and consumption of CFCs was phased out in developed countries by 2005. Production of some other halogenated compounds is being phased out for the same reasons. Halogenated compounds banned under the Montreal Convention are as follows:

- CFCs
- Halons (compounds containing bromine, fluorine and carbon; halons are used in fire extinguishers)
- Other fully halogenated CFCs
- Carbon tetrachloride
- Methyl chloroform
- Hydrochlorofluorocarbons (HCFCs)
- Methyl bromide (used as a fumigant)

Hydrochlorofluorocarbons

Hydrochlorofluorocarbons (HCFCs) are similar to CFCs but they are only partially halogenated and therefore retain some hydrogen. HCFCs have also been listed as ozone-depleting chemicals and their production is being phased out.

Hydrofluorocarbons

Hydrofluorocarbons are formed by partially fluorinating hydrocarbons. They retain some hydrogen and do not contain any chlorine. Suva 134 or 134a (CH_2FCF_3) is a popular replacement for Freon 12 for use as a refrigerant gas.

Perfluorocarbons

Perfluorocarbons (PFCs) are totally fluorinated hydrocarbons. They are very stable, have very good fire-suppressing properties but are not very biodegradable with a long life.

Hydrocarbons

Hydrocarbons (HCs) are naturally occurring organic substances. Generally they are stable and unreactive. Their negative point is their flammability and their ability to react with halogens.

Chemical names for the halogenated refrigerants are long and cumbersome to remember and therefore a numbering system was developed to identify the different refrigerants. The numbering system originally devised by DuPont (United States of America), has been adopted by the American Society of Heating, Refrigeration and Air Conditioning Engineers (ASHRAE) under its standard 34-1989. The refrigeration number is simply a way of pointing out their chemical structure and therefore their relative ozone depletion potential.

The number assigned to each refrigerant is related to its chemical composition as follows. The number consists of four digits (leading zeros are dropped), with each digit describing a character of the molecule.

1. The first digit is the number of carbon to carbon double bonds (in most cases zero).
2. The second digit is one less than the number of carbon (C) atoms.
3. The third digit is one more than the number of hydrogen (H) atoms.
4. The fourth digit is the number of fluorine (F) atoms.

Any spare atoms are assumed to be chlorine unless otherwise stated. When chlorine is replaced with bromine, the letter "B" is used in the number as in the case of R 13B1.

Example 1. Consider Freon 12 (one time popular but now a banned refrigerant). Its chemical name is dichlorfluoromethane and its chemical formula is $C.F_2.\ Cl_2$. This refrigerant is also known as CFC-12 or R-12, as follows:

First digit	= Number of double bonds	= 0
Second digit	= Number of carbon atoms –1	= 0
Third digit	= Number of hydrogen atoms +1	= 1
Fourth digit	= Number of fluorine atoms	= 2

The digits are numbered from right to left. The refrigerant number becomes 0012 or simply 12, ignoring leading zeros.

Example 2. Consider refrigerant gas R-22. Its chemical formula is $C.H.Cl.F_2$. This refrigerant number is computed, as follows:

First digit	= Number of double bonds	= 0
Second digit	= Number of carbon atoms –1	= 0
Third digit	= Number of hydrogen atoms +1	= 2
Fourth digit	= Number of fluorine atoms	= 2

Thus the refrigerant number is 0022 or 22.

Example 3. Consider the popular dry cleaning solvent perchloroethylene. Its chemical formula is $C.Cl_2{:}CCl_2$. This refrigerant number is computed, as follows:

First digit	= Number of double bonds	= 1
Second digit	= Number of carbon atoms – 1	= 1
Third digit	= Number of hydrogen atoms + 1	= 1
Fourth digit	= Number of fluorine atoms	= 0

Thus the refrigerant number is 1110.

Ordinary (nonazeotropic) mixtures are assigned numbers in the 400 series in order of their commercial introduction. Azeotropic mixtures are assigned numbers in the 500 series, in order of their commercial introduction.

Miscellaneous organic refrigerants are assigned arbitrary numbers in the 600 series. This includes hydrocarbon refrigerants that cannot be identified by the regular numbering system because they contain nine or more hydrogen atoms (because 9+1 cannot be represented by a single digit number).

Inorganic refrigerants are allocated to the 700 series, using the molecular weight prefixed by the number 7. Lowercase suffices are added to denote decreasing symmetry in isomers or to denote inorganic gases with the same molecular weight. To provide a similar three letters acronym, some natural refrigerants are sometimes referred to as "not in kind" or NIK refrigerants. The properties of some refrigerant gases are listed in Table 22-1.

AEROSOLS

The aerosol container has been termed as the finest packaging for its effectiveness, minimal waste, and convenience. Aerosol here is defined as a self-contained sprayable product in which the propulsion is provided by liquefied or compressed gas so that the product is self-dispensing. In 2004, it was estimated that about 11 billion aerosol units (excluding medical metered-dose inhalers) were sold worldwide. These are dispensed under pressure to create a spray or foam. They have made dispensing a multitude of products both easy and effective. The aerosol container is a sophisticated high-technology package in which a deodorant, insecticide, paint, or polish is dispensed in a safe and convenient manner. Another reason for using aerosol packaging is its ability for remote application and its penetrating ability.

The use of aerosol originated during World War II when U.S. troops used aerosol bombs to protect the troops from malaria in the tropical jungles of Southeast Asia. These bombs contained simply a solution of insecticide (mainly DDT) together with the high-pressure liquefied gas refrigerant Freon 12. Because of the pressure, which was over 70 lb/in^2, the liquid was packed in heavy steel containers. Soon it was found out that excellent results could be obtained using lower pressure propellant and the product could be packed in an inexpensive beer can type container.

Today, aerosol packaging is used in a very large number of consumer products such as the following:

- Household cleaners
- Disinfectants, deodorants
- Insect repellents
- Household insecticides
- Glass cleaners
- Paints and polishes
- Carpet shampoos
- Spray paints
- Industrial sealants and lubricants
- Personal care products such as shaving creams, hair sprays, deodorants, perfumes
- Medical inhalers

The aerosol package is a self-contained dispensing system with three elements:

- Product
- Propellant
- Container

PRODUCT

The product is in the form of a liquid, emulsion, or suspension. The product may be a blend of active ingredients like paint, cream, perfume, soap, disinfectant, and inactive ingredients, such as water.

PROPELLANT

Usually the propellant is the vapor of a liquid with a boiling point slightly lower than the ambient temperature. Thus inside the pressurized can, the vapor can exist in equilibrium with bulk liquid at a pressure higher than atmospheric pressure. Thus propellant can expel the contents without endangering the container, usually made of thin aluminum or tin-coated steel sheet. Because the aerosol container is under pressure, propellant exists mainly as a liquid but it will also be a gas in the head space. As the product is used up as the valve is opened, some of the liquid propellant evaporates and keeps the head space full of gas. Pressure in the container remains constant, and thus the spray performance remains constant throughout the life of the aerosol. An important property of aerosol propellant is its dissolving power; that is, its capacity to form a homogeneous liquid phase with other components of the aerosol formulation and the active ingredient. The most common aerosol propellants are low molecular weight hydrocarbons, dimethyl ether, carbon dioxide, nitrous oxide, nitrogen and refrigeration gases such as HFC-type hydrocarbons. Table 22-2 lists the properties of some popular propellants.

Liquified Petroleum Gas

Aerosol propellant-grade liquefied petroleum gas (LPG) consists of high-purity hydrocarbons such as propane, isobutene, n-butane, or a mixture of these with low boiling liquid hydrocarbons such as pentanes and isopentanes. These hydrocarbons have a low cost and low molecular weight but suffer from two serious drawbacks with regard to their use in aerosol compositions. The first drawback is their high flammability and the second is their low solvent power. LPG is used in household and industrial aerosols used for cleaning, insecticides, room fresheners, and some industrial products. The presence of water in the aerosol reduces the risk due to flammability of the hydrocarbons. Hydrocarbon propellants are preferred in formulations containing active ingredients in a state of emulsion or suspension.

TABLE 22-2 Properties of Common Aerosol Propellants

Compound	Formula	Molecular weight	Boiling point, °F	Vapor pressure at 100°F, lb/in^2	Specific gravity 60/60	Flammability limits, vol % in air Upper	Lower
Propane	C_3H_8	44.1	–43.75	188.60	0.5070	2.0	9.5
n-Butane	C_4H_{10}	58.12	31.08	51.71	0.5840	1.5	9.0
Isobutane	C_4H_{10}	58.12	10.78	72.58	0.5629	1.8	8.5
n-Pentane	C_5H_{12}	72.15	96.92	15.57	0.6311	1.4	8.3
Isopentane	C_5H_{12}	72.15	82.12	20.45	0.6247	1.3	8.0
Neopentane	C_5H_{12}	72.15	49.10	36.68	0.5974	1.4	7.5
Dimethyl ether	C_2H_6O	46.07	–12.71	122.28	0.6711	3.4	18.0
Carbon dioxide	CO_2	44.01	–109.26		0.8180		
Nitrogen	N_2	28.01	–320.45				
Nitrous oxide	N_2O	44.01	–127.26		0.8175		
HCFC-22	$CHClF_2$	86.5	–41.80	123.57*	1.2080		
HCFC-142b	CH_3CClF_2	98.5	14.00	29.58*	1.1230	6.7	14.9
HFC-152a	CH_3CHF_2	66.1	–13.00	64.107*	0.9110	3.9	16.9

*Vapor pressure at 70°F.

CFCs

Until recently, CFCs such as Freon 11 and Freon 12 appeared ideally suited for aerosol propulsion in term of stability, chemical inertness, compatibility with aerosol ingredients, lack of toxicity, and pressure-generating capacity. These were widely used as a mixture of these two gases. Freon 11 has good solvent power and Freon 12 is used to raise vapor pressure. But CFCs were banned in the late 1970s by various environmental agencies. The Montreal Protocol prohibited their use due to their adverse effects on the earth's ozone layer. Small quantities are still allowed in specialized applications such as metered-dose inhalers.

HCFCs

HCFCs differ from HFCs in that only some, not all hydrogen atoms in the parent hydrocarbon molecules have been replaced by chlorine or fluorine atoms. Solvents based on HCFC are still used because these are considered nonflammable, non-ozone-depleting and non-VOC compounds. Some aerosols (paints, solvents, degreasers, etc.) do contain VOCs, but their contribution to total VOC load emitted to the atmosphere is very small. Refrigeration gases such as HFCs 134a and 152s are used in many consumer products.

HFC 152a is noted for its exceptionally low odor and food solvency. It is used to make less flammable colognes and perfumes. Propellant used in the formulations is typically 2 to 10 percent.

Medical metered-dose inhalers use HFC 134a and HFC 227ea. HFC 134a is nonflammable and HFC 152a is less flammable than hydrocarbons. HFC 134a is the only nonflammable liquefied propellant available today. For marine and safety alarms, the propellant is the only chemical ingredient in the can.

Dimethyl Ether

Dimethyl ether is one of the most preferred propellants because it is a strong solvent and highly water soluble. It is highly flammable, but flammability is reduced by incorporating an aqueous phase in the solution. Dimethyl ether is commonly used in hair sprays, personal deodorants, room fresheners etc.

Nitrous Oxide

Nitrous oxide is approved as food additive and as an aerosol spray propellant. It is used in aerosol whipped cream canisters, cooking sprays, and sterilizing packaging of food items such as potato chips and other snacks, furniture polish, and in windshield and car lock deicer sprays.

Carbon Dioxide

Soluble compressed gases such as carbon dioxide are used as propellants in alcoholic systems such as water/alcohol disinfectants, deodorant sprays, bug killers, lubricants, air fresheners, perfumes, and other personal care products.

Nitrogen

Sterile saline solutions for rinsing contact lenses, long-range stream-type wasp and hornet killers, and injector-type engine cleaners etc. Carbon dioxide, nitrous oxide, and nitrogen are inexpensive but require special equipment to introduce them in aerosol cans.

CONTAINER

The final element is the container, which is usually a steel or aluminum can. The leakproof can protects the product from contamination and evaporation. An aerosol package is an airtight pressurized container. Pressing the actuator button opens the valve, pushing the product out of the valve. The system allows product to be delivered in a variety of ways; in a fine mist, a metered spray delivering just the right amount, a foam, or even a long spray.

Most aerosol containers are made of steel or aluminum, and empty containers can be recycled. Aerosols are generally formulated to a minimum pressure consistent with good operational performance. The maximum pressure limit for aerosol container can vary but is typically 140 to 150 lb/in^2.

CHAPTER 23
TRANSFORMER/ELECTRICAL INSULATING OILS

Transformer oils are well-refined hydrocarbon oils used as a coolant and insulating medium in oil-immersed electrical transformers. These oils are also used as insulating oils in circuit breakers, switches, oil-filled capacitors, tap changers, electrical enclosures, fuses, and other electrical devices. In the case of oil-immersed switches, rapid quenching of arc reduces contact erosion.

A transformer oil has two important functions to perform; to function as an electrical insulator and to act as a heat transfer medium for cooling the transformer. It should be able to perform these functions for 25 to 30 years. To get an oil to insulate against electricity is simple; all hydrocarbon oils that are free of water and sludge can do this. The real challenging task is to sustain these properties over a long period, which implies that no chemical changes at all should take place in the oil. Many transformers work around 160°F, and at this temperature, constant oxidation of the oil takes place, resulting in the formation of water and sludge. This, in turn, impedes the circulation of oil and thus adversely affects heat dissipation from the transformer. Thus oxidation stability of the oil is of vital importance.

Power transformers are generally located outdoors and exposed to the full force of extreme ambient temperatures. To act as an effective heat transfer medium for transformer cooling, oil viscosity has to be low at the transformer's operating temperature, and it should be sufficiently fluid at the lowest ambient temperature. Also, pour point should be sufficiently low so that transformer oil does not freeze at minimum ambient temperatures. The minimum ambient temperatures at which transformers function are diverse and may range from +20°F in warm climates to –70°F in arctic zones. Thus both the lowest temperature viscosity and the pour point of the oil are tailored for the transformer's geographic location.

PROPERTIES/SPECIFICATIONS

Properties of some commercial transformer oils manufactured by various companies are listed in Table 23-1. A major factor in the market acceptance of transformer oils is good oxidation resistance without the use of added inhibitors. Users of transformer oils have traditionally used uninhibited oils and are suspicious of oils containing inhibitors unless it is known that the base oil has good oxidation resistance. Inhibitors can be destroyed in the oil or volatilized by severe heat or extreme overloads. In cases where a transformer oil is formulated with inferior base stock inhibited with antioxidants, it can actually fail without warning. The significance of various properties and test methods used for their measurements are presented in the following paragraphs.

Color

The color of new transformer oil is generally accepted as an indication of its degree of refinement. For oil in service, an increase in the color number is indicative of contamination and/or deterioration. New oil typically has a color rating (ASTM D 1500) of 0.5 and is bright and clear.

TABLE 23-1 Properties of Transformer Oil

	Units	1	2	3	4
Color properties	ASTM D1500		L 0.5		L 0.5
Relative density		0.8854	0.889	0.8849	0.8938
Viscosity kinematic @ 0°C	cSt		66		
Viscosity kinematic @ 40°C	cSt	8.93	9.1	8.8	10.94
Viscosity kinematic @ 100°C	cSt		2.6	2.3	2.56
Viscosity index				58	
Pour point	°C	–57	–76	–47	–62
Flash point	°C	148	146	160	148
Sulfur, ASTM 1552	Wt %		0.07		
Acid number, ASTM D 974	mg KOH/g	0.02	0.01		0.014
Aniline point, °C			73		75.8
Demulsibility, ASTM D 1401	Minutes to pass		10		
Interfacial tension, ASTM D 971	Dynes/cm		45		42
Water, ASTM D 1533	ppmw	12	25		17
Inhibitor content, mass %	% mass			nil	0.002
PCB content	ppm	<1	nil	nil	
Dielectric strength ASTM D 877	kV	>30	35	50	>30
Power factor@60 hrz, ASTM D 924					
% at 25°C		0.00063	0.007		0.073
% at 100°C			0.26	0.05	
Oxidation stability, ASTM 2440					
Sludge, 72 h	Wt %		0.1		0.1
Total acid number, 72 h	mg KOH/g		0.3		0.1
Sludge, 164 h	Wt %		0.1		0.1
Total acid number, 164 h	mg KOH/g		0.4		

Specific Gravity

The specific gravity (SG) of mineral oil influences its heat transfer rates. In extremely cold climates, an upper limit of 0.895 is placed on the SG of mineral oil. If oil contains any moisture, the resulting ice from the freezing of the moisture in the oil-filled apparatus should not float on the surface (The SG of ice is approximately 0.915). Floating ice on oil can cause a flashover of conductors extending above the oil level. Oils of different SGs may not mix when added to each other, and precautions are taken to guard against this possibility.

Viscosity

Viscosity of an oil used as a coolant influences heat transfer rates and thus the temperature rise in a transformer. The viscosity of oil used in tap changers and circuit breakers influences the moving parts in these machines. To cool the transformer properly, electric insulating oil should have a viscosity as low as possible at the transformer operating temperature, so as to rapidly remove heat from the transformer core. Low-viscosity oils are particularly desirable in cold climates. Viscosity of electric insulating oil is chosen according to the lowest temperature to which the oil is exposed. Also the viscosity of oil at the lowest ambient temperature must not severely restrict heat transfer from the transformer core to the atmosphere. Although not a specification, the viscosity index (VI) of insulating oil is significant. Unlike in lubricating oils, where high VI oils are desired and are manufactured by removing aromatics from the feed, low VI oils are desirable for electric insulating service. A low VI causes the viscosity of the oil to fall quickly at transformer operating temperatures, thereby aiding heat transfer from core to atmosphere. The typical range of viscosities is as follows:

Temperature, °F	Viscosity, cSt
32	76 (maximum)
100	8–10
210	2–3

The VI is generally low (60 to 65), indicating a rapid fall in viscosity with the temperature.

Pour Point

A low pour point of transformer oil is important, particularly in cold climates, to ensure that oil circulates and serves its purpose as a cooling medium, apart from its insulating function. Pour point is also important for identifying the type of oil (naphthenic/paraffinic). The pour point of transformer oil is chosen according to climatic conditions of the region where it is to be used, typically varying between +20°F to –70°F from warm regions to arctic climates. The pour point of oil must be such that the oil must not freeze at the lowest ambient temperature. For example, for northern Europe or Canadian service, with winter temperatures falling to –40°F or more, an insulating oil of –76°F pour point would be suitable, whereas for tropical climates where the lowest ambient temperature never falls below 32°F, an oil with a 20°F pour point, representing the minimum winter temperature, would be adequate.

Neutralization Number (Acid Number)

The neutralization number of oil is a measure of the amount of acidic or alkaline material present. As oil ages in service, the acidity and thus neutralization number goes up. For oils in service, a high neutralization number indicates that oil is either oxidized or contaminated with materials such as varnish, paint, or other foreign matter. A severe increase in the neutralization number can be detrimental to insulation function. New oils typically have a neutralization number (ASTM D 974) of approximately 0.03 mg KOH/g or less.

Demulsibility

This test method provide a guideline for determining the water separation characteristics of oil subject to water contamination and turbulence. A 40-mL sample of oil and 40 mL of distilled water are stirred for 5 minutes at 54°C in a graduated cylinder. The time required for separation of emulsion thus formed is recorded. If complete separation does not occur after standing for 1 h, the volumes of oil, water, and emulsion remaining at the time are reported. Observations are recorded at 5-minute intervals until either (1) the product passes the water separability requirements or (2) the time limit for water separability is exceeded (usually 30 minutes maximum at 54°C and 60 minutes at 82°C). A demulsibility value (ASTM D 1401) of 10 to 15 minutes or lower indicates good water separation characteristics for the oil.

Aniline Point

The aniline point is the minimum equilibrium solution temperature for equal volumes of aniline and oil. The mixed aniline point is the minimum equilibrium solution temperature of a mixture of two volumes of aniline, one volume of sample, and one volume of n-heptane. The mixed aniline point is applicable to samples having an aniline point below the temperature at which aniline crystallizes from the mixture. Specified volumes of aniline and sample or aniline and n-heptane are placed in a test tube and mixed mechanically. The mixture is heated at a controlled rate until the two phases become miscible. The mixture is then cooled at a controlled rate, and the temperature at which the two phases separate is recorded as the aniline point or mixed aniline point. The aniline point or mixed aniline point is useful as an aid in the characterization of pure hydrocarbons and in the analysis of hydrocarbon mixtures. Aromatic hydrocarbons exhibit the lowest and paraffins, the highest. Naphthenes and olefins exhibit values that lie between those for paraffins and aromatics. The aniline point is often used to provide an estimate of the aromatic hydrocarbon content of mixtures. Aniline point (ASTM D 611, °F) of most commercial grades is seen to be in the range of 145 to 183°F maximum.

Interfacial Tension

Interfacial tension is very sensitive in detecting oil-soluble polar contaminants such as acid and sludges formed from the oxidation of transformer oil. Also when certain contaminants such as soaps, varnishes, and paints are present in oil, surface tension is decreased. For new oils, interfacial tension (ASTM D 971) should be 40 dynes/cm (minimum). For oils in service, a decreasing value indicates accumulation of contaminants or oxidation products, or both, which may impair transformer operation.

Interfacial tension is measured by measuring the force necessary to detach a planar ring of platinum wire from the surface of the liquid with higher surface tension, that is, upward from the water oil interface. To calculate the interfacial tension, the force so measured is corrected by an empirically determined factor that depends on the force applied. Interfacial tension is a reliable indicator of the presence of hydrophilic compounds in the oil.

Water contamination

A low water content is necessary to obtain and maintain required electrical breakdown strength and low dielectric losses in insulating oils. The water content of transformer oil is typically maintained below the 30 ppmw level. The sensitivity of electrical insulating oil properties to contaminants such as water and solid particles make it necessary to handle and store insulating oil with care so that the risk of moisture pickup and contamination is minimal. As a rule all shipments of electrical insulating oil are tested for dielectric strength before use. If the dielectric strength is low, oil is polished by filtration and dehydration. Many large transformers use custom-built equipment on sumps or reservoirs of transformer oil to reduce water and particulate contamination by filtering the oil through a fuller's earth bed. Drums of electrical insulating oil are stored indoors. If stored outside, the drums are placed on their side on racks to keep clear of any groundwater. The bungs are positioned low to prevent them from breathing moisture.

Sulfur

The presence of corrosive and thermally unstable sulfur compounds in oil, when present, can cause corrosion of copper and silver components in electrical machines.

Oxidation Inhibitor Content

Control of inhibitor content is important to maintain the long service life of insulating oil. For the inhibited class of transformer oils, the maximum concentration of antioxidants allowed is up to 0.3 wt %.

Dissolved Gas Analysis

Dissolved gas analysis (DGA) is recognized as an effective diagnostic tool for a routine checkup of transformers. A gas sample is extracted from the oil and nine key gas components (hydrogen, oxygen, nitrogen, methane, carbon monoxide, carbon dioxide, ethane, ethylene, and acetylene) are analyzed (ASTM D 3612). When a transformer is failing, hydrocarbons in the oil break down and give off these gases. A high concentration of these gases signals corona discharge, overheating, arcing, or cellulose insulation pyrolysis.

Furanic Compounds

Furanic compounds are generated as a by-product of the degradation of cellulosic materials such as insulating paper, wood etc. These compounds serve as indicators of insulation degradation. Because they are dissolved in oil, furanic compounds can be readily tested by high-performance liquid chromatography (HPLC). New oil should not contain any of these products.

The test (ASTM D 5837) is for five furanic compounds, 5 hydroxymethyl-2-furfural, furfuryl alcohol, 2-furfural, acetyl furan, and 5 methyl-2 furfural. The limit of 25 μg/g limits applies to each compound.

Polychlorinated Biphenyls Content

Polychlorinated biphenyls (PCBs) belong to a group of pollutants considered the most hazardous substance in nature because they do not biodegrade. Introduced in 1929, they became a big success for the chemical industry. These were extensively used in transformer oil as nonflammable cooling compounds in transformers and capacitors. Other PCB uses were as lubricants, hydraulic fluids, cutting oils, and many others. By the late 1970s, almost all countries banned the manufacture and use of these compounds after recognizing their disastrous effect on human beings and the environment. As a possible precaution against the use of recycled transformer oil containing PCBs, most specifications of transformer oil (ASTM D 4059) assure that new products do not contain detectable amounts of PCBs.

Dielectric Strength (ASTM D-877, D 1816)

Dielectric breakdown is the minimum voltage at which an electric flashover occurs in an oil. The dielectric breakdown strength of an insulating material is a measure of its resistance to electrical breakdown when an electric potential is applied across it. Immersing two electrodes in a sample of oil and then applying an AC voltage across the electrodes tests the electrical insulating oil. The voltage is increased in a specified manner until electrical breakdown occurs. The various tests used differ in electrode spacing and shape and rate of increase of voltage, and thus give different breakdown values for the same oil. The dielectric strength of insulating oil is reduced if there is water or solid particles in the oil. Also, oxidation products resulting from oil deterioration in service lowers the dielectric strength. Thus periodic check of the dielectric strength of an oil is a useful method to determine whether contamination or deterioration of oil has occurred or not. Dielectric strength of new oil on the average is between 35 and 45 kV.

Dielectric Dissipation Factor/Power Factor (ASTM D- 924)

Any two adjacent conductors form a capacitor. In an ideal capacitor, the phase difference between applied AC voltage and current is 90°, and the power dissipated is zero. If the dielectric between the conductors is less than ideal, the phase difference will be less than 90° and some power dissipation will occur. To keep this loss low, it is desirable to have the dielectric as near to ideal as possible.

For insulating oils the value of this characteristic is called the power factor or loss tangent (dissipation factor) and is expressed as a percentage at a specified temperature. These values are determined experimentally and represent trigonometric function of the angle of phase difference. With the particular function used a value of zero would represent a 90° phase difference and the ideal condition; therefore, low values are desirable.

Oxidation and the contamination of oil can cause power factor or a loss tangent to rise so determination of this property may provide useful information about a used electrical insulating oil. Because values vary with temperature, comparisons must be made at the same temperature. Power dissipation should not exceed 0.05 percent at 77°F and 0.30 percent at 212°F.

Impulse Breakdown Voltage (ASTM D 3300)

Impulse breakdown voltage is voltage at which electrical flashover occurs in an oil under impulse conditions. It indicates the ability of an oil to resist transient voltage stresses such as those caused by lightning strikes and high-voltage switching surges.

Oxidation Stability (ASTM D 2440)

The acid sludge test measures the oxidation resistance of an oil by determining the amount of acid/sludge products formed when tested under certain prescribed conditions. Oil that meets or exceeds the requirements tends to preserve its insulating system life and heat transfer characteristics. Good oxidation stability is the principal requirement for long service life (25 to 30 years) of transformer oils.

Doble Oxidation Test

Commercial transformer/insulating oils are tested by the Doble oxidation test, developed by the Doble Engineering Company,[1] USA. The method involves bubbling air through a known amount of oil held at a temperature of 203°F (95°C) in the presence of copper and iron and making two types of tests daily on a small sample of oil. One type of test is acidity measurement and the other is precipitation test in which one volume of oil is diluted with five volumes of pentane. The mixture is allowed to stand for at least 8 h, and the presence or absence of a sludge precipitate is noted. The end point of the Doble test is taken as the number of days of oxidation either before the acidity of the oil reaches 0.25 mg of KOH per gram oil or before a positive precipitation test for sludge is obtained.

An amplification of the Doble test is the "power factor valued oxidation" (PFVO) test. The test is conducted as just described, but the power factor of the oil is determined at every 2-h intervals throughout the oxidation period. A curve is obtained by plotting power factor against oxidation time. All commercial transformer oils show a sharp rise in the power factor in the early stages of oxidation. The power factor rapidly increases to a value considerably above 2 percent following which it decreases to a relatively low value and thereafter rises continuously. The time of oxidation before the curve crosses the 2 percent power factor is the Doble life of the transformer oil. The ideal transformer oil would be one that exhibits no increase in power factor throughout the test life.

The best of commercial oils typically have a life of about 3 days under Doble test conditions. Longer life could be obtained by adding an oxidation inhibitor to the oil, but the use of such additives traditionally has been considered unacceptable by transformer manufacturers and users.

TRANSFORMER OILS MANUFACTURE

Feedstocks for transformer oil manufacture are straight run light vacuum gas oil cuts from naphthenic crudes, with a boiling range of 460 to 775°F with ASTM 50 percent midboiling point in the

TABLE 23-2 Transformer Oil Typical Basestock Properties
Feed: Straight run distillate from naphthenic crude

Property	Units	Value
Boiling range	°F	460–775
API gravity		24.4
Flash point	°F	280
Viscosity, 100°F	SUS	55.3
Viscosity, 210°F	SUS	33.6
Nitrogen content	ppm	50
Sulfur	wt %	0.18
Viscosity-gravity constant*		0.855

*As per ASTM D-2501;

VGC = (10*G -1.0752* LOG (V-38))/(10-LOG(V-38))

Where: VGC = Viscosity gravity constant.
G = Specific gravity, 60/60°F.
V = Saybolt universal viscosity, 100°F.

range of 608 to 662°F. Viscosities are in the range of 50 to 65 SUS (7.5 to 12 cSt) at 100°F. Typical properties of the feed are shown in Table 23-2. Transformers and their operating conditions are diverse, and therefore transformer oil performance requirements are tailored to its operating conditions and geographic location.

Important manufacturing processes are as follows:

- Hydrocracking-Solvent dewaxing
- Hydrotreating-Clay contacting
- Sulfuric acid-Clay contacting

Hydrocracking-Dewaxing Process

Good quality transformer oils[2] can be produced from paraffinic oil sources by hydrocracking the heavy paraffinic oils such as vacuum gas oil, fractionating the hydrocracked petroleum hydrocarbons to recover a distillate boiling in the range of transformer oil (662 to 752°F). This stream is next dewaxed in a solvent dewaxing unit to a pour point of –6°F or lower. The dewaxed transformer oil can optionally be hydrofinished. Antioxidant and/or pour point depressant may be added. An example of an antioxidant additive used is 2-6 di-tertiary butyl para cresole. The use of such antioxidants, however, is limited by transformer oil specifications. ASTM D 3487 restricts type I oils to 0.08 percent oxidation inhibitor content maximum; type II oils are limited to a maximum of 0.3 percent by weight oxidation inhibitor. Pour point depressant are exemplified by Pearsell OA 100 A, an alkylated polystyrene. Typical dosage is being 0.1 to 1.0 percent.

Hydrotreating-Clay Contacting[3] Process

Feed to the process is straight run cut (662 to 752°F) from a crude distillation column processing low pour naphthenic crude. This feed is hydrotreated for decreasing the content of heterocyclic sulfur and nitrogen. Normal hydrodesulfurization catalysts such as Ni-Mo are used. Typical operating conditions are temperature 500 to 750°F and pressure 1200 lb/in^2. The sulfur and nitrogen content of the feed is substantially reduced. Also there is a partial saturation of polycyclic aromatic hydrocarbons such that UV absorptivity of hydrorefined product at 335 millimicrons (335 UVA) is below 0.04. Oxidation stability of transformer oil is improved by reducing sulfur and nitrogen content and by retaining polynuclear aromatics that contribute to 335 UVA and act as natural inhibitors (Fig. 23-1).

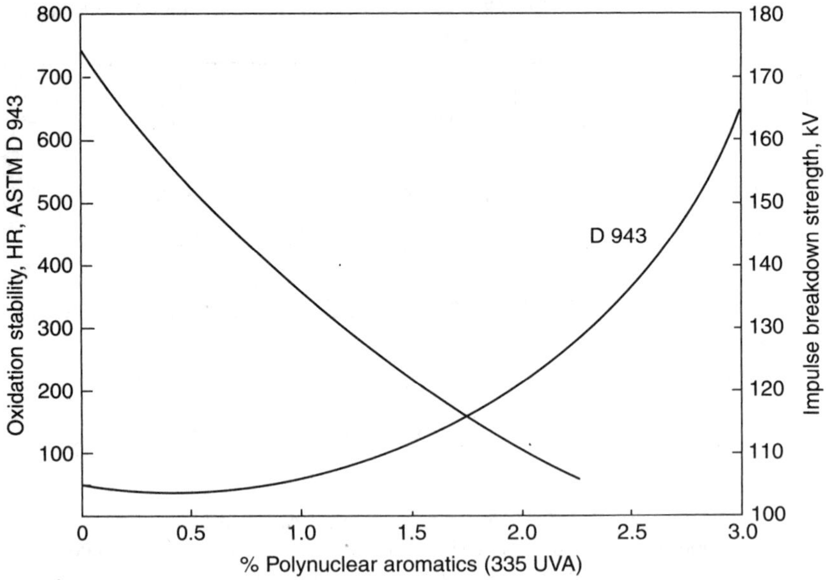

FIGURE 23-1 Effect of polynuclear aromatics on oxidation stability and impulse breakdown strength.

If hydrotreating severity is increased so that 335 UVA is less than 0.04, the impulse breakdown voltage of the resulting severely refined transformer oil may be as high as 175 kV. After hydrotreating, the transformer oil is contacted with an adsorbent such as attapulgite clay.

Acid and Clay Treating Process

In this process, feedstocks are treated with 2 to 3 volume percent oleum (20 percent). The sludge formed is allowed to settle for 48 h. After separation of sour oil from sludge, any free acid is neutralized with a sodium carbonate solution. The petroleum sulfonate formed is extracted with isopropyl alcohol. Acid-treated sulfonate-free oil is contacted with 2 percent by weight activated clay to yield the finished transformer oil. A simplified process flow diagram and material balance for an industrial plant is shown in Figs. 23-2, 23-3 and Table 23-3. Details of this process are described under white oil manufacture. Thus transformer oil feedstock may require only a single acid treatment, whereas white oils may require acid treatment four to six times. Raw sulfonate extract is sent to sulfonate recovery plant.

In another variant of the acid treating process,[4] feed, a naphthenic distillate, is treated with 1.5 vol % sulfuric acid of 99 percent strength. The charge is water washed and dried by air blowing. The oil is next heated to a temperature of 270°F, and sodium in an amount of 1 percent by weight is added and dispersed in oil by means of a high-speed mixer. Next isopropyl alcohol in an amount 0.5 vol % of oil is added, whereupon the reaction occurs and the temperature rises by 10°F over a time of about 6 minutes and thereafter drops. After about 30 minutes of mixing, carbon dioxide is bubbled into the mixture at a temperature of 245°F. This causes the temperature to rise during a period of 10 minutes, the amount of rise being somewhat greater than during treatment with sodium and alcohol. Contact with carbon dioxide is continued for about 10 minutes, after which the mixture is allowed to settle while cooling. Unreacted sodium and reaction product separate into a bottom layer and a clear transformer oil layer is decanted.

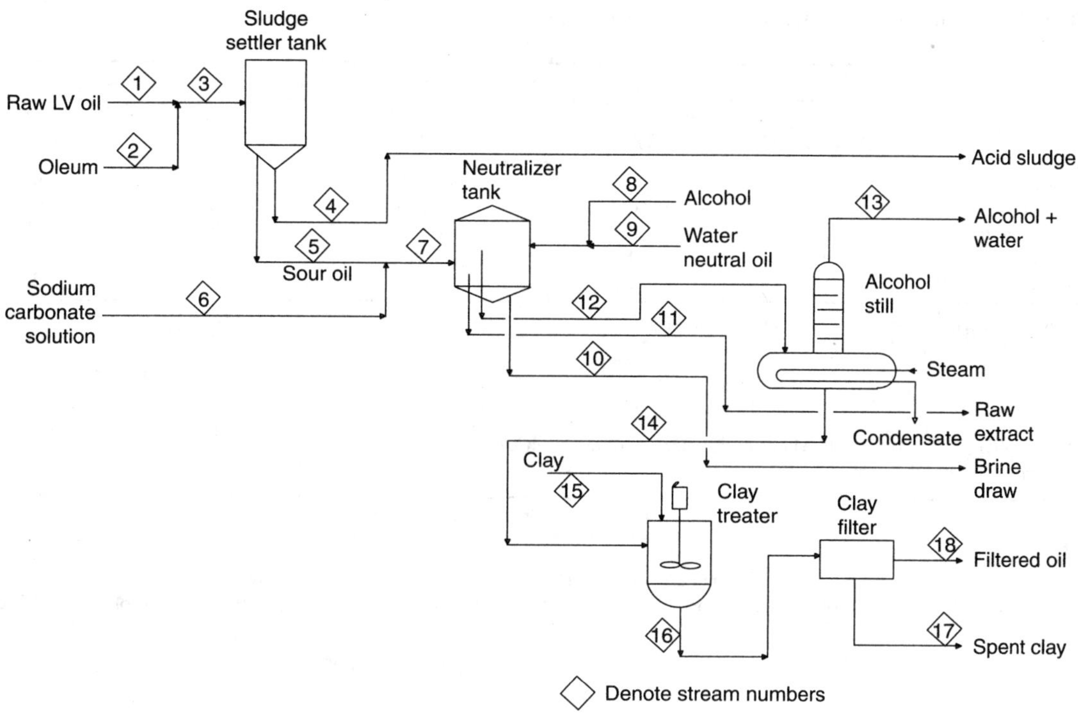

FIGURE 23-2 Transformer oil/white oil, material balance sream numbers.

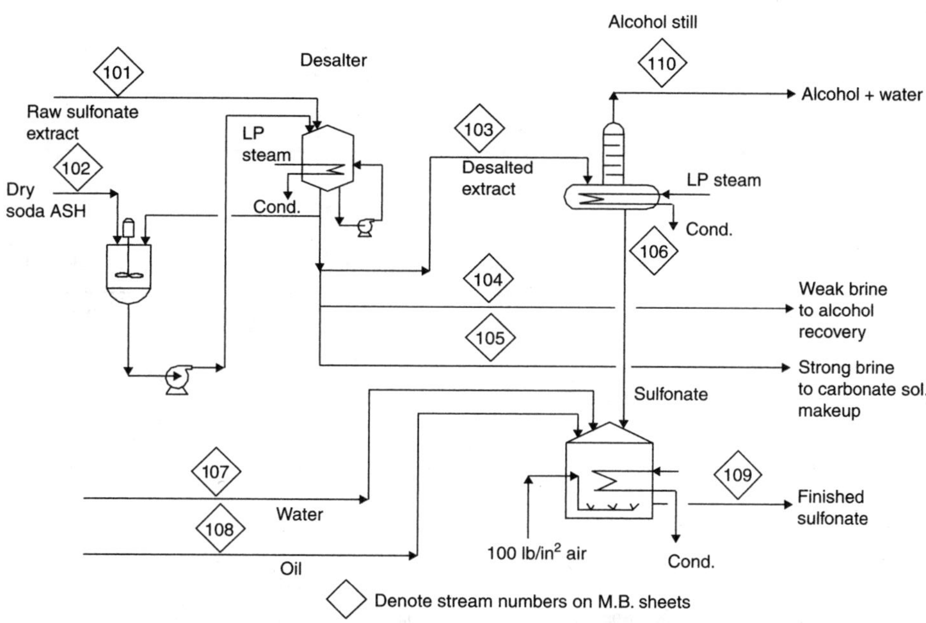

FIGURE 23-3 Sulfonate finishing materia balance stream numbers.

TABLE 23-3A Transformer Oil Material Balance: Acid Treating
Basis: 45,000–Gal batch

	Stream number and name*				
	1 Raw LV oil	2 Oleum	3 Acid oil	4 Sludge	5 Sour oil
Gal/Batch, 60°F	45,000	1350	46,350	2700	43,000
Lb/ Batch	330,000	21,400	351,400	33,800	317,600
Specific Gravity	0.87	1.9	0.91	1.5	0.874
Temperature, °F	80	80	110	110	110
Viscosity, cP, 100°F	12	29		10,000	12

*Stream numbers correspond to those in Fig. 23-2.

TABLE 23-3B Transformer Oil Material Balance: Neutralization and Extraction
Basis: Sour oil from acid-treating step

	Stream number and name							
	5 Sour oil	6 Carbonate solution	7 Oil + carbonate	10 Brine draw	11 Raw extract	12 Neutral oil	8 55% alcohol	9 Water
Gal/Batch, 60°F	43,000	6750	49,750	7500	4800	42,440	4500	900
Lb/Batch	317,600	66,900	384,500	73,570	38,400	311,190	32,900	7500
Specific Gravity	0.874	1.19	0.93	1.18	0.98	0.874	0.876	1.00
Temperature, °F	110	130	110	120	120	120	80	80
Viscosity, cP, 100°F	12	2		2	9	12	1	1
Components, lb								
CO_2 Gas		1750						
Sulfonates					5040			
Water				58,580	20,600	5160	16,800	
Alcohol				2460	9610	4030	16,100	
Salts				12,610	2040			
Oil					1650	301,000		

TABLE 23-3C Transformer Oil Material Balance: Oil Distillation and Cooling

	Stream number and name			
	12 Neutral oil	13 Water oil	14 Stripped alcohol	14 Unfiltered oil
Gal/Batch, 60°F	42440	1240	41200	41200
Lb/Batch	311,190	9190	301,000	301,000
Specific gravity	0.874	0.89	0.874	0.874
Temperature, °F	140	100	270	160
Viscosity, cP, operating temperature	7.5	1	2	5
Components, lb				
Water	5160	5160		
Isopropanol	4030	4030		
Oil	301,000		301,000	301,000

TABLE 23-3D Transformer Oil Material Balance: Oil-Clay Contacting and Filtering
Basis: 7500 Gal/Batch

	Stream number and name				
	14 Unfiltered oil	15 Clay	16 Oil-clay	17 Spent clay	18 Filtered
Gal/Batch, 60°F	7500		7550		7350
Lb/Batch	54,600	1092	55,692	2182	53,510
Specific gravity	0.874	2.5	0.885	1.31	0.874
Temperature, °F	80		160	160	160
Viscosity, cP, operating temperature	18				5
Components, lb					
Clay			1092	1092	
Oil			54,600	1090	

*Stream numbers correspond to those in Fig. 23-2.

REFERENCES

1. "ASTM Standards on Electrical Insulating Liquids and Gases" (December 1959): "Suggested Method of Test for Oxidation Characteristics of Mineral Transformer Oils.", pp. 307–313
2. U.S. Patent 5167847, "Process for Producing Transformer Oils from Hydrocracked Stock," issued December 1, 1992.
3. U.S. Patent 3839188, "Hydrorefined Transformer Oils and Process of Manufacture," issued May 22, 1968.
4. U.S. Patent 3406111, "Transformer Oils," issued October 15, 1968.

CHAPTER 24
WHITE MINERAL OILS

Mineral oils refined from petroleum crude oils are a complex mixture of straight and branched chain paraffinic, naphthenic hydrocarbons with 15 or more carbon atoms and boiling in the range of 600 to 1100°F. White mineral oils are colorless, clear, transparent, tasteless, nontoxic, and stable. They are odorless at room temperature with little odor after heating. White oils have a high degree of ultraviolet (UV) and color stability. The high purity white oils are essentially free of aromatics, unsaturated and polar compounds. Because of their innocuous and inert characteristics, these selectively refined hydrocarbon oils are used in a very large number of applications; in pharmaceuticals, cosmetics, plastics, and food processing. They are versatile because of their ability to lubricate, insulate, moisturize, protect, and act as a carrying agent.

White oils can be broadly classified into two broad categories; medicinal/food grades and technical grades. The medicinal grades of white oils are used in pharmaceutical preparations and cosmetics. Oils that come in direct food contact are refined to a high degree of purity to make them completely free from polynuclear aromatics as per UV absorption tests. These white oil grades are produced to higher standards of purity as reflected in their UV absorption test, color, odor, taste, and inertness and other tests to ensure that they are completely free from human carcinogens and conform to U.S. Food and Drug Administration (FDA) or European Pharmacopoeia regulations.

The technical grades of white oils are white mineral oils refined to a lesser degree as reflected in their color, odor, taste, and polynuclear aromatic content. These grades are meant to be used in nonfood/pharmaceutical uses. Processing required for their production is much less, and therefore cost of production is less compared with pharmaceutical grades.

PROPERTIES OF WHITE OILS

White oils are produced from both from naphthenic and paraffinic feedstocks. By careful selection of feedstocks (crude source and TBP cut), white oils are produced in a wide range of viscosity, specific gravity, volatility, pour point, and other properties to suit different end uses. Tables 24-1 and 24-2 show the typical properties of commercial white oils.

Color

The color of white oil is an indicator of its refining history. Typically, all food- and medicinal-grade white oils are water white and clear with a Saybolt color (ASTM D 156) of +30. Also, these grades are tasteless and odorless and have a high UV and color stability. Technical grades of white oils typically have a color rating between +20 and +30. Some medicinal- or technical-grade white mineral oils may contain up to 10 ppmw of an antioxidant such as alpha tocopherol (vitamin E), butylated hydroxy toluene (BHT), butylated hydroxyanisole (BHA), or other approved antioxidants.

Viscosity

White oils are produced in enormous range of viscosities, typically ranging from 10 to 120 cSt or more at 104°F, according to user industry requirements. A plant may basically produce two viscosity grades; lowest and highest viscosity grade. All other required grades are produced by blending these two grades in various proportions.

TABLE 24-1 Properties of Commercial White Oils

Property	Units	1	2	3	4	5	6	7	8	9
Color, Saybolt		+30	+30	+30	+30	+30	+30	+30	+30	+30
Density 68°F	kg/m^3	831	856	850	854	848	840	853	853	861
Viscosity, 68°F	cSt				31–35	34–38		60–66		
Viscosity, 68°F	SUS				146–164	160–178		279–306		
Viscosity, 100°F	SUS	53								186
Viscosity, 104°F	cSt	7.8	9.2	11.5	12.5–15.5	15–18.5	15	23–26	29.7	36
Viscosity, 104°F	SUS				68–79	77–92		111–124		
Viscosity, 212°F	cSt			3.14	3.1	3.5	3.38	4.5	5.25	
	SUS				36	38		41		
Flash point, COC, Min.	°F	310	331	338	313	338	356	356	372	395
Pour point, Max.	°F	20	–0.4	–0.4	10	5	–0.4	27	5	10
Sulfur, Wt %	Wt %				<0.001					
Neutral number					0	0		0		
Carbon distribution	Wt %									
	Aromatics				0	0		0		
	Naphthenes				37	35		34		
	Paraffins				63	65		66		
Initial boiling point	°F				536	626		554		
Final boiling point	°F				860	806		896		
Mol wt										

TABLE 24-2 Properties of Commercial White Oils

Property	Units	10	11	12	13	14	15	16	17
Color, saybolt		+30	+30	+30	+30	+30	+30	+30	+30
Density, 68°F	kg/m^3	859	861	861	863	866	885	875	872
Viscosity, 68°F	cSt			140–150	160–180	200–250			
Viscosity, 68°F	SUS			649–695	741–834	927–1158			
Viscosity, 100°F	SUS						325	370–410	
Viscosity, 104°F	cSt	39.5	44.8	48–51	56–59	65–77	70	72–79	180–240
Viscosity, 104°F	SUS			223–237	260–274	302–357			
Viscosity, 212°F	cSt	6.3	6.75	7.2	7.8	8.5 Min.			22
	SUS			49	51	54			
Flash point, COC, Min.	°F	379	385	383	392	428	392	475	
Pour point, Max.	°F	5	10.4	21	21	16	–0.4	15	
Sulfur, Wt %	Wt %								
Neutral number				0	0	0			
Carbon distribution	Wt %								
	Aromatics			0	0	0			
	Naphthenes			34	34	32			
	Paraffins			66	66	68			
Initial boiling point	°F			662	680	734			
Final boiling point	°F			977	986	1022			
Molecular weight									690

Pour Point

The pour point of white oil depends on the feedstock qualities. White oils produced from naphthenic feedstock have a lower pour point than those produced from paraffinic crudes. Typically, the pour point of white oils produced range between –0.4 and 27°F. Low pour point naphthenic-grade white oils find applications in hot melt adhesive and as air conditioners and refrigerator compressor lubricants. The pour point requirement of white oils is generally dictated by the lowest ambient temperatures prevailing in the geographic area of its use.

Polycyclic Aromatic Hydrocarbons

Polycyclic aromatic hydrocarbons (PCAHs) are nonbiodegradable and are known to be a hazardous substance. White oils used in pharmaceutical formulations or in/on food must be completely free from PCAHs. During manufacturing processes, these are targeted for complete removal, either by chemical treatment or by hydrogenation.

The concentration of PCAHs in white mineral oils is measured by the absorbance of UV light of different wavelengths in a cell of specified dimensions. This method is very sensitive and is able to detect the presence of very low concentrations of polycyclic aromatics, which is not feasible by conventional chemical analysis.

The UV absorbance of a mineral oil sample is determined by measuring the absorption spectrum of the undiluted liquid in a cell of known path length under specified conditions. The absorbance of liquids at specified wavelengths in the ultraviolet light is useful in characterizing petroleum products.[1]

The apparatus required is a spectrophotometer. It is equipped to handle liquid samples in the cell having sample path lengths up to 10 cm and capable of measuring absorbance in the spectral region from 220 to 400 nm. One or more pairs of fused silica cells having sample path lengths in the range from 0.1000 to 10.0 cm are required.

USP Specifications for White Oil

Regulation of white oil is by the U.S. Department of Health, Education and Welfare and the FDA. The applicable Codes of Federal Regulation (21 CFR) for white mineral oils are as follows:

Product	Direct use in food	Indirect food contact	Animal feeds
White mineral oils	172.878	178.3620	573.680
Light mineral oil NF	172.878	178.3620	573.680
Technical white mineral oils		178.3620	573.680

Alkaline and Acid Impurities

USP grades of white mineral oils must pass a test for the absence of alkalinity (phenolphthalein) and strong acid (methyl red). In this test, 5 mL of mineral oil is shaken with 20 mL of hot water at 90 to 95°C. The mixture is shaken for one minute. Aqueous layer is separated and tested for presence of acid or alkali. Upon addition of 0.1 mL phenolphthalein, the aqueous phase must not consume more than 0.1 mL of NaOH solution before turning red.

Sulfuric Acid Test

First, 5 mL of a white oil sample is mixed with 5 mL of 95.5 percent sulfuric acid. The contents are heated in a water bath to 70°C for 10 minutes. The contents are allowed to settle into two layers for 5 minutes. The color of the sulfuric acid layer is compared with the standard color solution.

When observed in transmitted light, the color of sulfuric acid may not be darker than the standard color solution or corresponding Lovibond reading.

Noack Volatility Test (ASTM D 5800/DIN 51581)

White mineral oils are also used as a lubricant in the textile and food processing industries. The Noack volatility test determines the evaporation loss of lubricant in high-temperature service. The more the oil vaporizes, the thicker and heavier the remaining oil becomes, contributing to poor circulation, reduced fuel economy, and increased oil consumption, wear, and emissions. A maximum of 15 percent evaporation loss is allowable.

Flash Point (COC) (ASTM D 93)

The flash point is the lowest temperature at which the application of small flame causes the vapor above the petroleum product to ignite when the product is heated under prescribed conditions. White mineral oils are generally derived from vacuum gas oils or bright stock (deasphalted vacuum resids). The flash point of these oils is generally high, in the range of 300 to 500°F.

Neutralization Number (ASTM D 664)

The neutralization number may be either an acid or a base number. It is the number of milligrams of potassium hydroxide required to neutralize all acidic constituents present in 1 g of the sample (acid number) or the number of milligrams of potassium hydroxide that would be equivalent to the amount of acid required to neutralize all basic constituents present in 1 g of the sample (base number).

New and used oils may contain acidic constituents that are present as additives or as degradation products formed during service. A wide variety of oxidation products and organic acids can contribute to the acid number.

The neutralization number of white mineral oils is of significance when used as a lubricating oil. The neutralization number of new white mineral oil should be close to zero, indicating the absence of corrosive constituents.

USES OF WHITE MINERAL OILS

Cosmetics and Personal Care Products

These personal care products may include baby oils, creams and lotions, suntan oils, sunscreens, hair products, makeup, makeup removers, absorption bases, depilatories, bath oils, emollients, and moisturizers. For these, only medicinal-grade white mineral oils are permissible. These are high-purity products with low irritancy, chemical inertness, and a consistent hydrophilic and lipophilic balance.

Pharmaceuticals

White mineral oils are used in laxative formulations, topical ointments, as gelatin capsule lubricants etc. Only medicinal-grade mineral oils are permissible for use, with compliance to major pharmacopoeias (FDA, European, etc.).

Food Contact (Medicinal and Technical Grades)

These may include bakery pan oils, divider oils, animal feed dedusting, mold release lubricants, egg coating, coating for fruit and vegetables, food packaging materials, food-grade lubricants and greases, and meat packaging. Here extremely low odor, low volatile residue, good color, heat and light stability, and chemical inertness are required.

Polymer Industries

White mineral oils are added to polystyrene (PS), PVC, polyolefin, thermoplastic elastomers etc. to improve and control the melt flow of the finished polymer. White oils are also used as internal and external lubricant in PS, PVC, PP, PE, and in many other polymer formulations.

Other Industrial Uses

White oils not intended for medicinal use are known as technical white oils, and as stated earlier, these grades need not pass a UV absorbance test to ensure the complete absence of polynuclear aromatics. These oils have many industrial applications in textile, chemical, and plastic industries. Their good color, nonstaining properties, and chemical inertness are highly desirable. Technical-grade white oils are used in textile machine lubricants, horticulture sprays, wrapping paper, for corrosion protection in the meatpacking industry, as a lubricant for watches, bicycles, and spindles, in adhesives, household cleaners, and polishes. Naphthenic white mineral oils with a low pour point are used in air conditioners and refrigerator compressor lubricants. White oils are used in the paper manufacturing process during the calendering operation and also as an antifoaming agent.

WHITE OIL MANUFACTURE

There are two basic methods to manufacture white mineral oils:

- Acid/Clay treatment process, which basically removes all aromatics and other reactive constituents from feed.
- Deep hydrotreating of feed, which saturates all aromatics to naphthenes and destroys all polynuclear aromatics. Clay or percolation treatment is sometimes also required for medicinal grades white oils produced by hydrotreating process.

Acid treating is the classical white oil manufacturing route. It also produces petroleum sulfonate, a valuable by-product. Due to environmental reasons, the use of the acid treating route is decreasing because the process produces a significant amount of acid sludge, which is difficult to dispose of.

Acid Treating Process

Feedstocks for white oils are vacuum distillates with a suitable viscosity, viscosity index, and pour point. Feed is treated with oleum, which reacts with aromatics and other reactive constituents of oil resulting in the formation of sulfonate, the petroleum hydrocarbons (approximately $C_{22\text{-}25}$) containing a highly polar—SO_2OH group and sludge. The sludge is separated after it settles down. The acidified oil is neutralized with aqueous sodium carbonate solution. The neutral oil is next processed to remove sulfonates formed in the oil during oleum treatment. Basically this involves extraction of sulfonates from oil with isopropyl alcohol and separating alcohol by distillation. This process sequence (oleum treatment, neutralization, and sulfonate extraction) is repeated a number of times until the oil broadly meets the specifications of the finished product. The number of times this processing sequence is repeated depends on the feedstock quality and the product specifications required. Medicinal-grade white oils may require this sequence of processing four to six times.

The finishing of white oils is done with either clay treatment or percolation filtration. In this latter treatment, the oil is passed through an activated bed of bauxite that removes color and odor from the product and improves taste.

PROCESS DESCRIPTION

Storage of Feedstocks

Referring to the process flow diagram in Fig. 24-1, raw materials, low- and high-viscosity white oils (LVN and HVN) are received from tank trucks in storage tanks V-050B and V-051. Oleum is received in tanks V-058 from pressurized tank trucks or from a pipeline in cases where oleum is manufactured close to the plant battery limits.

Isopropyl alcohol is received in tank V-049. Solid sodium carbonate or soda ash is received in bags and stored in the warehouse.

INTERMEDIATE PRODUCT STORAGE

The manufacturing process for low- and high-viscosity white oils follows these basic steps:

1. Oleum treatment, sludge settling
2. Neutralization and sulfonate extraction with alcohol
3. Alcohol recovery from oil, brine, and sulfonate

The preceding three steps may be considered to constitute a pass. Depending on feedstock properties, the three steps may have to be repeated four to six times or more to meet the final product specifications.

After every pass, the treated feedstock goes to an intermediate storage tank to awaiting the next processing step. Thus intermediate feedstocks must be properly identified by a proper nomenclature, as described next.

INTERMEDIATE PRODUCT NOMENCLATURE

Consider a four-pass white oil process:

- Raw feed oil for low-viscosity white oil is identified as LVN and for high viscosity white oil as HVN. This is the feed for the first pass.
- After the first pass (acid treatment, settling, extraction, etc.), the intermediate product goes to an intermediate storage tank named LV4. This in fact is feed for the second pass.
- After the second pass, the intermediate product is identified as LV3, the feed for third pass.
- After the third pass, the intermediate product is identified as LV2, the feed for the fourth pass.
- After the fourth pass, the intermediate product is identified as LV1. Now the feedstock processing is complete. LV1 is feed for the final finishing treatment, such as clay treatment or percolation filtration through a bauxite bed to improve the color, odor, and taste of the product.

Acid Treatment

Referring to the process flow diagram in Fig. 24-2, neutral oil is pumped from storage by transfer pumps to head tanks V-101A and B, which serve as feed tanks for the oil charging pumps. During acid treatment, a level controller maintains the level in the head tank by starting and stopping the transfer pump. At the end of the treatment, a batch meter in the oil line shuts off transfer pump P-101A and B when the desired amount of oil has been charged to the head tank. The rate of flow of oil from the charging pumps is measured by a turbine meter.

Simultaneously, oleum is pumped from storage through turbine meters. The oleum flow rate is manually adjusted to give the desired ratio of oleum to oil. The oil and oleum streams come together in premix chambers M-101/102. The premix chamber consists of a small length of two concentric pipes. The inner pipe is perforated with 1/8-inch holes. Acid flows through the inner pipe and next through the holes into the oil stream flowing through the outer pipe (see Fig. 24.7A). Oil and oleum are further mixed by mechanical line agitators M103/104. The mixture then flows to settling tanks V-102 A to E. When all the oil has been treated, the mixture is allowed to settle.

Viscosity contamination is minimized by using a separate head tank, charge pumps, charge lines, and mixers for high- and low-viscosity oils. The same viscosity oils are settled in the same tank wherever possible so as to minimize cleaning.

Sludge Separation

The bulk (88 to 95 percent) of the sludge separates from the oil during the first 24 hours of settling. The separated sludge is pumped from settlers to a storage tank that holds about 4 days of production. The sludge produced in treating transformer oil and first-pass low-viscosity white oil is very viscous and must be withdrawn every 8 hours during the settling period. If this is not done, the sludge will harden and cause considerable removal problems. At the end of the total settling time, the last portion of sludge is withdrawn to storage.

If laboratory tests indicate that the acidity of the settled oil is within specifications, the acidified, or "sour," oil is then pumped from the settler to neutralizing tank V-111 A to D. The exceptions are last-pass white oils, which require retreatment with acid before neutralization. Space for an emergency underground tank is provided that can serve as a reservoir for settler content in the event of a serious break or leak.

Oil Neutralization

Neutralization may be carried out either as a continuous or a batchwise operation as described next.

Continuous Neutralization

Referring to process flow diagram Fig. 24-3, following sludge separation, the next step in the process is neutralization. The primary chemical reaction is the neutralization of sulfonic acid with sodium carbonate (20 to 23°Be soda ash solution) to form sodium sulfonate. Several secondary reactions take place in varying degrees that consume additional sodium carbonate. These reactions are the neutralization of sulfuric acid and sulfur dioxide. The products of reaction are carbon dioxide, sodium sulfate, sodium sulfite, and water.

The neutralization process utilizes a pair of ratio proportioning pumps, P-111 A and B. Both the soda ash solution and acidic oil are pressurized simultaneously. The gear pumps, which are linked together, act like liquid meters. The ratio of soda ash solution to sour oil is maintained. Any variation in oil flow rate produces a corresponding change in the carbonate solution flow rate. Pressure control valves equalize the pressure of the two streams.

Sour oil and carbonate solution are mixed in premixer M-110 A and B. The premix chamber (see Fig. 24-7B) is similar in design to the acid-oil premix chamber.

The mixed stream next flows to inline mixer M-111 A and B, passes through degassers M-112 A to D and into the neutralizer. The carbon dioxide formed is separated in the degassers, which are essentially horizontal separator vessels for this service (Fig. 24-8).

Neutralizers are cone-bottomed tanks that have steam coils for heating. Mixing is accomplished by pump recirculation and by an air sparger. The primary function of the neutralizer (Fig. 24-9) is to separate neutral oil into three layers with the help of a solvent, isopropyl alcohol. Thus neutral oil is separated in a brine layer, sulfonate layer, and oil layer that are separated and withdrawn from nozzles at different levels in the neutralizer vessel.

Batchwise Neutralization

In the "batch" neutralization, the carbonate solution is pumped to the neutralizer first. Next, acidified oil is pumped in. Considerable foaming takes place, especially with first-pass high viscosity oils. Air is used to dissipate the foam and aid in mixing. High-level controllers that shut off the oil supply prevent the vessel from foaming over. Generally, longer loading times are experienced due to the level control of the foam. The volume of carbonate used depends on the acidity of the treated oil.

Sulfonate Extraction

After the batch has been charged and the correct alkalinity is established by laboratory tests, 55 percent isopropyl alcohol is charged to the neutralizer from the 55 percent alcohol tank. Depending on the type of oil, an initial solvent dose equal to 5 percent of the volume (vol %) of oil may be used to help break the emulsion of oil and brine formed as a result of the neutralization step. The entire mass is circulated for a few minutes and allowed to settle for 4 hours. After settling, the first brine draw is made and pumped to batch still charge tank V-141. Except for initial solvent addition, all oils are neutralized in a similar manner.

After the initial brine is removed, sodium sulfonate is extracted from the oil with an additional 55 percent isopropyl alcohol. This final solvent is metered into the batch with circulation and heating. The amount of alcohol depends on the type of oil and the amount of sulfonate present. Transformer oil and last-pass white oil contain very little sulfonate, and therefore a lesser amount of alcohol is required. Conversely, first-, second-, and third-pass white oils contain a larger amount of sulfonate, which requires a larger amount of alcohol.

After 1-hour circulation, the mass is settled for 8 hours. During this time the mass is heated to approximately 145°F to aid the extraction of the sulfonate.

During settling, three layers may be formed. The upper layer is oil, the middle layer is sulfonate extract, and the bottom layer is brine. Following the 8-hour settling, the brine layer, if present, is drawn into batch still charge tank V-141. Next the extract is pumped to raw extract storage tank V-120 or V-121.

Residual alcohol and salt remaining in the oil is reduced by adding 2 percent water by volume with circulation for 2 to 3 minutes maximum. The mass is settled for 4 hours, and the water layer is pumped to the batch still charge tank.

At this point in the process, the oil may be finished oil or an intermediate pass oil depending on the number of acid treatments that were previously applied. If the oil is finished oil, a small quantity of aqueous sodium hydroxide and a stabilizer are added and mixed for 30 minutes. Transformer oil, however, is considered finished oil, and only stabilizer is added.

BHT stabilizer or the equivalent is added to all last-pass oils. This helps prevent deterioration during distillation.

Alcohol Recovery from Oil and Brine

Referring to the process flow diagram in Fig. 24-5, at this stage, oil still contains 1 to 2 percent by volume alcohol, which must be removed and recovered. The oil is pumped through steam-heated exchangers E-140 A and B and E-141 A and B, and the vapors are allowed to flash off in flash drums V-140 A and B. The hot oil flows by gravity to oil rundown tanks V-143 A to D. Upon completion of the run, the oil is steam stripped for about an hour or until its temperature has dropped to about 220°F.

The oil is then air blown to remove residual alcohol and water. After air blowing, the oil is cooled by recirculating through water-cooled heat exchangers E-144 A and B and pumped to storage. If the oil is transformer oil or last-pass white oil, it is cooled to about 160°F and pumped to unfiltered oil storage tanks. Intermediate oils are further cooled to 100 to 110°F using refrigerated water. They are then pumped to their respective storage tanks or pumped directly to acid treatment.

The brine draws and water wash from neutralizers as well as weak brine draws from desalters are collected in batch still charge tank V-141. The alcohol concentration will vary from 3 to 15 percent depending on alcohol losses and the source of brine. It is important therefore to recover as much of this alcohol as possible. Feed is pumped to still V-142 and the mass is heated to about 214°F. Alcohol vapors are condensed in overhead condenser E-143 and flow to run down tank V-144 A to C. The still bottoms contain salts that are used to neutralize SO_2 and SO_3 fumes. The still bottoms are pumped to fume contact tanks.

Extract Desalting

Referring to the process flow diagram in Fig. 24-4, raw sulfonate extracts contain sodium carbonate, sodium sulfate, and other salts that are not removed in the neutralization step. Most of these remaining salts are removed from the raw extract in the desalting step by adding dry soda ash until a bottom brine layer separates. This brine layer is obtained by the addition of relatively small amounts of soda ash. In order to minimize soda ash consumption, the bulk of water in the extract is removed as a weak brine. The salt and other extraneous color bodies being very water soluble tend to remain dissolved in the brine layer and are thus separated.

Following the weak brine draw, additional dry soda ash is needed to form a strong brine layer. This strong brine is withdrawn and pumped to a mix tank where it is diluted with water and subsequently used for neutralization of acidified oils.

The raw extract is first pumped from storage to sulfonate desalter V-125 and heated by steam coils to 140°F. The desalter vessel (Fig. 24-10) is similar to the neutralizer, with a conical bottom, steam coils, and multiple product draw-off nozzles to separate the different layers. Dry soda ash is dissolved in small mixing vessel V-123 through which the extract is recirculated. After the first weak brine (15 to 18°Be) is formed, the mass is settled for an hour and then the brine is removed.

More soda ash is added until the strong brine has reached 34°Be concentration and the mass is recirculated for 1 hour. Strong brine is withdrawn after 2 hours and again after 12 hours of additional settling. Thus the total settling time is 14 hours. The desalted extract is recirculated and the alkalinity adjusted by the addition of 66°Be sulfuric acid. The volume of acid is determined by laboratory tests. A final brine draw is made before the extract is pumped out. The desalted extract is pumped to desalted extract storage tanks V-126 and V-127.

Extract Evaporation

The desalted adjusted extract is fed from storage to a continuous still (see Fig. 24-4). The still consists of preheater E-120, tubular heat exchanger E-122, and flash drum V-129. The sulfonate drum temperature in the flash drum bottom is measured and the still feed rate is manually adjusted to maintain this temperature. Alcohol and water vapors are flashed in the flash drum, cooled in the preheater, and condensed in overhead condenser E-121. The sulfonate flows to a rundown tank from the flash drum under level control.

Soda Ash Solution Preparation

In addition to soda ash solution obtained from the desalting step, additional soda ash solution is required. The solution is made up in diluting tank V-122 and stored in storage tanks outside battery limits. Bags of soda ash are manually dumped in slurry tank V-123.

SULFONATE BLENDING

Referring to the process flow diagram in Fig. 24-6, from rundown tank V-130 A and B, the sulfonate is pumped to "sulfonate adjust tank" V-160 where it is diluted from about 72 percent pure sulfonate to about 62 percent with oil and water. Mixing is accomplished with air agitation.

From the dilution tank the adjusted sulfonate is pumped to storage tanks V-161 or V-162, depending on the type of sulfonate. If the sulfonate is out of specs,. it is pumped to off-specs storage tank V-163 instead. From these storage tanks, custom blends can be mixed in final blending tank V-164 for shipment. The sulfonate rundown tank, storage tanks, and blend tanks are held at a temperature of 170 to 200°F.

Alcohol Make-Up

The alcohol recovered from various stills will average about 65 percent isopropanol. This is diluted with water to a 55 percent concentration for reuse and pumped to V-114, the 55 percent alcohol storage tank (Fig. 24-3). When 55 percent alcohol requires a make-up supply, fresh isopropanol is pumped from virgin alcohol storage tank V-059 (Fig. 24-1) and the proper proportion of water is added.

Product Blending and Loading

Referring to the process flow diagram in Fig. 24-6, five finished oil blending tanks of various sizes, typically ranging from 1.500 to 15,000 gallons, are provided for mixing oils to obtain custom blends. Mixing is done by pump recirculation.

Loading scales are provided for filling drums. A drum is placed on the scale and the proper weight of oil is added. When transformer oil is being loaded, each drum is flushed with about a gallon of oil to remove contaminants. This flush oil is pumped back to unfiltered storage. One mix tank and separate scales are provided for sulfonate handling and drumming.

Petrolatum Blending

Microcrystalline wax is received in drums. The drums are heated with a steam coil drum warmer until the wax has melted at about 150°F. While the drums are heating, the correct amount of white oil is pumped into one of two blend tanks, V-165 A and B, and heated to about 150°F. After the wax has melted, the drum contents are pumped into the blend tank and the air is agitated for about an hour. The blend is then pumped to product drums at the drum scale.

Finishing Treatment

Final purification of white oils is accomplished either by contacting the oil with clay or passing through a static bed of bauxite to remove impurities by adsorption. For certain feedstocks, both these treatments may be required. This process improves the product color, odor, and taste, which is very important for white oils for use in personal care and pharmaceutical products. Generally, percolation treatment is the preferred choice because of better quality product compared with that from clay treatment.

Clay Treatment

Referring to the process flow diagram in Fig. 24-12, from unfiltered oil storage tanks, the oil is pumped to clay mix tanks V-153 A and B, where it is treated with active clay to absorb any impurities left in the product. The treatment is performed batchwise in two vessels with mixers and steam coils. The clay (5 to 8 percent by weight [Wt %]) is added to the oil and agitated at 150 to 160°F for 1 hour. (Transformer oil requires about 2 percent weight clay.) The clay is separated from oil by passing it through two vertical leaf filters, F-150 A and B. From filters, the oil goes to a filter rundown tank, V-154 A and B, and after testing, to product storage tanks V-155, V-156, and V-157. A stabilizer is added in the rundown tank to the filtered oil. The clay cake is manually discharged from the filter and disposed off plot as landfill.

Percolation Filtration

In percolation filtration, the oil is passed through a static bed of bauxite to remove impurities by adsorption. This process removes coloring matter in oil and also some other constituents, resulting in a product that is colorless, odorless, and tasteless. The flow through the bed is continued until the effluent no longer meets the desired specifications. The bed is then drained of oil, washed with a solvent, and steamed. The bauxite is next removed from the bed and regenerated for reuse by burning off undesirable materials, in a rotary kiln, under controlled conditions.

Charge Stock

The charge stocks are low-viscosity white oils, high-viscosity white oils, and transformer oils, all of which have been acid treated, neutralized, solvent extracted, and stripped. These stocks are stored in V-150, V-151, and V-152. Tank V-204 (Fig. 24-13) is provided to store oil of intermediate viscosity, which is used for blending finished products.

Filter Preparation

Refer to the process flow diagram in Fig. 24-14. A percolation plant has multiple filters, for filtration of transformer oil and low- and high-viscosity white oils. A bauxite regeneration plant regenerates spent bauxite by burning off adsorbed hydrocarbons in a kiln.

The percolation vessel or filter is a cylindrical vessel loaded with activated bauxite from feed hoppers V-210A and V-210B. Feed enters from the top and the product exits through a bottom nozzle (Fig. 24-18). A medium-sized filter may hold 10 to 15 tons of bauxite. One hopper is used for fresh bauxite and the other for reactivated bauxite. Bauxite is transferred via conveyer P-210. The entire content of hopper F-210A is charged to a filter and make-up bauxite is added from V-210 B as needed to fill the vessel.

Valves are next set to feed white oil. For processing high-viscosity white oil, feed must go through feed preheater E-210. A number of charge pumps, G-901 to 908, are provided for high-viscosity white oil, for low-viscosity white oil, and for transformer oil. Generally, it is desirable to use a pump exclusively for a certain stock.

Pumping of oil to "soak" the filter is started. Soaking is a procedure used to wet all bauxite and displace all air from the bed to prevent channeling. Oil is charged to the filter as rapidly as possible to complete soaking within a 8-hour period. Oil flow is varied by manipulating the by-pass regulator at the charge pump. Pressure at the top of the filter is maintained at about 35 lb/in^2 by intermittent release of air and by manipulating the oil flow. When clear oil begins to flow from the air bleed valve, the oil feed is stopped and the filter allowed to soak for about 8 hours.

PERCOLATION

Feed to the filter is started, and the flow is adjusted to the required level. Pressure at the top of filter at maintained at approximately 35 lb/in^2. Valves are next set so that the white oil filtrate goes through a cartridge filter.

First, 5 to 10 gallons of oil from the filter are discarded to drain trough and sent to slop. When the filtrate effluent is tasteless, the flow is diverted to rundown tanks V-220 to 229 (see the process flow diagram in Fig. 24-16).Certain tanks must be used exclusively for specific oil to prevent cross-contamination. Necessary inhibitors are added as required. The filtrate is sampled every 8 hours and tested for appropriate specifications.

When the filter effluent goes off test or the design quantity of finished white oil (high or low viscosity) has been collected, the effluent is switched to "push oil" rundown tank V-227.

The effluent in push oil tank V-227 is collected until all the white oil has been displaced.

The effluent is sampled and tested every 8 hours. When the effluent goes off specification, the filter is shut down.

The product in the rundown tanks is tested and if approved, it is pumped to the appropriate storage tanks. Off specification material is pumped to V-228 or V-229 for reprocessing.

Shut Down, Washing, and Steaming

Oil flow to the filter is stopped and the effluent flow diverted to drain oil slop tank V-226. Oil in the filter is displaced with 30 lb/in^2 compressed air. When all the oil has been displaced, air pressure is bled off.

The spent bauxite is washed with naphtha to remove oil and then stripped with steam to remove naphtha from the filter in the following manner. First, "steaming naphtha" (recovered from previous stripping) is pumped from tank V-236, heated to 150 to 180°F, and fed to the filter for about 2 hours. This is followed by fresh (distilled) naphtha from tank V-237 at the same conditions for about 6 hours. All washings go to still feed tank V-230.

Then liquid in the filter is displaced with 30 lb/in^2 superheated steam, and the bed is purged with superheated steam for about 6 hours. Steaming is continued until the vapor from the filter no longer has a naphtha odor. The steam/naphtha vapors are condensed and collected in steam/naphtha tank V-236. After settling, the water layer is drawn from V-236 to a sewer. Wet naphtha (steaming naphtha) is used for the next wash cycle.

After steaming, the spent bauxite is discharged from the filter onto conveyer P-212 and conveyed to bucket elevator P-213 and elevated to kiln feed hopper V-240.

BAUXITE PROCESSING

Spent Bauxite Reactivation

Refer to Fig. 24-17. Rotary kiln H-240 is heated by burning natural gas. The combustion product flows through the kiln, providing the necessary temperature for bauxite regeneration.

When the kiln exhaust gases reach the required temperature, spent bauxite is fed by gravity to the kiln at a predetermined rate. The heat input to the kiln is regulated to maintain the temperature of bauxite leaving the kiln at 1100 to 1200°F. Bauxite flows through the kiln countercurrent to exhaust gases.

Hot bauxite from the kiln is elevated by bucket elevator P-214 and fed by gravity to bauxite cooler H-241. The bauxite is cooled to about 120°F by cooling water.

Cooled bauxite is fed to gravity separator M-210, and the high-density fraction is separated from the bulk of the bauxite. The high-density fraction is discarded to carts, which are used to transport this material to bucket elevator P-217. The high-density material is elevated to hopper V-241, where it is stored for periodical removal by trucks. Fines are also removed by separator M-210, collected by twin cyclones, and discharged to a drum for disposal.

The reactivated bauxite from the gravity separator is discharged onto conveyer P-215, raised by bucket elevator P-216, and discharged into reactivated bauxite feed hopper V-210A. This material is loaded into the next empty filter as described earlier.

NEW BAUXITE REACTIVATION

New bauxite is received in bags and stored next to the kiln. When the make-up bauxite feed hopper level permits, new bags of bauxite are transferred by forklift trucks or other suitable means to the base of elevator P-213. The bags are manually emptied into the boot of the elevator and transferred to kiln feed hopper V-240. Because the make-up new bauxite is only 2 percent of the total, new bauxite is processed through the kiln during periods when all the filters are on stream.

The new bauxite is processed through the kiln in the same manner as recycled/spent bauxite, except that the temperature of the bauxite leaving the kiln is maintained at about 800°F. The lower temperature is used because the purpose of this operation is only to dry the new bauxite.

NAPHTHA RECOVERY

Batch Distillation

Refer to Fig. 24-15. The naphtha/oil mixture is pumped from still feed tank V-230 to either of two direct gas-fired batch stills, V-231 or V-232. The batch is heated and naphtha distilled. A small amount of steam is sparged into the still to promote agitation. Vapors from the still are condensed and collected in still distillate rundown tank V-233.

When the liquid temperature in the still reaches about 400°F, the distillate is diverted into heavy ends tank V-234. Distillation is continued until the still liquid has a satisfactorily high flash point.

The oil residue in the still is circulated through a cooler, E-232, until the temperature of the residue is about 200°F. The residue is then pumped to bottoms storage tank V-235.

From the distillate rundown tank V-233, water is drawn after settling and distillate is pumped to distilled solvent storage tank V-237. After settling, water is drawn from the heavy ends rundown tank V-235, and the heavy ends are disposed of as fuel oil. The bottoms, depending on their quality, may be reprocessed or disposed of as fuel oil. Make-up naphtha can be delivered in tank trucks and unloaded directly into distilled naphtha tank V-237.

White Oils Processing Sequence

Refer to Fig. 24-11. The processing sequence described here is typical and the number of acid shots required and acid volumes required and other parameters is optimized by laboratory tests for a given feedstock and product specifications.

Low-Viscosity White Oil

Five times through the settler and four times through the neutralizer as follows:

First Pass: 4.5 vol % oleum (L.V-4), then neutralized

Second Pass: 6.8 vol % oleum (L.V-3), then neutralized

Third Pass: 9.0 vol % oleum (L.V-2), then neutralized

Fourth Pass: 9.0 vol % oleum (L.V-1), not neutralized

Fifth Pass: 9.0 vol % oleum (L.V-1-1), final neutralization

High-Viscosity White Oil

Six times through settler and four times through the neutralizer as follows:

First Pass: 4.6 vol % oleum (H.V-4), then neutralized

Second Pass: 4.5 vol % oleum (H.V-3), then neutralized

Third Pass: 9.4 vol % oleum (H.V-2), then neutralized

Fourth Pass: 9.0 vol % oleum, first shot (H.V-1), not neutralized
9.0 vol % oleum, second shot (H.V-1-1), not neutralized
9.0 vol % oleum, third shot (H.V-1-1-1), final neutralization

Where two or more shots are required, the oil is settled 24 hours between the shots and the sludge is removed each time. The oil is neutralized only after the last shot.

The material balance for a commercial white oil manufacturing unit producing low- and high-viscosity white oils is shown in Tables 24-6 and 24-7. The yield of white oils from the acid treating process is shown in Table 24-8. Typical batch cycle times for various steps are shown in Tables 24-11 to 24-16.

HYDROTREATING PROCESS

Traditionally, white oils have been produced by oleum/clay treatments of vacuum distillates or lubricating oils base stocks. In a conventional lube plant, lubricating oil base stocks of 85 to 100 viscosity index (VI) are produced by solvent extraction and the dewaxing of vacuum distillates of appropriate viscosity. Solvent extraction removes a part of the aromatics present in feed with a solvent like furfural; solvent dewaxing removes paraffin wax using a solvent (methyl ethyl ketone [MEK], methyl isobutyl ketone [MIBK], etc.) to separate wax and lower the pour point oil.

Acid treatment of feedstock removes aromatics by converting them into acid sludge or into petroleum sulfonates, which are extracted by isopropyl alcohol. The process is repeated a number of times until the aromatics are completely removed. Thus a part of the feedstock is converted to sludge or petroleum sulfonate and the yield of white oil is typically less than 66 percent by weight.

In the hydrotreating route, white oil is manufactured by deep hydrogenation of the base stock to achieve a near total saturation of aromatics. Noble metal and nickel on an alumina base has proven to be a most effective catalyst in achieving the trace product aromatics level demanded by white oils. Unfortunately, a nickel catalyst is more susceptible to poisoning by sulfur and nitrogen in feed. Hydrogenation is done in a two-stage process. The first stage must not only achieve its goal of maximum aromatic saturation but also must reduce feed contaminants (sulfur and nitrogen) to a ppm (parts per million) level. This is done with a nonprecious metal catalyst such as Ni-Mo or Co-Mo on alumina-type hydrogenation catalysts. The second stage uses a noble metal Pt-Pd on an alumina base to completely saturate the aromatics. With noble metal catalysts, the run length and catalyst life is much longer.

Depending on the feedstock, two process configurations are popular:

- In case feed is lube base stock from a conventional lube oil plant, white oils, both technical and medicinal grades, are made by a two-stage hydrogenation process. However, if the feed is vacuum distillate, it must undergo a dewaxing step to get the required pour point of white oil. The hydrotreating step does not change the pour point of the oil.
- In case the feed is from a lube hydrocracking plant, comprising lube hydrocracker, hydroisomerization, catalyst dewaxing, and hydrodewaxing units, medicinal-grade white oils can be produced by adding one more hydrotreating reactor to completely remove polynuclear aromatics and meet FDA white oil specifications.

Two-Stage Hydrotreating

The process flow diagram for a two stage hydrotreating plant for white oil manufacture is shown in Fig. 24-19. In the first stage, feed is desulfurized to completely remove sulfur, nitrogen and partially saturate aromatics. Desulfurization, denitrification, and aromatic saturation reactions take place over a nonprecious metal catalyst on an alumina base. The catalyst in the first stage of the hydrogenation operation can be any sulfur-resistant nonprecious metal hydrogenation catalyst such as Co, Mo, Ni, or W on an alumina base. The first-stage reactor operates at 650 to 715°F and a pressure of 2500 to 3000 lb/in^2. Hydrogen and feed are heated upstream of the first reactor and are separated downstream of the reactor into main product and by-product hydrogen sulfide and light hydrocarbons in a stripping column. The first-stage hydrotreating removes most of the sulfur and nitrogen in the feed, which are poison to the second-stage catalyst and partially saturate aromatics.

The partially hydrogenated oil from the first stage is then subjected to less severe operating conditions in the second-stage reactor. Average temperatures in the second-stage hydrogenation are 50 to 75°F lower than that of the first-stage reactor.[2] In the second reactor, feed and hydrogen is passed over a precious metal catalyst to achieve a very low level of aromatics, especially the polynuclear compounds. The second-stage catalyst is platinum group metals such as platinum, palladium, rhodium on an alumina base. The precious metal content of the second-stage catalyst is typically 0.6 percent by weight. Typical operating conditions, feed, and product properties are presented in Tables 24-3 and 24-4. Aromatics, and more particularly polynuclear aromatics, are completely saturated. Hydrogen and feed are heated upstream of the second reactor and are separated downstream of the reactor into main product and by-product light hydrocarbons in a stripping column As a result of hydrogenation, all aromatics, mono-, di-, and polyaromatics are almost completely removed, and the product meets UVA absorption specifications (ASTM D 2269).

HYDROPROCESSED BASE STOCKS

A block flow diagram of the process is shown in Fig. 24-20. Feed in this case is high pour waxy lube from a lube hydrocracking unit. Waxy lube is processed in a hydroisomerization/catalytic dewaxing unit where feed is treated with hydrogen in the presence of a dual functional noble metal/shape selective molecular sieve catalyst to hydroisomerize and selectively crack paraffins in the feed to produce low pour lube base stock. The low pour lube base stock is treated with hydrogen in a hydrofinishing unit, which saturates any olefinic hydrocarbons formed in the hydroisomerization/dewaxing stage. The lube base stock exiting from the hydrofinishing unit has a very low aromatic content and a very high VI (more than 130). This very high viscosity index lube (VHVI) lube is hydrogenated with a noble metal-alumina catalyst at high pressure (3700 lb/in^2) in a single-stage reactor to produce medicinal-grade white oils. Operating conditions are presented in Table 24-5.

PETROLEUM SULFONATES

Sodium petroleum sulfonates are produced as a by-product during the manufacture of white oil and transformer oils by the action of oleum on heavy petroleum distillates. During the process, a part of the feedstock is polymerized to tar, some of the feed is sulfonated to petroleum sulfonate, and this is extracted by the use of a solvent, isopropyl alcohol. The typical molecular weights of the product produced are 410 to 450 for light feeds and 550 to 650 for heavy feeds. The product is sold as approximately 60 percent active constituent, 30 percent oil, and the rest water and sodium sulfate produced during processing.

Sodium petroleum sulfonates are unique materials because of their dual action as surfactants and as rust inhibitors. The polar nature of the sulfonate end of the molecule functions as a typical anionic surfactant while the "hydrocarbon tail" provide the bridge to the nonpolar phase.

Sodium sulfonates have long been used in metal working fluids to act as emulsifiers and corrosion inhibitors. In the mining industry, sulfonates are used as flotation agents for various metals and minerals. The ability of the sulfonates to wet the surface of the ore particles renders the ore hydrophobic. This allows ore to adhere to air bubbles and move to the surface in froth.

Sodium sulfonates are often converted to calcium or barium sulfonates for use as motor oil and fuel oil additives for detergency and rust-inhibiting properties. Anticorrosion formulations such as preservative oils, automotive rust proofing are compounded with petroleum oils, waxes, petrolatums, and various synthetic materials. High molecular weight sodium sulfonate adheres to metal surfaces and provides an excellent moisture barrier. These are compatible with a wide range of solvents and thin film applications.

Other applications are in dry cleaning solvent, leather processing, printing inks, and oil well drilling fluids. In fact, demand for petroleum sulfonates is growing much faster than that of white oils, and petroleum sulfonate is now the main product and white oils have become the by-product of these units. The test method is ASTM D-3712 for all properties except color. For color, the sample is diluted to 6 percent active sulfonate in white mineral oil and measured by ASTM D 1500.

PETROLATUMS

Petrolatums function as carriers, emollients, lubricants, base ingredients, binders, protective coatings, plasticizers, waterproofing, release agents, and softeners. Some uses of petrolatums are described next

Petrolatums are homogeneous mixtures of oily and waxy long chain nonpolar hydrocarbons. Their hydrating properties set a standard against which other moisturizers are compared. Odorless and tasteless, they range in color from white to yellow and differ from one another in consistency and shear strength.

Personal care and pharmaceutical formulators often choose petrolatums as a formulation base. Petrolatums add lubricity and moisture resistance to lotions, creams, ointments, and hand cleaners. These meet FDA 21 CFR 172.880. Medium-consistency grades are used in many fine cosmetics and ointments. Softer grades are used in petroleum jellies and high-solid ointments. Blending and composition of petrolatums are shown in Tables 24-9 and 24-10.

Food processors rely on FDA grades for uses that range from baking to candy making candy packaging. In jar candles, the addition of petrolatums affects the crystallinity of wax, and gives a smoother, more pleasing finish. In metal polishes and buffing compounds, petrolatum protects against moisture and corrosion.

Petrolatums are often referred to as mineral jellies. Mineral jelly is an amorphous mixture of microcrystalline wax, mineral oil, and paraffin wax. Straight unblended petrolatums are a by-product of bright stock oil processing, the heaviest viscosity lube oil, and are mostly used as feedstock for microcrystalline wax production.

TABLE 24-3 White Oil by Hydrotreating Process Typical Operating Conditions

Property	Units	First stage	Second stage
Catalyst		Ni-Mo/Alumina base	Pt-Pd/Alumina base
Reactor pressure	lb/in^2	2,500	2,500
Temperature	°F	650	500
Space velocity	WHSV	0.25	0.25
Hydrogen/Feed ratio	SCF/BBL	2,500	2,500

TABLE 24-4 Two-Stage Hytrotreating Process for White Oil Manufacture Feed and Product Properties

Property	Units	Feed	Second-stage feed	Product
API gravity		17.7	22.8	23.9
Specific gravity		0.948	0.917	0.911
Yield	Vol %	100		90
Boiling range	°F	700–900	650–900	640–900
Viscosity, 100°F	SUS	900	536	503
Viscosity, 100°F	cSt	194.3	115.7	108.6
Sulfur content	ppmw	2,500	60	
Nitrogen	ppm	320	5	
Saybolt color				+30
Aromatics	ppmw			<100

TABLE 24-5 Approximate Operating Conditions for White Oil Manufacture Unconverted Hydrocracker Bottom Feed

Process condition	Units	ISOM-CDW reactor	Hydrofinishing reactor	Deep hydrotreating reactor
WABT	°F	530	500	485
Total pressure	lb/in^2	700	700	3,700
LHSV	HR^{-1}	1.0	1.5	0.5
H_2/Oil ratio	$N\ M^3/M^3$	850	850	400
H_2 consumption	$N\ M^3/M^3$	80	25	10

TABLE 24-6*a* Low-Viscosity White Oil Material Balance Acid Treating, Low-Viscosity Raw Oil (L.V-N)
Basis: 45,000 gal/batch

	Stream number and name				
	1 Raw LV oil	2 Oleum	3 Acid-oil	4 Sludge	5 Sour oil
gal/batch, 60°F	45,000	2,025	47,025	4,045	42,980
lb/batch	330,000	32,000	362,000	50,600	311,400
Specific gravity	0.879	1.9	0.918	1.5	0.873
Temperature,°F	80	80	110	110	110
Viscosity, cP, 100°F	12	29		10,000	12

Note: Stream numbers correspond to those shown in Figs. 23-2 and 23-3.

TABLE 24-6*b* Low-Viscosity First-Pass Oil (L.V-4A) Neutralization and Extraction

	Stream number and name							
	5 Sour oil	6 Carbonate solution	7 Oil + carbonate	10 Brine draw	11 Raw extract	12 Neutral oil	8 55% Alcohol	9 Water
gal/batch, 60°F	42,800	6,750	49,550	6,440	12,616	41,900	11,250	900
lb/batch	311,400	66,900	378,300	60,170	100,320	306,000	82,400	7,500
Specific gravity	0.873	1.19	0.915	1.12	0.96	0.872	0.876	1.00
Temperature, °F	110	130	110	120	120	120	80	80
Viscosity, cP, 100°F	12	2		2	8	12	1	1
Components, lb								
CO_2 gas			1,740					
Sulfonates					4,400			
Water		54,860		45,670	57,000	3,330	42,200	
Alcohol				6,100	31,500	2,600	40,200	
Salts		12,040		8,400	6,000			
Oil					1,420	300,070		

TABLE 24-6*c* Low-Viscosity First-Pass Oil (L.V-4) Oil Distillation and Cooling

	Stream number and name			
	12 Neutral oil	13 Water- alcohol	14 Stripped oil	14 Inter oil
gal/batch, 60°F	41,900	800	41,100	41,100
lb/batch	306,000	5,900	300,070	300,070
Specific gravity	0.872	0.89	0.872	0.872
Temperature, °F	135	100	270	110
Pressure, lb/in^2	75	Gravity	Gravity	30
Viscosity, cP, operating temperature	7.5	1	2	11
Components, lb				
CO_2 gas				
Sulfonates				
Water	3,330	3,330		
Isopropanol	2,600	2,600		
Salts				
Oil	300,070			

TABLE 24-6*d* Low-Viscosity Second-Pass Oil (L.V-4) Acid Treating

	Stream number and name				
	1 L.V-4 oil	2 Oleum	3 Acid and oil	4 Sludge	5 Sour oil
gal/batch, 60°F	45,000	3,060	48,060	5,350	42,710
lb/batch	327,500	48,400	375,900	66,900	309,000
Specific gravity	0.873	1.9	0.94	1.5	0.868
Temperature, °F	80	80	110	110	110
Viscosity, cP, 100°F	12	29		8,000	12

TABLE 24-6*e* Low-Viscosity Second-Pass Oil (L.V-3A) Neutralization and Extraction

	Stream number and name							
	5 Sour oil	6 Carbonate solution	7 Oil + carbonate	10 Brine draw	11 Raw extract	12 Neutral oil	8 55% alcohol	9 Water
gal/batch, 60°F	42,800	6,075	48,875	7,190	14,480	37,500	11,250	900
lb/batch	309,000	60,200	369,200	68,170	118,200	271,209	82,400	7,500
Specific gravity	0.868	1.19	0.906	1.14	0.98	0.863	0.876	1.00
Temperature, °F	110	130	110	120	120	120	80	80
Viscosity, cP, 100°F	12	2		2	10	12	1	1
Components, lb								
CO_2 gas			1,530					
Sulfonates					22,500			
Water		49,350		51,750	51,150	2,920	42,200	
Alcohol				6,420	31,500	2,280	40,200	
Salts		10,850		10,000	5,800			
Oil					7,250	266,000		

TABLE 24-6*f* Low-Viscosity Second-Pass Oil (L.V-3) Oil Distillation and Cooling

	Stream number and name			
	12 Neutral oil	13 Water-alcohol	14 Stripped oil	14 Inter oil
gal/batch, 60°F	37,500	700	36,800	36,800
lb/ batch	271,200	5,200	266,000	266,000
Specific gravity	0.863	0.8	0.863	0.863
Temperature, °F	135	100	270	110
Viscosity, cP, operating temperature	7.5	1	2	10.5
Components, lb				
CO_2 gas				
Sulfonates				
Water	2,920	2,920		
Isopropanol	2,280	2,280		
Salts				
Oil	266,000			

TABLE 24-6*g* Low-Viscosity Third-Pass Oil (L.V-3) Acid Treating

	Stream number and name				
	1 L.V-3 oil	2 Oleum	3 Acid and oil	4 Sludge	5 Sour oil
gal/batch, 60°F	45,000	4,050	49,050	6,075	42,975
lb/batch	325,000	64,000	389,000	80,900	308,100
Specific gravity	0.868	1.9	0.94	1.6	0.862
Temperature, °F	80	80	110	110	110
Viscosity, cP, 100°F	12	29		5,000	12

TABLE 24-6*h* Low-Viscosity Third-Pass Oil (L.V-2A) Neutralization and Extraction

	Stream number and name							
	5 Sour oil	6 Carbonate solution	7 Oil + carbonate	10 Brine draw	11 Raw extract	12 Neutral oil	8 55% alcohol	9 Water
gal/batch, 60°F	42,900	4,050	46,950	6,850	9,600	39,500	9,000	900
lb/batch	308,100	40,100	348,200	58,990	77,800	283,840	65,950	7,500
Specific gravity	0.862	1.19	0.89	1.03	0.97	0.862	0.876	1.00
Temperature, °F	110	130	110	120	120	120	80	80
Viscosity, cP, 100°F	12	2		2	10	12	1	1
Components, lb								
CO_2 gas			1,020					
Sulfonates					21,600			
Water		32,800		49,120	22,100	3,330	33,750	
Alcohol				6,200	23,400	2,600	32,200	
Salts		7,220		3,670	3,860			
Oil					6,840	277,910		

TABLE 24-6*i* Low-Viscosity Third-Pass Oil (L.V-2) Oil Distillation and Cooling

	Stream number and name			
	12 Neutral oil	13 Water-alcohol	14 Stripped oil	14 Inter oil
gal/batch, 60°F	39,500	800	38,700	38,700
lb/batch	283,840	5,900	277,910	277,910
Specific gravity	0.862	0.89	0.862	0.862
Temperature, °F	135	100	270	110
Viscosity, cP, operating temperature	7	1	2	10.5
Components, lb				
CO_2 gas				
Sulfonates				
Water	3,330	3,330		
Isopropanol	2,600	2,600		
Salts				
Oil	277,910			

TABLE 24-6*j* Low-Viscosity Fourth-Pass Oil (L.V-2): First Shot Acid Treating

	Stream number and name				
	1 L.V-2 oil	2 Oleum	3 Acid and oil	4 Sludge	5 Sour oil
gal/batch, 60°F	45,000	4,050	49,050	5,060	44,500
lb/batch	323,000	64,000	387,000	70,000	317,000
Specific gravity	0.862	1.9	0.947	1.66	0.855
Temperature, °F	80	80	110	110	110
Viscosity, cP, 100°F	12	29		2,000	12

TABLE 24-6*k* Low-Viscosity Fourth-Pass Oil (L.V-2): Second Shot Acid Treating

	Stream number and name				
	1 L.V-2 oil	2 Oleum	3 Acid and oil	4 Sludge	5 Sour oil out
gal/batch, 60°F	44,500	4,050	48,550	4,200	44,500
lb/batch	317,000	64,000	381,000	64,500	316,500
Specific gravity	0.855	1.9	0.94	1.85	0.853
Temperature, °F	80	80	110	110	140
Pressure, lb/in^2	30	40	45	50	50
Viscosity, cP, 100°F	12	29		1,000	12

TABLE 24-6*l* Low-Viscosity Fourth-Pass Oil (L.V-2) Neutralization and Extraction

	Stream number and name							
	5 Sour oil	6 Carbonate solution	7 Oil + carbonate	10 Brine draw	11 Raw extract	12 Neutral oil	8 55% alcohol	9 water
gal/batch, 60°F	44,500	3,040	47,540	5,180	3,856	44,200	4,500	900
lb/batch	316,500	30,100	346,600	44,030	30,500	311,680	32,900	7,500
Specific gravity	0.853	1.19	0.87	1.02	0.95	0.845	0.876	1.00
Temperature, °F	140	130	140	140	140	140	80	80
Viscosity, cP, 100°F	12	2		2	10	7	1	1
Components, lb								
CO_2 gas			790					
Sulfonates					690			
Water		24,690		35,910	18,300	3,750	16,800	
Isopropanol				3,420	9,750	2,990	16,100	
Salts		5,410		4,000	1,480			
Oil				700	280	305,000		

TABLE 24-6*m* Low-Viscosity Fourth-Pass Oil (L.V-1-1) Oil Distillation and Cooling

	Stream number and name			
	12 Neutral oil	13 Water-alcohol	14 Stripped oil	14 Unfiltered oil
gal/batch, 60°F	44,200	900	43,300	43,300
lb/batch	311,680	6,680	305,000	305,000
Specific gravity	0.845	0.89	0.845	0.845
Temperature, °F	140	100	270	160
Viscosity, cP, operating temperature	7	1	2	4.5
Components, lb				
CO_2 gas				
Sulfonates				
Water	3,750	3,750		
Isopropanol	2,930	2,930		
Salts				
Oil	305,000			

TABLE 24-6*n* Low-Viscosity White Oil Oil-Clay Contacting and Filtering

	Stream number and name				
	14 Unfiltered oil	15 Clay	16 Oil-clay	17 Spent clay	18 Filtered oil
gal/batch, 60°F	7,500		7,500		7,120
lb/batch	52,800	2,640	55,440	5,280	50,160
Specific gravity	0.845	2.5	0.877	1.27	0.845
Temperature, °F	80		160	160	160
Viscosity, cP, operating temperature	16		6		4.2
Components, lb					
Clay			2,640	2,640	
Oil			52,800	2,640	

TABLE 24-6*o* Low-Viscosity White Oil Desalting of Sulfonate Extract (Low-Viscosity White Oil Manufacture)

	Stream number and name				
	101 Raw extract	102 Soda ash	103 Desalted extract	104 Weak brine	105 Strong brine
gal/batch, 60°F	26,000		16,420	6,800	4,050
lb/batch	210,100	20,000	122,500	63,100	44,500
Specific gravity	0.97		0.9	1.12	1.32
Temperature, °F	80		140	140	140
Viscosity, cP, operating temperature	13		6	2	3
Components, lb					
Sulfonate	31,600		31,600		
Water	97,100		23,750	48,610	31,150
Isopropanol	60,100		57,050	3,050	
Salts	11,200		860	10,580	13,350
Oil	10,100		9,240	860	

TABLE 24-6*p* Sulfonate Finishing: Distillation and Adjusting of LV Extract

	Stream number and name					
	103 Desalted extract	104 Water-alcohol vapor	105 Sulfonate	106 Diluting water	107 Diluting oil	108 Adjusted sulfonate
gal/batch, 60°F	16,420	11,430	4,990	83	587	5,660
lb/batch	122,550	79,133	43,367	692	4,290	48,349
Specific gravity @ 60°F	0.9	0.83	1.04	1	0.876	1.026
Temperature, °F	100	200	320	80	80	190
Viscosity, cP, operating temperature	12		30	1	140	500
Components, lb						
Sulfonates	31,600		31,600			31,600
Water	23,750	22,083	1,667			2,359
Isopropanol	57,050	57,050				—
Salts	860		860			860
Oil	9,240		9,240			13,530

TABLE 24-7*a* High-Viscosity White Oil Material Balance Acid Treating, High-Viscosity Raw Oil (H.V-N)

	Stream number and name				
	1 Raw HV oil	2 oleum	3 Acid-oil	4 Sludge	5 Sour oil
gal/batch, 60°F	45,000	2,070	47,070	4,140	42,500
lb/batch	338,000	32,700	370,700	51,800	318,900
Specific gravity	0.904	1.9	0.947	1.5	0.901
Temperature, °F	80	80	110	110	110
Viscosity, cP, 100°F	120	29		10,000	120

Note: Stream numbers correspond to those indicated in Fig. 23-2 and Fig. 23-3.

TABLE 24-7*b* High-Viscosity First-Pass Oil (H.V-4A) Neutralization and Extraction

	Stream number and name							
	5 Sour oil	6 Carbonate solution	7 Oil + carbonate	10 Brine draw	11 Raw extract	12 Neutral oil	8 55% alcohol	9 water
gal/batch, 60°F	42,500	6,750	49,250	9,800	10,555	40,310	11,250	900
lb/batch	318,900	66,900	385,800	89,910	84,500	300,243	82,400	7,500
Specific gravity	0.901	1.19	0.94	1.09	0.96	0.894	0.876	1.00
Temperature, °F	110	130	110	130	130	130	80	80
Viscosity, cP, 100°F	120	2		2	10	120	1	1
Components, lb								
CO_2 gas			1,750					
Sulfonates					14,900			
Water		54,860		70,750	31,200	3,330	42,200	
Alcohol				8,300	29,300	2,600	40,200	
Salts		12,040		10,160	4,150			
Oil					4,950	294,310		

TABLE 24-7*c* High-Viscosity First-Pass Oil (L.V-4) Oil Distillation and Cooling

	Stream number and name			
	12 Neutral oil	13 Water-alcohol	14 Stripped oil	14 Inter oil
gal/batch, 60°F	40,310	800	39,510	39,510
lb/batch	300,240	5,900	294,310	294,310
Specific gravity	0.894	0.89	0.894	0.894
Temperature, °F	140	100	280	110
Viscosity, cP, operating temperature	33	1	4	78
Components, lb				
Water	3,330	3,330		
Isopropanol	2,600	2,600		
Salts				
Oil	294,310			

TABLE 24-7*d* High-Viscosity Second-Pass Oil (H.V-4) Acid Treating

	Stream Number and Name				
	1 H.V-4 oil	2 Oleum	3 Acid and oil	4 Sludge	5 Sour oil
gal/batch, 60°F	45,000	2,025	47,025	3,540	43,510
lb/batch	338,000	32,000	370,000	44,200	325,800
Specific gravity	0.901	1.9	0.944	1.5	0.898
Temperature, °F	80	80	110	110	110
Viscosity, cP, 100°F	120	29		8,000	120

TABLE 24-7*e* High-Viscosity Second-Pass Oil (H.V-3A) Neutralization and Extraction

	Stream number and name							
	5 Sour oil	6 Carbonate solution	7 Oil + carbonate	10 Brine draw	11 Raw extract	12 Neutral oil	8 55% alcohol	9 Water
gal/batch, 60°F	43,500	5,400	48,900	9,200	11,280	39,360	11,250	900
lb/batch	325,800	53,500	379,300	82,130	91,150	294,490	82,400	7,500
Specific gravity	0.898	1.19	0.93	1.07	0.97	0.899	0.876	1.00
Temperature, °F	110	130	110	130	130	130	80	80
Viscosity, cP, 100°F	120	2		2	10	120	1	1
Components, lb								
CO_2 gas			1,430					
Sulfonates					25,000			
Water		43,870		65,830	25,000	3,330	42,200	
Alcohol				8,400	29,200	2,600	40,200	
Salts		9,630		7,900	4,470			
Oil					7,480	288,650		

TABLE 24-7*f* High-Viscosity Second-Pass Oil (H.V-3) Oil Distillation and Cooling

	Stream number and name			
	12 Neutral oil	13 Water-alcohol	14 Stripped oil	14 Inter oil
gal/batch, 60°F	39,360	800	38,560	38,560
lb/batch	294,490	5,900	288,590	288,590
Specific gravity	0.899	0.89	0.899	0.899
Temperature, °F	140	100	280	110
Viscosity, cP, operating temperature	32	1	4	65
Components, lb				
CO_2 gas				
Sulfonates				
Water	3,330	3,330		
Isopropanol	2,600	2,600		
Salts				
Oil	288,560			

TABLE 24-7*g* High-Viscosity Third-Pass Oil (H.V-3) Acid Treating

	Stream number and name				
	1 HV-3 oil	2 Oleum	3 Acid and oil	4 Sludge	5 Sour oil
gal/batch, 60°F	45,000	4,230	49,230	6,350	42,880
lb/batch	337,000	66,800	403,800	84,500	319,300
Specific gravity	0.898	1.9	0.984	1.6	0.895
Temperature, °F	80	80	110	110	110
Viscosity, cP, 100°F	120	29		5,000	120

TABLE 24-7*h* High-Viscosity Third-Pass Oil (H.V-2A) Neutralization and Extraction

	Stream number and name							
	5 Sour oil	6 Carbonate solution	7 Oil + carbonate	10 Brine draw	11 Raw extract	12 Neutral oil	8 55% Alcohol	9 Water
gal/batch, 60°F	42,800	4,390	47,190	5,800	7,620	41,270	6,750	900
lb/batch	319,300	43,500	362,800	52,350	58,000	308,210	49,400	7,500
Specific gravity	0.895	1.19	0.92	1.08	0.91	0.895	0.876	1.00
Temperature,°F	110	130	110	130	130	130	80	80
Viscosity, cP, 100°F	120	2		2	10	120	1	1
Components, lb								
CO_2 gas			1,140					
Sulfonates					10,000			
Water		35,670		37,510	28,100	3,330	25,300	
Alcohol				8,170	13,330	2,600	24,100	
Salts		7,830		6,670	2,890			
Oil					3,580	302,280		

TABLE 24-7*i* High-Viscosity Third-Pass Oil (H.V-2) Oil Distillation and Cooling

	Stream number and name			
	12 Neutral oil	13 Water-alcohol	14 Stripped oil	14 Inter oil
gal/batch, 60°F	41,270	800	40,470	40,470
lb/batch	308,210	5,900	302,280	302,280
Specific gravity	0.895	0.89	0.895	0.895
Temperature, °F	140	100	280	110
Viscosity, cP, operating temperature	30	1	4	56
Components, lb				
Water	3,330	3,330		
Isopropanol	2,600	2,600		
Salts				
Oil	302,280			

TABLE 24-7*j* High-Viscosity Fourth-Pass Oil (H.V-2)- First Shot Acid Treating

	Stream number and name				
	1 H.V-2 oil	2 Oleum	3 Acid and oil	4 Sludge	5 Sour oil
gal/batch, 60°F	45,000	4,050	49,050	4,850	44,700
lb/batch	336,000	64,000	400,000	67,000	333,000
Specific gravity	0.895	1.9	0.978	1.66	0.892
Temperature, °F	60	80	90	90	90
Viscosity, cP, 100°F	120	29		2,000	120

TABLE 24-7*k* High-Viscosity Fourth-Pass Oil (H.V-1)-Second Shot Acid Treating

	Stream number and name				
	1 Sour oil in	2 Oleum	3 Acid and oil	4 Sludge	5 Sour oil out
gal/batch, 60°F	44,700	4,050	48,750	4,740	44,500
lb/batch	333,000	64,000	397,000	67,000	330,000
Specific gravity	0.892	1.9	0.977	1.7	0.889
Temperature, °F	90	80	120	120	120
Viscosity, cP, 100°F	120	29		1,000	120

TABLE 24-7*l* High-Viscosity Fourth-Pass Oil (H.V-1-1)-Third Shot Acid Treating

	Stream number and name				
	1 Sour oil in	2 Oleum	3 Acid and oil	4 Sludge	5 Sour oil out
gal/batch, 60°F	44,700	4,050	48,750	4,450	44,500
lb/batch	330,000	64,000	394,000	65,000	329,000
Specific gravity	0.889	1.9	0.974	1.75	0.886
Temperature, °F	120	80	140	140	140
Viscosity, cP, 100°F	120	29		500	120

TABLE 24-7*m* High-Viscosity Fourth-Pass Oil (H.V-1-1-1A) Neutralization and Extraction

	Stream number and name							
	5 Sour oil	6 Carbonate solution	7 Oil + carbonate	10 Brine draw	11 Raw extract	12 Neutral oil	8 55% Alcohol	9 Water
gal/batch, 60°F	44,500	2,700	47,200	5,400	6,730	44,200	6,750	2,700
lb/batch	329,000	26,800	355,800	50,190	51,510	326,000	49,400	22,500
Specific gravity	0.886	1.19	0.91	1.12	0.92	0.885	0.876	1.00
Temperature, °F	140	130	140	140	140	140	80	80
Viscosity, cP, 100°F	120	2		2	1	120	1	1
Components, lb								
CO_2 gas			730					
Sulfonates					1,220			
Water		21,980		39,000	30,210	3,750	25,300	
Isopropanol				4,170	17,000	2,930	24,100	
Salts		4,820		6,650	2,580			
Oil				370	500	319,320		

TABLE 24-7*n* High-Viscosity Fourth-Pass Oil (H.V-1-1-1) Oil Distillation and Cooling

	Stream number and name			
	12 Neutral oil	13 Water- alcohol	14 Stripped oil	14 Unfiltered oil
gal/batch, 60°F	44,200	900	43,300	43,300
lb/batch	326,000	6,680	319,320	319,320
Specific gravity	0.885	0.89	0.884	0.884
Temperature, °F	140	100	275	160
Viscosity, cP, operating temperature	24	1	3.5	20
Components, lb				
CO_2 gas				
Sulfonates				
Water	3,750	3,750		
Isopropanol	2,930	2,930		
Salts				
Oil	319,320			

TABLE 24-7*o* High-Viscosity White Oil Oil-Clay Contacting and Filtering

	Stream number and name				
	14 Unfiltered oil	15 Clay	16 Oil-clay	17 Spent clay	18 Filtered oil
gal/batch, 60°F	7,500		7,685		7,130
lb/batch	55,400	2,770	58,170	5,540	52,630
Specific gravity	0.886	2.5	0.925	1.36	0.886
Temperature, °F	80		160	160	160
Viscosity, cP, operating temperature	140		18		16
Components, lb					
Clay			2,770	2,770	
Oil			55,400	2,770	

TABLE 24-7*p* High-Viscosity White Oil Desalting of High-Viscosity Extract

	Stream number and name				
	101 Raw extract	102 Soda ash	103 Desalted extract	104 Weak brine	105 Strong brine
gal/batch, 60°F	26,000		17,510	6,490	4,050
lb/batch	206,850	22,680	135,750	49,080	44,500
Specific gravity	0.955		0.93	0.91	1.32
Temperature, °F	80		140	140	140
Viscosity, cP, operating temperature	15		7	2	3
Components, lb					
Sulfonate	44,300		44,300		
Water	73,700		28,900	24,930	31,150
Isopropanol	64,500		47,700	16,800	
Salts	10,200		1,350	6,900	13,350
Oil	14,150		13,500	650	

TABLE 24-7*q* Sulfonate Finishing: Distillation and Adjusting, HV Extract

	Stream number and name					
	103 Desalted extract	110 Water-alcohol	106 Sulfonate	107 Diluting water	108 Diluting oil	109 Adjusted sulfonate
gal/batch, 60°F	17,510	10,610	6,950	211	823	7,984
lb/batch	135,750	75,050	60,700	1,755	6,000	68,455
Specific gravity @ 60°F	0.91	0.85	1.05	1	0.876	1.034
Temperature, °F	100	200	320	80	80	190
Viscosity, cP, operating temperature	13		30	1	140	500
Components, lb						
CO_2 gas						
Sulfonates	44,300		44,300			44,300
Water	28,900	27,350	1,550			3,305
Isopropanol	47,700	47,700	—			—
Salts	1,350		1,350			1,350
Oil	13,500		13,500			19,500

TABLE 24-8 Yields of White Oil and Transformer Oil from Acid Treating Process Per Ton Feed

	Transformer oil	LV white oil	HV white oil
Feed			
Raw oil	1.0000	1.0000	1.0000
Oleum	0.0648	0.6421	0.7066
Soda ash	0.0398	0.1190	0.1160
Isopropyl alcohol	0.0024	0.0166	0.0163
Total	1.1242	1.7777	1.8812
Products			
Finished oil	0.8943	0.5666	0.6328
Sulfonate (62%)	0.0233	0.1868	0.2033
Acid sludge	0.1024	0.7739	0.8629
Salt and losses	0.0701	0.2504	0.1400
Total	1.1245	1.7777	1.8812
Minor raw materials per 1000 unit feed			
Sulfuric acid 66° Be	1.6900	1.6910	1.6910
Sodium hydroxide, flakes	0.0590	0.0594	0.0594

TABLE 24-9 Regular Petrolatum Blending (40% White Oil + 60% Wax)

Name of stream	LVWO	Wax	Blended petrolatum
gal/batch, 60°F	1,890	2,840	4,730
lb/batch	13,300	19,400	32,700
Specific gravity	0.845	0.82*	0.83
Temperature, °F	80	150	150
Viscosity, cP, 100°F	12	30	25
Components, lb			
LVWO	—	—	13,300
Wax	—	—	19,400

*Density of melted wax at 150°F.

TABLE 24-10 Special Petrolatum Blending (30% White Oil + 70% Wax)

Name of stream	HVWO	Wax	Blended petrolatum
gal/batch, 60°F	1,400	3,270	4,670
lb/batch	10,340	22,360	32,700
Specific gravity	0.886	0.82*	0.84
Temperature, °F	80	150	150
Viscosity, cP, 100°F	120	30	125
Components, lb			
HVWO	—	—	10,340
Wax	—	—	22,360

*Density of melted wax at 150°F.

TABLE 24-11 Batch Cycle Time Acid Treating
*Basis: 45,000 gal oil per batch**

	Feed name		Settling time, h
Transformer oil			
Transformer oil feed	TOFS†	First pass	48
L.V white oil			
Fresh feed	L.V-N†	First pass	48
	L.V-4	Second pass	48
	L.V-3	Third pass	36–48
L.V-2, first shot	L.V-2	Fourth pass	24
L.V-2, second shot	L.V-1	Fourth pass	48
H.V, white oil			
Fresh feed	H.V-N	First pass	48
	H.V-4	Seconnd pass	48
	H.V-3	Third pass	36–48
H.V-2, first shot	H.V-2	Fourth pass	24
H.V-2, second shot	H.V-1	Fourth pass	24
H.V-2, third shot	H.V-1-1	Fourth pass	48

*Sludge settling times are generally unaffected by batch size.
†Sludge draws are made every 8 h.

TABLE 24-12 Batch Cycle Time Neutralization and Extraction
Basis: Product (acidic oil) from settlers after sludge separation.

Activity	Hours
Load neutralizers*	5.50
Add initial solvent and circulate	0.50
Settle[2]	4.00
Brine draw	1.00
Final solvent addition	
(Heat and circulate)	1.50
Continue to circulate and heat	1.00
Settle†	8.00
Second brine draw (if any)	0.00
Sulfonate extract draw	2.00
Wash water addition and settling	4.00
Wash water draw	0.50
Add inhibitor and circulate	0.50
Pump out‡ sulfonate to still	12.50
Total	41.00

*Loading time for neutralizers is for adding sour oil and carbonate solution simultaneously through ratio pump.
†Settling time is independent of batch size.
‡Pumping time is a function of pumping rate and batch size.

TABLE 24-13 Batch Cycle Time Extract Desalting
Basis: 26,000 gal raw extract per batch

Activity	Hours
Load desalter (heat while loading)	3.00
Heat up and circulate	3.00
First dry soda ash addition and circulation	2.00
Settle	1.00
Weak brine draw	1.00
Second soda ash addition and circulation	3.00
Settling	2.00
First strong brine draw (if any)	1.00
Settle	12.00
Second strong brine draw (if any)	0.00
Circulate and heat	1.00
Sample and test	1.00
Add 60° Be sulfuric acid	2.00
Sample and test	1.00
Pump to storage	3.00
Total	36.00

Sulfonate finishing (solvent removal) and adjusting
Basis: 30,000 gal desalted extract

Activity	Hours
Continuous distillation*	50.00
Pump out sulfonate to adjust tank	4.00
Dilute with oil, water, and mix	5.50
Pump to storage	4.50
Total	64.00

*Assumed distillation rate of solvent at 10 Gal per minute.

TABLE 24-14 Batch Cycle Times Alcohol Recovery from Brine
Basis: 15,000 gal brine charge to batch still consisting of brine draws, water washes plus weak brine draws from desalter (3–15% isopropyl alcohol)

Activity	Hours
Load batch still	2.00
Heat up	7.00
Boil off alcohol	8.00
Natural cooling	5.00
Unload	2.00
Total	24.00

Oil blending and drumming
Basis: 15,000 gal oil per batch

Activity	Hours
Load oil	5.00
Mix by recirculation	6.50
Unload to drums	7.50
Total	19.00

Petrolatum blending
Basis: 4,500 gal mix per batch

Activity	Hours
Load oil (heat while loading)	0.50
Heat oil	1.00
Heat wax drums and load	16.00
Mix	1.00
Unload to drums	2.00
Total	20.50

TABLE 24-15 Batch Cycle Time Oil Distillation and Cooling
Basis: Product from neutralizers

Activity	Hours
Unload neutralizer to still	12.50
Steam strip	1.50
Circulate through cooler to 200°F	7.00
Air blow and continue to cool to 160°F	7.00
Circulate and cool to 110°F	7.00
Pump out	4.00
Total	39.00

Oil-clay contacting and filtering
Basis: 7,500 gal oil per batch

Activity	Hours
Load oil (heat while loading)	1.5
Heat and mix	3
Load clay (heat and mix while loading)	1.5
Heat and mix	1
Unload through filter	3
Remove cake and clean filter	1.5
Total	11.5

Inhibitor added to run down tank if white oil.

TABLE 24-16 Batch Cycle Time Bauxite Filtration
Basis: Filtration of 21,000 gal of white oil and 29,000 gal of transformer oil

White oil and transformer oil	
Activity	**Hours**
Fill	3.5
Soak	8.0
Run (white oil)	192.0
Push off (mixture)	15.0
Run (transformer oil)	273.0
Drain	4.0
Air blow	3.0
Naphtha wash	8.0
Steaming	6.0
Dump	3.5
Total	516.0

White oil only	
Activity	**Hours**
Fill	3.5
Soak	8.0
Run (white oil)	192.0
Push off (mixture)	20.0
Drain	4.0
Air blow	3.0
Naphtha wash	8.0
Steaming	6.0
Dump	3.5
Total	248.0

Bauxite reactivation *Basis: Burning rate 1.0 ton/hour*	
Activity	**Hours**
Fill feed bin	3.5
Unload product bin	3.5
Set up equipment	1.0
Burn	13.0
Total	21.0

Naphtha recovery *Basis: Distillation of wash from one filter cycle in two batches*	
Activity	**Hours**
Fill still	1.0
Heat to boiling temperature	5.0
Distill batch	4.0
Cool bottoms	2.0
Decant (naphtha/water)	2.0
Transfer naphtha	1.0
Transfer bottoms	1.0
Total	16.0

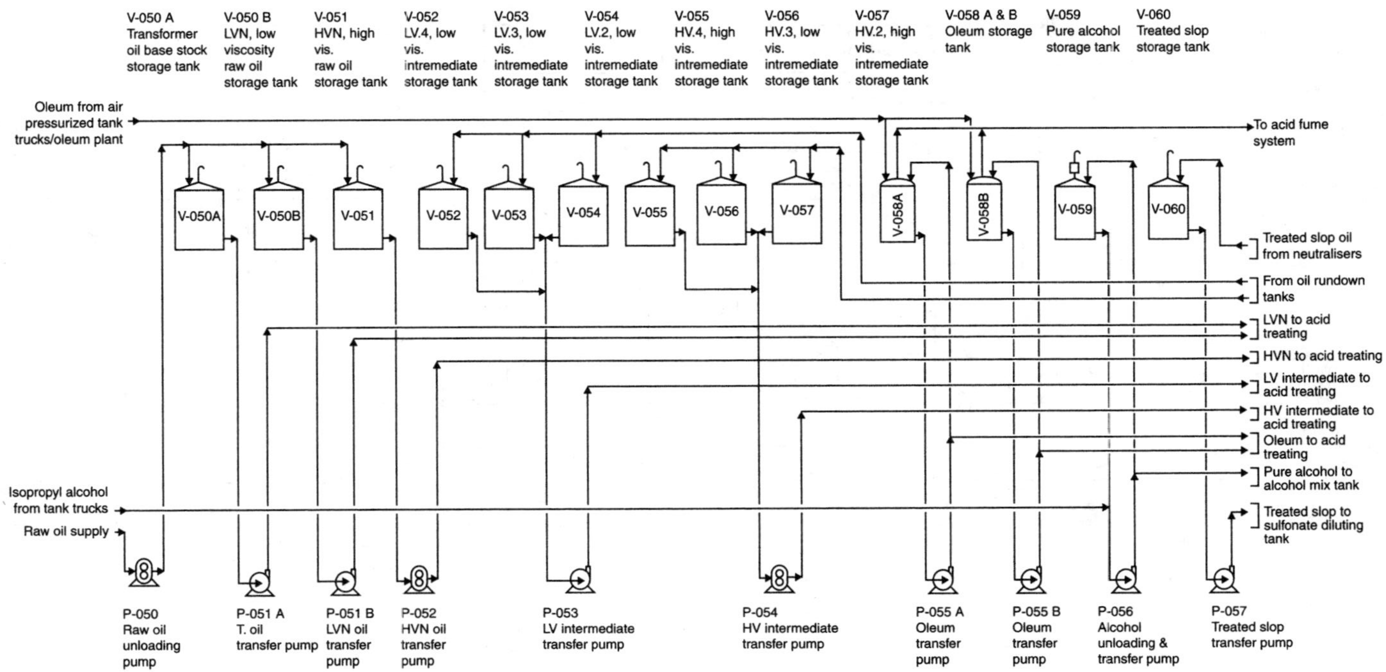

FIGURE 24-1 Classical white oil process (raw material and intermediate product storage).

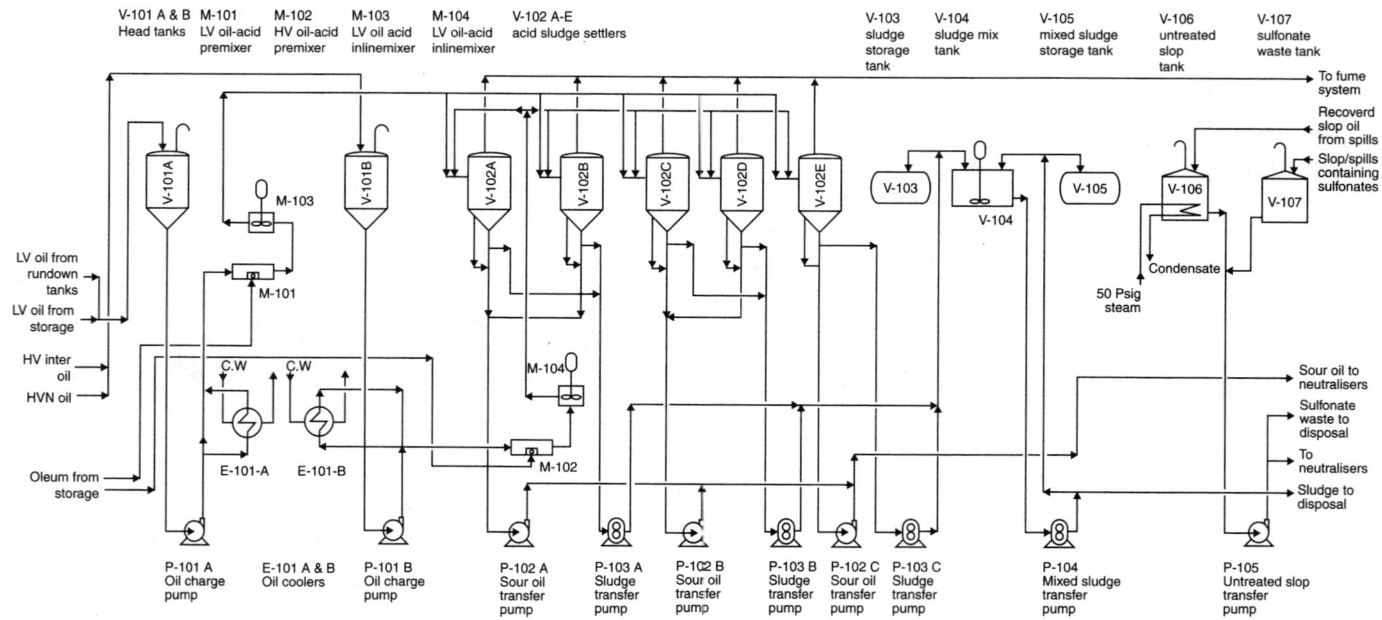

FIGURE 24-2 Transformer and white oil manufacture classical process; acid treating and sludge storage.

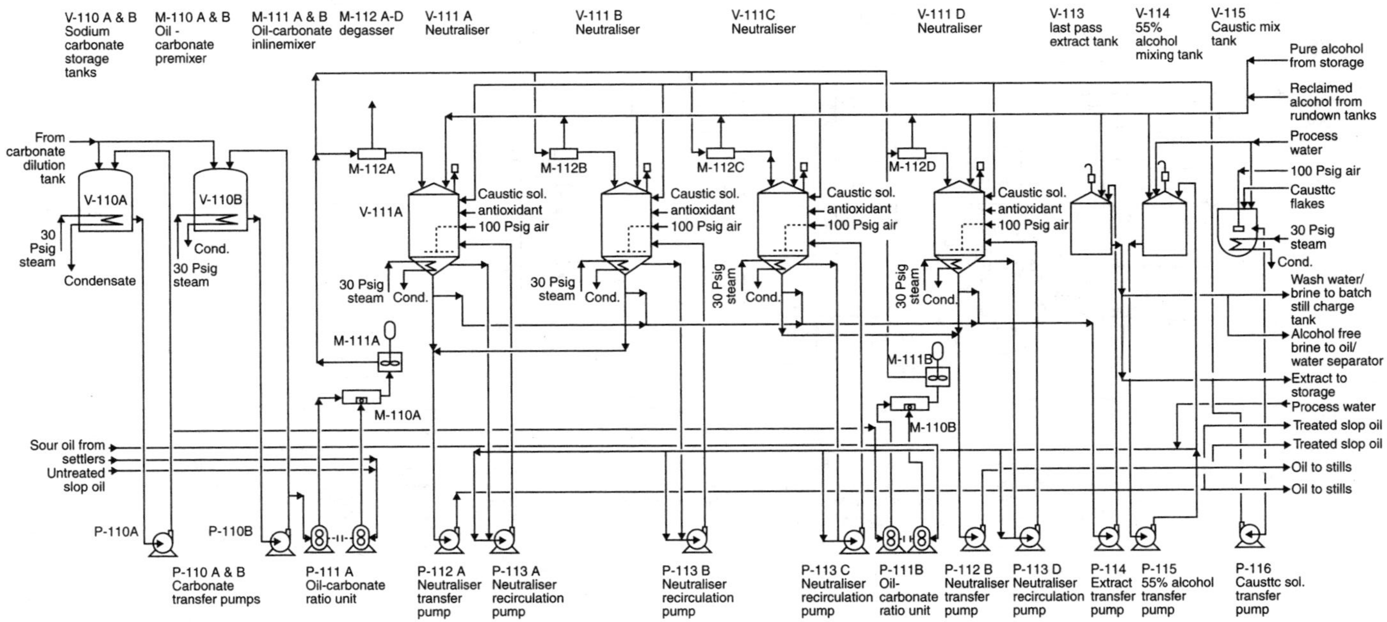

FIGURE 24-3 Classical white oil process (neturalization step).

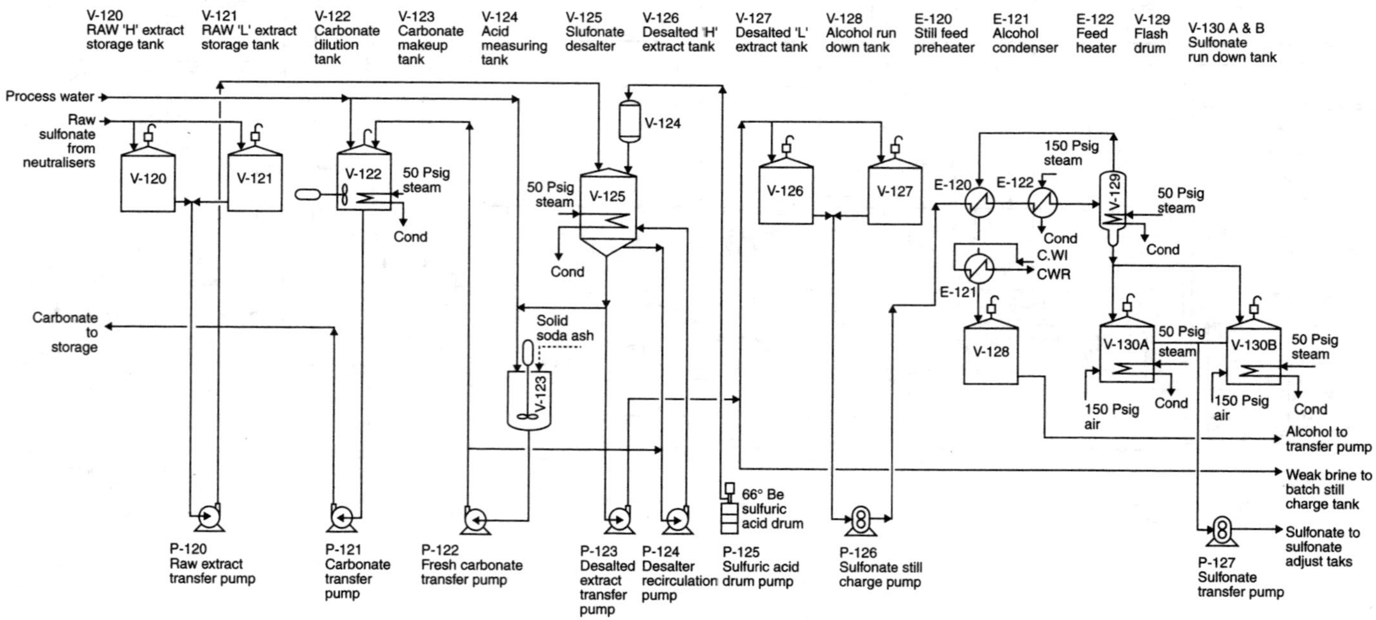

FIGURE 24-4 Classical white oil process (sulfonate finishing).

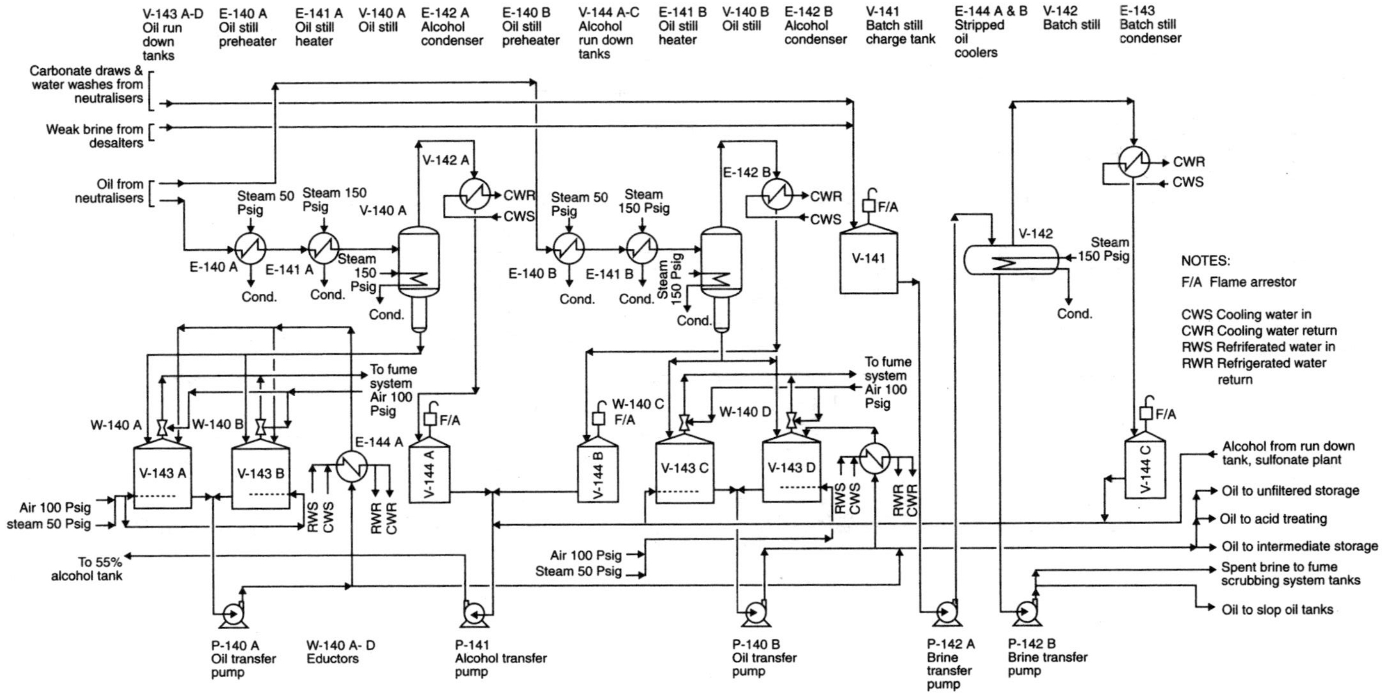

FIGURE 24-5 Classical white oil process (alcohol recovery from oil and brine).

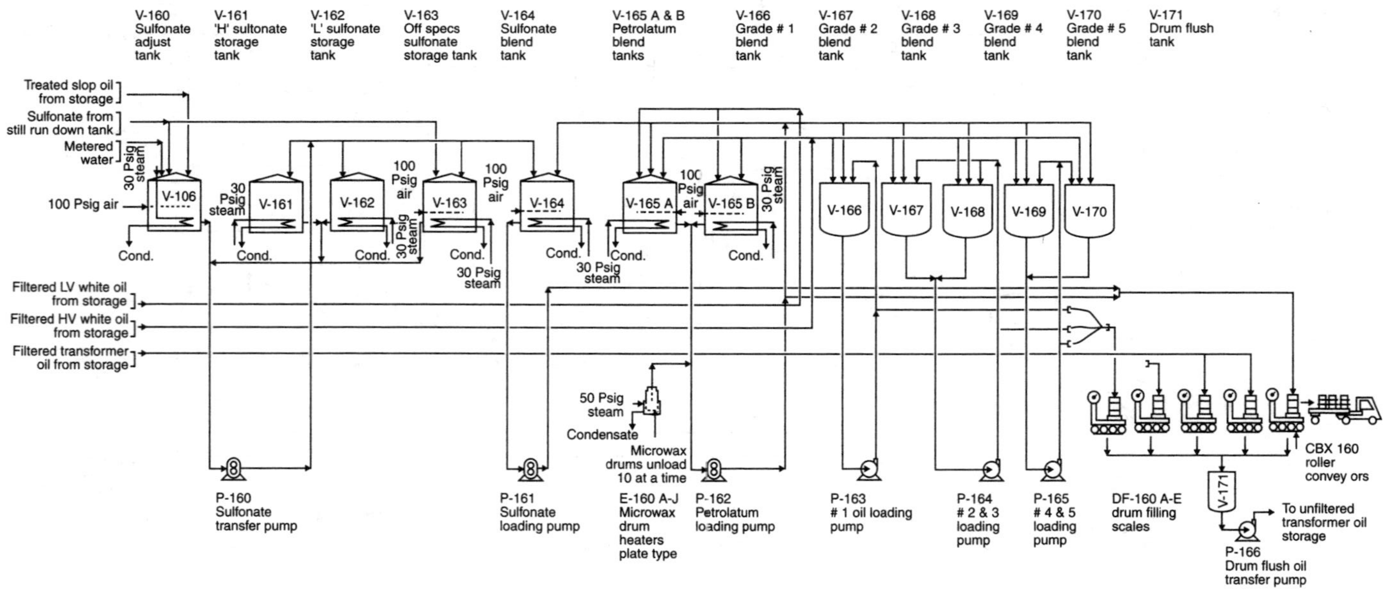

FIGURE 24-6 Classical white oil process (finished product blending and loading).

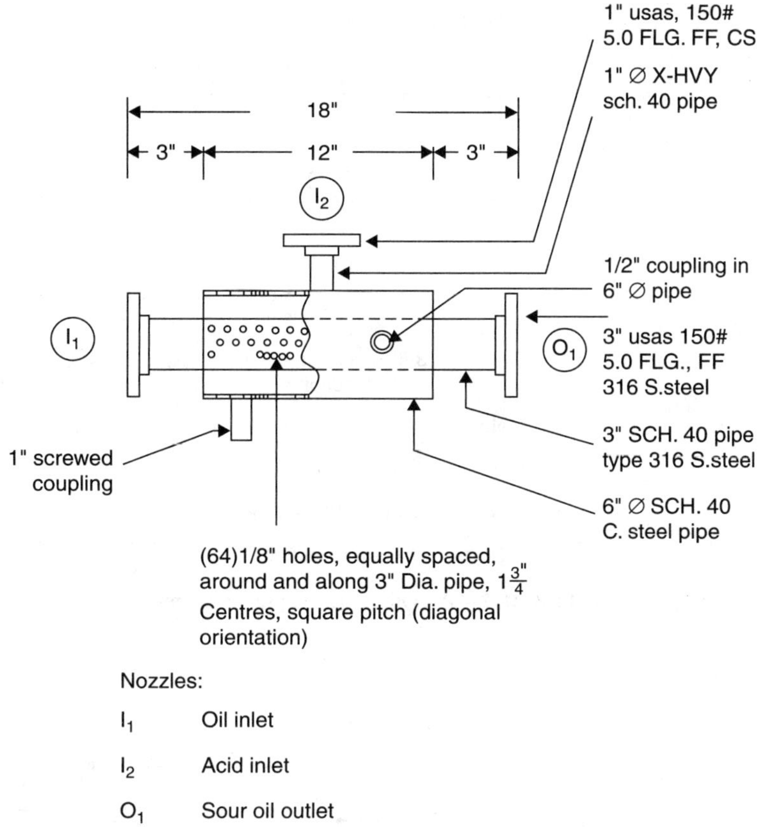

FIGURE 24-7A Acid-oil premix chamber.

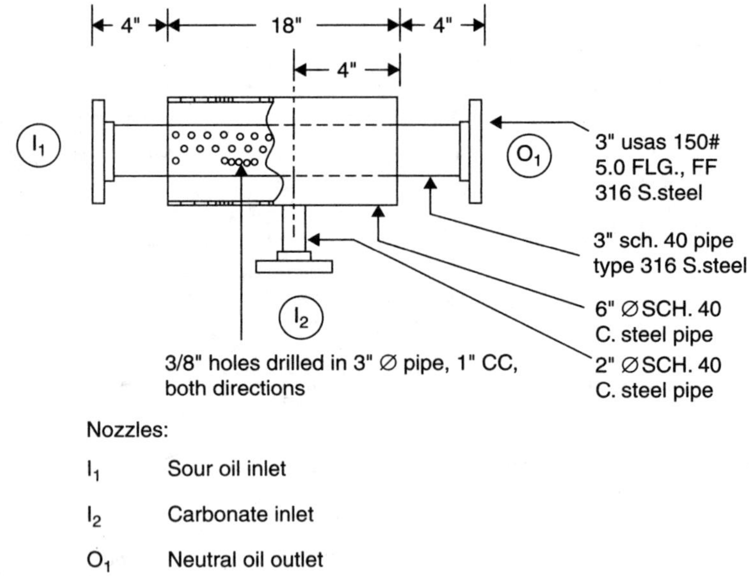

FIGURE 24-7B Oil-carbonate premix chamber.

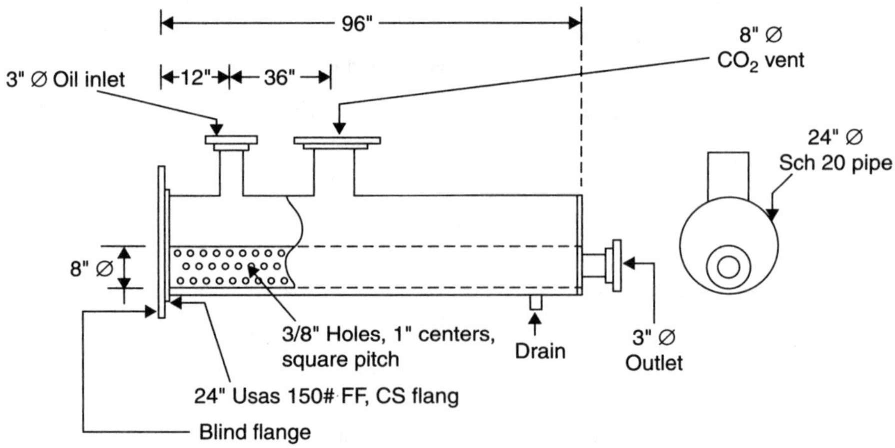

FIGURE 24-8 Degasser.

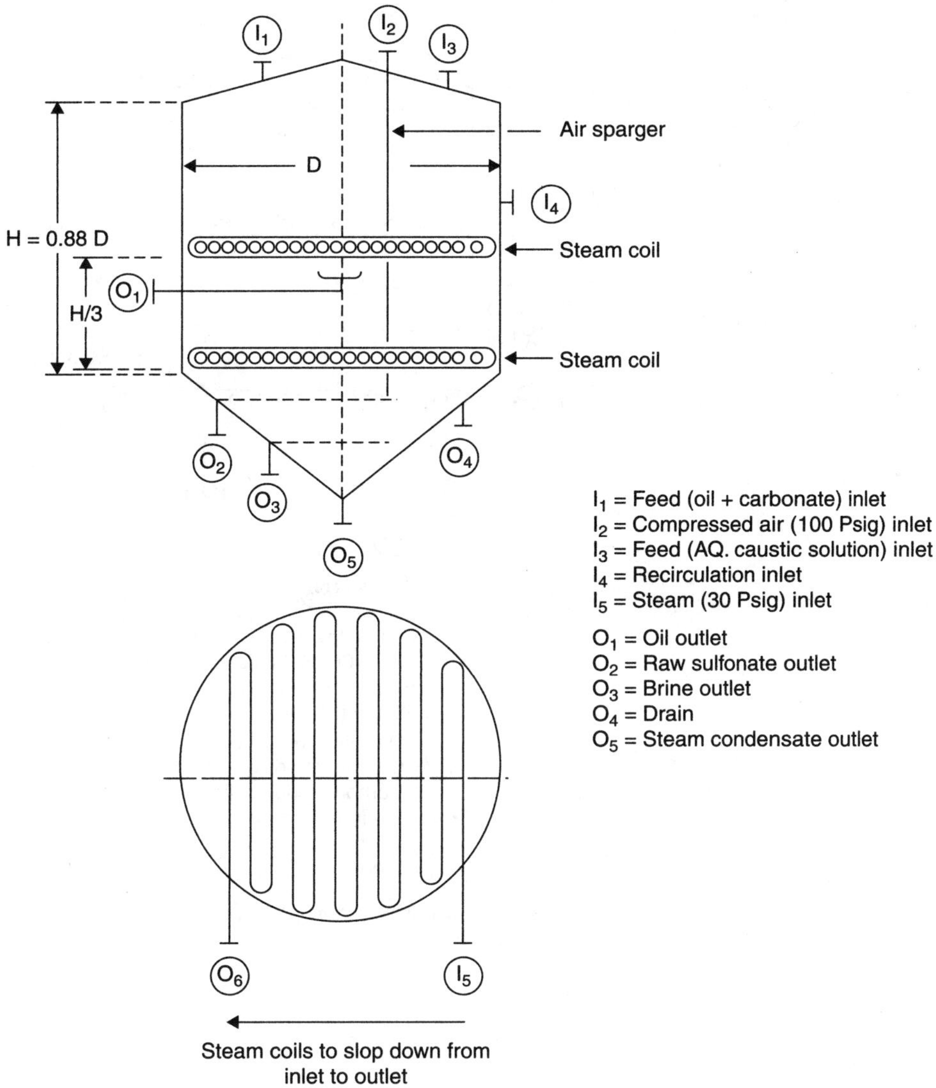

FIGURE 24-9 Classical white oil process (neutralizer vessel).

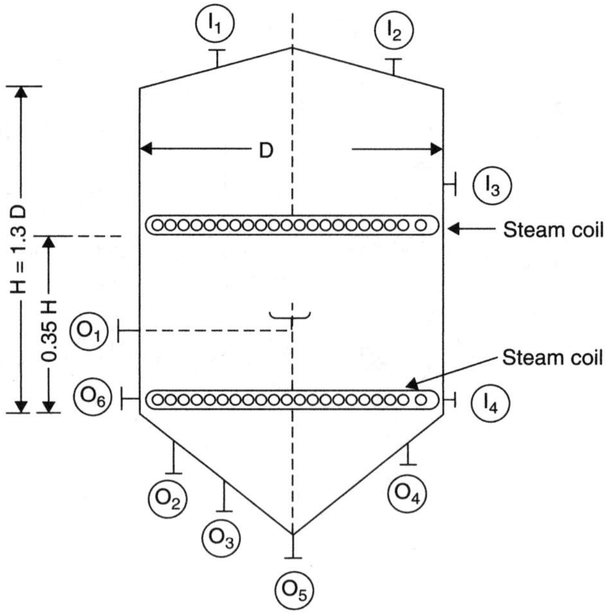

I_1 = Feed (Raw extract) inlet
I_2 = Feed (66° Be sulfuric acid) inlet
I_3 = Recirculation inlet
I_4 = Steam (30 Psig) inlet
O_1 = Desalted extract outlet
O_2 = Carbonate sol. outlet
O_3 = Carbonate sol. outlet
O_4 = Recirculation outlet
O_5 = Drain
O_6 = Steam condensate outlet

FIGURE 24-10 Classical white oil process (desalter vessel).

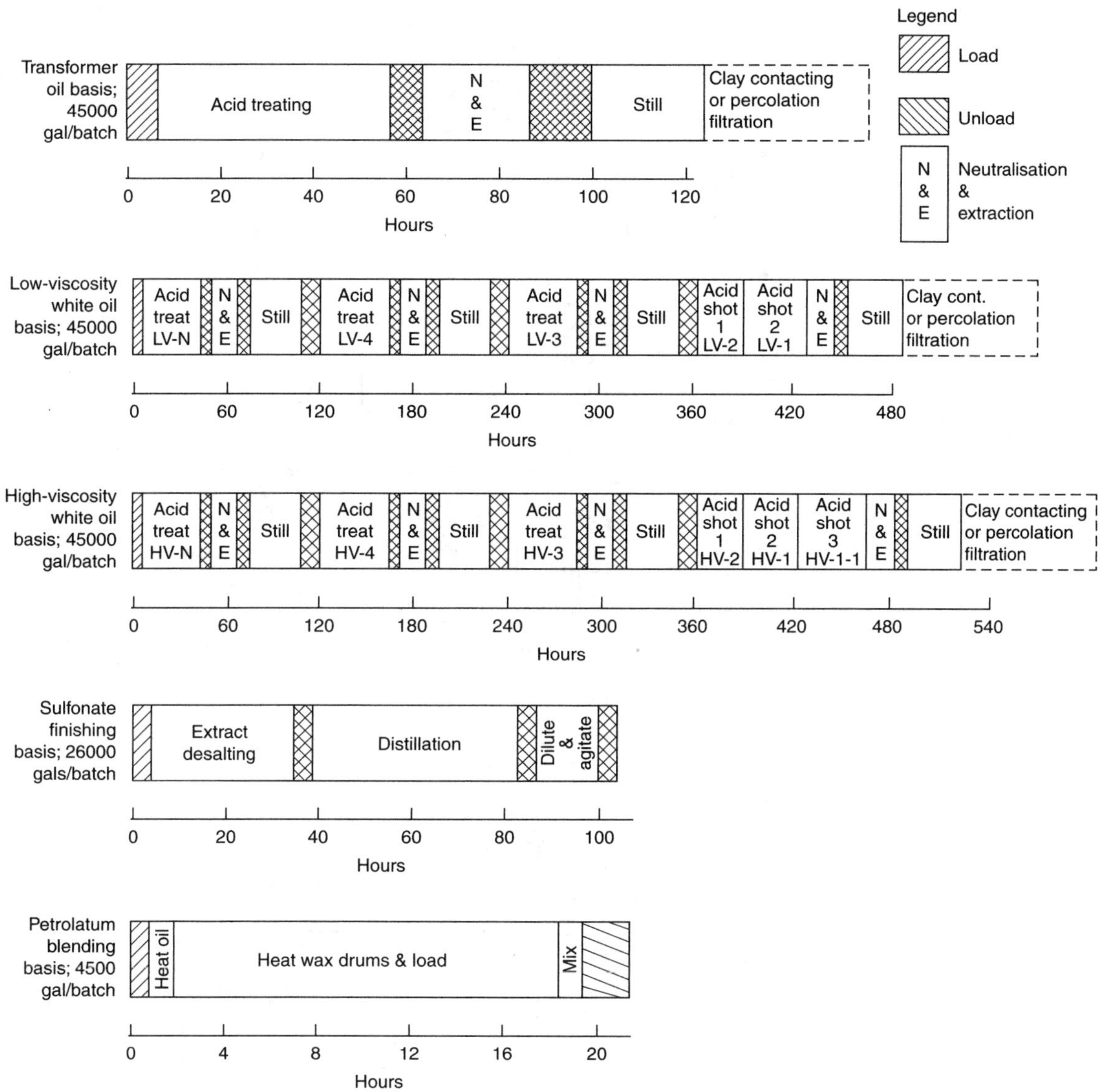

FIGURE 24-11 White oil manufacture: batch cycle times.

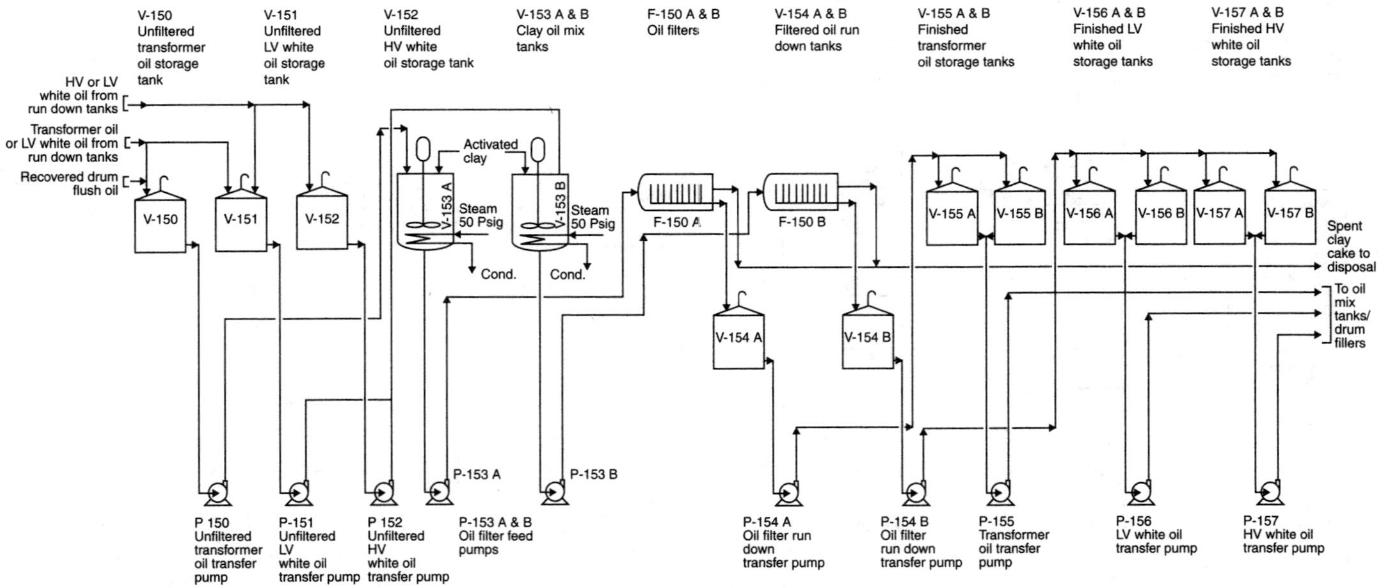

FIGURE 24-12 Classical white oil process (clay treatment).

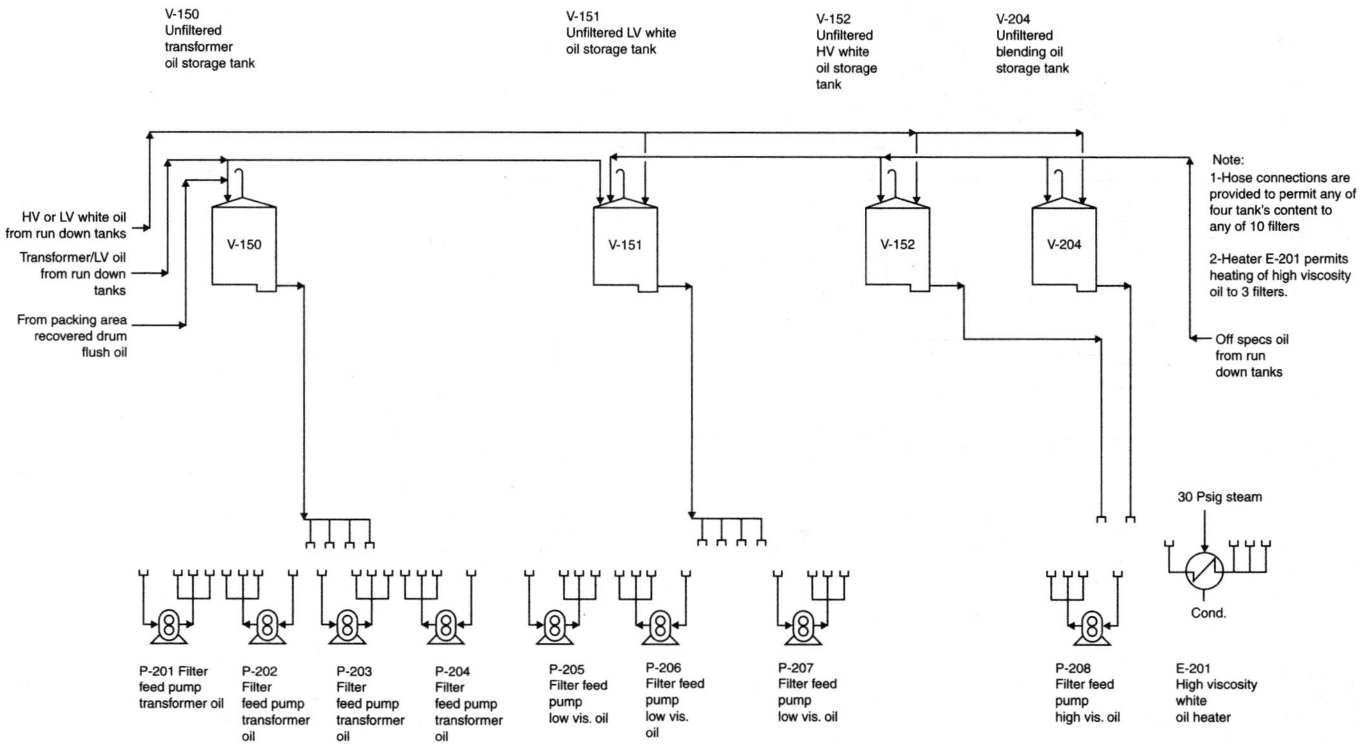

FIGURE 24-13 Classical white oil process (percolation-filtration; unfiltered oil storage tank).

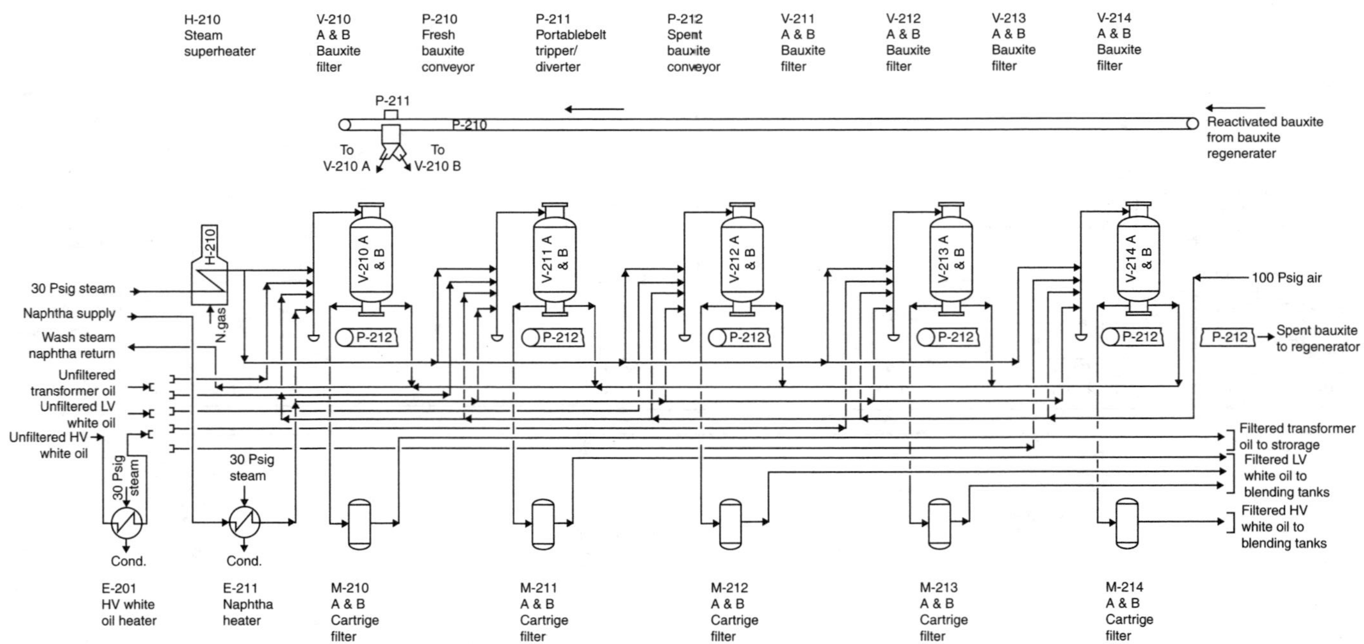

FIGURE 24-14 Classical white oil process (oil filtration/bauxite percolation).

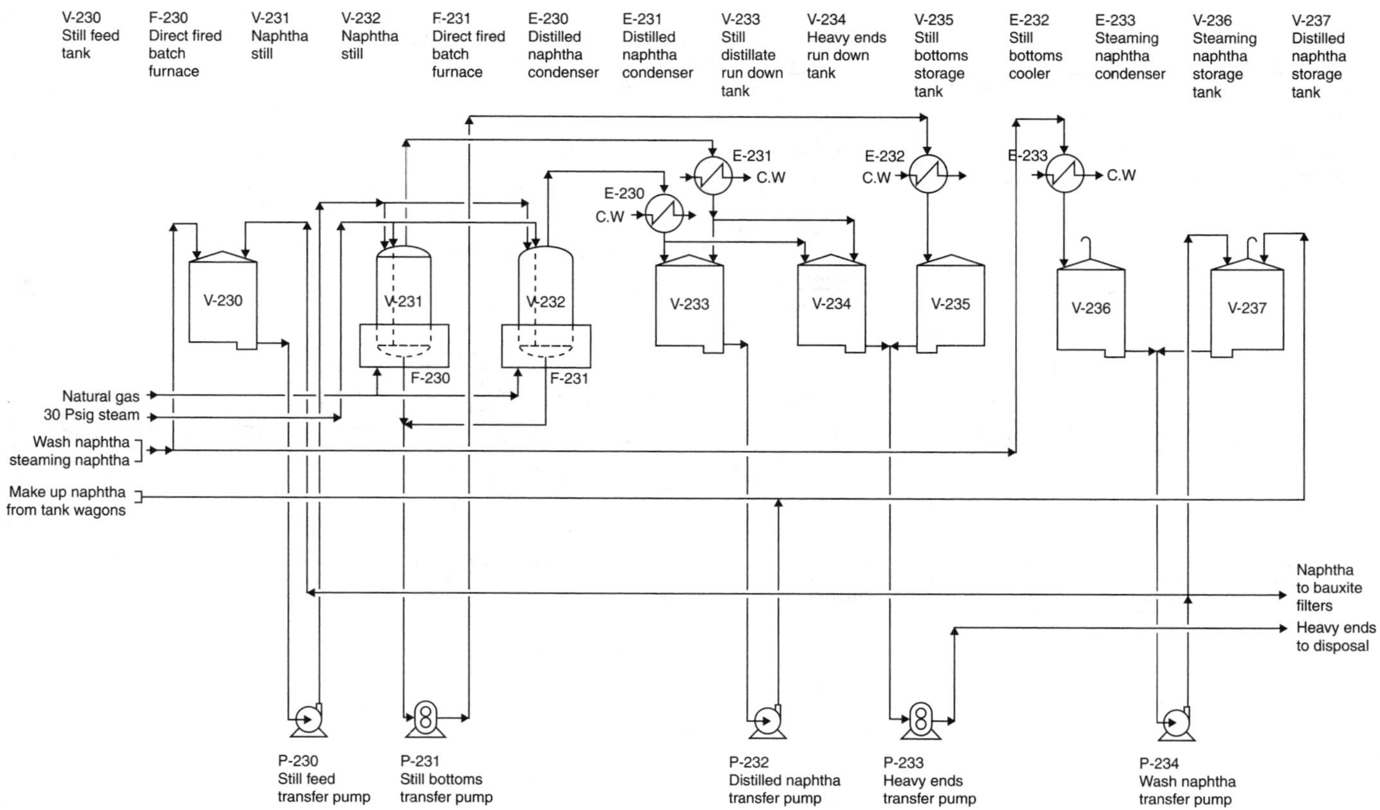

FIGURE 24-15 Classical white oil process (percolation-filtration clay treating plant; naphtha recovery).

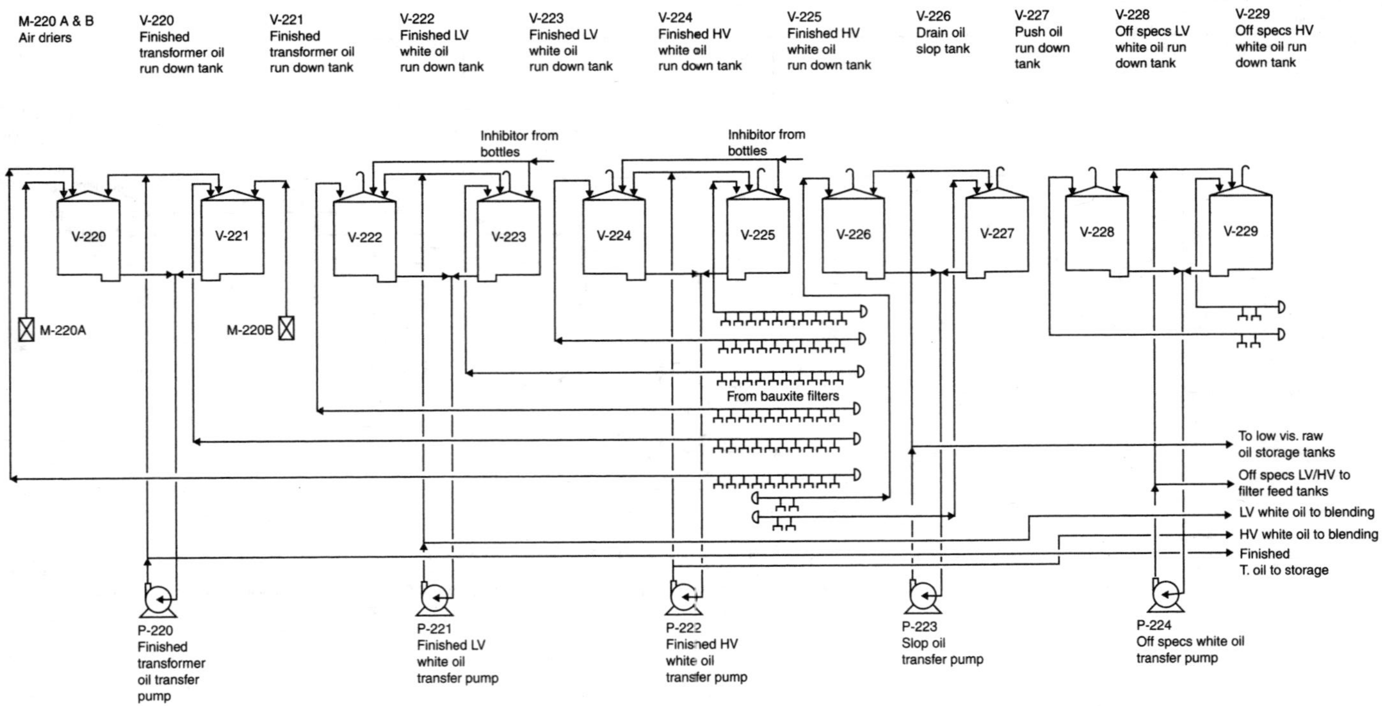

FIGURE 24-16 Classical white oil process (percolation-filtration clay treating plant; filtered oil rundown tanks).

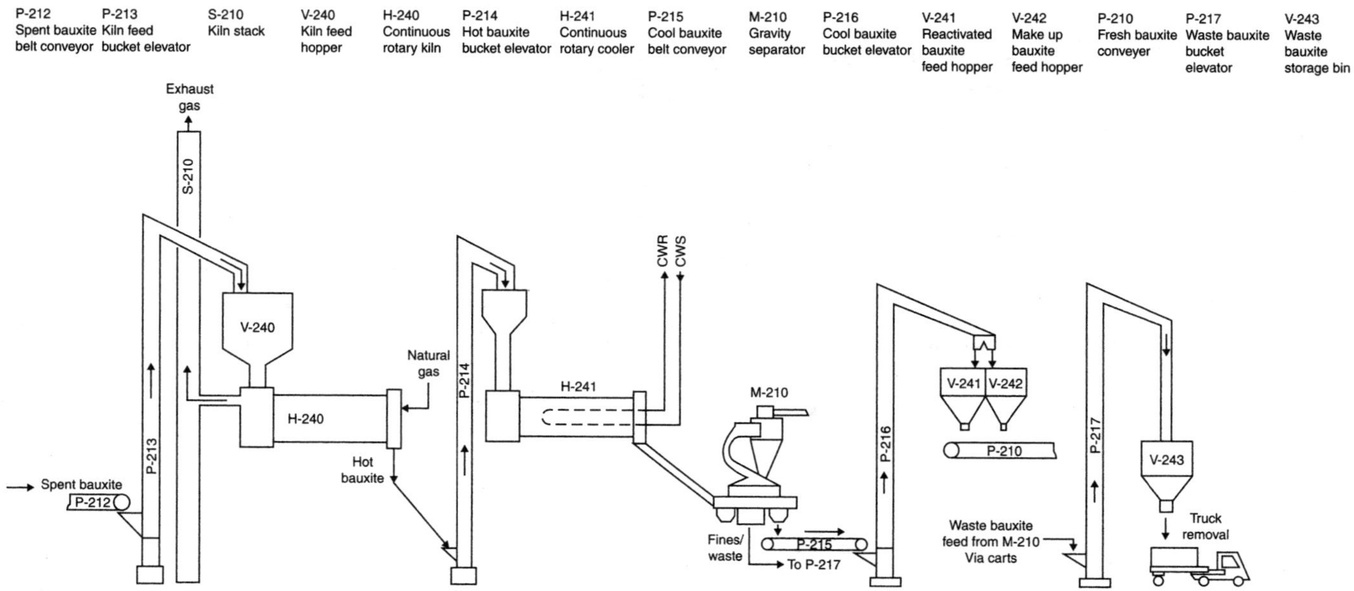

FIGURE 24-17 Classical white oil process (percolation-filtration clay treating plant; bauxite regeneration.

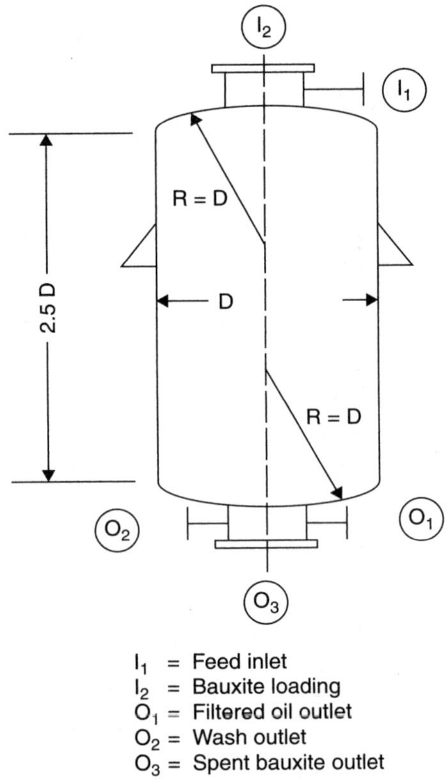

FIGURE 24-18 Bauxite filter vessel.

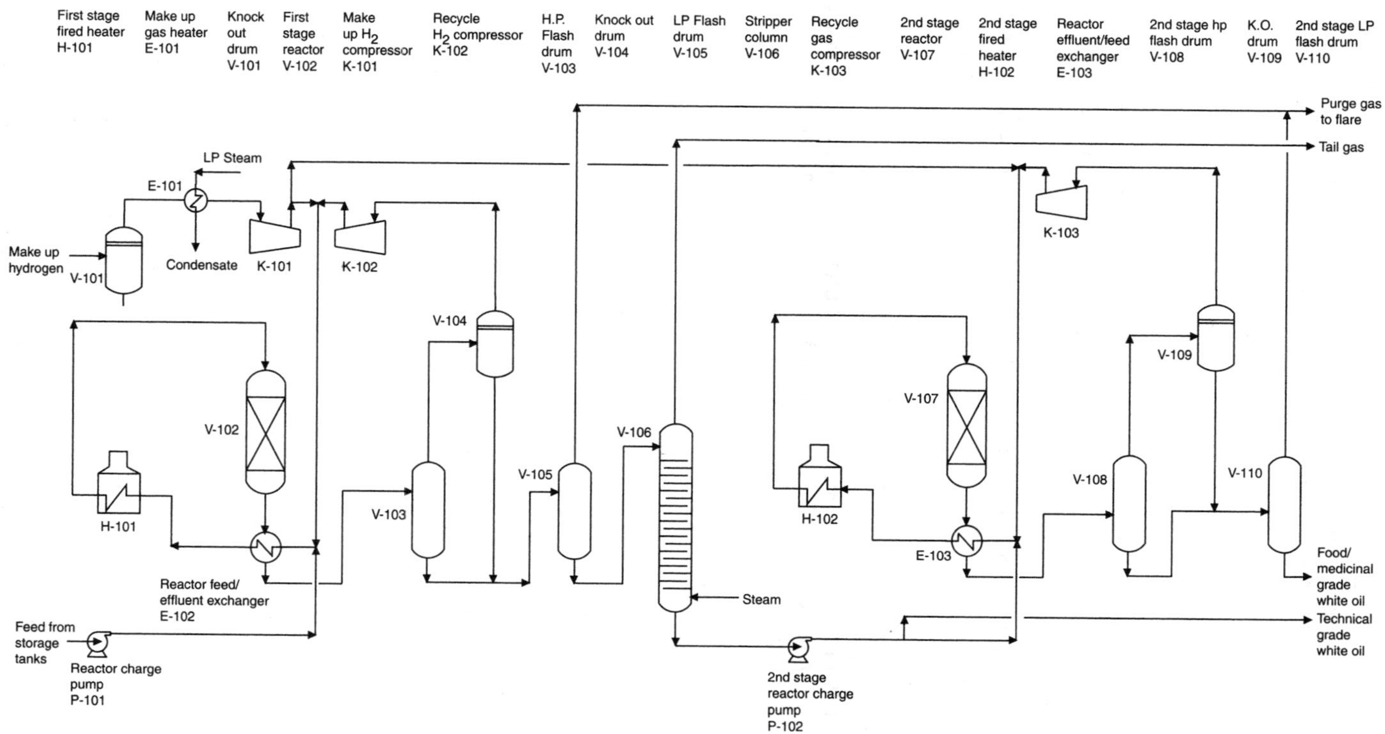

FIGURE 24-19 White oil (hydrotreating) process.

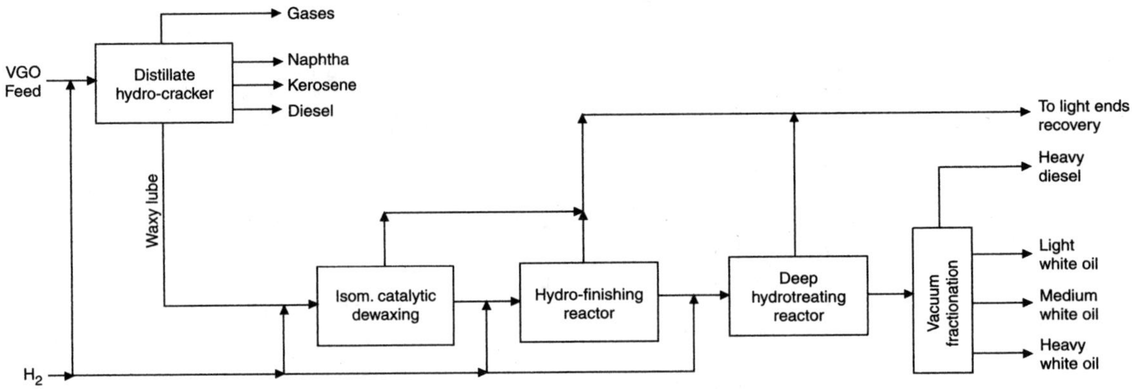

FIGURE 24-20 Manufacture of food-grade white oil: Hydrocracking/isomerization; catalytic dewaxing route.

REFERENCES

1. The PCAH test is described by J. W. Howard, E. O. Haenni, and F. L. Joe, *Journal of the Association of Official Agricultural Chemists* 48 (1965):304–315. Also see the UK mineral hydrocarbons in food regulations, 1966.
2. U.S. Patent 3459656, "Making White Oil by Two-Stage Catalytic Hydrogenation," issued August 5, 1969.

APPENDIX

TABLE A-1 Properties of Pure Hydrocarbons and Other Components

		Molecular weight	Normal boiling point, °F	Normal boiling point, °R	Critical temperature, °R	Critical pressure, lb/in²	Specific gravity
1	Hydrogen	2.060	–423.000	36.7	59.7	190.4	0.0707
2	Methane	16.043	–258.700	201.0	343.0	667.8	0.3000
3	Ethane	30.070	–127.500	332.2	549.8	707.8	0.3559
4	Propane	44.097	–43.700	416.0	665.7	616.3	0.5090
5	Isobutane	58.123	10.900	470.6	734.7	529.1	0.5653
6	n-Butane	58.123	31.100	490.8	765.3	550.7	0.5854
7	Isopentene	72.150	82.100	541.8	828.8	490.4	0.6274
8	n-Pentane	72.150	96.900	556.6	845.4	488.6	0.6323
9	Neopentane	72.150	49.100	508.8	780.8	464.0	0.5983
10	n-Hexane	86.177	155.700	615.4	913.4	436.9	0.6665
11	n-Heptane	100.204	209.100	668.8	972.5	396.8	0.6913
12	n-Octane	114.231	258.200	717.9	1023.9	360.6	0.7077
13	n-Nonane	128.258	303.400	763.1	1072.2	334.4	0.7243
14	n-Decane	142.285	345.400	805.1	1113.2	307.9	0.7372
15	n-Undecane	156.312	384.600	844.3	1149.8	285.1	0.7456
16	n-Dodecane	170.338	421.400	881.1	1184.8	264.5	0.7535
17	n-Tridecane	184.365	455.800	915.5	1216.4	249.8	0.7629
18	n-Tetradecane	198.392	488.400	948.1	1246.3	235.1	0.7643
19	n-Pentadecane	212.419	519.200	978.9	1272.2	220.4	0.7734
20	n-Hexadecane	226.446	548.300	1008.0	1297.1	205.7	0.9065
21	n-Heptadecane	240.473	575.800	1035.5	1320.1	191.0	0.9089
22	Ethylene	28.054	–154.700	305.0	508.2	729.8	0.5712
23	Propylene	42.081	–53.900	405.8	658.0	676.6	0.5200
24	1-Butene	56.107	20.700	480.4	755.3	583.0	0.6014
25	Cis -2-Butene	56.10/	38.700	498.4	784.0	610.0	0.6296
26	Trans -2-Butene	56.107	33.600	493.3	771.5	595.0	0.6121
27	Isobutene	56.107	19.600	479.3	752.2	580.0	0.6022
28	1,3 Butadiene	54.092	24.000	483.7	765.7	628.0	0.6282
29	1- Pentene	70.134	85.900	545.6	836.6	511.8	0.6466
30	Cis -2-Pentene	70.134	98.400	558.1	856.7	530.0	0.6619
31	Trans -2-Pentene	70.134	97.400	557.1	855.7	530.0	0.6541
32	2-Methyl -1-Butene	70.134	88.000	547.7	837.0	493.1	0.6573
33	3-Methyl -1-Butene	70.134	68.100	527.8	810.7	510.0	0.6335
34	2-Methyl -2 - Butene	70.134	101.400	561.1	847.8	493.1	0.6693
35	1-Hexene	84.161	146.200	605.9	907.3	455.4	0.6780
36	Cyclopentane	70.134	120.600	580.3	921.2	653.0	0.7615
37	Methylcyclopentane	84.161	161.200	620.9	959.0	548.9	0.7551
38	Cyclohexane	84.161	177.300	637.0	996.4	591.0	0.7835
39	Methyl Cyclohexane	98.188	213.700	673.4	1029.9	503.5	0.7760
40	Benzene	78.114	176.100	635.8	1011.9	710.4	0.8845
41	Toluene	92.141	231.100	690.8	1065.2	595.2	0.8757
42	O-Xylene	106.167	291.900	751.6	1134.7	541.6	0.8863
43	M-Xylene	106.167	282.400	742.1	1110.7	513.6	0.8707
44	P-Xylene	106.167	281.000	740.7	1109.3	509.2	0.8678
45	Ethylbenzene	106.167	277.100	736.8	1110.9	523.5	0.8758
46	Propadiene	40.060	–30.100	429.6	707.7	793.4	0.5997
47	1,2-Butadiene	54.090	51.530	511.2	799.2	652.7	0.6576
48	1,3 Butadiene	54.090	24.060	483.8	765.7	628.0	0.6273
49	Acetylene	26.040	–119.200	340.5	555.0	890.4	0.6150
50	Methyl Acetylene	40.050	–9.800	449.9	724.3	816.2	0.6212

°R = °F + 459.7

TABLE A-2 Conversion of Kinematic Viscosity to Uncorrected Saybolt Universal Viscosity at 100°F

Kinematic viscosity, mm^2/sec	Saybolt universal viscosity seconds	Kinematic viscosity, mm^2/sec	Saybolt universal viscosity, seconds	Kinematic viscosity, mm^2/sec	Saybolt universal viscosity seconds
1.82	32.0	7.0	48.8	16.6	83.8
1.84	32.1	7.2	49.4	16.8	84.6
1.86	32.1	7.4	50.1	17.0	85.4
1.88	32.2	7.6	50.7	17.2	86.2
1.90	32.3	7.8	51.4	17.4	87.0
1.92	32.3	8.0	52.1	17.6	87.8
1.94	32.4	8.2	52.7	17.8	88.7
1.96	32.5	8.4	53.4	18.0	89.5
1.98	32.5	8.6	54.1	18.2	90.3
2.00	32.6	8.8	54.7	18.4	91.1
2.10	32.9	9.0	55.4	18.6	92.0
2.20	33.3	9.2	56.1	18.8	92.8
2.30	33.6	9.4	56.8	19.0	93.6
2.40	34.0	9.6	57.5	19.2	94.5
2.50	34.3	9.8	58.1	19.4	95.3
2.60	34.6	10.0	58.8	19.6	96.1
2.70	35.0	10.2	59.5	19.8	97.0
2.80	35.3	10.4	60.2	20.0	97.8
2.90	35.6	10.6	60.9	20.2	98.7
3.00	36.0	10.8	61.7	20.4	99.5
3.10	36.3	11.0	62.4	20.6	100.4
3.20	36.6	11.2	63.1	20.8	101.2
3.30	36.9	11.4	63.8	21.0	102.1
3.40	37.3	11.6	64.5	21.2	102.9
3.50	37.6	11.8	65.3	21.4	103.8
3.60	37.9	12.0	66.0	21.6	104.6
3.70	38.2	12.2	66.7	21.8	105.5
3.80	38.6	12.4	67.5	22.0	106.3
3.90	38.9	12.6	68.2	22.2	107.2
4.00	39.2	12.8	69.0	22.4	108.1
4.10	39.5	13.0	69.7	22.6	108.9
4.20	39.8	13.2	70.5	22.8	109.8
4.30	40.2	13.4	71.2	23.0	110.6
4.40	40.5	13.6	72.0	23.2	111.5
4.50	40.8	13.8	72.7	23.4	112.4
4.60	41.1	14.0	73.5	23.6	113.2
4.70	41.4	14.2	74.3	23.8	114.1
4.80	41.8	14.4	75.1	24.0	115.0
4.90	42.1	14.6	75.8	24.2	115.9
5.00	42.4	14.8	76.6	24.4	116.7
5.20	43.0	15.0	77.4	24.6	117.6
5.40	43.7	15.2	78.2	24.8	118.5
5.60	44.3	15.4	79.0	25.0	119.4
5.80	44.9	15.6	79.8	25.2	120.2
6.0	45.6	15.8	80.6	25.4	121.1
6.2	46.2	16.0	81.4	24.6	117.6
6.4	46.9	16.2	82.2	24.8	118.5
6.8	48.1	16.4	83.0	25.0	119.4

(Continued)

TABLE A-2 Conversion of Kinematic Viscosity to Uncorrected Saybolt Universal Viscosity at 100°F (*Continued*)

Kinematic viscosity, mm^2/sec	Saybolt universal viscosity seconds	Kinematic viscosity, mm^2/sec	Saybolt universal viscosity seconds	Kinematic viscosity, mm^2/sec	Saybolt universal viscosity seconds
25.5	121.6	38.5	179.9	51.0	237.1
26.0	123.7	39.0	182.2	52.0	241.7
26.5	126.0	39.5	184.5	53.0	246.3
27.0	128.2	40.0	186.8	54.0	250.9
27.5	130.4	40.5	189.0	55.0	255.5
28.0	132.6	41.0	191.3	56.0	260.2
28.5	134.8	41.5	193.6	57.0	264.8
29.0	137.0	42.0	195.9	58.0	269.4
29.5	139.3	42.5	198.2	59.0	274.0
30.0	141.5	43.0	200.5	60.0	278.6
30.5	143.8	43.5	202.7	61.0	283.2
31.0	146.0	43.5	202.7	62.0	287.8
31.5	148.2	44.0	205.0	63.0	292.4
32.0	150.5	44.5	207.3	64.0	297.0
32.5	152.7	45.0	209.6	65.0	301.6
33.0	155.0	45.5	211.9	66.0	306.3
33.5	157.2	46.0	214.2	67.0	310.9
34.0	159.5	46.5	216.5	68.0	315.5
34.5	161.8	47.0	218.8	69.0	320.1
35.0	164.0	47.5	221.1	70.0	324.7
35.5	166.3	48.0	223.4	71.0	329.4
36.0	168.6	48.5	225.7	72.0	334.0
36.5	170.8	49.0	228.0	73.0	338.6
37.0	173.1	49.5	230.3	74.0	343.2
37.5	175.4	50.0	232.6	75.0	347.8
38.0	177.6				

For kinematic viscosity > 75 cSt at 40°C:
SUS = kinematic viscosity × 4.634.

TABLE A-3 Kinematic to Furol Viscosity Conversion

Kinematic viscosity, mm^2/sec	Furol vis. 122°F seconds	Furol vis. 210°F seconds	Kinematic viscosity, mm^2/sec	Furol vis. 122°F seconds	Furol vis. 210°F seconds	Kinematic viscosity, mm^2/sec	Furol vis. 122°F seconds	Furol vis. 210°F seconds
50	23.6	24.0	150	70.8	71.9	400	188.7	191.7
51	24.0	24.4	155	73.1	74.3	405	191.0	194.1
52	24.5	24.9	160	75.5	76.7	410	193.4	196.5
53	25.0	25.4	165	77.8	79.1	415	195.8	198.9
54	25.5	25.9	170	80.2	81.5	420	198.1	201.3
55	25.9	26.4	175	82.5	83.9	425	200.5	203.7
56	26.4	26.8	180	84.9	86.3	430	202.8	206.1
57	26.9	27.3	185	87.3	88.7	435	205.2	208.5
58	27.3	27.8	190	89.6	91.0	440	207.5	210.8
59	27.8	28.3	195	92.0	93.4	445	209.9	213.2
60	28.3	28.8	200	94.3	95.8	450	212.3	215.6
61	28.8	29.2	205	96.7	98.2	455	214.6	218.0
62	29.2	29.7	210	99.1	100.6	460	217.0	220.4
63	29.7	30.2	215	101.4	103.0	465	219.3	222.8
64	30.2	30.7	220	103.8	105.4	470	221.7	225.2
65	30.6	31.1	225	106.1	107.8	475	224.1	227.6
66	31.1	31.6	230	108.5	110.2	480	226.4	230.0
67	31.6	32.1	235	110.8	112.6	485	228.8	232.4
68	32.0	32.6	240	113.2	115.0	490	231.1	234.8
69	32.5	33.1	245	115.6	117.4	495	233.5	237.2
70	32.9	33.5	250	117.9	119.8	500	235.9	239.6
71	33.4	34.0	255	120.3	122.2	505	238.2	242.0
72	33.6	34.5	260	122.6	124.6	510	240.6	244.4
73	34.8	35.0	265	125.0	127.0	515	242.9	246.8
74	35.0	35.5	270	127.4	129.4	520	245.3	249.2
75	35.5	35.9	275	129.7	131.8	525	247.6	251.6
76	35.9	36.4	280	132.1	134.2	530	250.0	254.0
77	36.4	36.9	285	134.4	136.6	535	252.4	256.4
78	36.8	37.4	290	136.8	139.0	540	254.7	258.8
79	37.3	37.9	295	139.2	141.4	545	257.1	261.2
80	37.8	38.3	300	141.5	143.8	550	259.4	263.6
82	38.7	39.3	305	143.9	146.2	555	261.8	266.0
84	39.6	40.3	310	146.2	148.6	560	264.2	268.4
86	40.6	41.2	315	148.6	150.9	565	266.5	270.7
88	41.5	42.2	320	150.9	153.3	570	268.9	273.1
90	42.5	43.1	325	153.3	155.7	575	271.2	275.5
92	43.4	44.1	330	155.7	158.1	580	273.6	277.9
94	44.3	45.0	335	158.0	160.5	585	275.9	280.3
96	45.3	46.0	340	160.4	162.9	590	278.3	282.7
98	46.2	47.0	345	162.7	165.3	595	280.7	285.1
100	47.2	47.9	350	165.1	167.7	600	283.0	287.5
105	49.5	50.3	355	167.5	170.1	605	285.4	289.9
110	51.9	52.7	360	169.8	172.5	610	287.7	292.3
115	54.2	55.1	365	172.2	174.9	615	290.1	294.7
120	56.6	57.5	370	174.5	177.3	620	292.5	297.1
125	59.0	59.9	375	176.9	179.7	625	294.8	299.5
130	61.3	62.3	380	179.2	182.1	630	297.2	301.9
135	63.7	64.7	385	181.6	184.5	635	299.5	304.3
140	66.0	67.1	390	184.0	186.9	640	301.9	306.7
145	68.4	69.5	395	186.3	189.3	645	304.2	309.1

(*Continued*)

TABLE A-3 Kinematic to Furol Viscosity Conversion (*Continued*)

Kinematic viscosity, mm^2/sec	Furol vis. 122°F seconds	Furol vis. 210°F seconds	Kinematic viscosity, mm^2/sec	Furol vis. 122°F seconds	Furol vis. 210°F seconds	Kinematic viscosity, mm^2/sec	Furol vis. 122°F seconds	Furol vis. 210°F seconds
650	306.6	311.5	745	351	357	870	410	417
655	309.0	313.9	750	354	359	880	415	422
660	311.3	316.3	755	356	362	890	420	426
665	313.7	318.7	760	358	364	900	425	431
670	316.0	321.1	765	361	367	920	434	441
675	318.4	323.5	770	363	369	940	443	450
680	320.8	325.9	775	366	371	960	453	460
685	323.1	328.3	780	368	374	980	462	470
690	325.5	330.6	785	370	376	1000	472	479
695	327.8	333.0	790	373	379	1020	481	489
700	330.2	335.4	795	375	381	1040	491	498
705	332.5	337.8	800	377	383	1060	500	508
710	334.9	340.2	810	382	388	1080	509	518
715	337.3	342.6	820	387	393	1100	519	527
720	340	345	830	392	398	1150	542	551
725	342	347	840	396	403	1200	566	575
730	344	350	850	401	407	1250	590	599
735	347	352	860	406	412	1300	613	623
740	349	355						

For kinematic viscosity > 1300 mm^2/sec, calculate furol viscosity as below:
At 122°F furol viscosity = 0.4717 × kinematic viscosity
At 210°F furol viscosity = 0.4792 × kinematic viscosity

TABLE A-4 Volumes and Weights of Petroleum Products as a Function of API Gravity

API gravity	Specific gravity	Barrels/ metric ton	Barrels/ ton	Lb/gal	API gravity	Specific gravity	Barrels/ metric ton	Barrels/ ton	Lb/gal
0	1.0760	5.8455	5.9393	8.9798	31	0.8708	7.2235	7.3395	7.2667
1	1.0679	5.8899	5.9845	8.9120	32	0.8654	7.2680	7.3847	7.2223
2	1.0599	5.9344	6.0297	8.8453	33	0.8602	7.3124	7.4298	7.1784
3	1.0520	5.9788	6.0748	8.7795	34	0.8550	7.3569	7.4750	7.1350
4	1.0443	6.0233	6.1200	8.7147	35	0.8498	7.4013	7.5202	7.0922
5	1.0366	6.0677	6.1652	8.6509	36	0.8448	7.4458	7.5653	7.0498
6	1.0291	6.1122	6.2103	8.5880	37	0.8398	7.4902	7.6105	7.0080
7	1.0217	6.1566	6.2555	8.5260	38	0.8348	7.5347	7.6557	6.9666
8	1.0143	6.2011	6.3007	8.4648	39	0.8299	7.5791	7.7008	6.9258
9	1.0071	6.2455	6.3458	8.4046	40	0.8251	7.6236	7.7460	6.8854
10	1.0000	6.2900	6.3910	8.3452	41	0.8203	7.6680	7.7911	6.8455
11	0.9930	6.3345	6.4362	8.2866	42	0.8156	7.7125	7.8363	6.8060
12	0.9861	6.3789	6.4813	8.2289	43	0.8109	7.7569	7.8815	6.7670
13	0.9792	6.4234	6.5265	8.1719	44	0.8063	7.8014	7.9266	6.7285
14	0.9725	6.4678	6.5717	8.1158	45	0.8017	7.8458	7.9718	6.6903
15	0.9659	6.5123	6.6168	8.0604	46	0.7972	7.8903	8.0170	6.6527
16	0.9593	6.5567	6.6620	8.0057	47	0.7927	7.9347	8.0621	6.6154
17	0.9529	6.6012	6.7072	7.9518	48	0.7883	7.9792	8.1073	6.5785
18	0.9465	6.6456	6.7523	7.8986	49	0.7839	8.0236	8.1525	6.5421
19	0.9402	6.6901	6.7975	7.8462	50	0.7796	8.0681	8.1976	6.5060
20	0.9340	6.7345	6.8427	7.7944	51	0.7753	8.1125	8.2428	6.4704
21	0.9279	6.7790	6.8878	7.7433	52	0.7711	8.1570	8.2880	6.4351
22	0.9218	6.8234	6.9330	7.6928	53	0.7669	8.2014	8.3331	6.4002
23	0.9159	6.8679	6.9782	7.6430	54	0.7628	8.2459	8.3783	6.3657
24	0.9100	6.9123	7.0233	7.5939	55	0.7587	8.2904	8.4235	6.3316
25	0.9042	6.9568	7.0685	7.5453	56	0.7547	8.3348	8.4686	6.2978
26	0.8984	7.0012	7.1137	7.4974	57	0.7507	8.3793	8.5138	6.2644
27	0.8927	7.0457	7.1588	7.4501	58	0.7467	8.4237	8.5590	6.2314
28	0.8871	7.0901	7.2040	7.4034	59	0.7428	8.4682	8.6041	6.1987
29	0.8816	7.1346	7.2492	7.3573	60	0.7389	8.5126	8.6493	6.1663
30	0.8762	7.1790	7.2943	7.3117		1.0760	5.8455	5.9393	8.9798
61	0.7351	8.5571	8.6945	6.1343	81	0.6659	9.4461	9.5978	5.5569
62	0.7313	8.6015	8.7396	6.1026	82	0.6628	9.4906	9.6430	5.5309
63	0.7275	8.6460	8.7848	6.0712	83	0.6597	9.5350	9.6881	5.5051
64	0.7238	8.6904	8.8300	6.0401	84	0.6566	9.5795	9.7333	5.4796
65	0.7201	8.7349	8.8751	6.0094	85	0.6536	9.6239	9.7785	5.4543
66	0.7165	8.7793	8.9203	5.9790	86	0.6506	9.6684	9.8236	5.4292
67	0.7128	8.8238	8.9655	5.9488	87	0.6476	9.7128	9.8688	5.4043
68	0.7093	8.8682	9.0106	5.9190	88	0.6446	9.7573	9.9140	5.3797
69	0.7057	8.9127	9.0558	5.8895	89	0.6417	9.8017	9.9591	5.3553
70	0.7022	8.9571	9.1010	5.8603	90	0.6388	9.8462	10.0043	5.3311
71	0.6988	9.0016	9.1461	5.8313	91	0.6360	9.8906	10.0495	5.3072
72	0.6953	9.0460	9.1913	5.8027	92	0.6331	9.9351	10.0946	5.2834
73	0.6919	9.0905	9.2365	5.7743	93	0.6303	9.9795	10.1398	5.2599
74	0.6886	9.1349	9.2816	5.7462	94	0.6275	10.0240	10.1850	5.2366
75	0.6852	9.1794	9.3268	5.7184	95	0.6247	10.0684	10.2301	5.2134
76	0.6819	9.2239	9.3720	5.6908	96	0.6220	10.1129	10.2753	5.1905
77	0.6787	9.2683	9.4171	5.6635	97	0.6193	10.1573	10.3204	5.1678
78	0.6754	9.3128	9.4623	5.6365	98	0.6166	10.2018	10.3656	5.1453
79	0.6722	9.3572	9.5075	5.6097	99	0.6139	10.2463	10.4108	5.1230
80	0.6690	9.4017	9.5526	5.5832	100	0.6112	10.2907	10.4559	5.1008

TABLE A-5 Physical Properties of Gases

	Formula	Molecular weight	Boiling point, °F	Critical temperature, °F	Critical pressure, lb/in^2	Specific gravity, 60/60	Net heat of combustion, Btu/lb
Air		29.0	–317.83	–221.31	546.9	0.8748	
Ammonia	NH_3	17.0	–28.17	270.5	1635.67	0.6162	7998
Carbon Dioxide	CO_2	44.0	–109.26	87.91	1071	0.818	
Carbon Monoxide	CO	28.0	–312.61	–220.41	507.45		4344
Carbonyl Sulfide	COS	60.1	–58.27	222.17	920.85	1.02	3924
Chlorine	Cl_2	70.0	–29.25	291.2	1118.37	1.4274	0
Hydrogen Chloride	HCl	36.5	–121	124.7	1205.07	0.8463	337
Hydrogen Sulfide	H_2S	34.1	–76.5	212.45	1300	0.8014	7785
Nitrogen	N_2	28.0	–320.45	–232.51	493.1	0.8094	
Oxygen	O_2	32.0	–168.34	–181.43	731.4	1.1421	
Sulfur Dioxide	SO_2	64.1	13.96	315.68	1143.5	1.3946	0
Sulfur Trioxide	SO_3	80.1	–162.62	423.86	1190.38	1.9269	

TABLE A-6 Saturated Water and Steam Properties

Absolute pressure bar	Temperature, °C	Specific volume, m³/kg		Specific enthalpy, kJ/kg			Specific entropy, kJ/kg°K		
		Water	Steam	Water	Evaporation	Steam	Water	Evaporation	Steam
0.02	17.51	0.001001	67.01200	73.5	2460.2	2533.6	0.261	8.464	8.725
0.04	28.98	0.001004	34.80300	121.4	2433.1	2554.5	0.423	8.053	8.476
0.06	36.18	0.001006	23.74100	151.5	2416.0	2567.5	0.521	7.810	8.331
0.08	41.54	0.001008	18.10400	173.9	2403.2	2577.1	0.593	7.637	8.230
0.10	45.83	0.001010	14.67400	191.8	2392.9	2584.8	0.649	7.502	8.151
0.20	60.09	0.001017	7.64920	251.5	2358.4	2609.9	0.832	7.077	7.909
0.30	69.13	0.001022	5.22900	289.3	2336.1	2625.4	0.944	6.825	7.769
0.40	75.89	0.001027	3.99320	317.7	2319.2	2636.9	1.026	6.645	7.671
0.50	81.35	0.001030	3.24000	340.6	2305.4	2646.0	1.091	6.504	7.595
0.60	85.95	0.001033	2.73100	359.9	2293.6	2653.6	1.146	6.387	7.533
0.70	89.96	0.001036	2.36400	376.8	2283.3	2660.1	1.192	6.288	7.480
0.80	93.51	0.001039	2.08600	391.7	2274.0	2665.8	1.233	6.202	7.435
0.90	96.71	0.001041	1.86900	405.2	2265.6	2670.9	1.270	6.126	7.395
1.00	99.63	0.001043	1.69300	417.5	2257.9	2675.4	1.303	6.057	7.360
1.01325	100.00	0.001044	1.67300	419.1	2256.9	2676.0	1.307	6.048	7.355
1.50	111.40	0.001053	1.15900	467.1	2226.2	2693.4	1.434	5.790	7.223
2.00	120.20	0.001061	0.88500	504.7	2201.6	2706.3	1.530	5.597	7.127
2.50	127.40	0.001068	0.71800	535.4	2181.0	2716.4	1.607	5.445	7.052
3.00	133.50	0.001074	0.60500	561.4	2163.2	2724.7	1.672	5.319	6.991
3.50	138.90	0.001079	0.52300	584.3	2147.3	2731.6	1.727	5.212	6.939
4.00	143.60	0.001084	0.46200	604.7	2132.9	2737.6	1.776	5.118	6.894
4.50	147.90	0.001089	0.41300	623.2	2119.7	2742.9	1.820	5.034	6.855
5.00	151.90	0.001093	0.37400	640.1	2107.4	2747.5	1.860	4.959	6.819
6.00	158.80	0.001101	0.31546	670.4	2085.0	2755.5	1.931	4.827	6.758
7.00	165.00	0.001108	0.27268	697.1	2064.9	2762.0	1.992	4.713	6.705
8.00	170.40	0.001115	0.24026	720.9	2046.5	2767.5	2.046	4.614	6.660
9.00	175.40	0.001121	0.21482	742.6	2029.5	2772.1	2.094	4.525	6.619
10.00	179.90	0.001127	0.19430	762.6	2013.6	2776.2	2.138	4.445	6.583
11.00	184.10	0.001133	0.17739	781.1	1998.6	2779.7	2.179	4.371	6.550
12.00	188.00	0.001139	0.16321	798.4	1984.3	2782.7	2.216	4.303	6.519
13.00	191.60	0.001144	0.15114	814.7	1970.7	2785.4	2.251	4.240	6.491
14.00	195.00	0.001149	0.14073	830.1	1957.7	2787.8	2.284	4.181	6.465
15.00	198.30	0.001154	0.13167	844.6	1945.3	2789.9	2.314	4.126	6.441
16.00	201.40	0.001159	0.12370	858.5	1933.2	2791.7	2.344	4.074	6.418
17.00	204.30	0.001163	0.11664	871.8	1921.6	2793.4	2.371	4.025	6.396
18.00	207.10	0.001168	0.11033	884.5	1910.3	2794.8	2.398	3.978	6.375
19.00	209.80	0.001172	0.10467	896.8	1899.3	2796.1	2.423	3.933	6.356
20.00	212.40	0.001177	0.09955	908.6	1888.7	2797.2	2.447	3.890	6.337
21.00	214.80	0.001181	0.09490	919.9	1878.3	2798.2	2.470	3.849	6.319
22.00	217.20	0.001185	0.09066	930.9	1868.1	2799.1	2.492	3.809	6.301
23.00	219.60	0.001189	0.08678	941.6	1858.2	2799.8	2.514	3.771	6.285
24.00	221.80	0.001193	0.08321	951.9	1848.5	2800.4	2.534	3.735	6.269
25.00	223.90	0.001197	0.07992	961.9	1839.0	2800.9	2.554	3.699	6.254
26.00	226.00	0.001201	0.07687	971.7	1829.7	2801.4	2.574	3.665	6.239
27.00	228.10	0.001205	0.07403	981.2	1820.5	2801.7	2.592	3.632	6.224
28.00	230.00	0.001209	0.07140	990.5	1811.5	2802.0	2.611	3.600	6.211
29.00	232.00	0.001213	0.06894	999.5	1802.7	2802.2	2.628	3.569	6.197
30.00	233.80	0.001216	0.06663	1008.3	1794.0	2802.3	2.646	3.538	6.184
35.00	242.50	0.001235	0.05703	1049.7	1752.2	2802.0	2.725	3.398	6.123

(*Continued*)

TABLE A-6 Saturated Water and Steam Properties (*Continued*)

Absolute pressure bar	Temperature, °C	Specific volume, m³/kg		Specific enthalpy, kJ/kg			Specific entropy, kJ/kg°K		
		Water	Steam	Water	Evaporation	Steam	Water	Evaporation	Steam
40.00	250.30	0.001252	0.04975	1087.4	1712.9	2800.3	2.797	3.272	6.069
45.00	257.40	0.001269	0.04404	1122.1	1675.6	2797.7	2.861	3.158	6.019
50.00	261.90	0.001286	0.03943	1154.5	1639.7	2794.2	2.921	3.053	5.974
60.00	275.60	0.001319	0.03243	1213.7	1571.3	2785.0	3.027	2.863	5.891
70.00	285.80	0.001351	0.02737	1267.5	1506.0	2773.4	3.122	2.694	5.816
80.00	295.00	0.001384	0.02352	1317.2	1442.7	2759.9	3.208	2.539	5.747
90.00	303.30	0.001418	0.02049	1363.8	1380.8	2744.6	3.287	2.395	5.682
100.00	311.00	0.001453	0.01804	1408.1	1319.7	2727.7	3.361	2.259	5.620
110.00	318.00	0.001489	0.01601	1450.6	1258.8	2709.3	3.430	2.129	5.560
120.00	324.60	0.001527	0.01429	1491.7	1197.5	2698.2	3.497	2.003	5.500
130.00	330.80	0.001567	0.01280	1531.9	1135.1	2667.0	3.561	1.880	5.441
140.00	336.60	0.001611	0.01150	1571.5	1070.9	2642.4	3.624	1.756	5.380
150.00	342.10	0.001658	0.01343	1610.9	1004.2	2615.1	3.686	1.632	5.318
160.00	347.30	0.001710	0.00931	1650.4	934.5	2584.9	3.747	1.506	5.253
170.00	352.30	0.001770	0.00837	1691.6	860.0	2551.6	3.811	1.375	5.186
180.00	357.00	0.001840	0.00750	1734.8	779.0	2513.9	3.877	1.236	5.113
190.00	361.40	0.001926	0.00668	1778.7	691.8	2470.5	3.943	1.090	5.033
200.00	365.70	0.002037	0.00588	1826.6	591.6	2418.2	4.015	0.926	4.941
210.00	369.80	0.002202	0.00502	1886.3	461.2	2347.5	4.105	0.717	4.822
220.00	371.70	0.002668	0.00374	2010.3	186.3	2196.6	4.293	0.288	4.581
221.20	374.15	0.003170	0.00317	2107.4	0.00.0	2107.4	4.443	0.000	4.443

TABLE A-7 Physical Properties of Elements

Key (Sodium example): Atomic number 11; Atomic weight 22.9898; Boiling point °C 1330; Oxidation state 1; Melting point °C 97.8; Symbol Na; Density g/mL 0.97; Name Sodium.

Notes

1 Atomic weight in parentheses indicate most stable or best known isotope.

2 Density for gaseous elements are for liquids at boiling point.

Reference: Chemical engineers Handbook; Perry R.H et al Mcgraw hill, 1973

Group	Atomic number	Symbol	Name	Atomic weight	Oxidation state	Boiling point °C	Melting point °C	Density g/mL
IA	1	H	Hydrogen	1.0079	1	–252.7	–259.2	0.53
IA	3	Li	Lithium	6.939	1	892	180.5	0.071
IA	11	Na	Sodium	22.9898	1	1330	97.8	0.97
IA	19	K	Potassium	39.098	1	760	63.7	0.86
IA	37	Rb	Rubidium	85.47	1	688	38.9	1.53
IA	55	Cs	Cesium	132.9054	1	28.7	690	1.90
IA	87	Fr	Francium	(223)	1	–	(27)	–
IIA	4	Be	Beryllium	9.0122	2	2770	1277	1.85
IIA	12	Mg	Magnesium	24.312	2	1107	650	1.74
IIA	20	Ca	Calcium	40.08	2	1440	838	1.55
IIA	38	Sr	Strontium	40.08	2	1380	768	2.6
IIA	56	Ba	Barium	40.08	2	1640	714	3.5
IIA	88	Ra	Radium	40.08	2	1640	714	3.5
IIIB	21	Sc	Scandium	24.312	3	1107	650	1.74
IIIB	39	Y	Ytrium	88.905	3	2927	1509	4.47
IIIB	57	La	Lanthanum	138.91	3	3470	920	6.17
IIIB	89	Ac	Radium	(227)	3	–	1050	–
IVB	22	Ti	Titanium	47.90	4.3	3260	1668	4.51
IVB	40	Zr	Zircomium	91.22	4	3580	1852	6.49
IVB	72	Hf	Hafnium	178.49	4	5400	2200	13.1
IVB	104	Rf	Rutherfordium	–		–	–	
VB	23	V	Vanadium	50.942	5,4,3,2	3450	1900	6.1
VB	40	Nb	Niobium	92.906	5,3	3300	2468	8.4
VB	73	Ta	Tantalum	180.948	5	5425	2996	16.6
VB	105	Ha	Hahnium	–		–	–	
VIB	24	Cr	Chromium	51.996	6,3,2	2665	1875	7.19
VIB	42	Mo	Molybdenum	95.94	6,5,4,3,2	5560	2610	10.2
VIB	74	W	Wolfrom	183.85	6,5,4,3,2	5930	3410	19.3
VIIB	25	Mn	Mangnese	54.938	7,6,4,3,2	2150	1245	7.43
VIIB	43	Tc	Technetium	(98)	7	-	2140	11.5
VIIB	75	Re	Rhenium	186.2	7,6,4, 3,2,–1	5900	3180	21.0
VIII	26	Fe	Iron	55.847	2,3	3000	1536	7.86
VIII	44	Ru	Ruthenium	101.07	2,3,4,6,8	4900	2500	12.2
VIII	76	Os	Osmium	190.2	2,3,4,6,8	5500	3000	22.6
VIII	27	Co	Cobalt	58.933	2,3	2900	1495	8.9
VIII	45	Rh	Rhodium	102.905	2,3,4	4500	2500	12.4
VIII	77	Ir	Iridium	192.2	2,3,4,6	5300	2454	22.5
VIII	28	Ni	Nickel	58.71	2,3	2730	1453	8.9
VIII	46	Pd	Palladium	106.4	2,4	3980	1552	12.0
VIII	78	Pt	Platinum	195.09	2,4	4530	1769	21.4
IB	29	Cu	Copper	63.54	2,1	2595	1063	8.96
IB	47	Ag	Silver	107.870	1	2210	960.8	10.5
IB	79	Au	Gold	196.967	3,1	2970	1063	19.3
IIB	30	Zn	Zinc	65.38	2	906	419.5	7.14
IIB	48	Cd	Cadmium	112.4	2	765	320.9	8.65
IIB	80	Hg	Mercury	200.59	2,1	357	–38.4	13.6
IIIA	5	B	Boron	10.811	3	–	2030	2.34
IIIA	13	Al	Aluminum	26.9815	3	2450	660	2.70
IIIA	31	Ga	Gallium	69.72	3	2237	29.8	5.91
IIIA	49	In	Indium	114.82	3	2000	156.2	7.31
IIIA	81	Tl	Thallium	204.37	3,1	1457	303	11.85
IVA	6	C	Carbon	12.01115	±4,2	4830	3727	2.26
IVA	14	Si	Silicon	28.086	4	2680	1410	2.33
IVA	32	Ge	Germanium	72.59	4	2830	937.4	5.32
IVA	50	Sn	Tin	118.69	4,2	2270	231.9	7.30
IVA	82	Pb	Lead	207.19	4,2	1725	327.4	11.4
VA	7	N	Nitrogen	14.0067	±3,5,4,2	–195.8	–210	0.81
VA	15	P	Phosphorus	30.9738	±3,5,4	280	44.2	1.82
VA	33	As	Arsenic	74.922	3,5	613	817	5.72
VA	51	Sb	Antimony	121.75	±3,5	1380	630.5	6.62
VA	83	Bi	Bismuth	208.980	3.5	1560	271.3	9.8
VIA	8	O	Oxygen	15.9994	–2	–183	–218.8	1.14
VIA	16	S	Sulfur	32.064	±2,4,6	444.6	119.0	2.07
VIA	34	Se	Slenium	78.96	–2,4,6	685	217	4.79
VIA	52	Te	Tellurium	127.6	–2,4,6	989.8	449.5	6.24
VIA	84	Po	Polonium	210	2,4	-	254	(9.2)
VIIA	9	F	Fluorine	18.9984	–1	–188.2	–219.6	1.505
VIIA	17	Cl	Chlorine	35.453	±1,3,5,7	–34.7	–101.0	1.56
VIIA	35	Br	Bromine	79.909	±1,5	58	–7.2	3.12
VIIA	53	I	Iodine	126.904	±1,5,7	183	113.7	4.94
VIIA	85	At	Astatine	(210)	±1,3,5,7	–	(302)	–
VIIIA	2	He	Helium	4.0026		–268.9	–269.7	0.126
VIIIA	10	Ne	Neon	18.9984		–246	–248.6	1.20
VIIIA	18	Ar	Argon	39.948		–185.8	–189.4	1.40
VIIIA	36	Kr	Krypton	83.8		–152	–157.3	2.6
VIIIA	54	Xe	Xenon	131.3		–108.0	–111.9	3.06
VIIIA	86	Rn	Radon	(222)		(–61.8)	(–71)	–
	58	Ce	Cerium	140.12	3,4	3468	795	6.67
	59	Pr	Praseodymium	140.907	3,4	3127	935	6.77
	60	Nd	Neodymium	144.24	3	3027	1024	7.00
	61	Pm	Promethium	(147)	3	–	(1027)	–
	62	Sm	Samarium	150.35	3,2	1900	1072	7.54
	63	Eu	Europium	151.96	3,2	1439	826	5.26
	64	Gd	Gadolinium	157.25	3	3000	1312	7.89
	65	Tb	Terbium	158.924	3,4	2800	1356	8.27
	66	Dy	Dysprosium	162.50	3	2600	1407	8.54
	67	Ho	Holmium	164.930	3	2600	1461	8.80
	68	Er	Erbium	167.26	3	2900	1497	9.05
	69	Tm	Thulium	168.934	3,2	1727	1545	9.33
	70	Yb	Yttrium	173.04	3,2	1427	824	6.98
	71	Lu	Lutetium	174.97	3	3327	1652	9.84
	90	Th	Thorium	232.038	4	3850	1750	11.7
	91	Pa	Protactinium	(231)	5,4	–	(1230)	15.4
	92	U	Uranium	238.03	6,5,4,3	3818	1132	19.07
	93	Np	Neptunium	(237)	6,5,4,3	–	637	19.5
	94	Pu	Plutonium	(242)	6,5,4,3	3235	640	–
	95	Am	Americium	(243)	6,5,4,3	–	–	11.7
	96	Cm	Curium	(247)	3	–	–	–
	97	Bk	Berkelium	(247)	4,3	–	–	–
	98	Cf	Californium	249	3	2600	1407	8.54
	99	Es	Einsteinium	(254)	–	–	–	–
	100	Fm	Fermium	(253)	–	–	–	–
	101	Md	Mendelevium	(256)	–	–	–	–
	102	No	Nobelium	(254)	–	–	–	–
	103	Lw	Lawrencium	257	–	–	–	–

INDEX